SILICON PHOTC
HIGH-PERFORMANCE
COMPUTING AND BEYOND

SILICON PHOTONICS FOR HIGH-PERFORMANCE COMPUTING AND BEYOND

Edited by
Mahdi Nikdast
Sudeep Pasricha
Gabriela Nicolescu
Ashkan Seyedi
Di Liang

CRC Press is an imprint of the
Taylor & Francis Group, an informa business

First edition published 2022
by CRC Press
6000 Broken Sound Parkway NW, Suite 300, Boca Raton, FL 33487-2742

and by CRC Press
2 Park Square, Milton Park, Abingdon, Oxon, OX14 4RN

CRC Press is an imprint of Taylor & Francis Group, LLC

Library of Congress Cataloging-in-Publication Data
Names: Nikdast, Mahdi, editor. | Pasricha, Sudeep, editor. | Nicolescu, G.
(Gabriela), editor. | Seyedi, Ashkan, editor. | Liang, Di, editor.
Title: Silicon photonics for high-performance computing and beyond / edited by Mahdi Nikdast, Sudeep Pasricha, Gabriela Nicolescu, Ashkan Seyedi, Di Liang.
Description: First edition. | Boca Raton, FL : CRC Press, 2022. | Includes
bibliographical references and index.
Identifiers: LCCN 2021028854 (print) | LCCN 2021028855 (ebook) |
ISBN 9780367262143 (hbk) | ISBN 9781032122441 (pbk) | ISBN 9780429292033 (ebk)
Subjects: LCSH: Electronic digital computers–Circuits. | Photonics. |
Integrated optics. | Optoelectronic devices.
Classification: LCC TK7888.4 .S55 2022 (print) | LCC TK7888.4 (ebook) |
DDC 621.39/5–dc23
LC record available at https://lccn.loc.gov/2021028854
LC ebook record available at https://lccn.loc.gov/2021028855

ISBN: 978-0-367-26214-3 (hbk)
ISBN: 978-1-032-12244-1 (pbk)
ISBN: 978-0-429-29203-3 (ebk)

DOI: 10.1201/9780429292033

Typeset in Times
by MPS Limited, Dehradun

Contents

Section V Emerging Computing Technologies and Applications

Preface

Our daily lives depend heavily on efficiently moving and processing data generated by different applications, from social networks and online shopping to healthcare and educational applications, to emerging applications such as autonomous driving and artificial intelligence. The rapid growth in such data is putting increased pressure on high-performance computing (HPC) systems and interconnection networks, accelerating past the computing limits of Moore's Law. In particular, the conventional metallic interconnect has already reached its physical limit, requiring a shift to fiber-optic links to meet demands in bandwidth, power consumption and reach inside datacenters and supercomputers. Silicon photonics and optical interconnects offer a post–Moore's Law technological alternative, aiming to replicate the significant success of long-haul communications in Datacom applications but with a fraction of traditional solution costs and in orders of magnitude larger volume. Due to their inherent speed and energy benefits, optical interconnects are rapidly replacing electronics for data transmission at almost every scale of computing [1]. For example, silicon photonic transceivers present a low-cost and high-bandwidth solution for datacenters, and with the continuous increase in the global network traffic, the silicon photonic transceiver industry is expected to be worth US$3.6 billion in 2025 with 24 million units shipped [2]. As for high-performance computing systems and manycore architectures, where the performance is determined mainly by the communication efficiency among different resources (e.g., compute and memory), silicon photonics can potentially deliver the required communication bandwidths with scalable energy efficiency [3], [4].

In addition to Datacom applications, silicon photonics is paving the way towards enabling emerging computing paradigms, e.g., neuromorphic computing. It is now possible to use integrated photonic components to perform matrix-vector multiplication, the most time- and energy-intensive operation in deep neural networks (DNNs). This approach can reduce computation time from $O(N^2)$ to $O(1)$ by taking advantage of the natural parallelism of photonics [5]. Traditionally, these operations have relied on bulky optical components [6], but with the growing maturity of silicon photonics in recent years [7], optical interconnects and integrated photonic circuits can be implemented with CMOS-compatible manufacturing techniques to enable small-footprint, cost-effective, low-latency, and energy-efficient optical domain data transport and processing. Accordingly, integrated photonic neural networks with silicon-photonic-based deep learning accelerators [8] and optoelectronic arithmetic logic units for high-speed computing [9] are rapidly emerging. As a result, silicon photonics is shaping the future of not only Datacom and interconnect technology but also high-performance computing and emerging computing paradigms (e.g., artificial intelligence).

The purpose of this book on *"Silicon Photonics for High-Performance Computing and Beyond"* is to provide a comprehensive overview of the state-of-the-art in the field of silicon photonics and its applications with a focus on recent innovations for high-performance computing systems and interconnect networks. Towards achieving this goal, the book presents a compilation of 19 outstanding contributions, all from leading research groups and pioneers from both academia and industry in the fields of silicon photonics and high-performance computing. The contributions are grouped into five main sections. The first section focuses on the requirements and advances of interconnection networks in high-performance computing systems, while analyzing and comparing different interconnect technologies (e.g., optical versus electrical) in such systems. The second section discusses different challenges and improvements in the design, implementation, and integration of optical interconnects for HPC systems. The third section is composed of contributions that focus on novel design solutions and design-automation methodologies for silicon photonic devices and circuits as well as systems integrating silicon photonics. The fourth section includes contributions discussing novel materials and silicon photonic devices and integrated circuits for optical interconnects and communication in HPC systems. Finally, the last section focuses on the emerging applications of silicon photonics for neuromorphic computing, programmable optics, and integrated photonic neural networks.

All of the contributions have been carefully selected and organized to provide important and complementary discussions, from different perspectives and multidisciplinary groups, related to the design, analysis, optimization, and implementation of novel silicon photonic devices, circuits, and systems for high-performance computing systems and emerging applications. The editors themselves present a multidisciplinary team—from both academia and industry—with expertise in various areas, including silicon photonic devices and integrated photonics, interconnection networks, high-performance computing, computer architecture, and embedded systems.

Organization

This book is organized into five sections: (1) high-performance computing interconnect requirements and advances, (2) device- and system-level challenges and improvements, (3) novel design solutions and automation, (4) novel materials, devices, and photonic integrated circuits, and (5) emerging computing technologies and applications. In the following, we present an overview of the five sections along with a brief summary of each of the individual chapters.

High-Performance Computing Interconnect Requirements and Advances

The first part of this book discusses the shortcomings of conventional metallic interconnects in satisfying the aggressively increasing bandwidth and energy requirements of HPC and networked systems, and the promise of optical interconnects to address such requirements. The section starts with a discussion of optical carrier generation and modulation requirements, and challenges to realize Petabit per second (Pbps) data transmission for HPC systems, followed by the second chapter that reviews different laser modulation schemes for minimizing static power consumption in manycore and chip-multiprocessor systems. The promise of optical interconnects to boost the performance of rack-scale computing systems is discussed in the third chapter. Finally, the fourth chapter reviews the most advanced techniques developed in the HPC industry to move data over copper links, presenting a solid ground for comparing metallic and optical interconnects for HPC applications.

The chapter *Silicon Photonic Modulation for High-Performance* discusses the requirements and challenges for the generation of optical carriers for board-level optical interconnects. In particular, the need for optical interconnects in memory-intensive problems (e.g., training of convolutional neural networks for artificial intelligence, block chains, and edge computers for 5G data fusion) is motivated. This chapter presents calculations of the number of wavelength-division multiplexed (WDM) channels while considering present day optical modulator technology, and indicates that individual optical sources will need to generate tens of carriers suitable for multi-gigabit modulation and detection. Moreover, methods for producing dense-WDM channels are reviewed while recent results from the literature are discussed in light of simulation results in an effort to point at the most promising directions.

The chapter *Laser Modulation Schemes for Minimizing Static Power Dissipation* studies a wide variety of activity prediction and laser modulation schemes for CPUs, GPUs, multi-socket servers, and manycore processors. The chapter shows that the main challenge in designing an effective activity prediction and laser-modulation scheme is in deciding the nature of the relationship between the performance counters, the constants, and whether one uses a predictor of a single type or an ensemble of predictors, where all of these design choices together determine the accuracy of the final prediction. Moreover, the chapter presents a thorough overview of advantages and disadvantages of chip-scale optical interconnects for high-performance computing.

The chapter *Scalable Low-Power High-Performance Optical Network for Rack-Scale Computers* explores rack-scale optical network architectures with different path-reservation schemes and optical inter-chip networks. Moreover, the chapter presents a systematic analysis of such architectures and compares them with the most common architectures for high-performance computing systems (i.e., the Ethernet architecture). Results in the chapter show that a rack-scale optical network architecture with an optimized optical network switch and path-reservation scheme can considerably improve the system performance under the same energy consumption while achieving a better scalability than the state-of-the-art systems. Furthermore, it shows that the optical interconnects may become a potential alternative for the rack-scale computing systems.

The chapter *Network-in-Package for Low-Power and High-Performance Computing* explores data movement requirements in advanced computing systems. In particular, different from other chapters in this section and as an alternative to optical interconnects, the chapter reviews the most advanced techniques developed in the HPC industry to move data over copper links (wireline communication). The focus of this chapter is on Orthogonal Multi-Wire Signaling (OMWS) methods that exhibit low sensitivity to inter-symbol interference (ISI), and yet show a better data transfer rate compared to the conventional differential signaling.

Device- and System-Level Challenges and Improvements

The second part of this book highlights several device-level and system-level challenges in systems integrating silicon photonic integrated circuits and optical interconnects, as well as novel techniques to improve such systems' performance in the presence of device and link imperfections and failures. The first chapter discusses the thermal and fabrication-process variation sensitivities of silicon photonic devices and proposes novel solutions to compensate for the impact of such imperfections at the system level. The second chapter presents another novel approach to improve thermal resilience and communication performance in optical interconnection networks in manycore systems. The third chapter explores the voltage bias temperature induced (VBTI) aging effects in optical networks-on-chip, followed by the fourth chapter that shows how data approximation in optical interconnects can help improve the power and energy efficiency in systems integrating optical links.

The chapter *System-Level Management of Silicon-Photonic Networks in 2.5D Systems* discusses the thermal and fabrication-process variation sensitivities of silicon photonic devices, and presents some device-level techniques to mitigate their impact. Moreover, it motivates the need for system-level management techniques required to address the bandwidth-power trade-offs in optical links. The chapter presents frameworks for cross-layer modeling and simulation of optical links to account for the device-level characteristics. Leveraging these frameworks, the chapter presents runtime management techniques and system-level policies to achieve low-power operation of optical interconnects in several manycore system architectures.

The chapter *Thermal Reliability and Communication Performance Co-optimization for WDM-Based Optical Networks-on-Chip* proposes a co-optimization framework to improve the communication performance and thermal reliability in optical networks-on-chip(NoCs) for manycore systems. In particular, it presents a novel process-variation-tolerant optical temperature sensor design for accurate and efficient thermal monitoring in optical NoCs. Moreover, the chapter develops novel routing approaches to revolve both the communication conflict and the thermal susceptibility challenges in optical NoCs. Compared to the state-of-the-art, the proposed routing achieves better communication performance and reduces the energy overhead in the network.

The chapter *Exploring Aging Effects in Photonic Interconnects for High-Performance Manycore Architectures* presents a thorough frequency-domain analysis of voltage bias temperature induced (VBTI) aging effects on the performance of optical interconnects in manycore systems. The chapter discusses different modulation schemes, including on-off keying (OOK) and four pulse-amplitude modulation (4-PAM) and shows how using different modulation schemes can reduce signal degradations caused by aging-induced spectrum effects. Furthermore, the chapter analyzes system-level impacts of VBTI aging on several well-known optical network-on-chip architectures.

The chapter *Improving Energy Efficiency in Silicon Photonic Networks-on-Chip with Approximation Techniques* explores how using data approximation can help reduce power and energy consumption of laser power sources in optical networks-on-chip (NoCs) in manycore systems. In particular, the chapter proposes a novel framework to enable aggressive approximation during communication over optical links in optical NoCs. The proposed framework considers loss-aware laser power management and multilevel signaling to enable effective data approximation and energy-efficiency in chip-scale optical networks. Results in this chapter show that data approximation through optical interconnects can reduce laser power consumption and improve the overall energy-efficiency in optical networks-on-chip.

Novel Design Solutions and Automation

The third part of this book discusses the critical need for efficient design methods and novel design-automation solutions specific to silicon photonic device, circuit, and system design. The first chapter discusses the design automation and verification flow and its required innovations for silicon photonic devices and integrated circuits. The second chapter presents a physics-guided optimization approach to design photonic devices for high-performance computing systems, followed by the third chapter that discusses the system-level design process and automation in wavelength-routed optical networks-on-chip (NoCs).

The chapter *Automated, Scalable Silicon Photonics Design and Verification* motivates the urgent need for design-automation and verification solutions for integrated photonic circuits by comparing and discussing how the design procedure in electronic integrated circuits, as an example, has successfully evolved and advanced over years. In particular, the chapter presents a thorough discussion on process design kits (PDKs) for integrated silicon photonics, layout implementation, automated design methodologies (including schematic-driven layout and automated layout), and physical-layout verification, including layout versus schematic verification and design-rule checking (DRC).

The chapter *Inverse-Design for High-Performance Computing Photonics* introduces a physics-guided optimization approach to design silicon photonic devices for high-performance computing systems. The chapter begins by presenting a rough estimation of how improvements to the interconnect performance in a large computing system translate to improvements in the overall system performance. These results point to large system-level gains that could be accessed with optical interconnects, if better photonic components could be designed. Moreover, the chapter discusses the inverse-design method and its potential for designing photonic devices that are compact, highly efficient, and robust to fabrication and environmental factors, to realize efficient optical interconnects.

The chapter *Efficiency-Oriented Design of Wavelength-Routed Optical Network-on-Chip* discusses the system-level design process and the necessity for efficient design automation methods for optical networks-on-chip (NoCs) in multiprocessor systems. In particular, the chapter addresses the wavelength-routed optical NoC design problem while focusing on improving the efficiency of the network as well as the efficiency of the design process. The chapter covers two design automation strategies, including subtraction from fully connected router and design-template-based synthesis.

Novel Materials, Devices, and Photonic Integrated Circuits

The fourth part of this book includes chapters discussing innovations from materials, device structures, and integrated circuits in silicon photonics platform. The first chapter discusses recent research and product-development efforts to enable a disruptive dense wavelength-division multiplexing (DWDM) optical transceiver technology for HPC applications. The second chapter presents the theory and design fundamentals of integrated waveguide Bragg gratings, followed by the third chapter that focuses on principles and recent progress of silicon photonic orbital angular momentum (OAM) generators and multiplexers. Finally, the fourth chapter investigates the electro-optic properties of integrated barium titanate modulators and explores a new material system for building nonlinear optical devices.

The chapter *Innovative DWDM Silicon Photonics for High-Performance Computing* discusses a novel microresonator-based dense wavelength-division multiplexing (DWDM) transceiver architecture and recent progress on key silicon photonic building blocks, integration platforms, and several critical post-wafer fabrication steps for upcoming large-scale production. Innovations from materials, device structure, fabrication, testing, modeling, and packaging are covered in detail in this chapter. The chapter shows that a DWDM architecture backed by high-performance components and integration and testing solutions will be very competitive to other traditional coarse wavelength-division multiplexing (CWDM) approaches, and particularly shines in upcoming co-packaged optics applications in HPCs and high-end datacenters.

The chapter *Silicon Photonic Bragg Grating Devices* presents the theory and design fundamentals of integrated waveguide Bragg gratings and reviews recent research developments aimed at achieving complex Bragg grating based devices in silicon photonics. Moreover, the chapter discusses practical challenges in designing such Bragg gratings and discusses approaches to model and mitigate them. In addition to traditional Bragg gratings on a single waveguide, the chapter presents the contra-directional coupler, a Bragg grating based device that consists of two (or more) coupled waveguides that can be used to design multi-port optical add-drop filters.

The chapter *Silicon Photonic Integrated Circuits for OAM Generation and Multiplexing* reviews principles and recent progress of silicon photonic orbital angular momentum (OAM) generators and multiplexers, and discusses their future development for OAM-based space-division multiplexing (SDM) systems. The chapter reviews various types of OAM generators and multiplexers on the silicon platform with a careful comparison on their performance in terms of channel count per wavelength, bandwidth, polarization, and circuit complexity. Moreover, the chapter shows that OAM generators based on silicon photonic circuits can be integrated with many other photonic components on a single chip for large-scale integrated photonic systems.

The chapter *Novel Materials for Active Silicon Photonics* investigates the electro-optic properties of integrated barium titanate modulators. The barium titanate modulators show promise for realizing low-power, low-switching voltage modulators on chip, as a large effective electro-optic coefficient (100 pm/V) and small voltage device length product (2 V·cm) is demonstrated. Moreover, the chapter explores Gallium Nitride (GaN) and Aluminum Nitride (AlN) thin film as a new material system for building nonlinear optical devices that assist on-chip wavelength conversion and electro-optic modulation. The chapter shows that the fully CMOS-compatible AlN modulator is a promising candidate for electro-optic signal processing on a silicon photonics platform.

Emerging Computing Technologies and Applications

The last section of this book is devoted to novel applications of integrated silicon photonics for emerging computing paradigms from neuromorphic computing to integrated photonic neural networks. The first chapter discusses some recent advances in neuromorphic photonics, with emphasis on silicon photonics implementations. The second chapter reviews recent advances in optical logic computing and neural networks using photonic integrated circuits, as well as different challenges associated with realizing reliable and scalable optical computing systems. The third chapter proposes a phase-error- and loss-tolerant MZI-based optical processor for optical neural networks (ONNs), followed by the fourth chapter that presents a cross-layer optimized silicon photonic neural network accelerator.

The chapter *Neuromorphic Silicon Photonics* highlights recent advances in neuromorphic photonics, with emphasis on silicon photonics implementations. The chapter starts by discussing suitable neuronal models, with separate discussions on the implementation of linear weighted summation and nonlinear activation. Networking techniques to realize neural networks are also addressed. Finally, the chapter reviews three applications of neuromorphic photonics: ordinary differential equation (ODE) solving, model-predictive control, and intelligent signal processing.

The chapter discusses that ongoing investigations and progress in neuromorphic photonics enabled by silicon photonics promise to bring machine intelligence to unexplored regimes.

The chapter *Logic Computing and Neural Network on Photonic Integrated Circuit* discusses two optical computing paradigms, including digital computing and analog computing. Automated design methodologies are introduced to enhance different aspects, including scalability and robustness. The chapter further discusses different challenges and research opportunities to improve scalability and robustness in optical computing. In particular, it lists optical power loss, optical crosstalk, and manufacturing defects and process/environmental variations as the main obstacle to build complicated and efficient optical computing systems.

The chapter *High-Performance Programmable MZI-Based Optical Processors* presents the background principles of multiport programmable optical processors which are a mesh of 2 × 2 reconfigurable Mach–Zehnder interferometers (MZIs) in different topologies. It demonstrates how the unitary transformation matrix of a given application is decomposed for programming such MZI-based optical processors. Additionally, a phase-error- and loss-tolerant MZI-based optical processor for optical neural networks (ONNs) is investigated. The main goal of this chapter is to explore the design and implementation of more efficient and practical MZI-based optical processors that can better cope with inevitable fabrication-process and experimental imperfections.

The chapter *High-Performance Deep Learning Acceleration with Silicon Photonics* presents an innovative approach to designing high-performance deep learning accelerators with silicon photonics. The chapter motivates the use of cross-layer optimization to achieve complex design goals during deep learning accelerator design. Several cross-layer optimization solutions are investigated and integrated together, including device-level engineering for resilience to process variations and thermal crosstalk, circuit-level tuning enhancements for inference latency reduction, and architecture-level organization to enable higher resolution, better energy-efficiency, and improved throughput. The resulting cross-layer optimized deep learning accelerator, called CrossLight, is shown to achieve significant gains in energy-per-bit and performance-per-watt, compared to several state-of-the-art electronic and photonic deep learning accelerator platforms.

REFERENCES

[1] D. Thomson, A. Zilkie, J.E. Bowers, T. Komljenovic, G.T. Reed, L. Vivien, D. Marris-Morini, E. Cassan, L. Virot, M. Fédéli, J.M. Hartmann, “Roadmap on silicon photonics,” *Journal of Optics, vol. 18, no. 7*, p. 073003, 2016.

[2] Silicon photonics: Datacom, yes, but not only. [Online]. Available: http://www.yole.fr/iso_upload/News/2020/PR_SI_PHOTONICS_MarketGrowth_YOLE_GROUP_Apr2020.pdf.

[3] S. Rumley, D. Nikolova, R. Hendry, Q. Li, D. Calhoun, K. Bergman, “Silicon photonics for exascale systems,” *IEEE Journal of Lightwave Technology, vol. 33, no. 3*, pp. 547–562, 1 Feb 2015.

[4] S. Pasricha, M. Nikdast, “A survey of silicon photonics for energy efficient manycore computing,” *IEEE Design and Test, vol. 37, no. 4*, pp. 60–81, Aug 2020.

[5] Q. Cheng, J. Kwon, M. Glick, M. Bahadori, L.P. Carloni, K. Bergman, “Silicon photonics codesign for deep learning,” *Proceedings of the IEEE*, 2020.

[6] N.H. Farhat, D. Psaltis, A. Prata, E. Paek, “Optical implementation of the Hopfield model,” *Applied Optics, vol. 24, no. 10*, pp. 1469–1475, May 1985.

[7] The promise of silicon photonics. [Online]. Available: https://physicsworld.com/a/the-promise-of-silicon-photonics/. Accessed: August 2021.

[8] F. Sunny, E. Taheri, M. Nikdast, S. Pasricha, “A survey on silicon photonics for deep learning,” *ACM Journal of Emerging Technologies in Computing Systems (JETC)*, vol. 17, no. 4, article no. 61, pp. 1–57, 2021.

[9] Z. Ying, C. Feng, Z. Zhao, S. Dhar, H. Dalir, J. Gu, Y. Cheng, R. Soref, D. Pan, R.T. Chen, “Electronic-photonic arithmetic logic unit for high-speed computing,” *Nature Communication, vol. 11*, article no. 2154, 2020.

Editors

Mahdi Nikdast is with the Department of Electrical and Computer Engineering at Colorado State University (CSU), Fort Collins. Prof. Nikdast received his Ph.D. in electronic and computer engineering from The Hong Kong University of Science and Technology (HKUST), Hong Kong, in 2014. From 2014 to 2017, he was a postdoctoral fellow at McGill University and Polytechnique Montreal, Quebec, Canada, where he was a member of the Photonics System Group and Heterogenous Embedded System Lab. He is the director of Electronic-PhotoniC System Design (ECSyD) Laboratory at CSU. His research interests include various topics related to integrated photonics, interconnection networks, and high-performance computing systems.

Prof. Nikdast has authored and coauthored numerous papers in refereed journals and international conference publications. He has edited a book on *Photonic Interconnects for Computing Systems: Understanding and Pushing Design Challenges*, published by River Publishers in 2017. Prof. Nikdast has served as a reviewer for many journals as well as on the technical program committee (TPC) of numerous international conferences. He is a co-founder of the International Workshop on Optical/Photonic Interconnects for Computing Systems (OPTICS workshop) and the North American Workshop on Silicon Photonics for High-Performance Computing (SPHPC Workshop). Prof. Nikdast serves as an associate editor for IEEE Transactions on Very Large Scale Integration Systems (IEEE TVLSI). He was the recipient of various awards, including the Second Best Project Award at the AMD Technical Forum and Exhibition (AMD-TFE 2010, Taiwan), the Best Paper Award at the Asia Communications and Photonics Conference (ACP 2015, Hong Kong), the Best Paper Award at the Design, Automation, and Test in Europe (DATE) Conference and Exhibition (DATE 2016 – Test Track, Germany), the Best Paper Award Finalist at ACM Great Lakes Symposium on VLSI (GLSVLSI 2018, USA), the Best Paper Honorable Mention Award at ACM Great Lakes Symposium on VLSI (GLSVLSI 2020, China), and the prestigious NSF CAREER award in 2021. Prof. Nikdast is a senior member of the IEEE.

Sudeep Pasricha received the B.E. degree in electronics and communication engineering from Delhi Institute of Technology, India, in 2000, after which he spent several years working for STMicroelectronics, India/France, and Conexant, USA. He received his Ph.D. degree in computer science from the University of California, Irvine, in 2008. He joined Colorado State University (CSU) in 2008 where he is currently a Walter Scott Jr. College of Engineering professor in the Department of Electrical and Computer Engineering. He is also the chair of computer engineering and the director of the Embedded, High Performance, and Intelligent Computing (EPIC) Laboratory. His research broadly focuses on software algorithms, hardware architectures, and hardware-software co-design for energy-efficient, fault-tolerant, real-time, and secure computing. These efforts target multi-scale computing platforms, including embedded and IoT systems, cyber-physical systems, mobile devices, and datacenters.

Prof. Pasricha has published more than 200 papers in peer-reviewed journals and conference publications that have received 7 best paper awards and 6 best paper nominations. He has filed for multiple patents, and co-authored several books and book chapters. His contributions have been recognized with several awards for research and service excellence, including the George T. Abell

Outstanding Research Faculty Award, IEEE-CS/TCVLSI Mid-Career Research Achievement Award, IEEE/TCSC Award for Excellence for a Mid-Career Researcher, AFOSR Young Investigator Award, ACM Technical Leadership Award, and ACM SIGDA Distinguished Service Award. He is currently the vice chair of ACM SIGDA, a senior associate editor for the ACM Journal of Emerging Technologies in Computing, and an associate editor for the ACM Transactions on Embedded Computing Systems, IEEE Transactions on Computer-Aided Design of Integrated Circuits and Systems, IEEE Consumer Electronics, and IEEE Design & Test of Computers. He also serves as the chair of the steering committee of IEEE Transactions on Sustainable Computing. He has served as general chair and technical program chair of 12 conferences, steering and organizing committee member of 40 conferences, and technical program committee member of 100+ conferences. He is a senior member of the IEEE and distinguished member of the ACM.

Gabriela Nicolescu is a full professor at Polytechnique Montréal, Candra in the Department of Software and Computer Engineering. Her research interests are in the fields of automatic design for secure IoT and she is the director of the Heterogeneous Embedded Systems Laboratory. Prof. Nicolescu obtained her B.Eng. degree in electrical engineering from UPB (Polytechnic University Bucharest) in 1998 and her Ph.D. degree in 2002 from INPG (Institut National Polytechnique de Grenoble) France. She is a member of several Technical Program Committees (DAC, MPSoC, ICCAD, etc.). She is one of the founders of the International Workshop on Optical/Photonic Interconnects for Computing Systems (OPTICS workshop), the North American Workshop on Silicon Photonics for High Performance Computing (SPHPC Workshop), and (Francophone school on heterogeneous embedded systems) FETCH winter school. She co-authored papers awarded for best papers in DATE and ISSS-CODES. Her papers were candidates for best papers in several other conferences. She is member of RESMIQ (Regroupement Stratégique en Microélectronique du Québec) and represents Polytechnique Montréal in RESMIQ Board of Directors. She is also founder of a start-up providing security solutions, member of the directors board for IN-SEC-M Canada and CyberNB Canada security industry clusters. She edited 6 books and authored and co-authored more than 200 journal articles and papers in international conference proceedings.

Ashkan Seyedi received a dual bachelor's degree in electrical and computer engineering from the University of Missouri-Columbia and a Ph.D. from University of Southern California working on photonic crystal devices, high-speed nanowire photodetectors, efficient white LEDs, and solar cells. With over ten years of industry experience, Dr. Seyedi has been working on developing high-bandwidth, efficient optical interconnects for GPUs, CPUs and exascale high-performance computing systems.

Di Liang is currently a distinguished technologist and research manager at Hewlett Packard Labs in Hewlett Packard Enterprise. Dr. Liang leads the advanced R&D of silicon and compound semiconductor integrated photonics for high-speed communication, high-performance computing and many emerging applications. His research interests include silicon and III-V photonics, heterogeneous and monolithic integration, and nanofabrication technology. He has (co)authored more than 250 journal and conference papers and 5 book chapters with over 5,800 google citations, and was granted by 47 patents with another 55+ pending. He received his B.S. degree in optical engineering from the Zhejiang University, China, and Ph.D. degree in electrical engineering from the University of Notre Dame, USA. He has been invited to serve as TPC chairs/members over 30 times in a number of international conferences. He is a fellow of OSA, senior member of IEEE, and associated editor of OSA *Photonics Research* and IEEE *Journal of Quantum Electronics*.

Contributors

Geun Ho Ahn
Stanford University
Stanford, California, USA

Simon Bilodeau
Princeton University
Princeton, New Jersey, USA

Ray T. Chen
University of Texas at Austin
Austin, Texas, USA

Shixi Chen
The Hong Kong University of Science and Technology
Hong Kong, China

Yuxuan Chen
Laval University
Québec City, Québec, Canada

Sai Vineel Reddy Chittamuru
Micron Technology
Boise, Idaho, USA

Lukas Chrostowski
The University of British Columbia
Vancouver, British Columbia, Canada

Ayse K. Coskun
Boston University
Boston Massachusetts, USA

Tom Daspit
Mentor, a Siemens Business
Wilsonville, Oregon, USA

Chenghao Feng
University of Texas at Austin
Austin, Texas, USA

Jun Feng
The Hong Kong University of Science and Technology
Hong Kong, China

John Ferguson
Mentor, a Siemens business
Wilsonville, Oregon, USA

Simon Geoffroy Gagnon
McGill University
Montréal, Québec, Canada

Mustafa Hammood
The University of British Columbia
Vancouver, British Columbia, Canada

Yingtao Hu
Hewlett Packard Labs
Palo Alto, California, USA

Chaoran Huang
Princeton University
Princeton, New Jersey, USA

Jared Hulme
Hewlett Packard Labs
Palo Alto, California, USA

Nicolas A. F. Jaeger
The University of British Columbia
Vancouver, British Columbia, Canada

Ajay Joshi
Boston University
Boston Massachusetts, USA

Venkata Sai Praneeth Karempudi
University of Kentucky
Lexington, Kentucky, USA

Geza Kurczveil
Hewlett Packard Labs
Palo Alto, California, USA

Odile Liboiron Laouceur
McGill University
Montréal, Québec, Canada

Mengchu Li
Technical University of Munich
Munich, Germany

Mengquan Li
Nanyang Technological University
Singapore

Di Liang
Hewlett Packard Labs
Palo Alto, California, USA

Thomas Ferreira de Lima
Princeton University
Princeton, New Jersey, USA

Weichen Liu
Nanyang Technological University
Singapore

Sagi Mathai
Hewlett Packard Labs
Palo Alto, California, USA

Alan Mickelson
University of Colorado at Boulder
Boulder, Colorado, USA

Asif Mirza
Colorado State University
Fort Collins, Colorado, USA

Subhsish Mitra
Stanford University
Stanford, California, USA

Aditya Narayan
Boston University
Boston, Massachusetts, USA

Gabriela Nicolescu
Polytechnique Montréal
Montréal, Québec, Canada

Mahdi Nikdast
Colorado State University
Fort Collins, Colorado, USA

David Z. Pan
University of Texas at Austin
Austin, Texas, USA

Sudeep Pasricha
Colorado State University
Fort Collins, Colorado, USA

Paul R. Prucnal
Princeton University
Princeton, New Jersey, USA

Leslie A. Rusch
Laval University
Québec City, Québec, Canada

Maithem Salih
University of Kufka
Najaf, Iraq

Smruti Ranjan Sarangi
Indian Institute of Technology (IIT)
Delhi, India

Ulf Schlichtmann
Technical University of Munich
Munich, Germany

Omar El Sewefy
Mentor, a Siemens business
Cairo, Egypt

Ashkan Seyedi
Hewlett Packard Labs
Palo Alto, California, USA

Bhavin J. Shastri
Queen's University
Kingston, Ontario, Canada

Wei Shi
Laval University
Québec City, Québec, Canada

Farhad Shokraneh
McGill University
Montréal, Québec, Canada

Jinhie Skarda
Stanford University
Stanford, California, USA

Peng Sun
Hewlett Packard Labs
Palo Alto, California, USA

Febin P. Sunny
Colorado State University
Fort Collins, Colorado, USA

Armin Tajalli
The University of Utah
Salt Lake City, Utah, USA

Ishan G. Thakkar
University of Kentucky
Lexington, Kentucky, USA

Bassem Tossoun
Hewlett Packard Labs
Palo Alto, California, USA

Rahul Trivedi
Stanford University
Stanford, California, USA

Alexandre Truppel
Technical University of Munich
Munich, Germany

Tsun-Ming Tseng
Technical University of Munich
Munich, Germany

Jelena Vuckovic
Stanford University
Stanford, California, USA

Tony Wu
Stanford University
Stanford, California, USA

Chi Xiong
IBM T.J. Watson Research Center
New Haven, Connecticut
New York, New York, USA

Jiang Xu
The Hong Kong University of Science and Technology
Hong Kong, China

Zhoufeng Ying
University of Texas at Austin
Austin, Texas, USA

Jinsung Youn
Hewlett Packard Labs
Palo Alto, California, USA

Mohamed Youssef
Mentor, a Siemens business
Wilsonville, Oregon, USA

Yuan Yuan
Hewlett Packard Labs
Palo Alto, California, USA

Jiaxu Zhang
The Hong Kong University of Science and Technology
Hong Kong, China

Zheng Zhao
University of Texas at Austin
Austin, Texas, USA

Zhidan Zheng
Technical University of Munich
Munich, Germany

Section 1

High-Performance Computing Interconnect Requirements and Advances

1 Silicon Photonic Modulation for High-Performance Computing

Maithem Salih[1] *and Alan Mickelson*[2]
[1]University of Kufa
[2]University of Colorado at Boulder

CONTENTS

1.1 INTRODUCTION

Optical transmission technology became the choice for long haul telecommunication in the 1980s. The demonstration of low loss optical fiber and the room temperature semiconductor laser in 1970 fueled the telecommunication revolution that has increased link rates from Megabits per second (Mbps) in the 1960s to the hundreds of terabits per second (Tbps) aggregate rates now transmitted by 200 fiber undersea cables (Winzer, Neilson, & Chraplyvy, 2018).

Much of the story of optics in data communication has been that of standards. IEEE 802.3z Gigabit Ethernet (GbE) standard of 1998 leveled the playing field for networking in data centers (Frazier, 1998). The follow on 1999 standard for GbE in twisted pairs entrenched multimode fiber not only in high level rack-to-rack interconnects in warehouse scale data centers but also as the backbone of choice for networks of desktop computers. The climb up the bandwidth ladder from there has not been quite at the rate of Moore's law (doubling every two years) but is a factor 400 here in 2019 (Shalf, 2019). Notable along this path of doubling every 4 years was the 2006 jump to 10 GbE that retained VCSEL sources and multimode transmission, the original GbE combination. The jumps to higher rates were not so much harnessed by standards as by need. The 40 and

DOI: 10.1201/9780429292033-1

100 Gbps single mode standards of 2010 and 2013 were more milestones along a path than changes in paradigm. The tipping point in 100 Gbps transceivers that occurred circa 2015 required not only single mode fiber and laser sources but also four-channel coarse wavelength division multiplexing (J.-W. Kim, Kim, & Kim, 2019). The 100 Gbps transceiver also marked the first real commercial viability of silicon photonic (SiP) devices necessary for compact multiplexing/demultiplexing (Mux/Demux) and detection (Yu et al., 2019). In 2019, we are on the verge of widespread deployment of the 400 Gbps data center transceiver (Zhong, Mo, Grzybowski, & Lau, 2019).

More important than the speed of present-day transceivers is the versatility of the components that have become ubiquitous through transceiver development. The developments of the last five years have put integrated and semiconductor laser optics together with electrical and digital drivers through a cost test that the telecommunications of the preceding 40 years had not. Telecommunications right of way is so expensive that transceivers are in some sense "free". Repeater huts are so large that size is not an issue. The cost, thermal, and volumetric constraints of the data center transceiver package have prepared the technology for the transition to the board.

In what follows, we will discuss extant issues with the transition of optics to the board. Silicon photonics (SiP) offers compact and cost effective solutions for modulation, multiplexing/demultiplexing and detection. Sources remain problematic. Optical interconnect is most advantageous for the highest of bandwidth densities. The highest of modulation rates will still require the densest of dense wavelength division multiplexing (DWDM) to exceed electrical bandwidth densities by the requisite factors to drive the paradigm shift to on-board optical interconnect. The number of required color channels is large.

The focus in much of what follows then is the generation of optical channels. The distributed feedback lasers that have driven transceiver technology forward are too large and expensive to package to be used one per color (WDM) channel. Generation of multiple channels from single sources will be the rule rather than the exception. Discussion will turn to comb generators, methods of producing large numbers of equally spaced harmonics from high-power semiconductor lasers. The discussion will include both review and original simulations.

The presentation will follow from this introduction to a discussion of one particular problem that has the requirements for an optical interconnection solution, the memory hard problem. The third section will then discuss the device (component) needs for a peta bit per second (Pbps) optical interconnection. The fourth section then will discuss some salient features of wavelength division multiplexing in the context of optical interconnection. Methods of generating multiple channels from single sources will then be discussed in light of the detection noise problem to motivate later focus on comb generation. The discussion then turns to combs as generated by both nonlinear optics as well as RF modulation. Some discussion is given to the state of the art of commercially available comb sources. The last section summarizes. The summary includes some speculation as to future direction.

1.2 A BOARD PROBLEM WITH AN OPTICAL SOLUTION

Optical transmission technology has some positive and some negative attributes. Optics modulates, transmits point to point at near the speed of light, and detects at tens of Gbps rates. Loss is small, generally due to linear scattering. Integrated optics (here we use silicon photonics (SiP) as the specific realization) can fan out as well as perform essentially lossless wavelength division multiplexing and demultiplexing with little overhead. Optics, however, do not switch well nor perform any kind of contention resolution.

Warehouse scale data centers are large on the scale of meters. The interconnection problem is to deliver packets from one rack to another without allowing processing elements within a rack to be ever idle. There will be latency from distance. The interconnection is a mesh of transceivers with significant computing power residing at the points that the transceivers plug into racks. The computing power and inherent latency allow that there can be threading and virtualization. In fact, a data center can be a software-defined network (SDN). In this way, electronics can schedule and optics can transmit.

Board level interconnections are short on the scale of meters. Latency is crucial. There is no time for error correcting codes embedded in optical streams. Oftentimes, architects can mass cache memory about processing elements. If computational and storage patterns are sufficiently repeatable, there may be no need for different processors to interconnect at all if there are enough cache levels and processing power. There is a growing class of problems where this is not true, the class of the memory hard problems.

Caching of memory is only useful if one uses the cached bits repeatedly. Problems that require storage of ever-longer recursions cannot avail of cache. Such problems require a large volume of memory that cannot be easily accessible. Some problems have quite predicable interconnection, for example, cellular automata. Pieces of a small puzzle can be solved independently. Pieces of a large puzzle can be solved in parallel with infrequent but predictable communication between pieces. Other problems require globally unpredictable interconnection, for example, training of neural networks. Such training algorithms require all to all interconnections that range from the sparse to the dense, with the only constancy being randomness.

Electrical interconnection does not apply well to memory hard problems. Optical interconnection can, if the optical network can be kept contention free. Our focus here, though, is more on size than contention. Electrical interconnection is already being used for training in deep AI. However, there is a limit to the size of the problem. In what follows, we want to see how optics can be used to scale to ever larger memory hard problems, well beyond the scale to which electrical interconnection can compete. Our focus will be on using ever more complex circuits in SoI.

1.3 PETABYTE OPTICAL INTERCONNECTION

Recent success in deep learning has inaugurated a renaissance in the application of artificial neural networks. The training of a convolutional neural network is an archetypical memory hard problem. The time taken to train a convolutional neural network (CNN) can be a significant number of orders of magnitude greater than the classification time. The real power of a CNN, though, is continual improvement through adaptation. The requirement is repeated training as new data becomes available. The retraining of autonomous vehicles can become both a time and energy barrier to real-world application.

Even before the recent success in deep learning, it was known that graphical processing units (GPUs) could be used as training accelerators. Their use is presently ubiquitous. The use of 8 NVIDIA Tesla P40 accelerators is said to increase training speeds by as much as a factor of 140 for the state of the art processors now in 2019. These present GPU boards operate at 12 Tera floating-point operations per second (12 TFlops), 47 tera operations per second (47 TOps), and have bus rates of up to 2.8 tera bits per second (Tbps). Scaling to ever-larger CNNs, however, poses a problem. The quoted rates require that the electrical connections be short, that is, at most, the length from a surrounding accelerator unit to a processor.

Let's say we would like to train a CNN that consists of N_g groups where a group may consist of an accelerated processor with associated GPUs, either external or internal to a chip. Let's assume we have attached optical modulators to a number N_c of electrical contact points within a group. We will further assume that each of the N_c modulators can be coded with a different color of a dense wavelength division multiplexing (DWDM) table. Such an interconnection would be N_g times larger than the single group interconnection that was discussed above. The larger CNN should allow for much more complex problems to be addressed.

Can an optical interconnection achieve the complexity and bandwidth necessary to accommodate larger CNNs? A 2017 paper (Khanna et al., 2017) notes that already in 2016, silicon photonic circuits had been demonstrated with 10^4 circuit elements. It was noted in the same paper that complexity was doubling each 12 months indicating that circuits with more than 10^5 devices (circuit elements) should presently exist. Optical bandwidth is plentiful. At 1,550 nm, 1 nm is equivalent to roughly 125 GHz. At a bit per Hz (roughly achievable by on off coding), the 300 nm from 1,400 nm to 1,700 nm should contain roughly 37.5 THz of bandwidth that could be used to code a 37.5 Tbps in signals. With tens of spatial channels running in parallel, cut through bandwidths in excess of a petabit per second (Pbps) should be obtainable. The question is can this be achieved with 10^5 of today's components (devices).

Silicon photonic optical modulators presently achieve rates of 50 Gbps with on-off keying. These rates can be detected with silicon photonic Si-Ge detectors at bit error rates (BERs) smaller than 10^{-9} if power levels exceed circa 25 mW. Multiplexer-demultiplexer pairs with more than 512 channels have been demonstrated in SiP as well. A silicon photonic interconnection if on a large enough substrate could then connect all of the N_c connection points of N_g groups. If the SiP substrate were not large enough, interposers could be used to expand the number of groups.

An $N_g \times N_c$ all to all interconnection would require that each of the N_g groups would connect all of its N_c channels to all of the N_g groups including its own. Each of the groups could then electrically connect the N_c colors to each of their own N_c colors. Connecting each of the N_g to all of the N_g will require fan out and likely amplification. Both fan out and amplification have been demonstrated.

The cut through bandwidth of an $N_g \times N_c$ interconnection with each of the N_c points transmitting at 50 Gbps would be 50 $N_g \times N_c$. For $N_g = N_c = 64$, this number would slightly greater than 2 Pbps. This interconnection would be N_g as large as a present-day electrical interconnection on a single accelerated processor and thereby able to address more complex problems in AI. The question of feasibility is really of cost and size as the components have been demonstrated but the complexity not.

1.4 MULTIPLEXING

Figure 1.1 illustrates a typical wavelength division multiplexing (WDM) system. WDM is a crucial element in the achievement of the Pbps rates as was noted above. In fact, the Pbps requires three types of multiplexes, time division multiplexing (TDM), space division multiplexing (SDM), and wavelength division multiplexing (WDM). This combination of multiplexing is also the basis of the telecommunications network.

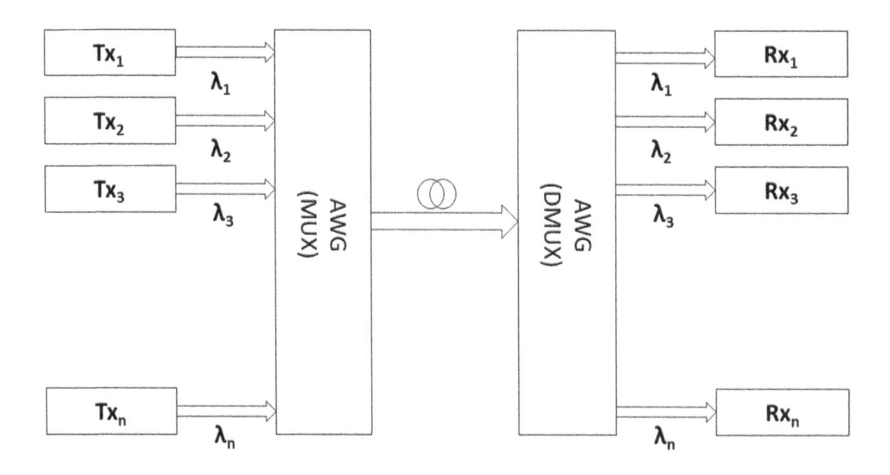

FIGURE 1.1 An illustration of a WDM system.

1.4.1 Time and Space Division Multiplexing

Implementation of time division multiplexing (TDM) was the first step in the creation of the digital telecommunications network. A single digitized voice channel requires roughly 64 kbps, that is, a 4 kHz bandwidth for human voice sampled at the Shannon rate of twice the 4 kHz bandwidth (8 kbps) and digitized into 256 (2^8) levels. Optical fiber first became necessary when city trunk lines that generally span two kilometers required 45 kbps rates in the mid- to late 1970s. The ducts in which cables lay could fit only so many cables or fibers, that is, the ducts were limited in the degree of space division multiplexing they could support. Multimode fiber was smaller than cable and higher bandwidth (at the time (1975), 100 Mbps/km) and thus optical communication was born.

The bandwidth of single channels of silicon photonics waveguide interconnection at present is limited by the modulators that insert information on the optical carriers rather than by waveguide dispersion. Modulators can operate up to roughly 50 Gbps at present. Clock rates on CMOS chips have saturated at roughly 4 GHz. That is, the highest rates of digital transmission are at a rate of 8 Gbps. Most on chip rates are much slower. Serialization, deserialization units (SERDES) on chips can boost these rates significantly, for example, up to the 50 Gbps rates that modulators can sustain. Even at 50 Gbps rates, a Pbps

1.4.2 Wavelength Division Multiplexing

The prism dates back to the ancient Romans and well embodies the basic principle of WDM. Most implementations of WDM are simply guided wave versions of splitting and combining with prisms. Channels of different colors propagating in the same direction can be either placed on top of each other or spread out to propagate in spatially separated parallel channels. A problem with circa pre-1990 guided wave implementations of WDM was amplifying the multiplexed signals without demultiplexing.

In the late 1980s, the telecommunication rates peaked out around 2.5 Gbps. When erbium doped optical amplifiers with greater than 4 THz amplification bandwidths in the 1,550 nm band became available, wavelength division multiplexing (WDM) became the telecommunication transmission standard.

Silicon is transparent from its bandgap wavelength of 1.1 micron out to roughly 10 microns. Present-day low-cost optical sources are more available in the range of 1.4 to 1.7 micron, a 300 nm wavelength range. A 50 GHz channel has a width of 0.4 nm at a center wavelength of 1.5 microns. With a channel guard band of one channel width, almost 400 50 GHz channels could be provisioned into the 1,400–1,700 nm wavelength range. Arrayed waveguide gratings multiplexer-demultiplexer pairs with 512 channels have been demonstrated indicating that the hardware for inserting 37.5 Tbps of information bandwidth between 1.4 and 1.7 microns can be produced that can multiplex such information densities onto a single silicon waveguide.

A question arises as to the quantity of silicon photonic real estate that would be taken up by multiplexing, demultiplexing and detection in a Pbps interconnection. A study of a number of papers in the literature (Cheung, Su, Okamoto, & Yoo, 2014; Fukazawa, Ohno & Baba, 2004; Paiam & MacDonald, 1998; Weimann et al., 2014) that used arrayed waveguide gratings (AWGs) for multiplexing and demultiplexing indicated that an N channel interconnection required 400 N^2 $micron^2$. Each of the N^2 outputs of the guides would require another roughly 200 $micron^2$ for detectors or another 50% in area. Other configurations such as a ring based system grows only linearly in N (Bogaerts et al., 2012). The loss of a ring arrangement, however, grows more rapidly with N. Other arrangements such as those that result from constrained optimization (Su, Piggott, Sapra, Petykiewicz & Vučković, 2017) Su 2017 result in more compact devices but have not been scaled to anywhere near the size of AWG arrangements.

In the above, we considered an archetypical 2 Pbps interconnection as a 64 chip interconnect in which 64 pins on each of the 64 chips would be all to all interconnected. Such an interconnection would require one modulator per pin and one multiplexer per chip in order to transmit. Each chip would need to be transmitted to each other, requiring one 1 to 64 fan-out per chip. On receive, each chip would require 64 incoming channels, 64 demultiplexers (one for each incoming channel) and then 64^2 detectors. The highest component counts are for the demultplexers and detectors.

A chip receiving from all other pins and chips would then have 64 incident waveguides, each multiplexed with 64 color (color) channels. Sixty-four WDMs would require $400 \times 64 \times 64^2$ $\text{microns}^2 = 1.05\ \text{cm}^2$. This is large but the receiving chip is circa 2 cm × 2 cm of 4 cm^2 so the multiplexers would fit under the receiving chip. Even with another 50% of the area for detectors (maybe 100% including pads), the demultiplexing arrangement would fit underneath a usual size processor chip. Other arrangements could be more compact.

A problem we did not address above is the generation of optical carriers. In our archetypical Pbps interconnection, we will require 64 modulators, one per electrical pin. Lasers are fabricated in InGaAsP. They cannot be (satisfactorily) grown monolithically on silicon. They must be bonded. If sources are not frequency (wavelength) locked, they also would all require precision temperature control in order to work in a DWDM system. Such a solution is not tenable. Slicing up the spectrum from a single broadband optical source results in carriers that have a fixed wavelength relation, eliminating individual wavelength control. Broadband sources, though, are noise sources. There are receiver noise consequences for slicing a noise signal. Techniques for generating wavelength combs from a single laser output, though, are becoming more advanced with time. We discuss and simulate some of these sliced as well as comb techniques in the next two sections.

1.5 GENERATING MULTIPLE OPTICAL CARRIERS

In Figure 1.1, each transceiver (TXi) is fed by a different wavelength. Here, we consider two different ways to carry that out. In the following section, we will elaborate on the presently more promising of those techniques, that of comb generation.

1.5.1 Optical Spectral-Slicing

Light-emitting diodes (LEDs) in the 1,550 nm band have 3 dB linewidths as broad of 70 nm. Super-luminescent LEDs (SLEDs) may have linewidths ranging from 40 to 100 nm. If a 50 Gbps information stream requires 0.4 nm, then 50 Gbps streams spaced by 1 nm could be used for transmission if the lines were to keep their relative locations in wavelength space. Slicing techniques could allow for this (Lee, Chung, & DiGiovanni, 1993). That is, an arrayed waveguide grating (AWG) with fixed spacing would always leave the center wavelengths and guard bands unaffected so long as the total linewidth of the source were greater than the sampling width of the AWG. If the AWG split a 64 nm linewidth input into 64 channels, 0.4 nm channels, 64 carriers would be generated for later multiplexing. This would be an example of spectral slicing of high brightness single spatial mode (perfect spatial coherence) sources (Akimoto, Kani, Teshima, & Iwatsuki, 2003; Han, Kim, & Lee, 1999; S.-J. Kim, Han, Lee, & Park, 1999). Indeed, an SLED can be generated in a spatially coherent single spatial mode and the splitting could take place almost losslessly in a single mode AWG.

Now, N is not limited to 64 nor is the data rate to 50 Gbps. N could be larger. A limitation, though, is the received power. A 100 Gbps receiver bandwidth results in an effective thermal noise ($kT\Delta f$ electrical) of 10 μW optical. A BER rate of 10^{-12} (without error correction) will require then circa 80 μW of optical power. If an optical link loss budget is 9 dB, then each of the channels will require a greater than 1 mW carrier taking into account the 3 dB modulation loss (Personick, 1973).

Choices for broadband sources include LEDs, erbium doped fiber amplifier EDFA, and super-luminescent light emitting diode SLED. An LED doesn't have enough power to generate even a few channels at 1 μW (Reeve et al., 1988). However, both the EDFA and the SLED have a problem that when sliced, the noise increases (sampling from a finite population or a Gaussian standard deviation depends on the number of samples). The problem is the relative intensity noise (RIN) that arises from splitting signals from a single source (Dravnieks & Spolitis, 2017; Hung, Lee, Sung, Hsu, & Lee, 2017; Kilkelly, Chidgey, & Hill, 1990; Lee et al.,1993; Wagner & Chapuran, 1990). The splitting noises are often classified as mode partition noise and amplifier spontaneous emission noise.

Mode partition noise is exhibited by sources that switch modes slowly enough that those modes can be observed independently. A multimode laser may exhibit tens of different spikes. The output power (integrated over the lines) may be quite constant and determined by the pump. If one tries split each mode into a different channel, one finds those modes are amplitude coupled. The amplitude of each spectral spike varies rapidly and randomly. When the modes hop so randomly that they appear as a continuum as in an LED or an SLED, we refer to the noise as spontaneous emission as there is no noticeable effect of the cavity on the noise spectrum.

Amplified spontaneous emission (ASE) noise is noise that is generated by a spontaneous emission source when the gain in an amplifying region is so strong than many of the individual photons events are stimulated. A fiber laser below threshold may exhibit constant wavelength integrated power but large wavelength dependent relative intensity noise (RIN). When one uses a set of rings or an arrayed waveguide grating (AWG) to split the spectrum, each spectral slice exhibits significant RIN. It is sometimes called spontaneous-spontaneous beat noise, which consists of a dc part arising from the beat between the same optical frequency components and an ac part due to the beat between the different frequency components (Giles & Desurvire, 1991; Olsson, 1989). Thus, when an ASE source is used as a WDM light source, we may consider the dc ASE power I_{ASE}^2, as carrier and the time-varying ac part I_{sp-sp}^2, as noise. Passing the ASE spectrum through a strongly gain saturated semiconductor optical amplifier (SOA) can limit the wavelength peaks and flatten the gain spectrum. This effect is similar passing the signal through a flat-topped passband filter. However, any form of filtering has a cost in both power and bandwidth.

1.5.2 Noise in a Slice from an SLED

The signal-to-noise ratio (SNR) of a spectral slice at the receiver from an incoherent source such as an EDFA or SLED is given by (Lee et al., 1993)

$$SNR = \frac{I_{ASE}^2}{I_{sp-sp}^2 + I_{shot}^2 + I_{ckt}^2} \tag{1.1}$$

where I_{shot}^2 and I_{ckt}^2 are the noise power produced by the ASE shot noise and the receiver electronics, respectively. In a SLED source, I_{sp-sp}^2 is dominates the electrical noise and limits the total transmission capacity. As shown in Lee et al. (1993) when one neglects the electrical noise, one can write Eq. (1.1) in the form (where $SNR = I_{ASE}^2/I)_{sp-sp}^2$)

$$SNR \approx \frac{BW_O}{BW_E} \tag{1.2}$$

where BW_O is the optical bandwidth of the channel and BW_E is the electrical bandwidth, with $BW_E = 0.7B$, and B is the bit rate. Thus, the Q-factor at the receiver is given by (Personick, 1973)

$$Q \approx \sqrt{SNR} \tag{1.3}$$

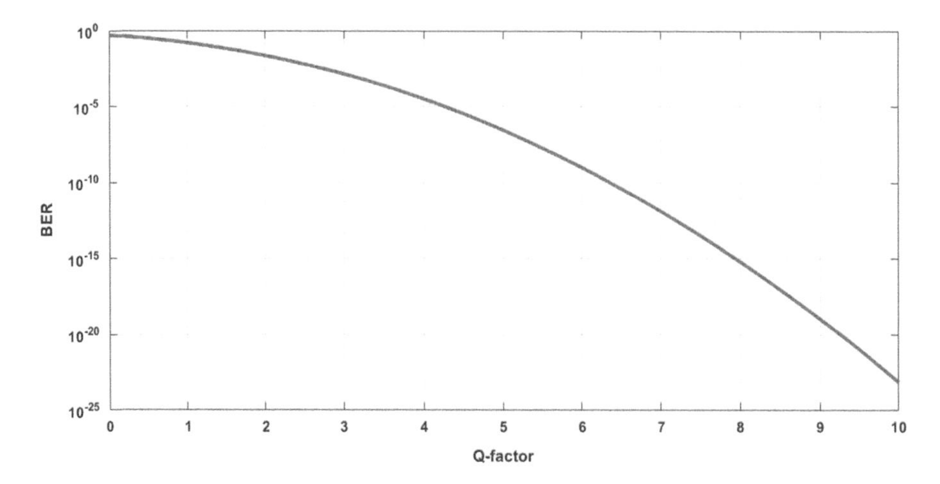

FIGURE 1.2 Bit error rate (BER) vs. Q-factor.

Q = 3 is required for BER = 10^{-3} and (Q = 8) for BER = 10^{-15}, as shown in Figure 1.2. That means BW_O needs to be from 6 to 40 times wider than BW_E. This is serious limitation if one desires 100 channels from a source. For example, a 1 nm (@1,550 nm center wavelength) spectral width is roughly a 125 GHz bandwidth, then 100 nm (@1,550 nm) results in rough 12.5 THz.

One hundred channels would require converting all of the bandwidth (zero channel space) to modulation. That needs $BW_O = BW_E$ which would result in a BER = 10^{-2}. Such an error rate cannot be corrected. Forward error correction (FEC) requires BER = 10^{-3} and results in latency of microsecond or more. This is useable for data centers but not HPCs. For an error rate of 10^{-15} requires a Q of 8 or a channel spacing of roughly $BW_O = 40B$. This would be low latency (HPC) but only 100 Gbps per source or roughly a single channel of SLED or EDFA.

For the present, let's consider 1 nm slice of spectral width in order to discuss the trade-offs between bandwidth, power, noise, and dispersion. Some of competing effects include:

1 A thicker (in terms of nm) slice carries more power (to generate higher BER) and also will average over a larger spectral area to minimize relative intensity noise (RIN).
2 A thicker line has a dispersion penalty. There is a limit with how thick a line can be used. The limit we set for ourselves during our first transceiver design was 1 nm in order to gather enough power for a 2 km link. Some numbers
   a One nm at 1,500 nm at 1,550 nm is very close to 125 GHz. There would be no dispersion penalty for a 125 Gbps modulation rate for 1 nm. There is no dispersion penalty for lower rates.
   b At 1,300 nm, 1 nm subtends more Hz, and, in fact, is close to 175 GHz. Again, there would be no dispersion penalty at a modulation bandwidth of 175 GHz if we were using 1 nm at 1,300 nm. (175 GHz is really not quite the same as 175 Gbps. I used 1 Hz equals 1 bps above. 1 bps really requires more than 1 Hz.)
   c 0.3 nm at 1,300 nm would be around 50 GHz. We did not use this previously because of the power penalty.

The next plot illustrates the SNR of 50 Gbps modulated slices of 1, 3.5, and 10 nm SLED slice as a function of increasing SLED power. This is compared with the SNR of a 50 Gbps modulated comb tooth (to be discussed in the next section). The two SNRs are identical at small signal strength but sharply diverge at higher received power. The SNR of the comb tends to infinity,

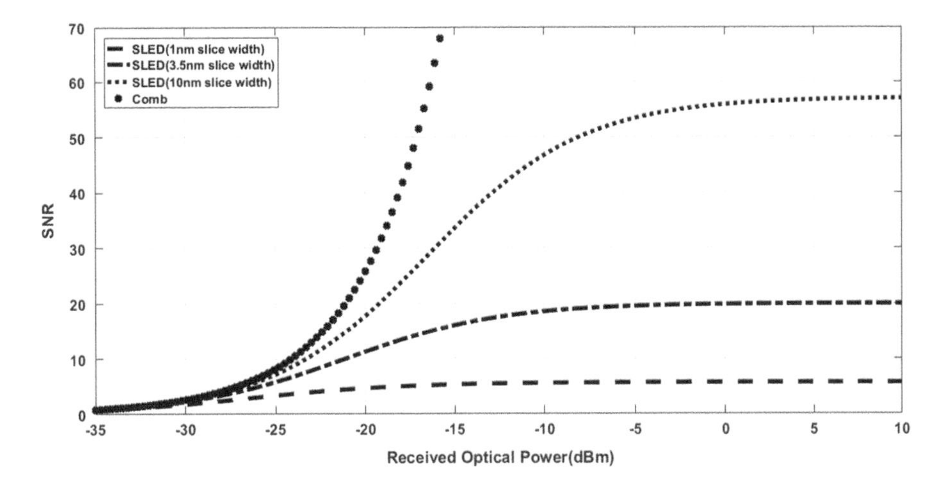

FIGURE 1.3 Signal-to-noise ratio (SNR) as a function of the received optical power of a 50 Gbps modulated sliced of 1, 3.5, and 10 nm SLED slices (dashed, dash-dot, and dotted, respectively) vs. a 1 nm comb tooth (asterisk) modulated with 50 Gbps.

reaching 60 (about Q = 8, and BER = 10^{-15}) at about -16 dBm optical received power, whereas the SNR of the sliced SLED converges to about 2.5, 9, and 31, respectively (Figure 1.3).

We see that slicing is problematic. The problem stems from the spectrum consisting of noise without any carrier so to speak. Any modulation of noise corresponds to rapidly turning the noise on and off. When one modulates a narrowband source, however, one adds sidebands to the delta like carrier. There is no excess noise. A long line of spike-like carriers in frequency space is referred to as a frequency or wavelength comb.

1.6 OPTICAL COMB GENERATION

An optical frequency comb (OFC) is defined to be a spectrum that consists of a series of equally spaced discrete lines. Such spectra are used in applications that range from microwave photonics (Shao et al., 2015) and millimeter wave and THz generation (Thorpe, Balslev-Clausen, Kirchner, & Ye, 2008) to optical transceivers and interconnection (Vujicic et al., 2015).

1.6.1 Properties of an Optical Comb

In the time domain, a comb is any pulse shape limited to a time period T centered on zero convolved the delta-like pulse train

$$f(t) = \sum_{n=-\infty}^{+\infty} \delta(t - nT) \tag{1.4}$$

The convolving factor Fourier transforms to the factor

$$F(\omega) = \sum_{n=-\infty}^{+\infty} \delta(\omega - n\frac{2\pi}{T}) \tag{1.5}$$

In reality the comb will be limited to a finite number of repeat periods depending on the bandwidth of the equipment that generates the signal. The repetition in both time and frequency, though, leads to useful mathematical properties (Deseada Gutierrez Pascual & Barry, 2017).

A comb signal may be generated from a coherent laser source. The spontaneous-spontaneous beat noise I^2_{sp-sp} of such a source is negligible because of the large optical bandwidth but the electrical noise is significant. The SNR of a single tooth of a comb at a receiver is then given by (Kaneko et al., 2006)

$$SNR = \frac{electrical\ power}{electrical\ noise\ power} = \frac{Rs^2P^2R}{4kTBW_E} \tag{1.6}$$

where Rs is the responsively of the detector, P is the received optical power, R is the load resistance that generates the electrical thermal noise ($4kTBW_E$), k is Boltzman's constant, T is temperature, and BW_E is the electrical bandwidth. As one increases the optical power, the SNR increases without bound. If one generates a comb and then slices it, each slice carries away the coherent optical power but no appreciable noise power assuming the teeth of the comb are uncoupled. The noise in each of the channels is dominated by the thermal noise of the receiver circuit.

The choice of channel spacing for modulation of a low coherence, high brightness source fixes the maximum SNR that can be achieved, independent of power level at the detector. For a coherent source, increasing the power to the detector always improves the SNR. That is what would be predicted if strong coherent sources, such as laser diodes, were used to generate carriers.

Driving a free running oscillator at its oscillation frequency can produce harmonic spiking. Providing positive feedback to an optical oscillator at a multiple of the round trip time can produce spiking at cavity harmonics. In a Ti:Sa laser, spiking at a multiple of the cavity round trip time can be produced by carefully controlled feedback (Huang, Jiang, Leaird, & Weiner, 2006). Driving a laser with an external electrical signal at the relaxation oscillation frequency can produce tens of harmonics. What we focus on here are mechanisms that are most applicable on-chip comb generation.

1.6.2 Generating Combs from Lasers and RF Sources

Mode-locked lasers (MLLs, such as a Ti:Sa as mentioned above) are comb generators (Lundberg et al., 2018). Compact integrated micro-resonators may also generate such combs. Optical frequency comb (OFC) generation can also be carried out by purely electro-optic modulation (Deseada, Gutierrez, Pascual, & Barry, 2017). Electro-optic modulation is the most natural for on-chip generation in silicon photonic chips fabricated in a foundry with a multi-user process design kit (PDK). Mode-locking generally requires bonding a cavity to the on-chip structure in an addition to a source laser.

1.6.2.1 *Single-Spaced Optical Frequency Comb Generators*

A narrow line continuous wave source modulated with a set of overdriven modulators (Dou, Zhang, & Yao, 2012; Fujiwara et al., 2003; He, Pan, Guo, Zhao, & Pan, 2012; Metcalf, Torres-Company, Leaird, & Weiner, 2013; Sakamoto, Kawanishi, & Izutsu, 2007; Shang, Li, Ma, & Chen, 2015; Tran, Song, Song, & Seo, 2019; Veselka & Korotky, 1998; Wu, Supradeepa, Long, Leaird, & Weiner, 2010) generate combs that exhibit the characteristics of the RF drive signals. Figure 1.4 illustrates two systems of electro-optic modulators used for comb generation. In one case, an intensity modulator (IM) followed by a cascade of phase modulators (PMs). In the other

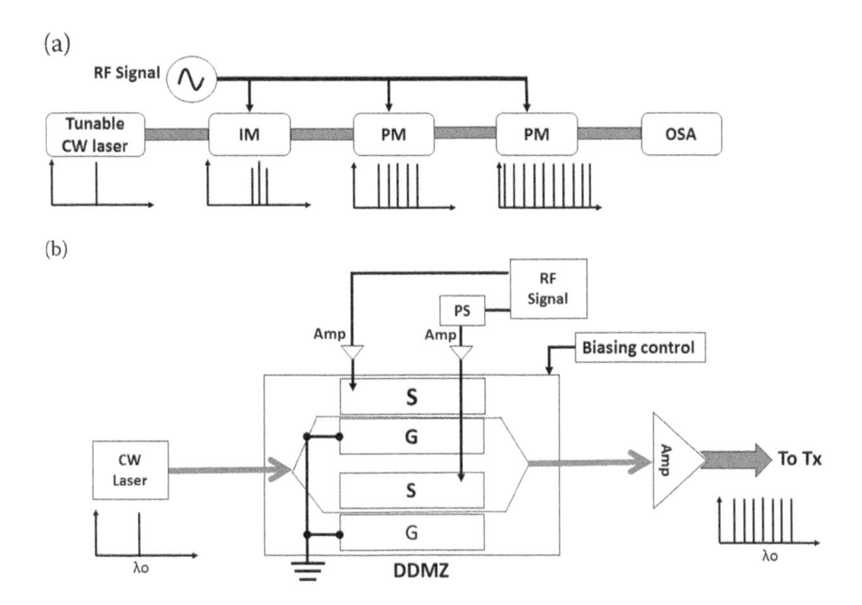

FIGURE 1.4 Two types of electro-optic modulator-based comb generators. (a) Using a cascade of intensity and phase modulators (IM-PMs cascade). (b) Using a dual-drive Mach-Zehnder modulator (DDMZM).

(Tran et al., 2019), a single stage of a conventional dual-drive Mach-Zhender (MZ) modulator (Sakamoto et al., 2007) is used.

In both cases, the RF drive signal operates at the frequency of the tooth spacing. The dual-drive MZs are optically phased to subtract but electrically tuned to be $\pi/2$ out of phase in the two arms. This phasing subtracts to produce, to first order in the modulation depth, a single sideband. With other bias schemes, such an arrangement can be used to generate multiple sidebands. Spectral flattening is important (Weimann et al., 2014). Flattening is obtained when the flattening condition (Sakamoto et al., 2007)

$$\Delta A \pm \Delta\theta = \frac{\pi}{2} \tag{1.7}$$

where $\Delta A = (A_1 - A_2)/2$, A_1 and A_2 are amplitudes of RF signals, is satisfied. $2\Delta A$ is the peak-to-peak phase difference induced in each arm. $\Delta\theta = (\theta_1 - \theta_2)/2$, where $2\Delta\theta$ is the dc bias difference between the arms. A cascade of intensity and phase modulators can be used to generate flat-top spectral lines. In such a case, the pulse shape is produced by an intensity modulator before a periodic linear chirp is applied to the phase (Tran et al., 2019). A comb with many with roughly equal amplitudes can result from a cascade of IMs and PMs as shown in Figure 1.5.

The spectral profiles shown in Figure 1.5, is quite flat. Using such a technique, it should be possible to convert a one watt source to 200 lines of 5 mW per line. A spacing of 50 GHz would then result in 10 Tbps of bandwidth for transmission. A thousand 1-watt sources could generate a comb that could be modulated with a Pbps of information. The 5 mW used for each comb peak would allow for a 20 dB loss budget and still result in 50 μW per channel. 50 μW result in a Q of greater than 8 for a BER less than 10^{-15} s this would require no error correction. The resulting bandwidth of about each comb tooth is still too narrow to allow for guard bands about the information stream.

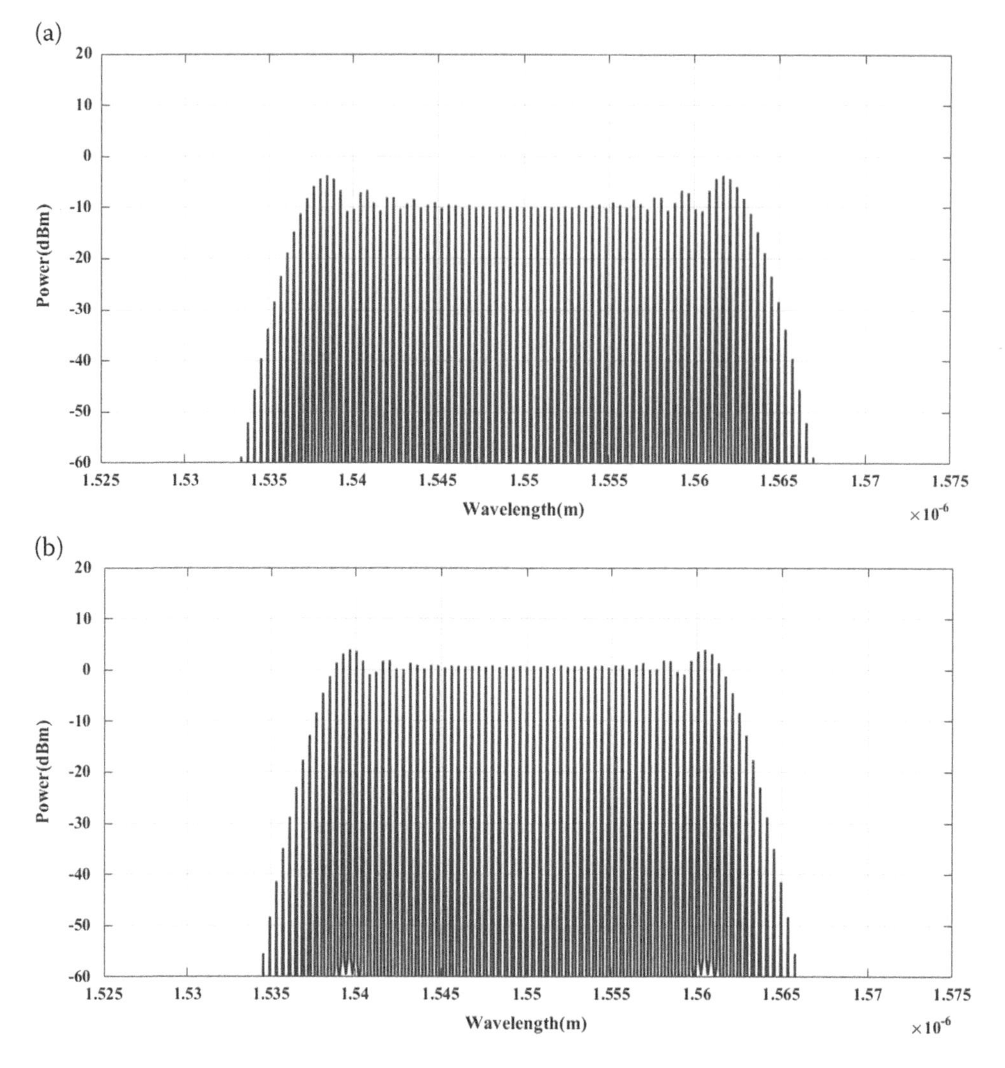

FIGURE 1.5 Spectral profile of an optical frequency comb signal with 48 lines and 50 GHz space that's generated by (a) DDMZM, (b) IM-PMs cascade.

1.6.2.2 *Double-Spaced Comb Generation*

In Sakamoto (2017) and Sakamoto and Chiba (2017), flat optical combs are generated with a set of phase modulators. In double-frequency-spaced optical comb generation, four phase modulators (PMs) are used, as shown in Figure 1.6. A single stage dual-parallel Mach-Zhender (MZ) modulator (DPMZM) could replace these four PMs, where each arm is driven by an RF signal as illustrated below

$$\begin{aligned} RF1 &= A_1 \sin(wt + \varphi_1) + \theta_1 \\ RF2 &= A_2 \sin(wt + \varphi_2) + \theta_2 \\ RF3 &= A_3 \sin(wt + \varphi_3) + \theta_3 \\ RF4 &= A_4 \sin(wt + \varphi_4) + \theta_4 \end{aligned} \tag{1.8}$$

where the amplitudes are $A_1 = A_2$ and $A_3 = A_4$, the initial phases are $\varphi_1 = \varphi_3 = 0$, and $\varphi_2 = \varphi_4 = \pi$,

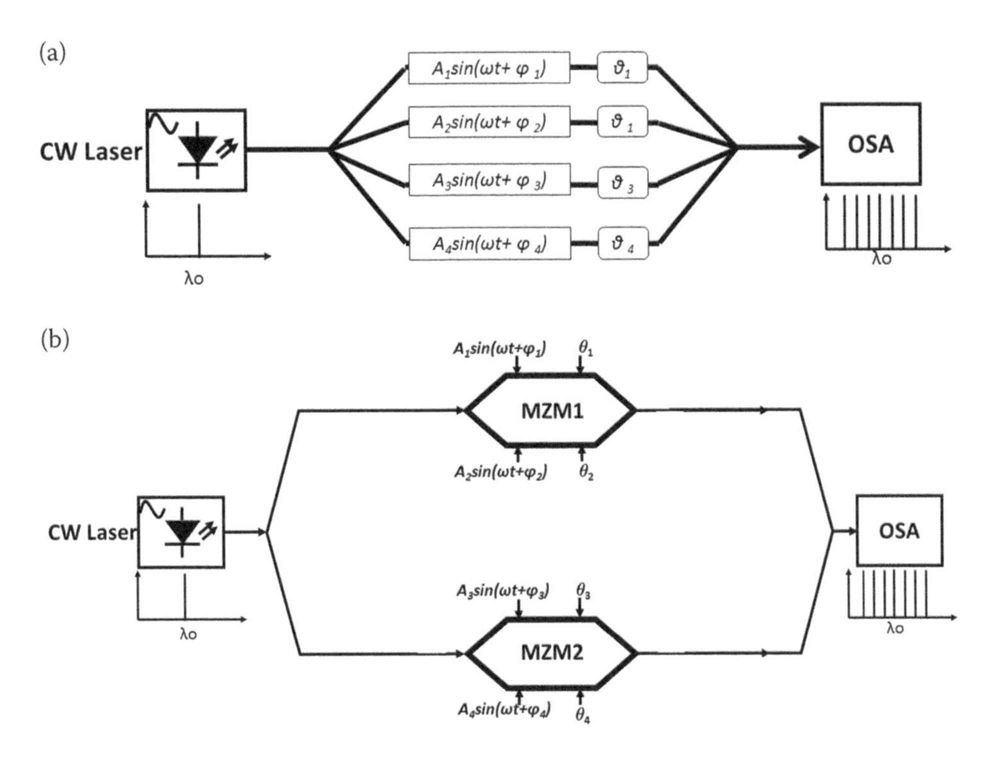

FIGURE 1.6 Basics of a double-frequency-spaced optical comb generator using (a) four phase modulators, (b) a single dual-parallel Mach-Zehnder modulator (DPMZM).

and the phased which induced by biasing voltages are $\theta_1 = \frac{-\pi}{2}$, $\theta_2 = \frac{\pi}{2}$, $\theta_3 = 0$, and $\theta_4 = -\pi$. With keeping the flatness condition between both upper and lower MZs as shown in Eq. 1.7 flat-top comb with double-space frequency could be generated, as shown in Figure 1.7a.

Also, we could apply same of above conditions on four driven PMs in four IM/PM cascade system (Tran et al., 2019) and then combine the outputs, as shown in Figure 1.7, to get an OFC with double-space frequency, as shown in Figure 1.7b.

1.6.3 General Model of a Comb Generator

A conclusion to be made from the previous sections is that it is not possible space channels from a super-luminescent source close enough together to satisfy the needs of a high-performance computer. With an electro-optic comb, one can set the RF spacing to that required for the information streams.

An important consideration with the combs is the spacing. If we modulate at BW_E Gbps (where $BW_E = 0.7B$, and B is the information bandwidth (Lee et al., 1993)), then the teeth must be spaced by at least $BW_O = 2BW_E$ to accommodate the double sidebands of an intensity modulator. If a modulator at $BW_E = 50$ Gbps (requires 35 GHz of bandwidth to pass) on a line at center frequency f_{opt}, it will require a resulting spectrum from f_{opt} - $0.7B$ to f_{opt} $+0.7B$ clear to pass the signal. If our spacing then is $2BW_O$ and we have two lines, we will have a composite spectra from $f_{opt} - BW_E$ to $f_{opt} + BW_E$ and from $f_{opt} + 2BW_O - BW_E$ to $f_{opt} + 2BW_O + BW_E$. We need $f_{opt} + BW_E < f_{opt} + 2BW_O - BW_E$. So, BW_E has to be greater than BW_E. This will give us a $0.6B$ guard rail between the lines (using $BW_E = 0.7B$). Figure 1.8 illustrates the above.

A double-spaced OFC generator is a promising solution for high-speed optical transceivers.

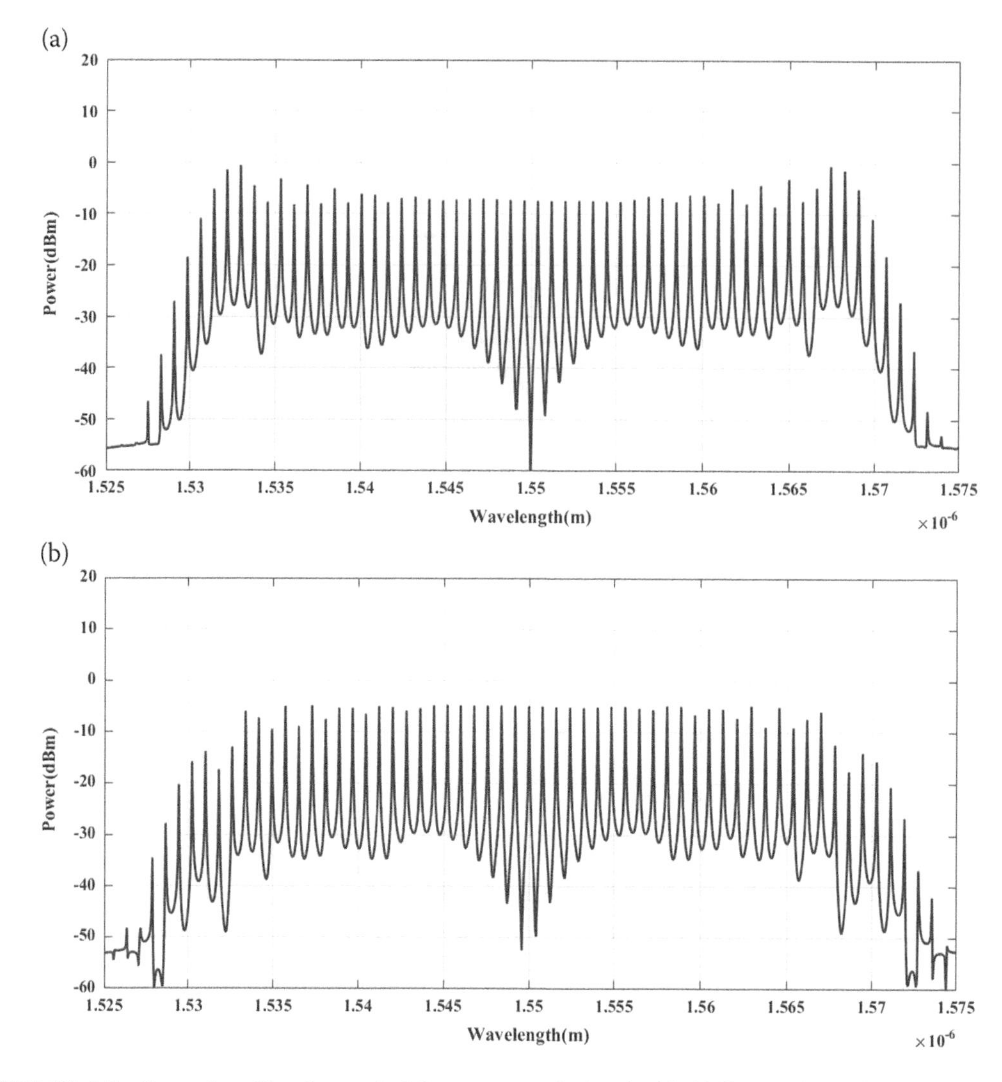

FIGURE 1.7 Spectral profile of an optical frequency comb signal with 30 lines and 100 GHz space that's generated by (a) DPMZM, (b) two lines of IM-PMs cascade.

1.7 MODELING A TRANSMITTER IN A HIGH-SPEED OPTICAL INTERCONNECTION

In this section, we will present a model of an optical transmitter in a high-speed optical interconnection using an optical frequency comb (OFC).

The optical field of a temporally and spatially coherent source can be expressed in the form

$$E_o(t) = E_0 e^{jw_0 t} \tag{1.9}$$

where E_0 is the amplitude of the optical field and w_0 is a central radian frequency of this optical field. In modeling one of the double-space frequency comb generation techniques, especially the combination of four IM/PM cascade lines, the optical field in Eq. (1.9) will be divided into four equal intensity (one quarter in intensity is one half in amplitude) parallel dual drive conventional IMs each one is driven by two RF signals, where the output of each IM (assuming a 3 dB intensity modulation loss) will be

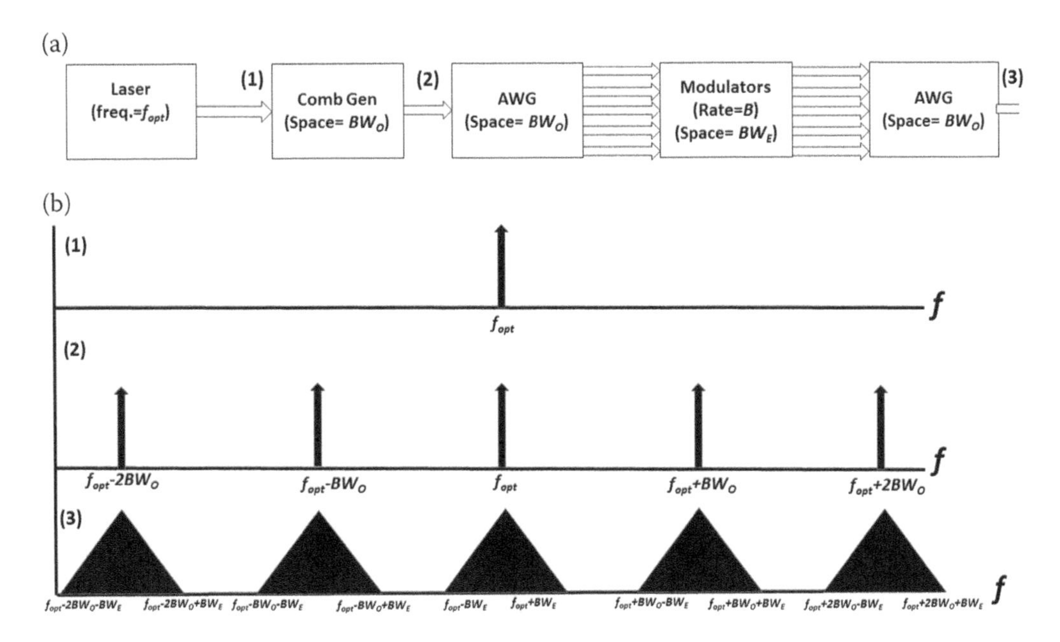

FIGURE 1.8 (a) A block diagram of combing a signal from a laser source with BW_O spectral space, then each tooth in the comb is modulated at B. (b) The spectral profile of the signal according to the location as indicated in (a).

$$E_{IMx} = 0.25E_{in}(t)\left(1 + \exp(j\pi\left(\frac{V_1 \cos(wt) + V_2 \cos(wt)}{V_\pi}\right)\right) \tag{1.10}$$

where V_1 and V_2 are the amplitude RF signals, w is the corresponding angular frequency of the RF signals, and V_π is the required applied voltage to change the phase of an optical field by π. The output of each IM (Eq. 1.10) is modulated by a phase modulator. The phase modulators are driven by RF signals with the same amplitude and corresponding angular radio frequency (w) but shifted and biased as illustrated in Eq. 1.8 (Section 1.6). The combination of the output fields of these four PMs is appeared as an optical comb signal (with roughly 40 peaks) with a double-space frequency ($2w$ = 100 GHz) and 5 dBm output power with 1 dB flatness variation as illustrated in Figure 1.7(b).

As illustrated in Figure 1.9(a), an AWG could be modeled as a demultiplexer, as in Tsao and Lin (2004), to spread out these comb lines in individual channels. Each line then is carried by information, which is modeled as a Pseudo Random Binary Stream (PRSB) at 50 Gbps rate via a MZM, which is the maximum rate of the available recent modulators, as presented in Figure 1.9(b).

After impressing the information streams on the lines but before coupling them to the interconnection fiber or waveguide, the modulated comb peaks are multiplexed again, using the model of 64 × 64 AWG as in Tsao and Lin (2004), as shown in Figure 1.10.

Figure 1.10(b) presents the spectral profile of a multiplexed comb signal with about 40 modulated lines. The spectrum represents about 2 Tbps information rate and occupies about 30 nm of a spectral width. Each line has a greater than -10 dBm output power level. If we increase the input optical power of the CW laser by 10 dB (from 13 to 23 dBm), the comb signal will exhibit a capacity of 2 Tbps. So if we assume that the loss budget through this interconnect is not tdoes not exceed 16 dB, a 10^{-15} BER could be achieved without any error correction. The system is illustrated as in Figure 1.11.

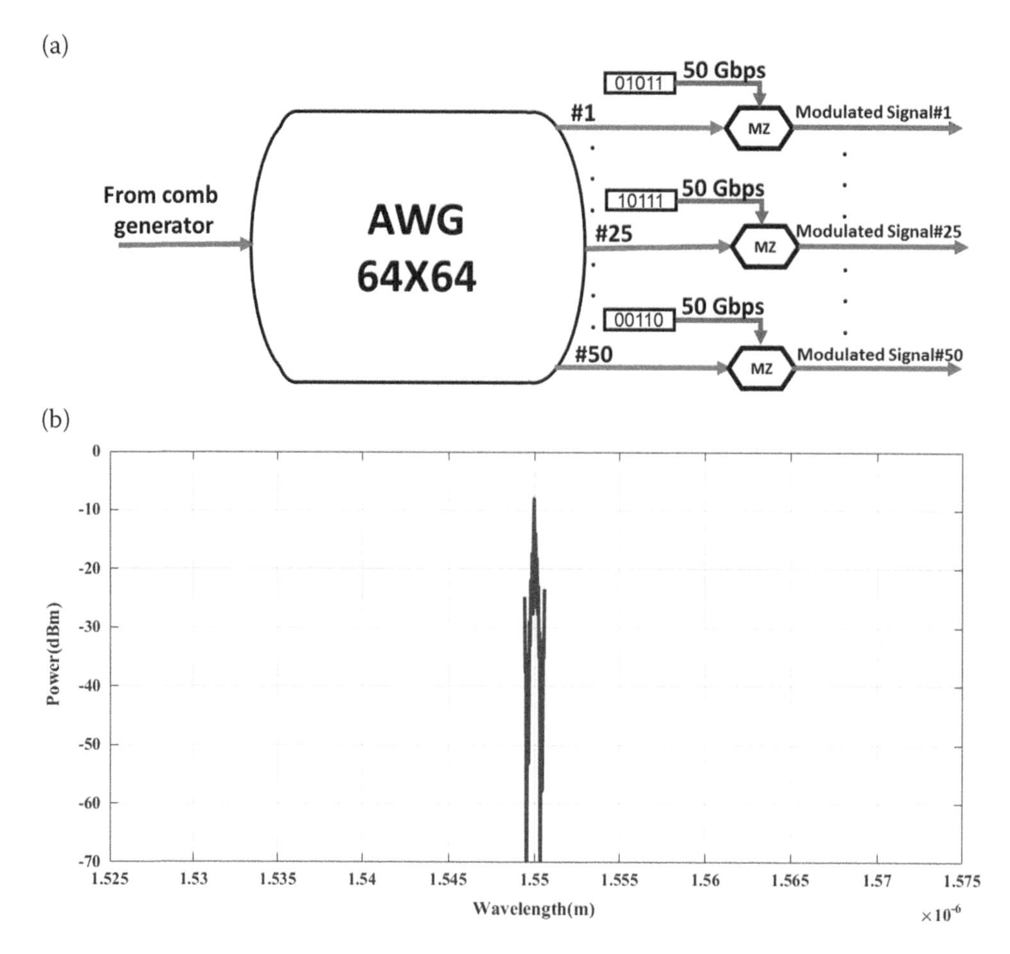

FIGURE 1.9 (a) Demultiplexing of the comb signal using 64 x 64 AWG and applying data on each line @ 50 Gbps. (b) The spectrum of a modulated signal by a 50 Gbps MZ modulator.

1.8 SUMMARY

Silicon photonics (SiP) is enabling a new generation of guided wave optics. To the present, guided wave optics has been the technology of choice for telecommunication interconnections. SiP has found its first commercial applications in data communication, specifically in the data center transceiver boxes. The demonstrated potential of SiP, however, is much greater than as a multiplexing and demultiplexing board.

Dense optical interconnections can be realized by attaching lasers on SiPs that are bonded to CMOS processors. Memory hard problems (training of convolutional neural networks for AI, block chains, edge computers for 5G data fusion) are becoming ever more ubiquitous. Scaling such problems to numbers of processors larger than those integrable on a single chip whole require complex optical interconnection.

Modest scale memory hard problems may well require interconnection with peta bit per second cut through bandwidth. Such system can be achieved by combining SiP and CMOS. In order to make the optics more cost effective, better techniques of optical carrier generation are necessary.

Optical frequency combs were first developed for precision measurement. The present-day need is for cost effectiveness. Frequency combs generated by RF sources operating in the same

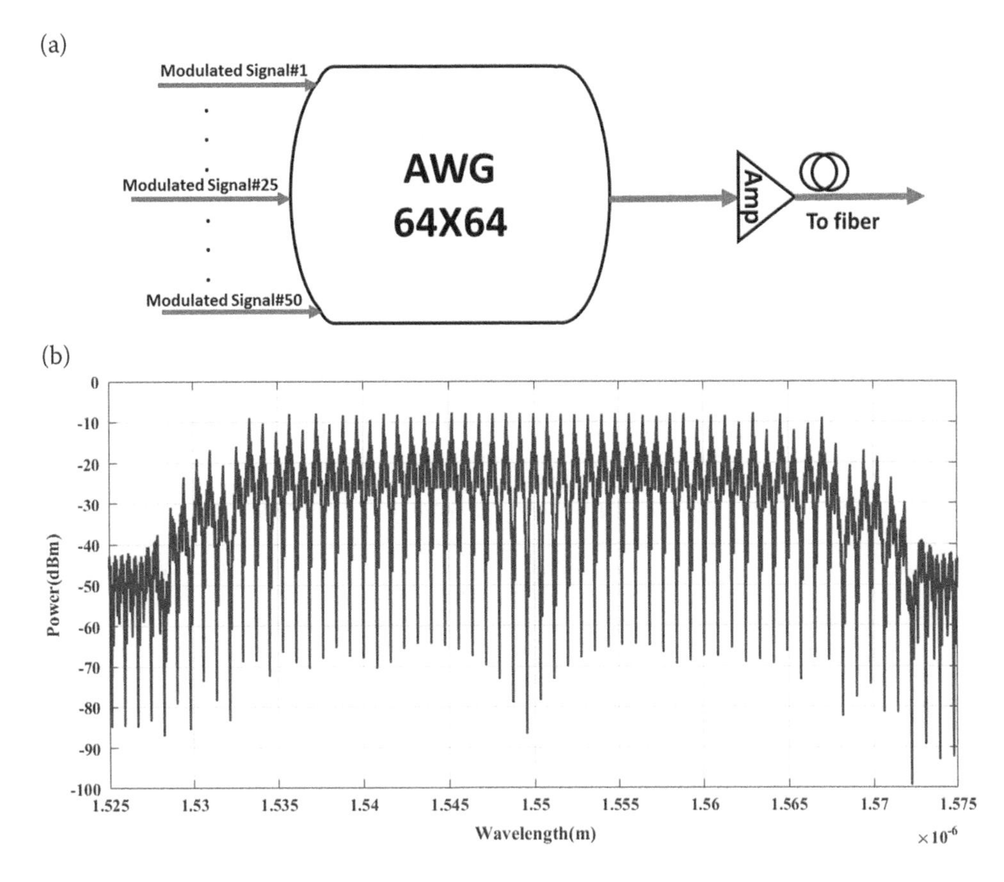

FIGURE 1.10 (a) Multiplexing of modulated signals using a 64 x 64 AWG. (b) The spectrum of a multiplexing signal modulated by information.

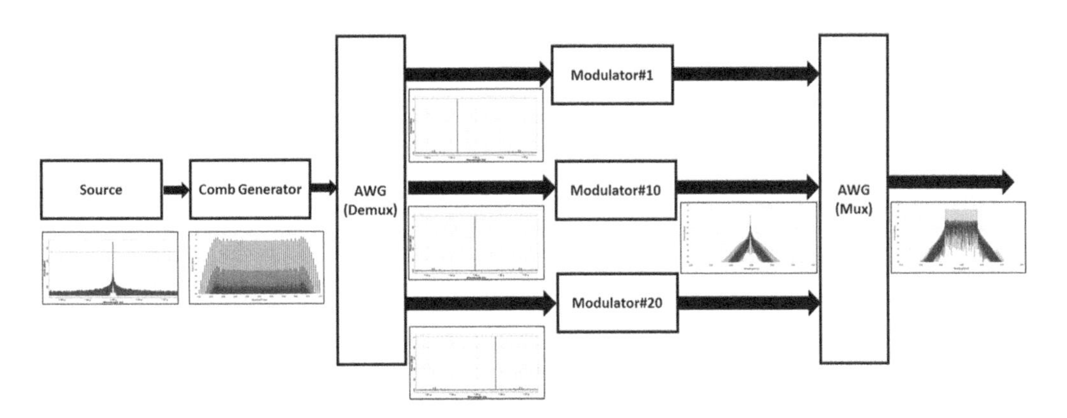

FIGURE 1.11 A general block diagram of a high-speed transmitter using a comb generator with the spectrum of the signal at each stage.

frequency regime as are present-day optical modulation rates, offer a viable solution to minimizing both cost and power dissipation in the SiP-based optical interconnection.

A number of promising techniques for generating and modulating combs presently exist. It is an exciting time to be in optical communication. The future is bright.

REFERENCES

Akimoto, K., Kani, J., Teshima, M., & Iwatsuki, K. (2003). Super-dense WDM transmission of spectrum-sliced incoherent light for wide-area access network. *Journal of Lightwave Technology*, *21*(11), 2715–2722.

Su, Logan, Piggott, Alexander Y., Sapra, Neil V., Petykiewicz, Jan, & Vučković, Jelena (2017). Inverse Design and Demonstration of a Compact on-Chip Narrowband Three-Channel Wavelength Demultiplexer. ACS Photonics, 5, 301–30510.1021/acsphotonics.7b00987.

Bogaerts, W., de Heyn, P., van Vaerenbergh, T., de Vos, K., Kumar Selvaraja, S., Claes, T., ... Baets, R. (2012). Silicon microring resonators. *Laser and Photonics Reviews*, *6*(1), 47–73. 10.1002/lpor.201100017.

Cheung, S., Su, T., Okamoto, K., & Yoo, S. J. B. (2014). Ultra-compact silicon photonic 512 × 512 25 GHz arrayed waveguide grating router. *IEEE Journal on Selected Topics in Quantum Electronics*, *20*(4), 310–316. 10.1109/JSTQE.2013.2295879.

Deseada Gutierrez Pascual, M., & Barry, L. P. (2017). *Development and Investigation of Optical Frequency Combs for Photonic Communication Systems* (September).

Dou, Y., Zhang, H., & Yao, M. (2012). Generation of flat optical-frequency comb using cascaded intensity and phase modulators. *IEEE Photonics Technology Letters*, *24*(9), 727–729. 10.1109/LPT.2012.2187330.

Dravnieks, K., & Spolitis, S. (2017). Demonstration of scalable spectrum-sliced optical WDM-PON access system. *Progress in Electromagnetics Research Symposium*, 2425–2430. 10.1109/PIERS.2017.8262158.

Frazier, H. (1998). The 802.3 z gigabit ethernet standard. *IEEE Network*, *12*(3), 6–7.

Fujiwara, M., Teshima, M., Kani, J. I., Suzuki, H., Takachio, N., & Iwatsuki, K. (2003). Optical carrier supply module using flattened optical multicarrier generation based on sinusoidal amplitude and phase hybrid modulation. *Journal of Lightwave Technology*, *21*(11), 2705–2714. 10.1109/JLT.2003.819147.

Giles, C. R., & Desurvire, E. (1991). Propagation of signal and noise in concatenated erbium-doped fiber optical amplifiers. *Journal of Lightwave Technology*, *9*(2), 147–154.

Han, J.-H., Kim, S.-J., & Lee, J.-S. (1999). Transmission of 4 × 2.5-Gb/s spectrum-sliced incoherent light channels over 240 km of dispersion-shifted fiber with 200-GHz channel spacing. *IEEE Photonics Technology Letters*, *11*(7), 901–903..

He, C., Pan, S., Guo, R., Zhao, Y., & Pan, M. (2012). Ultraflat optical frequency comb generated based on cascaded polarization modulators. *Optics Letters*, *37*(18), 3834. 10.1364/ol.37.003834.

Huang, C. B., Jiang, Z., Leaird, D. E., & Weiner, A. M. (2006). High-rate femtosecond pulse generation via line-by-line processing of phase-modulated CW laser frequency comb. *Electronics Letters*, *42*(19), 1.

Hung, H. W., Lee, Y. L., Sung, C. W., Hsu, C. H., & Lee, S. L. (2017). 10-GB/S bidirectional WDM-PON transmission using spectrum-sliced ASE light sources. *Journal of the Chinese Institute of Engineers, Transactions of the Chinese Institute of Engineers, Series A/Chung-Kuo Kung Ch'eng Hsuch K'an*, *40*(1), 93–99. 10.1080/02533839.2016.1271288.

Kaneko, S., Kani, J. I., Iwatsuki, K., Ohki, A., Sugo, M., & Kamei, S. (2006). Scalability of spectrum-sliced DWDM transmission and its expansion using forward error correction. *Journal of Lightwave Technology*, *24*(3), 1295–1301. 10.1109/JLT.2005.863304.

Khanna, A., Chen, Y., Novack, A., Liu, Y., Ding, R., Baehr-Jones, T., & Hochberg, M. (2017). Complexity scaling in silicon photonics. *Optical Fiber Communication Conference*, Th1B-3. Optical Society of America.

Kilkelly, P. D. D., Chidgey, P. J., & Hill, G. (1990). Experimental demonstration of a three channel WDM system over 110 km using superluminescent diodes. *Electronics Letters*, *26*(20), 1671–1673.

Kim, J.-W., Kim, S.-H., & Kim, D.-S. (2019). Implement of 100-Gbps optical transceiver firmware for optical communication systems. *2019 Eleventh International Conference on Ubiquitous and Future Networks (ICUFN)*, 581–583. IEEE.

Kim, S.-J., Han, J.-H., Lee, J.-S., & Park, C.-S. (1999). Intensity noise suppression in spectrum-sliced incoherent light communication systems using a gain-saturated semiconductor optical amplifier. *IEEE Photonics Technology Letters*, *11*(8), 1042–1044.

Fukazawa, T., Ohno, F., & Baba, T. (2004). Very compact arrayed-waveguide-grating demultiplexer using Si photonic wire waveguides related content low loss intersection of Si photonic wire waveguides. *Japanese Journal of Applied Physics*, *43*(5), 673–675. 10.1143/jjap.43.l673.

Lee, J. S., Chung, Y. C., & DiGiovanni, D. J. (1993). Spectrum-sliced fiber amplifier light source for multichannel WDM applications. *IEEE Photonics Technology Letters*, *5*(12), 1458–1461. 10.1109/68.262573.

Lundberg, L., Karlsson, M., Lorences-Riesgo, A., Mazur, M., Torres-Company, V., Schröder, J., & Andrekson, P. A. (2018). Frequency comb-based WDM transmission systems enabling joint signal processing. *Applied Sciences (Switzerland)*, *8*(5), 718–742. 10.3390/app8050718.

Metcalf, A. J., Torres-Company, V., Leaird, D. E., & Weiner, A. M. (2013). High-power broadly tunable electrooptic frequency comb generator. *IEEE Journal on Selected Topics in Quantum Electronics*, *19*(6), 231–236. 10.1109/JSTQE.2013.2268384.

Olsson, N. A. (1989). Lightwave systems with optical amplifiers. *Journal of Lightwave Technology*, *7*(7), 1071–1082.

Paiam, M. R., & MacDonald, R. I. (1998). A 12-channel phased-array wavelength multiplexer with multimode interference couplers. *IEEE Photonics Technology Letters*, *10*(2), 241–243.

Personick, S. D. (1973). Receiver design for digital fiber optic communication systems, I. *Bell System Technical Journal*, *52*(6), 843–874.

Reeve, M. H., Hunwicks, A. R., Zhao, W., Methley, S. G., Bickers, L., & Hornung, S. (1988). LED spectral slicing for single-mode local loop applications. *Electronics Letters*, *24*(7), 389–390.

Sakamoto, T. (2017). Double-frequency-spaced optical comb generation technique based on quad-parallel phase modulators. *2016 IEEE Photonics Conference, IPC 2016*, 613–614. 10.1109/IPCon.2016.7831252.

Sakamoto, T., & Chiba, A. (2017). Multiple-frequency-spaced flat optical comb generation using a multiple-parallel phase modulator. *Optics Letters*, *42*(21), 4462. 10.1364/ol.42.004462.

Sakamoto, T., Kawanishi, T., & Izutsu, M. (2007). Asymptotic formalism for ultraflat optical frequency comb generation using a Mach-Zehnder modulator. *Optics Letters*, *32*(11), 1515. 10.1364/ol.32.001515.

Shalf, J. (2019). HPC Interconnects at the End of Moore's Law. *2019 Optical Fiber Communications Conference and Exhibition (OFC)*, 1–3. IEEE.

Shang, L., Li, Y., Ma, L., & Chen, J. (2015). A flexible and ultra-flat optical frequency comb generator using a parallel Mach-Zehnder modulator with a single DC bias. *Optics Communications*, *356*, 70–73. 10.1016/j.optcom.2015.07.065.

Shao, T., Shams, H., Anandarajah, P. M., Fice, M. J., Renaud, C. C., van Dijk, F., … Barry, L. P. (2015). Phase noise investigation of multicarrier sub-THz wireless transmission system based on an injection-locked gain-switched laser. *IEEE Transactions on Terahertz Science and Technology*, *5*(4), 590–597.

Thorpe, M. J., Balslev-Clausen, D., Kirchner, M. S., & Ye, J. (2008). Cavity-enhanced optical frequency comb spectroscopy: application to human breath analysis. *Optics Express*, *16*(4), 2387–2397.

Tran, T. T., Song, M., Song, M., & Seo, D. S. (2019). Highly flat optical comb generation based on DP-MZM and phase modulators. *Electronics Letters*, *55*(1), 43–45. 10.1049/el.2018.6454.

Tsao, S.-L., & Lin, Y.-H. (2004). 64x64 silicon-on-insulator arrayed waveguide grating mux/demux for optical interconnections. *Photonic Devices and Algorithms for Computing VI*, *5556*, 288–296. International Society for Optics and Photonics.

Veselka, J. J., & Korotky, S. K. (1998). A multiwavelength source having precise channel spacing for WDM systems. *IEEE Photonics Technology Letters*, *10*(7), 958–960. 10.1109/68.681283.

Vujicic, V., Calo, C., Watts, R., Lelarge, F., Browning, C., Merghem, K., … Barry, L. P. (2015). Quantum dash mode-locked lasers for data centre applications. *IEEE Journal of Selected Topics in Quantum Electronics*, *21*(6), 53–60.

Wagner, S. S., & Chapuran, T. E. (1990). Broadband high-density WDM transmission using superluminescent diodes. *Electronics Letters*, *26*(11), 696–697.

Weimann, C., Schindler, P. C., Palmer, R., Wolf, S., Bekele, D., Korn, D., … Koos, C. (2014). Silicon-organic hybrid (SOH) frequency comb sources for terabit/s data transmission. *Optics Express*, *22*(3), 3629. 10.1364/oe.22.003629.

Winzer, P. J., Neilson, D. T., & Chraplyvy, A. R. (2018). Fiber-optic transmission and networking: the previous 20 and the next 20 years [Invited]. *Optics Express*, *26*(18), 24190. 10.1364/oe.26.024190.

Wu, R., Supradeepa, V. R., Long, C. M., Leaird, D. E., & Weiner, A. M. (2010). Generation of very flat optical frequency combs from continuous-wave lasers using cascaded intensity and phase modulators driven by tailored radio frequency waveforms. *Optics Letters*, *35*(19), 3234. 10.1364/ol.35.003234.

Yu, H., Doylend, J., Lin, W., Nguyen, K., Liu, W., Gold, D., … Ghiurcan, G. A. (2019). 100Gbps CWDM4 silicon photonics transmitter for 5G applications. *Optical Fiber Communication Conference*, W3E-4. Optical Society of America.

Zhong, K., Mo, J., Grzybowski, R., & Lau, A. P. T. (2019). 400 Gbps PAM-4 signal transmission using a monolithic laser integrated silicon photonics transmitter. *2019 Optical Fiber Communications Conference and Exhibition (OFC)*, 1–3. IEEE.

2 Laser Modulation Schemes for Minimizing Static Power Dissipation

Smruti Ranjan Sarangi

CONTENTS

2.1 INTRODUCTION

In the last decade, we have seen a gradual slowing down of the Moore's law. As of 2020, companies have already started to fabricate transistors with 7 nm technology. Over the next few years, we will reach the limit in terms of silicon transistor scaling. Henceforth, it will be very difficult to increase the number of transistors as we have been doing for the last 50 to 60 years. Hence, it is necessary to incorporate newer technologies such that we can continue to increase the performance and computational throughput of modern processors. Up until now, most of these gains were coming from faster, smaller, and power efficient transistors. However, this trend will end in the next few years. It will be necessary to incorporate newer technologies into a conventional silicon chip.

One of the key components that can be optimized further is the on-chip network (also referred to as the *NoC*). Even though the on-chip network has seen great innovations over the last few decades. Most of these have been rather incremental and have not been able to overshadow the gains

DOI: 10.1201/9780429292033-2

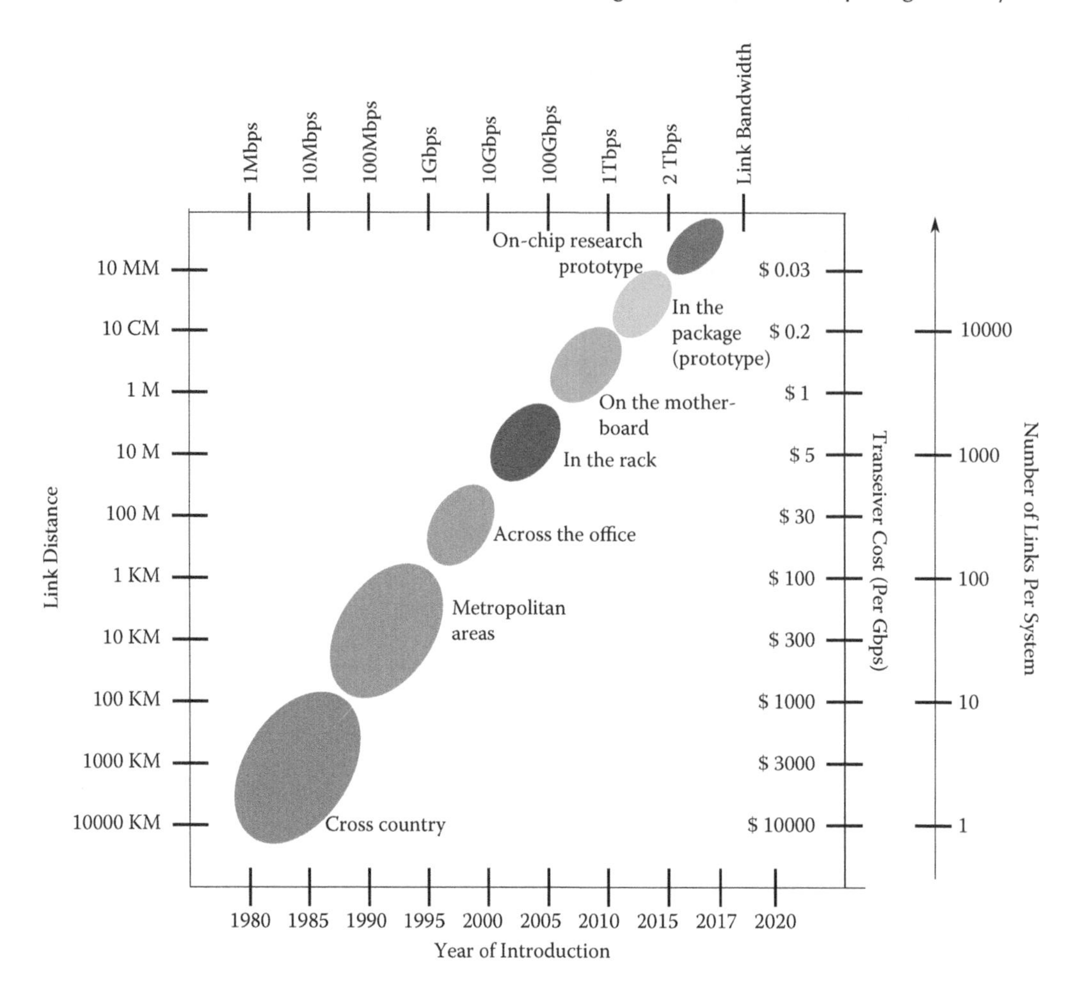

FIGURE 2.1 Evolution of optical networks in the last 40 years (Source: Bashir, 2019a).

due to improvements in transistor technology. All of this is set to change if non-conventional interconnects become commonplace. There are two worthy candidates in this area: on-chip photonic networks and wireless interconnects. On-chip photonics, particularly, is very promising because optical communication has been around for a very long time. Optical communication is gradually getting miniaturized, it is being used to connect servers and data centers, servers across racks, and even experimental versions are now available for on-board optical communication. Many of these components including transceivers that can be manufactured by commercial fabs and can also be integrated as accelerators in general-purpose chips.

Figure 2.1 shows the evolution of optical networks in the last 40 years. We started with large cross-country networks; from 2010 onwards, we moved towards rack scale optical networks, and now we are ultimately approaching board-level optical networks. It is widely believed that in the next few years, optical networks will be used to implement on-chip NOCs.

2.2 PROS AND CONS OF ON-CHIP OPTICAL NETWORKS

Let us briefly review the pros and cons of on-chip optical networks. The biggest advantages are in terms of latency, bandwidth, and power consumption. The typical signal transmission latency of an electrical wire is 35 ps/mm, whereas for an optical wire it is 7 ps/mm. However, the real advantages are not in terms of the raw signal transmission latency, but they lie in the fact that regardless of the distance, we typically use a similar amount of optical power to send the signal. This is starkly different from electrical networks, where, as the length of the interconnect increases, we need proportionately additional power to charge all the capacitors along the way. Furthermore, in a typical unbuffered line, the signal propagation latency is proportional to the square of the length, primarily because of parasitic resistances and capacitances. These problems are not there in an optical network where the losses in transmission are rather low, even inside a chip with numerous bends and crossings. Hence, it is possible to send signals over large distances within the chip without incurring large power losses.

Let us see how basic physics determines the topologies of the different kinds of networks. In electrical networks, we cannot have a very long lines; it is necessary to insert buffers along the way. Furthermore, since we cannot have crossings between electrical lines, these buffers need to have the capacity to route messages to different destinations by making them take turns. This is why such buffers need to be much smarter and need to be more sophisticated, and thus instead of simple buffers, they have taken the form of complex *5-stage routers* in electrical NoCs. The router is a fairly complex piece of hardware that supports complicated virtual channel allocation and arbitration logic; as of 2020, state of the art routers have multi-cycle delays that can vary from 2 to 5 cycles. The main source of delays in electrical networks is actually the time it takes to traverse these routers. The time spent in traversing just the electrical links is comparatively much lower. If we have an *8 x 8* chip with 64 cores, then we need to traverse 15 routers while sending a message from one corner of the chip to the opposite corner. Assuming a 2D mesh network, this can easily take somewhere between 30 and 75 cycles. The response will also take a similar amount of time. Hence, in such large chips, electrical networks become a performance bottleneck. Additionally, the power consumption of electrical networks is often as high as 20 to 25% of the overall power budget of the chip. This is prohibitive in terms of both performance and power consumption.

In an optical NoC, all of these problems are not there. Given that the length of the interconnect within the chip does not determine the transmission power to a great extent (it is a weak sub-linear function), we can afford to have very long optical interconnects in the chip. This has two advantages.

1. We can afford to create a large, snakelike buses that touch all the cores and cache banks. Such long buses are typically point-to-point links, one-to-many broadcast links. We can afford such interconnection structures primarily because the transmission power is mostly independent of the length of the bus.
2. We need not have costly on-chip routers. Given that we can tolerate crossings to a certain extent, there is no need to spend 80% of the transmission time in just traversing through routers. We can just have long point-to-point links or buses.

Optical networks have one more advantage, which is that in a single bus we can actually transmit signals using 32 to 64 different wavelengths. This is known as dense wavelength division multiplexing, abbreviated as DWDM. This allows us to tremendously increase the bandwidth that can be supported by a single optical bus.

There are several important and subtle points to note here. It is true that the raw transmission latency of an optical network is five times more than that of an electrical network. However, superior propagation characteristics of such networks do not arise from this fact, instead they arise from the fact that we actually do not need a message to traverse multiple routers on the way. This is because we can afford to have very long point-to-point links without a concomitant increase in transmission power, as well as we can leverage the power of dense wavelength division multiplexing.

However, optical networks of the unique challenges. The first is that the fabrication technology for optical networks is a mature. Many of the devices that are used to create on-chip optical networks need to further improve and mature before they can be mass produced.

This is a very active area of research and a lot of advances are being made in this field. However, we will look at computer architecture aspect in this section. We need to understand that the basic physics of photons introduces some fundamental limitations to the power efficiency of such networks.

Photons are *bosons* that follow the Bose-Einstein statistics. An important property of such bosons is that the Pauli's exclusion principle does not apply to them. This means that any number of bosons can occupy the same quantum state. In comparison, electrons, protons, and neutrons are *fermions*. This means that the Pauli's exclusion principle applies to them – two fermions cannot occupy the same quantum state. This is why we can make matter with electrons, protons and neutrons. Because we cannot pack two fermions in the same quantum state, we can store such particles, they are associated with a notion of mass and volume, and most importantly, we can create capacitors that can store charge. Such capacitors are associated with the notion of a potential, which essentially quantifies the number of electrons that are stored across the plates of the capacitor. In comparison, we cannot create capacitors that store photons. This is primarily because of the fact that they are bosons and cannot be stored. However, there is an advantage of being a boson. It is that we can transmit different signals at different wavelengths, and they are guaranteed to not interfere with each other. This is why we can realize dense wavelength division multiplexing; this would not have been possible with electrical transmission, primarily because the individual signals would interfere with each other.

Let us focus on the fact that we cannot create capacitors to store photons as compared to electrons. First, let us understand the mechanisms for power consumption in traditional electrical circuits. Different elements within the circuit have associated capacitance values which can arise either because of real capacitors or because of parasitic capacitances. In either case, there is some degree of charge storage. Whenever, we need to change the state of the circuit, we basically need to charge or discharge capacitors. The current that is required to charge and discharge capacitors needs to be drawn from the power supply. The total power dissipation of the circuit is determined by this current. In most circuits, including CMOS circuits, we typically try to minimize this current, primarily because large parts of the circuit can hold onto their charge (or their state). This is why when we are dealing with electrical circuits, we only count the number of voltage transitions. Power is only dissipated when we transition from state 0 to 1 or state 1 to 0. In comparison, since photons are not inherently associated with any form of storage, we cannot transmit signals and operate our communication circuits in the same manner. It is necessary for a light source such as an on-chip or off-chip laser to continuously supply photons. Because we do not have the notion of a photonic potential, in this case, the state basically refers to the presence or absence of photons. If a detector is detecting the presence of light at a given wavelength, we can infer a logical 1, or else we can infer a logical 0.

This is the fundamental difference between transmission in electrical and optical on-chip networks. In optical on-chip networks, it is necessary to run the lasers all the time, primarily because we make all the decisions based on the presence or absence of light. We cannot rely on the fact that some photons are stored somewhere in the circuit and hence we may not transmit power. This is because we do not have an analog of a photonic potential.

Even though optical transmission is supposed to be power efficient when considered in isolation, but continuously having the lasers on is a significant source of power dissipation. This is also referred to a static power dissipation because most of the time we are actually wasting light, it is not being used to transmit valid signals. Such static power dissipation has been measured to be around 80 to 90% of the overall power consumption of an optical on-chip network. This can be almost as high as the total power budget of the chip and can seriously reduce the feasibility of having such on-chip optical networks. Hence, we need to accurately *predict* the activity of the on-chip optical network, and also *modulate* the intensity of the lasers such that they supply just enough power to transmit messages – no power is wasted.

We thus need to create a very complicated set of technologies, where we accurately monitor the network activity, predict network activity in the future, and use this to modulate the on-chip and off-chip lasers. If we are too conservative and keep lasers on most of the time, then performance will suffer, primarily because we will not have enough power to transmit messages. If we are lenient, and lasers are turned on most of the time, this will lead to a large amount of static power dissipation. Hence, laser modulation schemes occupy a very central position in the design of on-chip optical networks. This is the area of study in this chapter.

2.3 BACKGROUND

2.3.1 Optical Devices

Let us provide a brief background of on-chip optical communication systems. Figure 2.2 shows a typical diagram of an on-chip optical communication system. We have a light source: it can either be an external light source such as an off-chip laser, or an on-chip laser. In general, external light sources are preferred because lasers have a very low wall clock efficiency. It is of the order of 20 to 30%. Hence, it is much better that the additional heat is dissipated outside the package such that it can be removed easily. However, in some scenarios, particularly when we desire faster modulation, we typically use on-chip lasers such as VCSEL lasers.

For off-chip lasers, most of the time, commercially available directly modulated lasers (DML) are used such as Fitel FOL15DDBA and Finisar DM 80. These lasers have even been fabricated by many researchers and provide very high modulation rates at multi-GHz speeds (J.-R. Burie, 2010). The main issue in using such lasers is their thermal stability. However, it has been addressed to a large extent by Fukamachi et al. (T. Fukamachi, 2010). They demonstrated a thermally stable and a very efficient DML laser showing a stable operation up to 100 degrees Celsius with a modulation speed of 25 Gbps. Moreover, an array of DML lasers can be used to create a tunable laser with different levels of optical power. For on-chip optical communication InGaAsP/InP lasers are used that emit light 1,550 nm. Sometimes, InP/InGaAsP lasers are used that emit light at 1,310 nm.

For on-chip lasers, we can use any wavelength division multiplexing compatible laser with a fast switching time. Prominent examples of such lasers are VCSEL lasers that have a typical switching time of 100 ps. Their main problem is their low wall clock efficiencies and the fact that they dissipate all the heat inside the chip.

It is necessary to use a modulated, such as a micro-ring resonator are shown in Figure 2.3, to modulate the optical signal. In this case, modulation, the optical signal means either turning it on turning it off. As shown in Figure 2.3, we can see that if there micro-ring resonator is in the resonant state, then the signal is fully absorbed and it does not propagate through the waveguide (channel to carry the optical signal). However, if it is not in the resonant state, then you can allow signals to propagate. The resonance of ring can be changed by slightly charging the material of the ring. This uses the electro-optic effect a change effective refractive index of the material. This can change the resonance properties of the ring resonator.

The signal subsequently propagates through the waveguide. Note that waveguides can have crossings and can of beans. At this point, there is some loss in the signal strength. Finally, we use photodetector to detector to detect the presence or absence of an optical signal. Its threshold value is typically 36 μW. This lets the receiver know whether we are transmitting a logical zero or a logical one.

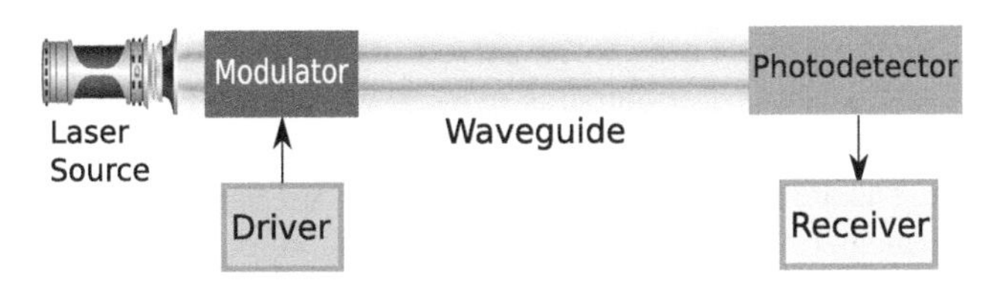

FIGURE 2.2 A diagram of a typical optical communication system (Source: Bashir, 2019).

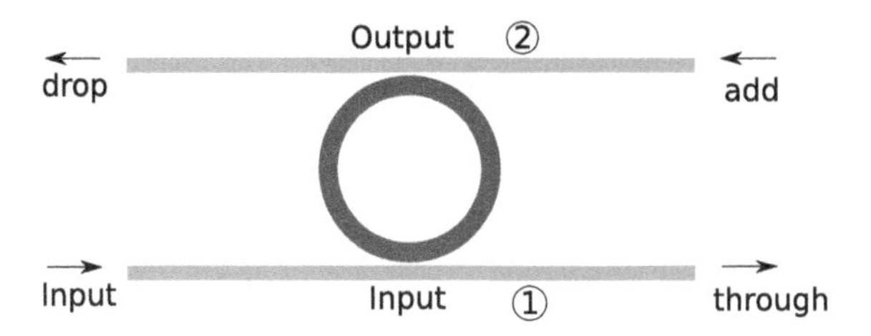

FIGURE 2.3 A micro-ring resonator (Source: Bashir, 2019).

2.3.2 Optical Buses

There are three kinds of commonly used optical buses. Let us define the term, *optical station*, which refers to any element within the chip that can either send a message or receive a message. This includes course, cash banks, directory elements, and memory controllers. Let us assume a total of N optical stations.

1. SWMR buses: These are single-writer, multiple-reader buses. Here, each optical station is connected to the rest of the $N-1$ optical stations using a single bus. The writing station can broadcast the value and any subset of the reading stations can read the value. They use a beam splitter such as a Y junction to split a part of the signal and read it. The split ratios of the beam splitters need to be tuned such that there is a minimal amount of power loss. Peter et al. (Eldhose Peter, 2015) have proposed an algorithm based on dynamic programming that yields optimal solutions.

   Modern buses typically use reservation assisted schemes that was originally proposed in Firefly (Pan Y. P., 2009). Here, all the receiving stations are by default turned off, and we add a new waveguide called a reservation waveguide. Before sending a message, the sender sends a short message on the *reservation waveguide* to identify the set of receivers that need to turn themselves on. Subsequently, once the receivers have turned themselves on, the message is sent and the receivers read the message correctly.
2. MWSR buses: This is a multiple-writer single-reader bus. Each bus has a single reading station and $N - 1$ writing stations. A process of arbitration is required to decide the writing station. Once, a station gets the permission to write, it can write on the bus. One advantage of this scheme is that it does not require beam splitters that reduce the intensity of the signal and also entail additional circuitry to configure them.
3. MWMR buses: This approach that was originally proposed in Flexishare (Pan Y. J., 2010) combines the SWMR and MWSR schemes. It tries to achieve the best of both worlds by supporting multiple readers and multiple writers on each bus. This does require beam splitting and limited support for arbitration; however, if operated and configured correctly it can be more efficient than both.

2.4 LASER MODULATION IN CPUS

2.4.1 Probe

The most important problem that needs to be solved in this case is that we need to effectively modulate the lasers such that its power output is reduced when there is no network activity. One of the first proposals in this area was Probe (Zhou, 2013). This is a 64-core architecture, where 4 cores are grouped into a tile. The architecture is a regular grid of 4 x 4 tiles. Each tile has a dedicated off-chip laser that has a dedicated voltage regulator to set the output power. Furthermore, each tile can broadcast a message using SWMR buses in the x and y directions. The design uses tunable beam splitters that are based on MMI devices.

We shall focus on the methods to predict the laser power in the future. The authors propose to use two metrics: link utilization and buffer utilization. The link utilization (L) for a given link is given as follows:

$$L = \frac{\sum_{t=1}^{T} active\,(t)}{T}$$

Here, T is the length of an epoch; we assume that time is divided into epochs. The function *active* (t) indicates if at time instant t, a flit was transmitted on the link or not. On similar lines, we can define the buffer utilization.

$$B = \frac{\sum_{t=1}^{T} occupancy\,(t)/BufSize}{T}$$

Here, *occupancy*(t) is the number of flit buffers occupied in the flit queue at time t. *BufSize* refers to the total number of buffers. This information is used to implement two predictors.

The first predictor is used when the traffic variation is low. Let us show an example for the link utilization.

$$L_{pred} = \frac{L_{past*}3 + L_{current}}{4}$$

We have a similar equation for the buffer utilization as well. The aim is to give a higher weightage for past values and give a lesser weightage to current values.

The second predictor is used when the variation in the traffic is high. In this case, the link or buffer utilization levels are divided into five buckets. This is done by normalizing them to lie between zero and one; the buckets are then created uniformly: 0 to 0.2, 0.2 to 0.4, and so on. Next, the authors consider the history of the last five epochs; this is used to index a pattern table that stores the predicted link or buffer utilization. This is similar to a regular pattern history-based branch predictor.

Similar to the tournament branch predictor, there is an array of saturating counters that chooses between the two predictors. This automatically adjusts itself based on amount of variation in the traffic.

Every laser modulation algorithm has a reconfiguration phase that typically happens once every epoch. In most laser modulation schemes the reconfiguration happens at the beginning of the epoch. At the beginning of an epoch all the hardware counters that collect the link and buffer utilization statistics send their data to a local controller (specific to each tile). Subsequently, the controller predicts the link and buffer utilization for the current epoch. The predicted link or buffer utilization is compared with two thresholds α and β. Consider the case for link utilization. If the predicted link utilization, L_{pred} is less than the lower threshold α, then the bandwidth is decreased, if it is between α and β, the status quo is maintained, and finally if it is more than β, the bandwidth is increased. This information is sent by the local controller is to a global controller. The global controller then configures the network accordingly: it powers of ring resonators, tunes the tunable splitters, and computes the power output of the laser. Once this is done, it sends an acknowledgment to all the local controllers and sends messages to the off-chip lasers to modulate their output power. We typically have an array of output lasers where we can turn off any subset of them. For example, if we have 32 lasers and only 17 of them are on, then the power output is 17/32 times the maximum power.

2.4.2 ColdBus

As compared to Probe, ColdBus (Peter, 2015) uses a very different kind of prediction mechanism. It does not rely on link or buffer utilization. The basic insight is as follows. If we consider a system with private L1 caches and a shared L2 cache, then most of the network traffic is accounted for by L1 misses. The network traffic can either be because of coherence messages, or it can be because of messages sent between the L1 cache banks and L2 cache banks. Given that the instruction cache miss rates are very low, just predicting L1 misses can give us a very accurate indication of the amount of on-chip network traffic.

For most benchmarks, it was observed that most of the L1 cache misses were accounted for by a very small minority of instructions. These were mostly irregular memory accesses or *reduce* operations where data is read from sister caches and a final value is computed. Given that a very small number of instructions are actually involved in the majority of the misses, we can design a predictor based on the program counter of the instruction. Akin to a branch predictor, the authors of ColdBus designed a predictor that takes the last 10 bits of the PC and uses 3-bit saturating counters. For the saturating counters, the threshold for deciding if a PC will lead to a miss was set to 5. The accuracy was roughly 90% for a suite of Splash (S. C. Woo, 1995), Parsec (C. Bienia, 2008), and Parboil benchmarks (Stratton, 2012).

There is a problem in using the program counters for prediction. We don't know when the load or store in consideration will actually be sent to the memory system. We at least need to predict the epoch in which the load or store will be sent to the memory system. Hence, we need a dedicated epoch predictor. ColdBus solves this problem as follows: there is 1,024-entry epoch predictor for loads and a similar-sized table for stores that are indexed with the last 10 bits of the PC. Each entry in this table contains the number of epochs that a memory instruction takes to be sent to the L1 caches after it has been decoded. This table can be used to find in which epoch the given instruction will most likely be sent to the L1 cache. Furthermore, if it is predicted to miss in the L1 cache, we can then accurately predict the epoch in which there will be network traffic, arising out of this miss. This process is made slightly more accurate by storing the values for the last five epochs. The weighted average is used as the prediction.

ColdBus has one more innovation – an extra waveguide. The *extra waveguide* is an additional waveguide that carries a little bit of additional contingency power, which can be used by any station that needs power to transmit messages but does not have enough power available. For using the extra waveguide, a dedicated arbitration mechanism is required. We transmit a given number of tokens on the extra waveguide, where one token corresponds to power that can be used to send one packet. In all such prediction schemes, two thresholds are used: an upper threshold and a lower threshold. If the requirement is more than upper threshold, then the number of tokens is decreased, if it is between the thresholds, the number of tokens is kept the same, and finally if the requirement is below the lower threshold, the number of tokens is decreased.

Akin to Probe, here at the end of an epoch, all the predictions are sent to a global controller. Given that a single array of lasers is used, the global controller predicts the power requirement of the entire chip, calculates the configurations of all the ring resonators and beam splitters, and prepares all the messages that need to be sent. In the beginning of the next epoch, the entire system is reconfigured to operate at the new power level. During this time, the system remains quiescent.

2.4.3 PShaRe

ColdBus and Probe are exclusively laser modulation schemes. However, for effective power management, laser modulation needs to be combined with other approaches that manage the inefficiencies in accurately predicting the laser power in future epochs. We need to understand that regardless of our efforts, the accuracy will never be 100% because of information theoretic reasons.

Let us discuss the PShaRe (Bashir, 2019) scheme, which is the most comprehensive solution as of 2020. Figure 2.4 shows the relative laser power consumption of an ideal scheme as compared to Probe and ColdBus. An ideal scheme is defined as a scheme where the lasers turn on exactly when required, and then they turn off. This is clearly a hypothetical scheme, nevertheless it establishes a lower bound on the power consumption. We observe that both Probe and Coldbus are quite inefficient in terms of laser power consumption. The ideal scheme consumes 3-8X lower power. The authors of PShaRe tried to reduce this as much as possible. Furthermore, they realized that beyond a certain point, it is not possible to reduce the power anymore. Hence, they proposed a method to *reuse* the wasted power.

The key insights in this work is that we need separate networks for coherence and non-coherence messages. The traffic patterns of both these message types differ significantly, and this is a much better idea to have separate networks for them. The second insight is that there is a tremendous amount of imbalance between the traffic generated and received by different stations. Hence, we often end up over predicting, which is detrimental to us in terms of power consumption. Finally, the last insight is that the predicted value is often not a linear function of the values of performance counters as has been assumed in prior work. Instead, it is much better to use a nonlinear predictor such as a neural network that can capture such effects.

2.4.3.1 Basic Architecture

Figure 2.5 shows the architecture of the PShaRe system. In this representative system there are 32 cores and 32 cache banks. Each optical station is connected to a tile that contains two cores and two cache banks. In this architecture, there are 16 power waveguides and 16 data waveguides. Each power waveguide carries monochromatic light at 1,550 nm, and subsequently a comb splitter

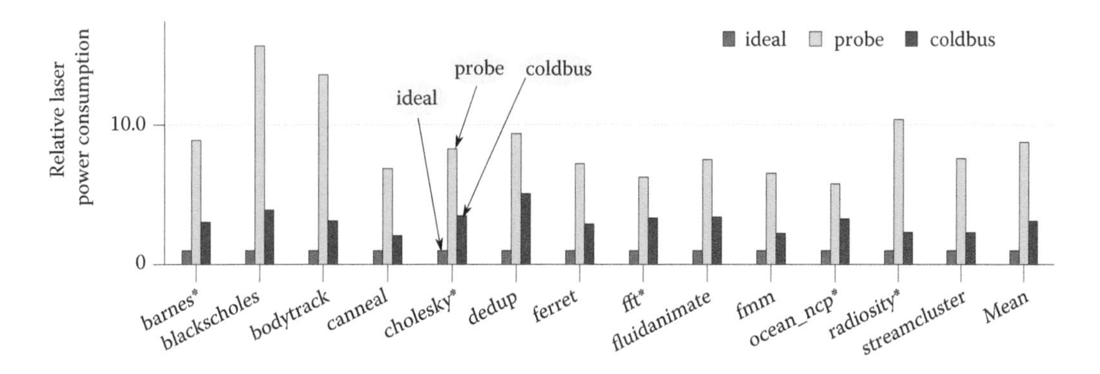

FIGURE 2.4 Relative laser power consumption (Source: Bashir, 2019).

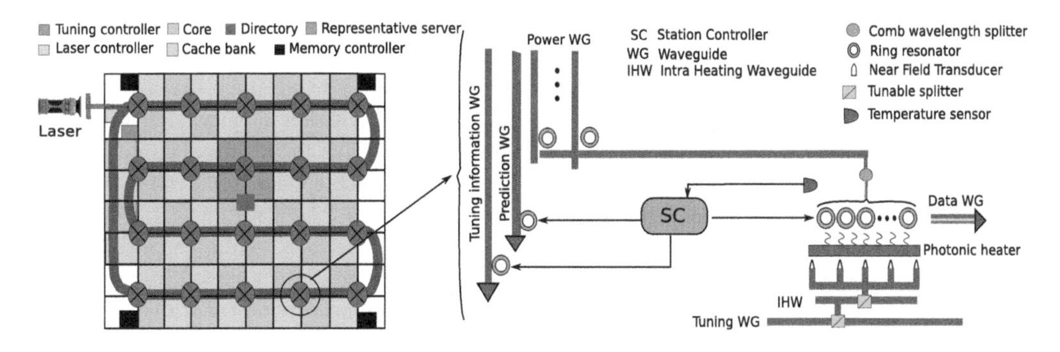

FIGURE 2.5 PShaRe architecture (Source: Bashir, 2019).

splits it into 64 equally spaced wavelengths. Each data waveguide is used to send messages using DWDM multiplexing. This is an MWMR architecture where any station can use a data waveguide for both reading and writing. An arbitration mechanism is thus required for getting a power and data waveguide allocated. Note that for sending a message on the ith data waveguide, it is necessary to acquire the rights to source power from the ith power waveguide. There are additional waveguides for sending prediction information, and for reusing wasted power (the tuning waveguides).

2.4.3.2 Prediction

PShaRe divides time into fixed size durations called *epochs* and predicts the laser power in advance for the next epoch. For each station, the prediction is binary. A value of '1' indicates that it is expected to transmit a message in the next epoch, and a value of '0' indicates complete inactivity during the next epoch. False positives are wasteful in terms of power, because the laser remains on when no messages are actually sent. Likewise, false negatives are hurtful in terms of performance because of the unavailability of laser power. Each optical station is associated with a separate neural network-based predictor. Each neural network is trained using the values of 14 performance counters that include miss rates, network activities, and waiting times. This is done at the beginning of an epoch only if there is a misprediction, and the prediction is performed at the end of every epoch. During the training phase, the optical station provides all the inputs that were used for predicting laser power in the previous epoch and the desired output. The ANN (artificial neural network) is trained iteratively until convergence. For the prediction, the ANN is run at the end of every epoch. The need for ANN-based predictors arose because of the nonlinear relationship between the values of different performance counters and the predicted output.

2.4.3.3 Reconfiguring the Network

After the predictions are made by the optical stations, each station sends this information to the laser controller at the end of an epoch through a separate waveguide called the prediction waveguide. Each station uses a specific wavelength in the prediction waveguide to indicate its requirement (active low signaling). The predictions are sent to the laser controller, which collates the predictions and retunes the network at the beginning of the next epoch. The optical stations also send this information to their respective representative servers, which are dedicated hardware units that perform arbitration.

2.4.3.4 Tuning Network

In PShaRe, the ring resonators are divided into two groups: one group includes the ring resonators attached to the data waveguides (roughly 85%) and the remaining ring resonators belong to the other group (power waveguides and arbitration logic). The resonators in these groups are tuned separately. Note that tuning is required to bring all the resonators to the same temperature because the resonant wavelength is a function of temperature. Our idea is to use the unused optical power to partially tune the ring resonators attached with the data waveguides. Note that all the ring resonators will still need to be connected to conventional microheaters. The objective is to decrease the power supplied to the microheaters by *reusing* wasted optical power. Since heat is the lowest form of energy, conversion of optical power to heat is easy and has a very high efficiency. Resonators attached to the data waveguides are co-located, and thus they have a similar thermal profile. This property allows us to place heaters at the center of such clusters (size: 64 resonators), and target all of them at once.

A separate set of waveguides called the tuning waveguides (TW) are used to carry the unused optical power. At the end of the power waveguides, the unused power is coupled into these waveguides. A cascaded set of Y-junctions is used to combine the unused power and route it into the tuning waveguides (lengths adjusted appropriately for nullifying phase delay). As light travels through the TWs, the attached splitters split some portion of the light and send it towards each optical station. Within each station, the optical power is guided through a very small waveguide called the intra-heating waveguide (IHW) that distributes it to all the photonic heaters. We use near field transducers (NFT) to focus the light on the plasmonic material embedded in the heaters for maximum conversion efficiency.

2.4.4 BigBus

PShaRe, Probe, and ColdBus are all multicore designs. Let us now consider designs that have upwards of 500 cores such as BigBus (Janibul Bashir, 2019).

2.4.4.1 Architecture

In this architecture, there are 768 2-issue in-order cores and 256 cache banks. We cannot use existing solutions based on SWMR, MWSR, and MWMR waveguides because they don't scale. Figure 2.6 shows the architecture of BigBus, where the cache banks are in the center and the codes are on the periphery. In this case, each tile contains either four cores or four cache banks. We then create a cluster of tiles. Sixteen clusters (4 x 4) form a P-Cluster, and 16 P-Clusters form an O-cluster. This is a 2-level hierarchical design that has four O-Clusters at the highest level. As we can see from the figure, all the stations within an O-Cluster are connected via a serpentine-shaped optical link called an O-Link. We have a dedicated waveguide to connect all the O-Links. Furthermore, all the cache banks are connected with a dedicated link known as the CB-Link.

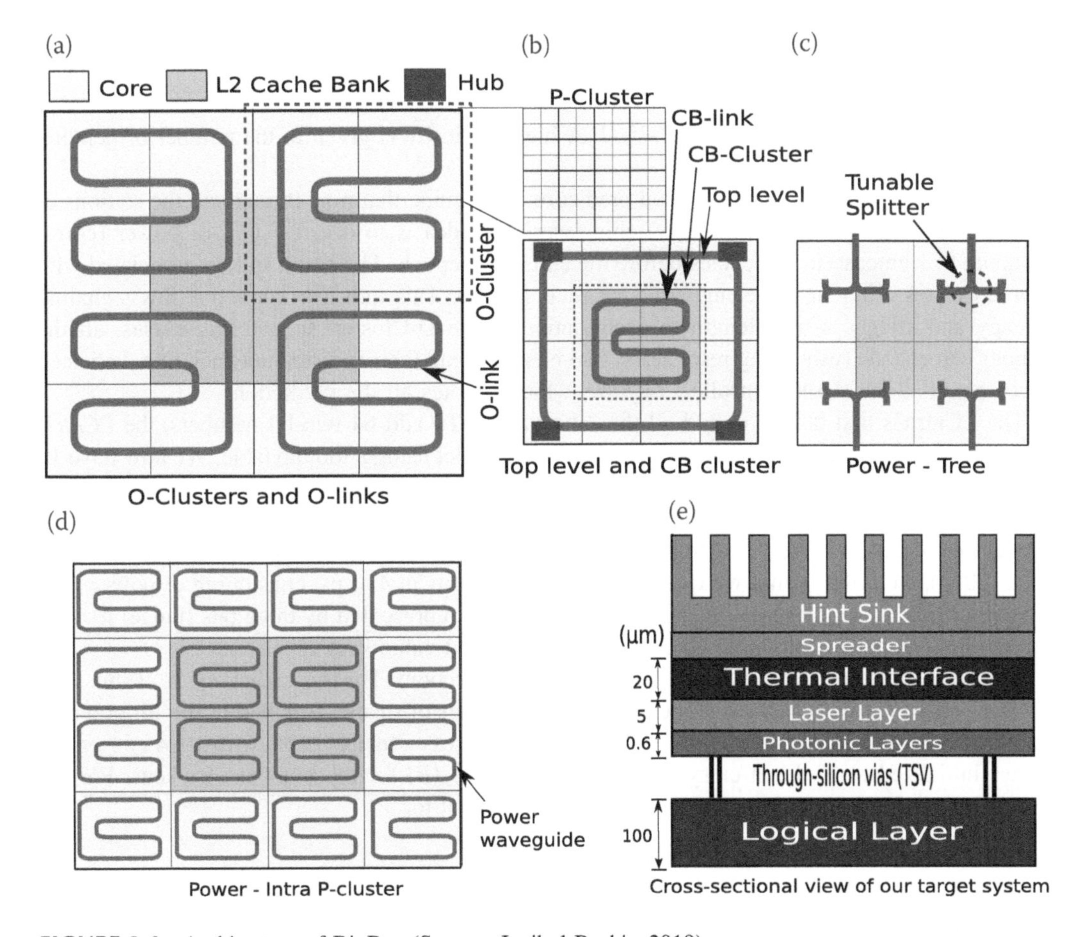

FIGURE 2.6 Architecture of BigBus (Source: Janibul Bashir, 2019).

We have four off-chip lasers – one for each O-Cluster. This power is then distributed to the individual P-Clusters using power waveguides. We assume that power is shared between the stations in a P-Cluster and not across stations in different P-Clusters. For each P-Cluster, we circulate certain number of tokens (1 token = power of 1 unicast) on the token waveguide corresponding to that P-Cluster. We use a maximum of 16 tokens on the token waveguide. A station can grab tokens from only the token waveguide corresponding to its P-Cluster. This scheme is similar to that used in PShaRe.

2.4.4.2 Prediction

The number of tokens that need to be transmitted on the token waveguide have to be accurately determined for each P-Cluster. We do not want to send a lot of tokens (power wastage), nor too few tokens (too much of contention).

A novel prediction scheme was designed to consider the current trends in traffic along with historical values. Activity prediction is done in two stages. In the first stage, each station decides whether to increase or decrease tokens based on a function that has two inputs: wait time (T) and the number of pending events (N) at that station. The output of this function, F, is 0, 1, 2, or 3. Here we give priority to the number of pending events over the wait time. The rules are shown below in decreasing order of precedence for setting the value of $F(T, N)$.

3 $N \geq Tp$

2 $T \geq Tw \text{ or } \frac{Tp}{2} \leq N < Tp$

1 $\frac{Tw}{2} \leq T < Tw \text{ or } N < \frac{Tp}{2}$

0 $T < \frac{Tw}{2}$

BigBus used $Tp = 8$, and $Tw = \text{epoch size}/2$. A higher priority is given to the number of pending events to improve the performance.

In Figure 2.6, we assume a dedicated prediction waveguide that runs through all the stations of an O-Cluster. It is powered by the off-chip laser. The idea is to divert 1 unit of power (corresponding to 1 unicast) to it 9 cycles before the end of the epoch. The comb splitter associated with it produces 64 different wavelengths (1 for each station). This unit uses an active-low signaling strategy and assigns a wavelength to each station in the O-Cluster. In two half-cycles, all the stations can encode (using ring resonators) two bits and send their information back to a dedicated structure called the (laser controller) *LCntrlr*, which collates all the predictions.

The LCntrlr's first task is to add all the 2-bit values. To add 64 two-bit numbers, the *LCntrlr* first adds sets of four 2-bit numbers (16 such sets) in parallel using a lookup table. We now have 16 partial sums, where each sum is at the most 12. Now, another lookup table is used to again add two partial sums at a time to generate 8 partial sums. We now have 8 partial sums, each being 5 bits. A 3-level tree of adders is then used to add them. The final output is 8 bits. Using numbers from Herr et al. (2013), we observe that we can finish all the additions in 450 ps, consuming of power. The top three MSB bits of the sum are extracted. They can be represented by three bits (range: [0…7]) and are then mapped to the set [–3 … 4]. Let the mapped value be V.

Now it is time to take historical information into account. BigBus maintains a 10-bit shift register, where the ith bit contains the MSB of the number of tokens supplied in the ith previous epoch. The 10-bit value in this shift register is used to access a history table with 1024 entries (all entries initialized to 0). Each entry contains a base value (B) for the number of tokens. For the current epoch, let the predicted value be P, which is equal to.

$$P = B + V$$

The history table is updated with P.

This is the most complicated prediction mechanism that we have seen up till now. However, given the extreme power efficiency that is required in such large networks, it was necessary for the authors to design such a complicated and elaborate scheme. If we look at the scheme carefully, we will see that there are two distinct components in it. The first is that we access the history table using an index that is composed of the MSB bits of the tokens supplied in the previous epochs. The value read from the table is like a base value for the current prediction, which is also equal to the predicted number of tokens the last time the same pattern was encountered. Furthermore, we add the mapped value V to it because this is like a perturbation over the base value. It takes care of instantaneous needs. Figuring out the exact formulae and the constants requires a fair amount of engineering effort; this is representative of the design processes that are required to design optical networks for such large systems.

2.5 LASER MODULATION IN MULTI-SOCKET SYSTEMS

Let us now turn our attention to multi-socket systems that have a large number of cores that are distributed across multiple sockets on a server. They use a shared memory-based paradigm. We will discuss the Nuplet (Bashir, 2017) design in this space. In *Nuplet*, each chip in such a socket is called a *chiplet*.

First, note that there is an intra-chip optical network based on ColdBus and an inter-chip optical network. The inter-chip optical network is an MWMR bus, which has its own set of lasers to provide power. Since it is an MWMR network, there is a need to first arbitrate for the bus and then send a message once it is allocated to a station. Each chiplet has a dedicated station for sending messages to other chiplets. It is known as the inter-chiplet optical station or ICOS. The laser modulation mechanism for inter-chiplet messages is as follows.

Akin to BigBus and PShaRe, there are inter-chiplet data, token, power, and prediction waveguides. The token waveguides are used for arbitration. In this case, the aim is to predict the number of messages that will be sent across the chiplets. This will determine the number of tokens that will be circulated in the token waveguide in the next epoch.

In the prediction phase, two values are calculated. Each station sends a 2-bit value to the laser controller that corresponds to the number of messages that it sent to the ICOSs for transmission to other chiplets. The laser controller adds all of these values, and computes the sum S. Similarly, all the ICOSs send the average number of pending events in their queues to the controller. The controller adds these numbers, and computes the sum P.

The controller has a power request table, PRT, which is indexed by the 6-bit sum S. Each entry of this table contains half the maximum number of tokens. Let the sum for the previous epoch be S' and for the current epoch be S. The formula for computing the number of tokens for the current epoch is as follows.

Predicted Number of Tokens	Condition	Action
$PRT[S'] - 1$	$(S < S')$ *and* $(16 \le P < 32)$	–
$PRT[S']$	$(S \ge S')$ *and* $(16 \le P < 32)$	–
$PRT[S'] - 1$	$P < 16$	$PRT[S'] = PRT[S'] - 1$
$PRT[S'] + 1$	$P \ge 32$	$PRT[S'] = PRT[S'] + 1$

Let us understand the logic behind these equations. Akin to BigBus, here also priority is given to the number of *pending messages* at the ICOS stations. The first two cases correspond to situations where the number of pending messages has an intermediate value (between 16 and 32). In such cases, we look at the traffic generated from the optical stations. If the traffic in the current

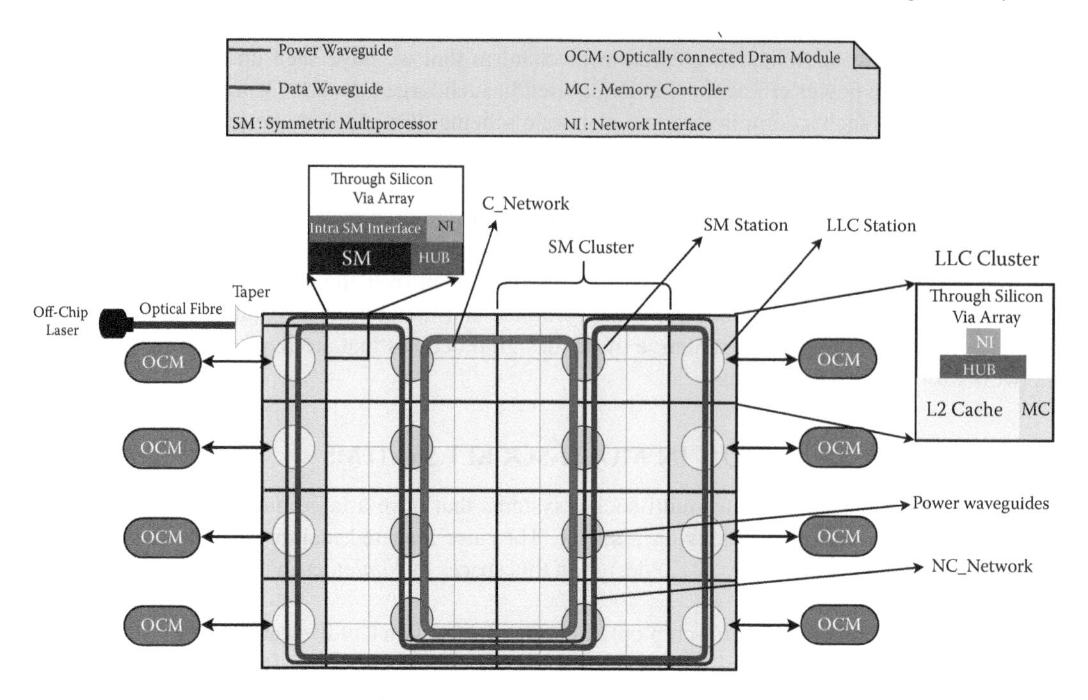

FIGURE 2.7 Architecture of an optical NoC for a GPU (Source: Bashir, 2020).

epoch is less than the traffic in the previous epoch, then the predicted number of tokens is one less than $PRT[S']$. However, if the traffic in the current epoch is more than the traffic in the previous epoch, then we do not predict more tokens because the number of pending messages still has an intermediate value.

Next, let us consider the last two cases where we exclusively look at the number of pending messages. In this case, the number of pending messages is either lower than 16 or more than 32. In the first case, we predict $PRT[S'] - 1$, and also update the value in the PRT table. In the second case ($P \geq 32$), we predict $PRT[S'] + 1$, and also update the value in the PRT table.

The notion of using $PRT[S']$ as a base value is inspired from earlier prediction approaches that rely on a degree of hysteresis and historical information.

2.6 LASER MODULATION IN GPUS

Let us now extend these schemes for a GPU. A GPU is somewhat similar to a manycore processor; however, this is a very superficial similarity. A GPU is constructed very differently and its workloads are mostly sensitive to memory bandwidth and relatively less sensitive to latency. Keeping this in mind, let us describe the GPUOpt (Bashir, 2020) scheme in this section.

2.6.1 Architecture

Figure 2.7 shows the design of the optical NoC (uses 2.5D integration with two stacked layers – optical and silicon). The figure shows the optical NoC, optically connected global memory modules, and other optical components. The chip has two separate layers: logical layer: containing SMs, L2 banks (last level cache), and memory controllers (MC); and the photonic layer containing optical components. The logical layer is divided into 16 clusters: 8 *SM_Clusters* and 8 *LLC_Clusters*.

Each *SM_Cluster* has 8 SMs where each SM contains a private L1 instruction and data cache. Each *LLC_Cluster* has an L2 bank along with a memory controller. The intra-cluster communication is done electrically, whereas, for inter-cluster communication, we incorporate a separate silicon photonics layer underneath the logical layer. The optical layer has optical stations – one for each cluster – and these stations are connected using an optical bus. The clusters are connected to their respective optical stations using through-silicon vias (TSVs). Note that the optical stations attached to *SM_clusters* are called *SM_stations*, whereas the stations connected to *LLC_clusters* are called *LLC_stations*.

There are two optical buses in this system. The first is an SWMR-based bus called the *C_Network* to connect the *SM_stations* with each other. This network contains SWMR links with each link composed of 1 waveguide carrying 64 wavelengths. All the coherence traffic is routed through this network. Then there is an MWSR *NC_Network* for non-coherence traffic. Each link in the *NC_network* is assumed to be carrying 128 wavelengths: it is 2 waveguides wide. Similar to BigBus and PShaRe, there are 16 power waveguides, and a dedicated waveguide for power tokens.

2.6.2 Prediction Mechanism

Every *SM_station* uses the *Restr_Pred* module to predict the laser power requirement in the next epoch. *Restr_Pred* uses a multivariable function (Ψ) to make its prediction. This function takes three inputs: messages received in the current epoch (M_R), messages sent in the current epoch (M_S), and the waiting time at a station (W), and produces a 1-bit output to be sent to a Laser Controller (*L_Cntrlr*) at the end of every epoch. The output indicates if power is required or not in the next epoch. Ψ is defined as follows

In this function, R_T and W_T are threshold values, α and β are constants (all empirically determined). The explanation is as follows.

F_1 means that if a station has received a lot of messages, and sent a few, it is most likely to transmit in the next epoch, because it needs to send responses. F_2 suggests that if a lot of messages have been waiting, then they will be sent in the next epoch, and finally, F_3 means that if a station has been very quiescent in the current epoch, it has a higher likelihood of being active in the next epoch. Note that these are GPU-specific observations and do not hold for other platforms.

Akin to *SM_stations*, *LLC_stations* also send a 1-bit prediction to the laser controller based on the following rules. They use the *Flex_Pred* module.

In Figure 2.9, P_E is the number of pending events at the *LLC_station* in the current epoch. The intuition for G_1 is similar to F_1 (see Figure 2.8), and that for G_2 is similar to that of F_2. G_3 considers the ratio P_E/M_R and predicts a '1' if it is more than a threshold, β. This means that there

$$\Psi(\mathcal{M_R}, \mathcal{M_S}, \mathcal{W}) = \begin{cases} 1 & \underbrace{(\mathcal{M_R} \geq R_T \wedge \mathcal{M_S} \leq \alpha * R_T)}_{F_1} \vee \underbrace{(\mathcal{W} \geq W_T)}_{F_2}, \\ & \vee \underbrace{(\mathcal{M_R} \leq R_T \wedge \mathcal{M_S} < \alpha * M_R)}_{F_3}, \\ 0 & otherwise. \end{cases}$$

FIGURE 2.8 The prediction function used by Restr_Pred (Source: Bashir, 2020).

$$\zeta(\mathcal{M_R}, \mathcal{P_E}, \mathcal{W}) = \begin{cases} 1 & \underbrace{(\mathcal{M_R} \geq \alpha * R_T)}_{G_1} \vee \underbrace{(\mathcal{W} \geq \alpha * W_T)}_{G_2}, \\ & \vee \underbrace{(\mathcal{P_E} \geq \beta * \mathcal{M_R})}_{G_3}, \\ 0 & otherwise. \end{cases}$$

FIGURE 2.9 The logic used by Flex_Pred (Source: Bashir, 2020).

are more events at the station other than the expected number of responses to reads, and these events need to be sent in the next epoch.

At the end of every epoch, *SM_stations* and *LLC_stations* send their 1-bit recommendations to the *L_Cntrlr* through a separate waveguide called the prediction waveguide. This process is initiated 8 cycles before the end of the epoch.

In the reconfiguration phase, the first job of *L_Cntrlr* is to collect sixteen 1-bit predictions sent by the 16 optical stations. The *L_Cntrlr* adds the eight 1-bit predictions received from the 8 *SM_stations* to calculate the amount of power required by the *SM_stations* in the next epoch. Let this sum be equal to v; if the sum was 8, it is set to 7 such that it remains within 3 bits.

The eight 1-bit predictions received from the 8 *LLC_stations* are used to index a 256-entry table called the Power Table (PT). Each entry of the PT stores the predicted number of power units that will be used by the *LLC_stations* in the next epoch. Let the value read from the table be w (3 bits).

The PT is updated at the end of every epoch. Let *PT_index_prev* denote the 8-bit number created using the eight 1-bit predictions received by the *L_Cntrlr* in the previous epoch. Let P be the optical power used in the current epoch and S be the optical power units required in the next epoch (sum of 1-bit predictions), then the PT table is updated as follows:

$$PT\left[PT\_index\_prev\right]\begin{cases} P+1, & S \geq P \\ P, & \frac{P}{2} \leq S < P \\ P-1, & default \end{cases}$$

The intuition is the same as similar formulae where if the power required is within a certain range then status quo is maintained otherwise the corresponding prediction is either incremented or decremented.

For the SMs the authors assume $P = S$, which means that the system reacts immediately and modulates the power requirements. L1 miss rates are more frequent, and thus faster reactivity is beneficial. Additionally, if we underestimate the power, the additional latency will not affect us much given that GPUs are relatively intolerant to changes in latency. However, in the case of LLC banks, we make the system far less reactive particularly for reducing the number of tokens. We always prefer to have the same number of tokens (or more) even if the expected number reduces by up to 50%, because we wish to prioritize this traffic from the point of view of bandwidth and latency.

2.6.3 Reconfiguration

Finally, the *L_Cntrlr* adds the two 3-bit numbers (v and w) to determine the number of lasers to be turned on in the off-chip laser array. This 4-bit sum is sent to the off-chip laser array for turning on the required number of lasers in the next epoch. It is assumed that the off-chip laser is connected using a fast optical link and it is located at a distance of 2 cm from the chip. Thus, it takes less than a cycle to reach the off-chip light source.

2.7 CONCLUSION

In this chapter, we looked at a wide variety of activity prediction and laser modulation schemes for CPUs, GPUs, multi-socket servers, and manycore processors. The basic concepts are similar: we have an array of performance counters, we combine their values using either linear methods or nonlinear methods based on neural networks, subsequently we make an ad hoc prediction which is compared against two thresholds – an upper threshold and a lower threshold. If it is less than the lower threshold, then we reduce the power output, and likewise if it is higher than upper threshold, then we increase the laser power output. If the predicted value lies between the two thresholds, then we maintain status quo. The main challenge in designing an effective activity prediction and laser modulation scheme is in deciding the nature of the relationship between the performance counters, the constants, and whether we use a predictor of a single type or we use an ensemble of predictors. All of these design choices together determine the accuracy of the final prediction. This chapter captures the state of the art as of 2020; it is expected that in the years to come, as optical networks become more commonplace, these techniques will substantially mature and see practical applications.

REFERENCES

Bashir, J. A. (2017). NUPLet: A photonic based multi-chip NUCA architecture. *International Conference on Computer Design (ICCD)* (pp. 617–624). IEEE.

Bashir, J. A. (2019a). A survey of on-chip optical interconnects. *ACM Computing Surveys (CSUR)*, 1–34.

Bashir, J. A. (2019b). Predict, share, and recycle your way to low-power nanophotonic networks. *ACM Journal on Emerging Technologies in Computing Systems (JETC)*, *16*(1), 1–26.

Bashir, J. A. (2020). GPUOPT: Power-efficient photonic network-on-chip for a scalable GPU. *ACM Journal on Emerging Technologies in Computing Systems (JETC)*, *17*(1), 1–26.

C. Bienia, S. K. (2008). The PARSEC Benchmark Suite: Characterization and architectural implications. *Proceedings of the 17th International Conference on Parallel Architectures and Compilation Techniques*.

Eldhose Peter, S. R. (2015). Optimal power efficient photonic SWMR buses. *2nd Workshop on Silicon Photonics, along with the HiPEAC Conference*. Amsterdam.

Herr, Anna Y, Herr, Quentin P, Oberg, Oliver T, Naaman, Ofer, Przybysz, John X, and Borodulin, Pavel, Shauck, Steven B. (2013). An 8-bit carry look-ahead adder with 150 ps latency and sub-microwatt power dissipation at 10 GHz. *Journal of Applied Physics*, *113*(3), 033911.

J.-R. Burie, G. B.-R.-M. (2010). Ultra high power, ultra low RIN up to 20 GHz 1.55 um DFB AlGaInAsP laser for analog applications. *OPTO, International Society for Optics and Photonics*.

Janibul Bashir, E. P. (2019). BigBus: A scalable optical interconnect. *ACM Journal of Emerging Technologies in Computing Systems (ACM JETC)*, *15*(1), 1–24.

Pan, Y. J. (2010). Flexishare: Channel sharing for an energy-efficient nanophotonic crossbar. *The Sixteenth International Symposium on High-Performance Computer Architecture (HPCA)*.

Pan, Y. P. (2009). Firefly: Illuminating future network-on-chip with nanophotonics. *Proceedings of the 36th Annual International Symposium on Computer Architecture* (pp. 429–440).

Peter, E. A. (2015). ColdBus: A near-optimal power efficient optical bus. *International Conference on High Performance Computing (HiPC)*. IEEE.

S. C. Woo, M. O. (1995). SPLASH-2 programs: Characterization and methodological considerations. *Proceedings of the 22nd International Symposiumon Computer Architecture* (pp. 24–36).

Stratton, J. A.-J.-W.-M. (2012). Parboil: A revised benchmark suite for scientific and commercial throughput computing. Center for Reliable and High-Performance Computing.

T. Fukamachi, K. A. (2010). Recent progress in 1.3-um uncooled InGaAlAs directly modulated lasers. *International Semiconductor Laser Conference (ISLC)* (pp. 189–190). IEEE.

Zhou, L. A. (2013). Probe: Prediction-based optical bandwidth scaling for energy-efficient NoCs. *Seventh IEEE/ACM International Symposium on Networks-on-Chip (NoCS)*. IEEE.

3 Scalable Low-Power High-Performance Optical Network for Rack-Scale Computers

Jun Feng, Jiaxu Zhang, Shixi Chen, and Jiang Xu

CONTENTS

3.1 INTRODUCTION

Large-scale applications like scientific computing, big data applications and artificial intelligent applications such as weather prediction and social network analysis are demanding more processing cores and computation power to handle the increased computation tasks and large data sets [1]. The high-performance computers and cloud computing systems keep evolving to fulfill such demands which push computing systems towards exascale systems [2]. Scale-up and scale-out models are two common solutions towards the efficient exascale systems [3]. Scale-up model improves computation power by integrating more cores into a single chip server. Since the trend of Moore's Law on semiconductor technology is slowing down, it is becoming more difficult to integrate more cores into a single chip server. On the other hand, scale-out model can lower single server requirements and link together more distributed chips to bring more computation power. But it asks for efficient data movement among distributed chips.

With the rack increasingly replacing the individual server as the standard unit to build large-scale computing systems, rack-scale computing systems are promising choices for future exascale systems [4]. And its computation and communication resources are closely coupled for efficient data movement within rack. However, interconnection is currently the bottleneck between the excessive computation power and large volume data sets, and this problem may become even worse for the closely coupled rack-scale computing systems because of the high latency and limited bandwidth [5]. It is estimated that 80% of the traffic generated by servers stays within the rack in cloud computing, and the disaggregated computing and memory resources in the rack tend to generate more traffic [6]. Furthermore, the energy consumption of rack-scale interconnection

DOI: 10.1201/9780429292033-3

occupies a quarter of the overall system energy and is expected to grow when the system scale becomes large [7]. Rack-scale interconnects are required to provide low-latency, high-bandwidth, and low-power consumption for local and remote data access, which is essential to build rack-scale computing systems with higher performance and better energy efficiency.

The traditional electrical interconnects cannot meet the energy consumption requirements of rack-scale high-performance computing systems. Optical interconnects therefore become an alternative to replace traditional electrical network fabrics for high-performance computing systems thanks to their superiority in bandwidth and latency of transmission and switching, and attenuation and power consumption at high data rate [8]. However, there are three main disadvantages of existing optical network architectures. Firstly, we use circuit switching in optical networks, which requires path reservation before data transmission. The path reservation stage might degrade the system performance and lower the network resource utilization especially when the data transmission covers both on-chip and off-chip domains [9]. Secondly, separate off-chip and on-chip networks are considered in previous studies, which neglects the performance gap between on-chip and off-chip domains [10]. Rack-scale computing system targets large-scale systems, but these designs mainly target at small-scale systems. Thirdly, optical interconnects have different characteristics from electrical interconnects especially in bandwidth and switching methods [11]. But most communication procedures such as application level connection establishment for the inter-chip traffic mainly focus on electrical domain. These methods and designs cannot be used in optical network architectures directly because they are inefficient and need better optimizations.

In this chapter, we first explore different rack-scale computing system architectures including rack-scale optical network (RSON) architecture with different path reservation schemes and optical inter-chip networks. Then we systematically analyze these architectures and compare with the most-commonly used architecture for high-performance computing systems: the Ethernet architecture. Experimental results show that the RSON with a floorplan optimized delta optical network (FODON) switch and preemptive chain feedback (PCF) scheme can improve system performance by up to 5x under the same energy consumption which maintaining better scalability than the state-of-the-art systems, and the optical interconnects may become a potential alternative for the rack-scale computing systems.

The rest of this chapter is organized as follows. Section 3.2 gives the related work about the existing optical network architecture and its control. The scalable low-power high-performance rack-scale optical network architecture in this work is described in detail in Section 3.3. Section 3.4 presents quantitative comparisons between an Ethernet-based rack-scale system and the RSON architecture with different control schemes and inter-chip networks. Finally, Section 3.5 concludes this work.

3.2 RELATED WORK

To deal with the performance and energy challenges for manycore processors and high-performance computing systems, optical network architecture is introduced for both intra-chip and inter-chip communications [12]. Hybrid electrical and optical on-chip architectures were explored to use electrical links for local communications and optical links for global communications to take advantages of both electrical and optical interconnects [13]. All-optical intra-chip networks were also investigated that wavelength division multiplexing (WDM) was adopted to boost the bandwidth [14]. Either passive or active routing is used in these designs for a trade-off between scalability and control complexity. Inter- and intra-chip network was studied in [13], but they did not provide details about how to connect these two nor how they interact with each other. Different kinds of passive switching for optical interconnects at the top of the rack were presented in [15], but they suffered from the low flexibility and scalability issues. Rack-scale inter-chip optical network architectures have also been extensively investigated in [16]. The optical top-of-rack (ToR) switch is critical to construct all-optical rack-scale inter-chip networks. A ToR switch was experimentally demonstrated with 75 ns end-to-end delay [17]. However, the scalability of the switch radix and bandwidth could be potential issues for the increasing communication demand within the rack.

The control flow for both intra-chip and inter-chip networks was introduced to improve system performance. For example, explicit congestion notification was proposed for multi-domain network congestion management [18]. However, when the number of server nodes inside the rack grows, the scalability of the passive switching will become an issue. Other techniques, like resource allocation, congestion control and so on have been investigated widely, but these techniques are also limited to small-scale systems [13]. For optical path reservation, the prebooking reservation mechanism was proposed [19], but it is still inefficient and targets at small-scale systems, which need further attention and optimization.

3.3 ARCHITECTURE

The RSON is an optical intra-rack communication architecture that targets at achieving high performance and energy efficiency for rack-scale computing systems [20]. This multi-domain optical network contains both on-chip and off-chip networks. Thus, the design of intra-chip optical network (or optical network-on-chip (ONoC)) and inter-chip optical network, along with the optical inter-node interface and optical transceiver is required. Here for the inter-chip optical network, a FODON switch is designed with a deterministic routing algorithm and PCF is used as the control scheme and communication flow optimization to improve the overall system performance and energy efficiency.

3.3.1 RSON Architecture Overview

The structural overview of the RSON system is shown in Figure 3.1. The RSON architecture uses a low-loss high-radix multi-stage optical switch fabric to fully connect multiple electrical/optical hybrid network server nodes. For each server node, this hybrid electrical network-on-chip (ENoC) and ONoC can take advantage of both electrical and optical interconnects to provide sufficient computing power and memory space to fulfill the application demands. There are multiple core clusters on each server node, which can provide sufficient computation power and keep better scalability while increasing the number of cores.

The ONoC is designed to connect the on-chip memory controllers and the optical inter-node interfaces via the ONoC interface. It can provide high-speed data movement bypass for memory access between on-chip and off-chip domains. This will dramatically mitigate the burden on the legacy ENoC with little hardware overhead. This architecture can realize real zero-copy remote direct memory access (RDMA) communication between the local and remote server node's memory, enabled by the multi-domain circuit switching. No buffer, E/O and O/E conversion is needed in the middle stage, avoiding extra power and latency overhead.

3.3.2 FODON Switch Design

The multi-stage optical switch used in the RSON is a FODON, whose architecture is shown in the gray region of Figure 3.1. Since the ToR integrated optical switch is expected to support high radix with low signal loss to connect the increasing number of more server nodes and other components, this optical fabric shows an unprecedented advantage on loss and is promising for high-radix switch networks [21].

The FODON architecture contains silicon waveguides and optical fibers, and it uses micro-resonators as the basic switching element to switch optical signals between data channels. The silicon waveguides are designed as the data channels to connect the input/output ports. The optical fabric also contains an input switch stage, middle switch stage, and output switch stage. The optical fibers are used to connect the ONoC via the optical inter-node interface and the input/output stage switch in the optical switch fabric. Moreover, to reduce the optical crossing- and passing-incurred signal loss, the floorplan is optimized so that the average port-to-port loss is limited and maintains good scalability.

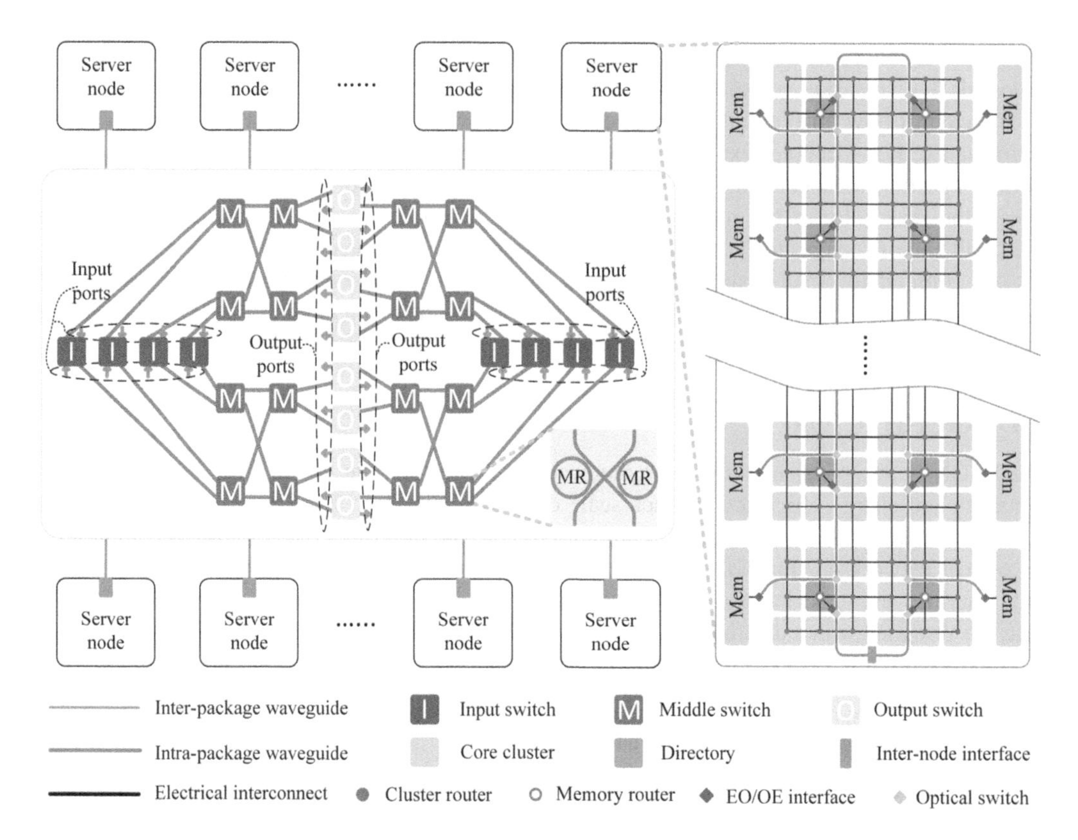

FIGURE 3.1 Overview of RSON structure. The inter-chip network with multi-stage optical switch fabric connects server nodes, and the ONoC is designed to connect the on-chip memory controller and the optical inter-node interface.

Off-chip laser bank is used for each ONoC interface to provide optical power source. The off-chip laser shows lower power consumption compared to on-chip laser source but induces higher latency on controlling the lasers. However, the control latency for the off-chip laser can be hidden because the laser turn-on and turn-off can be done in advance with preemptive reservation mechanism. Edge coupling is used to couple the optical signal into waveguides. To improve the bandwidth, multiple wavelengths are multiplexed into a single fiber/waveguide by wavelength division multiplexing (WDM).

The switch controls the optical path actively to set up the connection between any input and output port pair and when the number of input/output ports and interconnect bandwidth increases, passive switching maintains better scalability. By configuring the status of micro-resonators, the path can be changed under the particular path requirements. Thus, the optical switch fabric only needs to handle data transmission requests and corresponding path reservation. The circuit switching method for the optical switch fabric helps reduce buffer area and power consumption because the data transmission just passes through the switch without buffering. Internal signal forwarding is implemented between input and output control ports, which is essential for control messages and path reservation across multiple optical domains.

3.3.3 Preemptive Chain Feedback Control Scheme

For optical communications, circuit switching is used for data transmission and the path will be reserved and occupied until the data transmission is completed, which will result in unfairness to other waiting messages and large turnaround and response latency because the ongoing message transmission will occupy the path for a long time. Then periodical resource rescheduling is proposed in the RSON architecture to periodically reschedule the optical resources such as the input/output ports so that the long-term blocking and unfairness issues can be avoided due to long message transmission. Long messages will be divided into multiple small blocks for transmission and reschedule the waiting small message requests in a given period *T*, which gives the interval in which the reschedule should be triggered. When the channel is occupied by small message, the controller can schedule multiple messages into a single period *T* because the small message will not occupy the channel for too long. Then the reschedule mechanism will not be triggered so that unnecessary reschedule overhead will be reduced. Moreover, the reschedule mechanism will not be triggered when the channel is free, both long and short messages can make full use of the current resource. Thus, the reschedule mechanism will not affect normal scheduling process, which will only introduce neglected overhead.

The PCF scheme [22] allocates network resources and manages the requests contention with the help of a multi-cell reservation window. The resources are the input and output ports of optical devices, senders or receivers. The chain feedback mechanism is devised to convey the reservation information between neighboring hops in a responsive way to reduce the overhead. Figure 3.2 gives an example of the scheme between neighboring resources. Specifically, the network path is allocated to a communication request only for a given time period *T*. In the preemptive reservation window, multiple cells (blue box in Figure 3.2) stand for a set of continuous time slots to be preemptively allocated to corresponding communication requests. Here, we set *C* cells under each reservation window and each cell corresponds to the rescheduling period *T*. There is a list of requests that are assigned to each cell in the reservation window, which is shown in the red part of Figure 3.2. Each cell holds necessary information to record the preemptive reservation status. In each cell, we use *C_id* as the index for each cell and the *id* number is incremented when the clock moves to the next time period, which is synchronized across all resources. We use the local clock source in each domain to realize the synchronization and the synchronization process is conducted at the initialization stage. There exists propagation delay between neighboring resources and we calculate the proper offset at the starting time to compensate for the propagation delay, which is important to make sure that the *C_id* in different resources keep the consistent pace. This synchronization process is very important for the multidomain path reservation mechanism that make sure the correct data transmission in each cell at all resources.

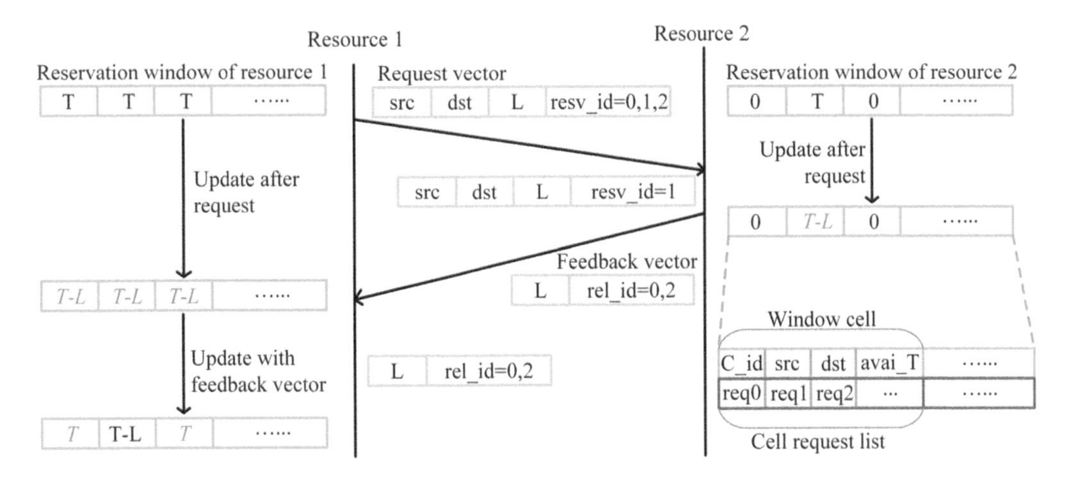

FIGURE 3.2 Example of PCF between neighboring resource 1 and 2 together with the structure of window cell.

To explain the meaning of preemptive reservations, the scheduled request will reserve the time slot of each resource preemptively, which means other requests cannot reserve the corresponding time slot unless it is released due to reservation failure. The resources at each time slot can only be shared by communication flows that use the same resource set. In circuit switching, this restriction is reduced to that only messages with the same source and destination can be scheduled to the same cell, which ensures that the flows in each cell can use resources out of order, such as that in electrical packet or cut-through switching. This can bring two benefits: by omitting the request sequence of the reservation window of the intermediate resources, the path reservation process can be simplified; and future communication requests can be immediately scheduled for the reservation period with the same source and destination. This is essential for scheduling small messages in time and improving resource utilization. It is sufficient to use just the source and destination information to identify suitable cells.

Regarding the reservation, the earliest cell will be preemptively scheduled to the incoming requests in a round-robin fashion. It should be mentioned that, only at the sender's interface, each cell of the reservation window needs to keep a list of issued path reservation requests. This is necessary to correctly inject data into the network once the reservation completes. The request vector propagates to the next hop resource via the optical control channel at the time when the reservation at the current resource is successful. Once the reservation process fails at the current resource, it will generate the corresponding feedback vector to release the related reservations at previous resources across the requested path. The feedback vector is the key component in the multidomain path reservation mechanism that can release the unused resourced more efficiently and improve the network resource utilization. As described in the former section, for inter-node data transmission in RSON, it usually contains three critical resources reservations: local ONoC domain, optical inter-node network, and remote ONoC domain. Then to reserve the required resources for a single time slot may result in a low possibility, which can even lead to low success rate of path reservation and too many reserved packets in the network. To solve this shortcoming, the reservation window works in a greedy manner to reserve cells. If there are free cells, greedy reservation is simply achieved by allowing twice the number of required resource cells to be reserved for each request. Additional resource reservation makes it possible to prepare any cell in a multi-domain network. Together with the responsive chain feedback mechanism, it can achieve a good reservation speed without causing high resource utilization overhead.

For simplicity, only the available time is marked in each window cell. The detailed information of the window cell is shown on the right-bottom side of Figure 3.2. The request vector and feedback vector propagate along the routing path to reserve or release resources like a chain. They can effectively coordinate the resource allocation and release among different network domains. The operations of the request vector and feedback vector between resource 1 and 2 are as shown in Figure 3.2. The request vector firstly flows from resource 1 to 2, carrying the request source, destination, length, and *C_id* to reserve. Cells 0, 1, and 2 are the cells to be reserved in this stage. The successful reservation will update corresponding resource cells and forward the request vector to the next hop. Upon failed reservation of intended cells, the feedback vector will be triggered. The reservation window of resource 2 shows that it only has a free cell in cell 1, which means cell 0 and 2 are occupied by other communication flows. The feedback vector will carry this information and release the corresponding preemptively reserved time slots in previous resources.

3.4 EVALUATIONS

In this section, we conduct evaluations on various rack-scale computing system architectures with different control schemes. The most common interconnection solutions in industry and academia for high-performance computing systems are the traditional server nodes with an Ethernet switch-

TABLE 3.1
Basic architecture configurations

Item	Configuration
Core cluster	4, 4 cores/cluster
Coherence protocol	Directory-based MOSI
L1 Instruction $	32 KB/core, private
L1 Data $	32 KB/core, private
L2 $	512 KB/cluster, cluster shared
LLC	2.5 MB/slice
Memory bandwidth	480 GBps/port
Processing core frequency	4 GHz
Fabrication technology	7 nm

connected rack (Ethernet for short). Two different control schemes, the generic handshake-based path reservation control scheme (Handshake for short) and the PCF scheme, will be used in the RSON. For the inter-chip network, three different multi-stage optical switches, Benes, Fat-tree and FODON, will be compared.

3.4.1 Evaluation Setup and Methodology

We use the simulator JADE [23] as the platform to model and functionally implement the RSON with the two different control schemes and different inter-chip networks along with its counterpart, the Ethernet architecture [24]. We consider both the network features and corresponding impacts on computing and memory systems. Thus, the processor microarchitecture, cache subsystem and memory, besides the interconnect itself, need to be considered. Some basic architecture configurations are shown in Table 3.1 below.

The on-chip electrical link bit width is 128, which is consistent with mainstream designs. The optical interconnect bandwidth is set to be 100, 200, 400, and 800 Gbps, enabled by WDM with 25 Gbps per wavelength, and the length of the link that is used to connect the server node and inter-chip optical switch is set to 1, 2, 3, 4, and 5 m. These are both carried out for a 128-node system unless otherwise stated. Furthermore, we use the large machine learning applications Cifar, AlexNet, and ONMT, which are included in the COSMIC benchmarks.

3.4.2 Performance Evaluation

Firstly, we present the overall system performance results under different system architectures. Figure 3.3 shows the performance comparison under different system scales when the optical inter-chip interconnect bandwidth is 800 Gbps and the link length is one meter. The performance of the RSON architecture with the PCF control scheme under the three multi-stage optical switch scales best among all the architectures since the corresponding performance improves fastest when the scale becomes larger. The contention for the network resources becomes worse when there are more server nodes within the rack and this problem may become even worse for the Ethernet architecture as its inter-chip network uses electrical interconnects. The electrical interconnects perform worse than the optical interconnects, especially in network latency and energy consumption. The PCF scheme can preemptively reserve the resources and release the contention, improving the performance by 99% and 19% when compared with the Ethernet architecture and the RSON architecture with the handshake scheme.

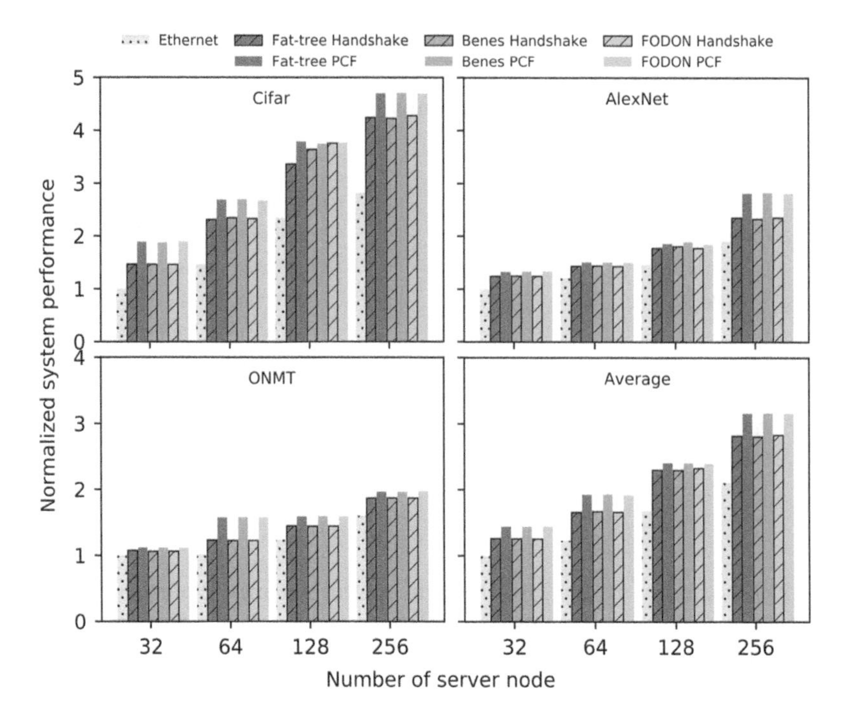

FIGURE 3.3 Normalized performance comparison under different system scales for different applications and the average case.

The performance under different interconnect bandwidths is presented in Figure 3.4. The inter-chip optical interconnect bandwidth shows transmission capacity, which means a larger bandwidth presents a larger transmission volume within the same time. Two kinds of traffic are involved in the inter-node communication: send/receive and RDMA. The message size of these two kinds of traffic is different where the send/receive operation communicates between the host process's and remote process's internal buffer with small messages and the RDMA operation transfers large messages between the local node's memory and remote node's memory. Therefore, the larger interconnect bandwidth improves the inter-node communications, especially for larger scale with more nodes. Also, the long messages of the RDMA operations are the main parts of the inter-node communications. The results show that the RSON architecture with the PCF scheme and FODON switch can achieve 98% and 15% higher performance than the Ethernet architecture and the RSON with the handshake scheme respectively. Furthermore, the larger interconnect bandwidth can help improve the system performance and the RSON architecture with the PCF scheme can obtain the highest performance. These results demonstrate that a larger system scale and interconnect bandwidth can bring higher system performance.

3.4.3 System Energy Consumption

The energy consumption is an important metric to evaluate a high-performance and low-power computing system. Here, we first present Figure 3.5 to show the system energy breakdown for the average case. The energy consumption of the processor and cache memory subsystems are conducted by the McPAT and CACTI tools integrated in JADE. To measure the optical power, we use the minimum laser power that needs to be generated to compensate for the optical devices' signal losses, and the corresponding device loss values used in the evaluation are listed in Table 3.2 below.

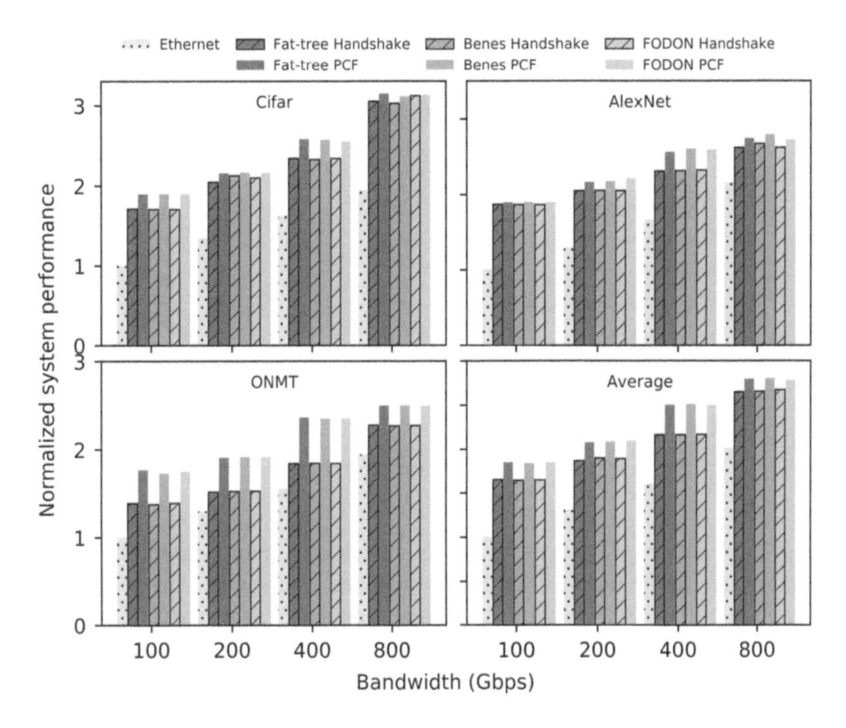

FIGURE 3.4 Normalized performance comparison under different interconnect bandwidths for different applications and the average case.

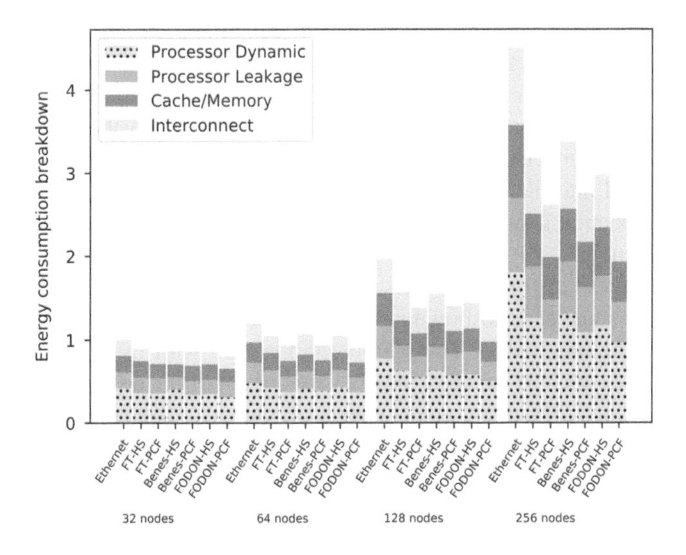

FIGURE 3.5 The energy breakdown for the average case. Here, FT is short for Fat-tree and HS stands for the handshake scheme.

Figure 3.5 shows that the system needs more energy consumption with a larger architecture scale. The energy consumption of the Ethernet architecture grows much faster compared with that of the RSON architecture. The energy consumed by the processor cores occupies around 60% of the whole system energy consumption. Moreover, the RSON architecture with the PCF control scheme and FODON optical switch achieves the lowest system energy consumption among these architectures because optical interconnects are superior in latency, bandwidth and power

TABLE 3.2
Parameters for optical interconnects

Item	Configuration
MR insertion loss	1 dB
MR passing loss	0.2 dB
Edge coupling loss	2 dB
Waveguide crossing loss	0.5 dB
Waveguide bending loss	0.1 dB/90°
Waveguide propagation loss	1 dB/cm
Photodetector sensitivity	-14.2 dBm
Laser efficiency	0.33

consumption at high data rates and the FODON switch has better scalability and lower power consumption compared to the Fat-tree and Benes switches. Even at the 256-node scale, the system energy consumption of the RSON architecture with the FODON switch and PCF control scheme is reduced by 26% when compared with the same architecture with handshake scheme. It is also 12% less than the system energy consumption of the RSON architecture with the Benes switch and PCF control scheme and 65% less than that of the Ethernet architecture.

3.4.4 Optical Energy Efficiency

To measure optical interconnects and the control subsystem, the optical energy efficiency is also an important factor. Here we define the energy efficiency as the number of bits in a transmission under the energy consumption. Higher energy efficiency means that a larger number of bits are transmitted with the same energy consumption. For fair comparison with the RSON architecture, the inter-chip network interconnects and the RDMA corresponding interconnects of the Ethernet architecture are calculated into the interconnect energy efficiency. (For simplicity, we treat them all as optical energy efficiency.) We also consider the electrical control logic power in the interconnect energy efficiency.

The optical interconnect energy efficiency under different interconnect bandwidths for different applications and the average case is shown in Figure 3.6. The interconnect energy efficiency grows with the increase of the bandwidth. The RSON architecture with the FODON switch and PCF control scheme achieves the best energy efficiency among all of the architectures, especially when the bandwidth is 800 Gbps. The interconnect energy efficiency of the Ethernet architecture is much lower, and though the PCF scheme brings power overhead because of the reservation window and feedback mechanism, it can still improve the interconnect energy efficiency. Figure 3.7 shows the interconnect energy efficiency under different link lengths. Although the energy efficiency decreases with the link length, the RSON architecture can transmit more bits under the same energy consumption. The FODON switch performs best along with the PCF control scheme. The results also show that the improved network performance by the PCF scheme can well compensate for the induced power overhead.

3.4.5 System Performance per Energy

Another important factor in evaluating large-scale and high-performance computing systems is the system performance per energy (PPE) from the perspective of power consumption and energy efficiency. A larger PPE means that better system performance is achieved or a greater

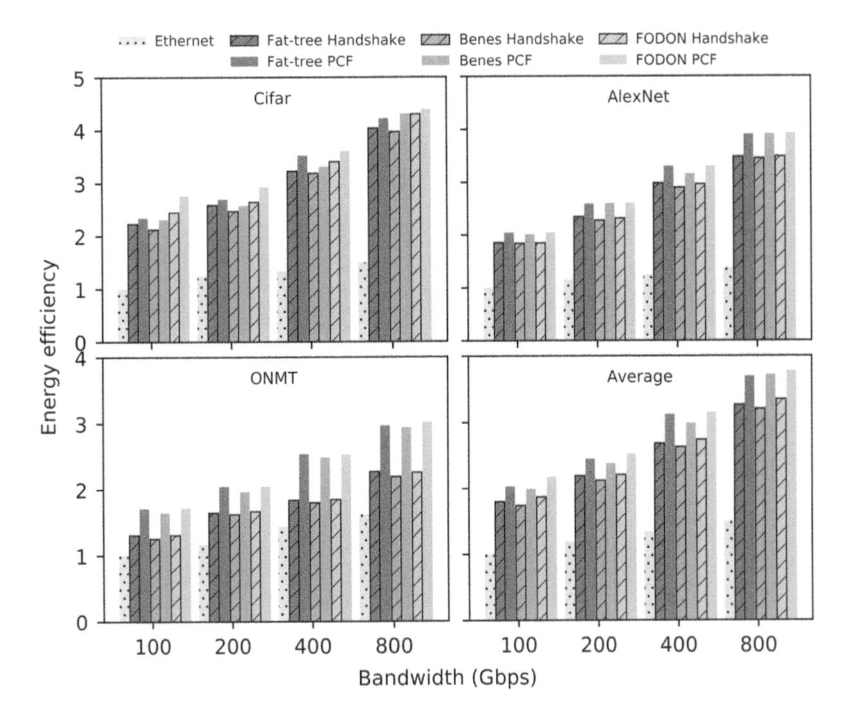

FIGURE 3.6 Normalized optical interconnect energy efficiency under different interconnect bandwidths for different applications and the average case.

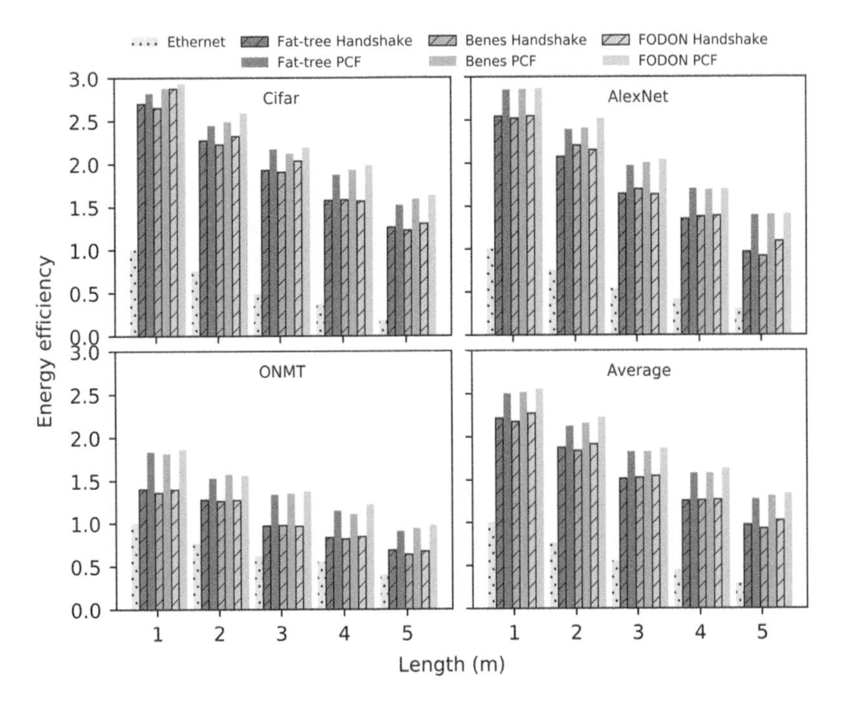

FIGURE 3.7 Normalized optical interconnect energy efficiency under different link lengths for different applications and the average case.

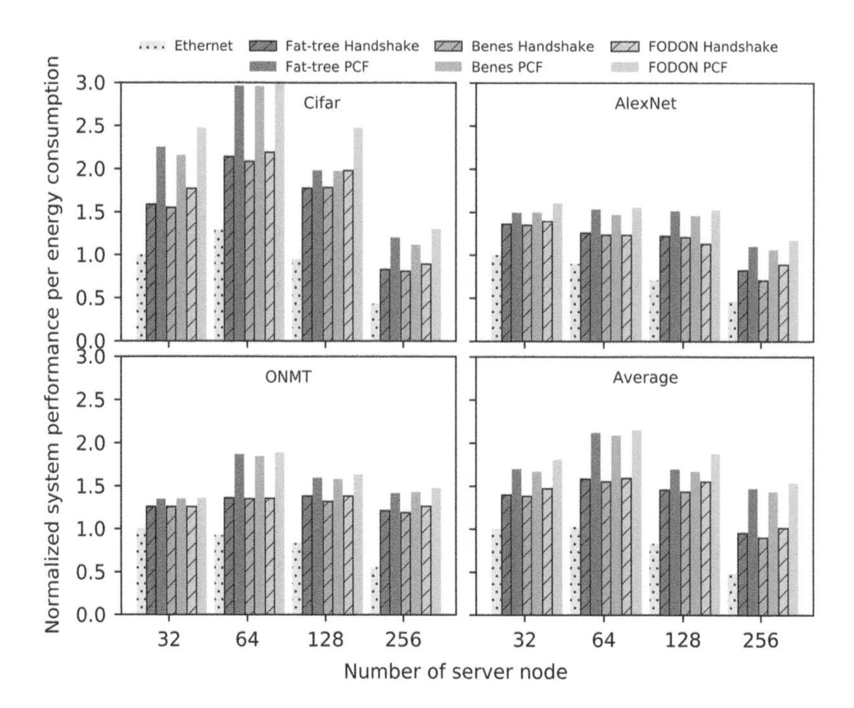

FIGURE 3.8 Normalized system performance per energy consumption under different system scales for different applications and the average case.

number of instructions are executed under the same energy consumption. We have conducted PPE evaluations under different conditions. Firstly, Figure 3.8 shows the PPE under different system scales. The Ethernet architecture presents much smaller PPE values when compared to the RSON architecture, and the trend of the decrease is much more rapid. This is because the electrical interconnects consume much more energy than the optical interconnects. As mentioned in the above section, electrical interconnects suffer from low system performance and may worsen this issue as the system scale become large. When the RSON architecture is used, the FODON switch and the PCF control scheme show the largest PPE, and the change of the PPE value is also very small, which demonstrates the better scalability, low power consumption, and high performance of the RSON architecture with the PCF control scheme and FODON switch.

Secondly, we present Figure 3.9 to show the PPE under different optical interconnect bandwidths. The PPE values become larger with higher optical interconnect bandwidth. When the interconnect bandwidth is 800 Gbps, the RSON architecture with the PCF scheme and FODON switch achieves the largest PPE value, which is around 56% and 18% lower than the values of the Ethernet architecture and RSON architecture with the handshake scheme. From the previous section, we know a higher interconnect bandwidth means more transmissions over the interconnects. What's more, the higher bandwidth can not only reduce the energy consumption of the interconnection but benefits the whole system by reducing the system energy and the latency to offer better communication service of the overall system.

Thirdly, we also evaluate the PPE under different link lengths. Here, a link is used to connect the server node and the inter-node switch. Since fiber is the most common link used for long distance optical transmission, a change of the link length will have a great effect on the whole system, especially a large-scale system. Figure 3.10 gives the result that as the link length increases, the system PPE will decrease correspondingly. Interestingly, the Ethernet architecture with the electrical interconnects consumes more energy with the same system performance when

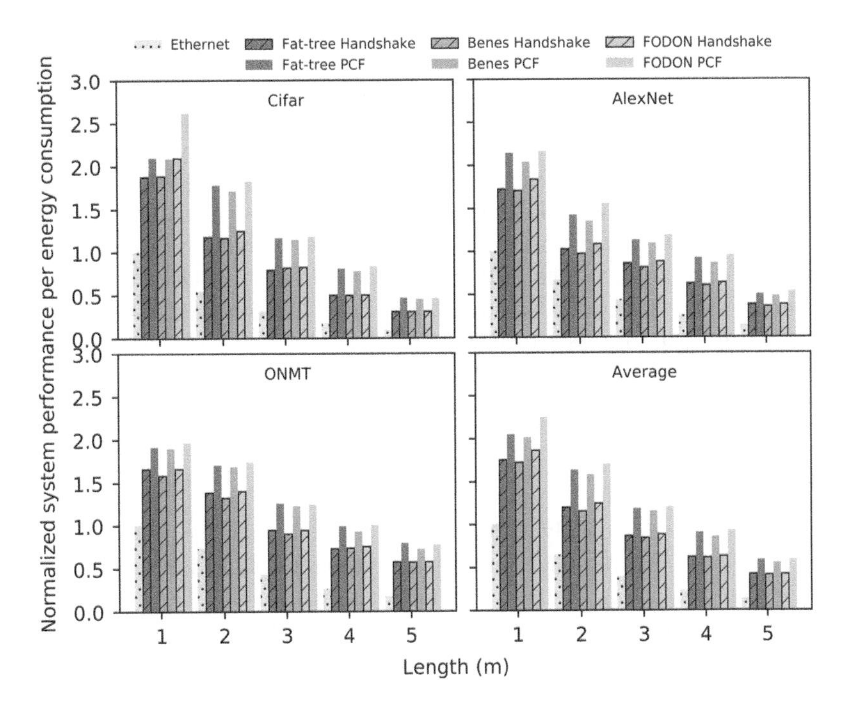

FIGURE 3.9 Normalized system performance per energy consumption under different interconnect bandwidths for different applications and the average case.

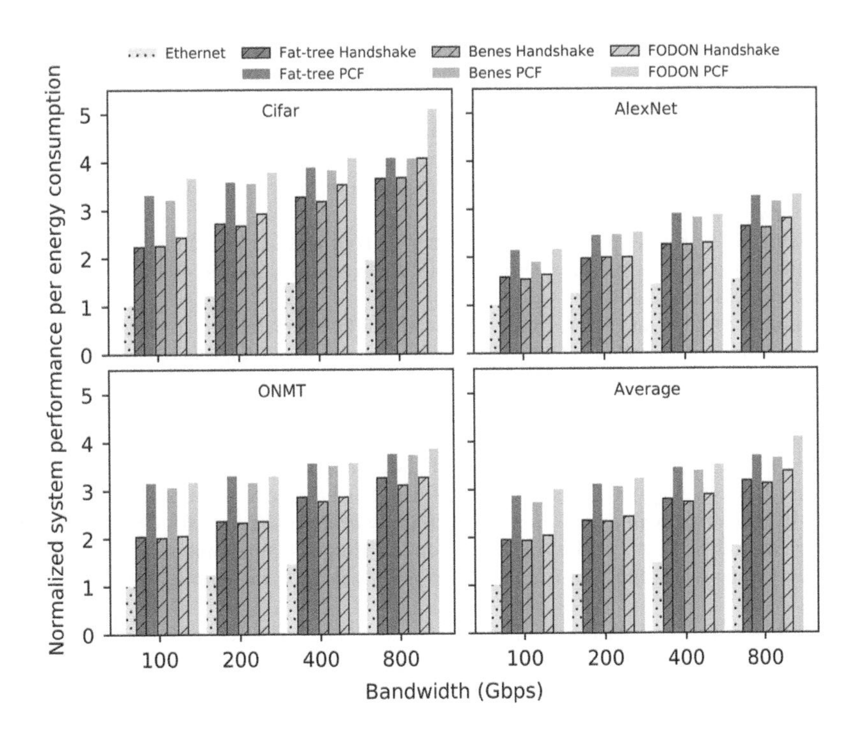

FIGURE 3.10 Normalized system performance per energy consumption under different link lengths for different applications and the average case.

compared to the RSON architecture. Although the system PPE of the RSON architecture also decreases with the increase of the link length, its decreasing trend is much slower. In fact, the link length has an impact mainly on the propagation delay. For the electrical interconnects, a longer link length may result in higher latency in the network and higher energy consumption, which will severely degrade the system performance at the same time. Then the PPE of the Ethernet architecture decreases exponentially with the link length. However, the optical interconnects show their advantage in terms of low latency and energy consumption. Despite the increase in latency, the optical interconnects are stable on the PPE. The fiber used in the RSON can maintain high performance under the same energy consumption, and we can also conclude from the results that the optical fiber performs better on long distance transmission within rack than the electrical interconnects. The RSON architecture with the PCF scheme and FODON switch achieves the largest PPE compared with the handshake scheme and the other two optical switches.

3.5 CONCLUSIONS

This work conducts systematic analyses and comparisons between the RSON architecture with different path reservation schemes and optical inter-chip networks and the most commonly used architecture for high-computing systems, the Ethernet architecture. The results show that the RSON architecture with the FODON switch and PCF control scheme can achieve higher performance with better scalability and lower energy consumption.

REFERENCES

[1] C. Minkenberg, "Interconnection network architectures for high performance computing," IBM Research Technology Report, Zurich, 2013.

[2] S. Rumley, D. Nikolova, R. Hendry, Q. Li, D. Calhoun and K. Bergman, "Silicon photonics for exascale systems," *Journal of Lightwave Technology*, vol. 33, pp. 547–562, 2015.

[3] R. Azimi, T. Fox, W. Gonzalez and S. Reda, "Scale-out vs scale-up: A study of ARM-based SoCs on server-class workloads," *ACM Transactions on Modeling and Performance Evaluation of Computing Systems*, vol. 3, no. 2376-3639, p. 4, 2018.

[4] C. Barthels, S. Loesing, G. Alonso and D. Kossmann, "Rack-scale in-memory join processing using RDMA," in *Proceedings of the 2015 ACM SIGMOD International Conference on Management of Data*, New York, 2015.

[5] C. Berger, M. A. Kossel, C. Menolfi, T. Morf, T. Toifl and M. L. Schmatz, "High-density optical interconnects within large-scale systems," in *Proceedings Volume 4942, VCSELs and Optical Interconnects*, Bruges, 2003.

[6] T. Benson, A. Akella and D. A. Maltz, "Network traffic characteristics of data centers in the wild," in *Proceedings of the 10th ACM SIGCOMM conference on Internet measurement*, New York, 2010.

[7] T. Hoefler, "Software and hardware techniques for power-efficient HPC networking," in *Computing in Science & Engineering*, 2010.

[8] N. Kirman, M. Kirman, R. K. Dokania, J. F. Martinez, A. B. Apsel, M. A. Watkins and D. H. Albonesi, "Leveraging optical technology in future bus-based chip multiprocessors," in *39th Annual IEEE/ACM International Symposium on Microarchitecture (MICRO'06)*, Orlando, FL, 2006.

[9] S. Pasricha and N. Dutt, "ORB: An on-chip optical ring bus communication architecture for multiprocessor systems-on-chip," in *Asia and South Pacific Design Automation Conference*, Seoul, South Korea, 2008.

[10] X. Wu, J. Xu, Y. Ye, Z. Wang, M. Nikdast and X. Wang, "SUOR: Sectioned undirectional optical ring for chip multiprocessor," *ACM Journal on Emerging Technologies in Computing Systems*, vol. 10, no. 1550-4832, p. 4, 2014.

[11] S. Bahirat and S. Pasricha, "UC-PHOTON: A novel hybrid photonic network-on-chip for multiple use-case applications," in *11th International Symposium on Quality Electronic Design*, San Jose, CA, USA, 2010.

[12] Y. Pan, P. Kumar, J. Kim, G. Memik, Y. Zhang and A. Choudhary, "Firefly: Illuminating future

network-on-chip with nanophotonics," in *Proceedings of the 36th annual international symposium on Computer architecture*, New York, NY, USA, 2009.
[13] X. Wu, J. Xu, Y. Ye, X. Wang, M. Nikdast, Z. Wang and Z. Wang, "An inter/intra-chip optical network for manycore processors," *IEEE Transactions on Very Large Scale Integration (VLSI) Systems*, vol. 23, pp. 678–691, 2015.
[14] S. Koohi and S. Hessabi, "All-optical wavelength-routed architecture for a power-efficient network on chip," *IEEE Transactions on Computers*, vol. 63, pp. 777–792, 2014.
[15] J. Chen, Y. Gong, M. Fiorani and S. Aleksic, "Optical interconnects at the top of the rack for energy-efficient data centers," in *IEEE Communications Magazine*, 2015.
[16] S. Rumley, M. Glick, G. Dongaonkar, R. Hendry, K. Bergman and R. Dutt, "Low latency, rack scale optical interconnection network for data center applications," in *39th European Conference and Exhibition on Optical Communication (ECOC 2013)*, 2013.
[17] P. Andreades, Y. Wang, J. Shen, S. Liu and P. M. Watt, "Experimental demonstration of 75 ns end-to-end latency in an optical top-of-rack switch," in *OSA Technical Digest (online) (Optical Society of America)*, 2015.
[18] InfiniBand Trade Association, "InfiniBand Architecture Specification Volume 1 Release 1.3," 2012.
[19] H. Kong and C. Phillips, "Prebooking reservation mechanism for next-generation optical networks," *IEEE Journal of Selected Topics in Quantum Electronics*, vol. 12, no. 1558-4542, pp. 645–652, 2006.
[20] P. Yang, Z. Pang, Z. Wang, Z. Wang, M. Xie, X. Chen, L. H. K. Duong and J. Xu, "RSON: An inter/intra-chip silicon photonic network for rack-scale computing systems," in *Design, Automation & Test in Europe Conference & Exhibition (DATE)*, Dresden, Germany, 2018.
[21] Z. Wang, Z. Wang, J. Xu, P. Yang, L. H. K. Duong, Z. Wang, H. Li and R. K. V. Maeda, "Low-loss high-radix integrated optical switch networks for software-defined servers," *Journal of Lightwave Technology*, vol. 34, no. 1558-2213, pp. 4364–4375, 2016.
[22] P. Yang, Z. Wang, Z. Wang, J. Xu, Y.-S. Chang, X. Chen, R. K. V. Maeda and J. Feng, "Multidomain inter/intrachip silicon photonic networks for energy-efficient rack-scale computing systems," *IEEE Transactions on Computer-Aided Design of Integrated Circuits and Systems*, vol. 39, no. 3, pp. 626–639, 2019.
[23] R. K. Vivas Maeda, P. Yang, H. Li, Z. Tian, Z. Wang, Z. Wang, X. Chen, J. Feng and J. Xu, "A fast joint application-architecture exploration platform for heterogeneous systems," in *Embedded, Cyber-Physical, and IoT Systems: Essays Dedicated to Marilyn Wolf on the Occasion of Her 60th Birthday*, Springer International Publishing, 2020, pp. 203–232.
[24] J. Feng, Z. Wang, Z. Wang, X. Chen, S. Chen, J. Zhang and J. Xu, "Scalable low-power high-performance rack-scale optical network," in *IEEE/ACM Workshop on Photonics-Optics Technology Oriented Networking, Information and Computing Systems (PHOTONICS)*, Denver, CO, USA, 2019.

4 Network-in-Package for Low-Power and High-Performance Computing

Armin Tajalli

CONTENTS

4.1 DATA MOVEMENT IN ADVANCED COMPUTING SYSTEMS

Modern high-performance computing (HPC) systems rely on high-speed links to transfer data among different computing, processing, and memory units. The higher the bandwidth, the higher the processing power. In addition to the data transfer bandwidth, the energy that is consumed to move the data is critically important. While the core computing power has been drastically reduced using advanced nano-scale technologies, energy dissipated for data movement has not been reduced with the same pace. As a result, designing power-efficient and high-speed serial data links are becoming more and more important.

Figure 4.1 shows an advanced computing system, where a 3D structure is employed to minimize the communication channels among different chips. Shorter the distance, lower the consumption and higher the communication speed. Figure 4.2 illustrates the evolution of the HPC technology [1]. Integrated circuit (IC) technology scaling together with progress of circuits and architectures have steadily pushed the signal processing capabilities forward [2]. For example, multi-core processor architecture replaced the traditional single-core systems around 2005 in order to mitigate the problem of heat and power density that saturated the clock speed. This paradigm shift in architecture led into development of the Network-on-chip (NoC) concept [3]. The NoC technology enables high BW data movement among different cores sitting inside a multi-core processor chipset. Such an architecture can substantially improve the processing performance, and simultaneously keep the power consumption and heat generation under control. By increasing the number of cores per chip (or, equivalently the number of cores per IC die), yield, and the corresponding excess cost manifested themselves as the new physical barriers to realize processors with higher performance, and keep pace with the Moore's Law. To keep the silicon area of multi-core processors within a reasonable size, improve the yield [4], and reduce design and manufacturing complexities, Multi-Chip Module (MCM) systems were proposed. The possibility of co-

DOI: 10.1201/9780429292033-4

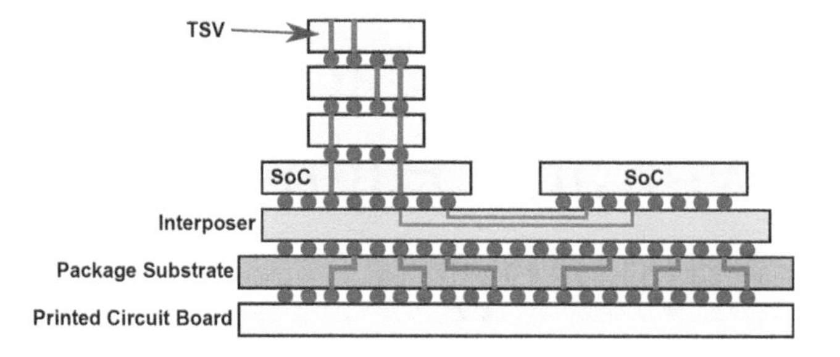

FIGURE 4.1 Example of a multi-layer high-performance integrated system with different types of chip-to-chip (C2C) communication mechanisms to transfer data between different units.

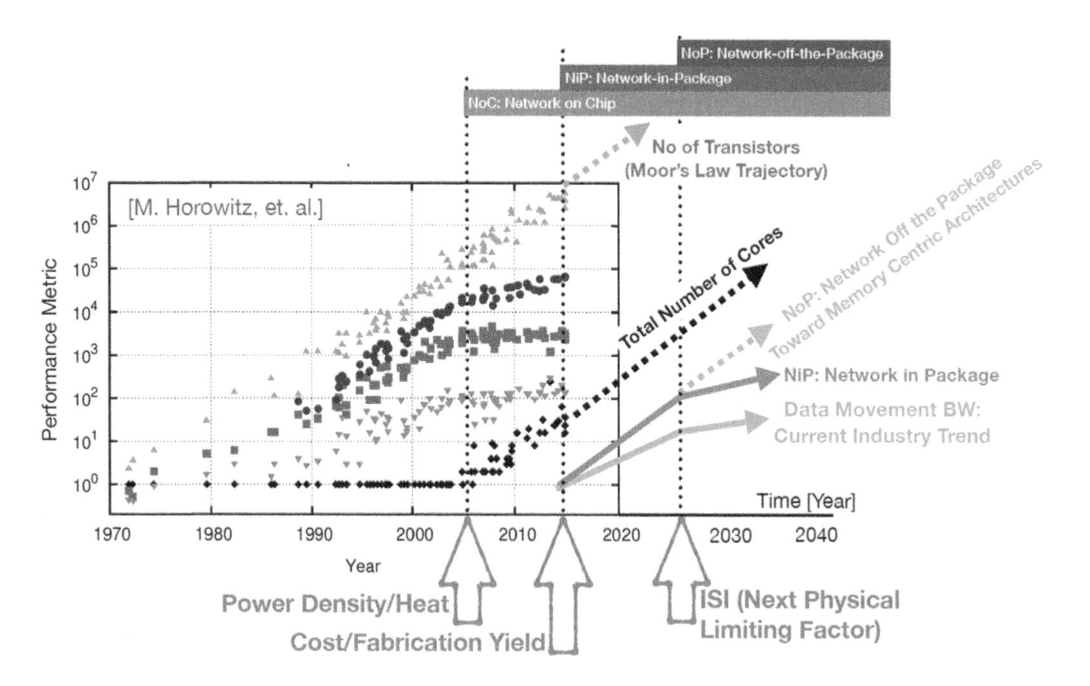

FIGURE 4.2 Evolution of the high-performance computing systems [1], and its implications for high-speed data movement.

packaging different dies, which are potentially fabricated in different technologies, has made the MCM technology very prevalent in today's electronics industry. In addition, the MCM architecture provides more design flexibility, at lower cost to implement more advanced System-on-Chip (SoC) embodiments. Today, MCM systems are considered as miniaturized printed circuit boards (PCBs), which rely on very high-speed network-in-package (NiP) capable of transferring data between different units residing on the same package substrates (see Figure 4.2).

Considering modern integrated systems, such as the one depicted in Figure 4.1, performance of the HPC systems depends directly on the speed and robustness of their Input and Output (I/O) links. The data communication BW is more critical in the emerging and growing applications such as Artificial Intelligence (AI), and Machine Learning (ML). In this type of system, large sizes of data is required to be continuously stored, transferred, and processed [5–8]. Thus, high-speed and low-power data movement has become a vital factor in such applications. As a consequence, high bandwidth communication among packages (Network-off-the-Package, or NoP) now is becoming one of the main topics of research, and a key requirement in the HPC industry. The main goal for

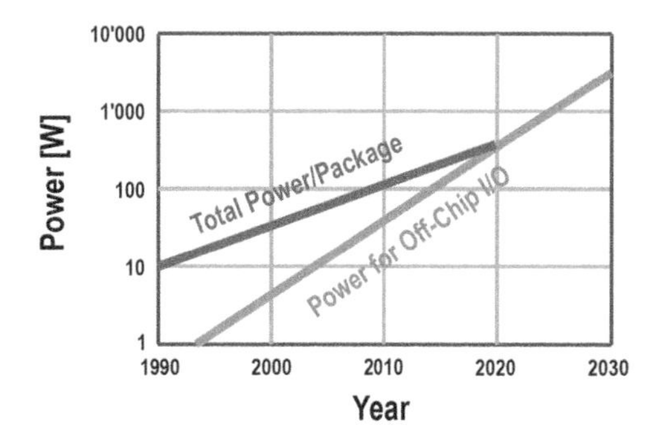

FIGURE 4.3 Power dissipation in processors: comparing processing power with respect to the I/O component [13].

the NoP systems is to transfer huge sizes of data (tens of Tb/s) among different SoCs, memory units, and processors at a very low level of energy dissipation, as well as small latency. To support a high data transfer BW, advanced computing systems employ high-performance serial data transceivers [4,9–12]. Complex circuit topologies and transceiver architectures, as well as advanced equalizing schemes are employed in such communication systems to compensate for the channel loss and inter-symbol interference (ISI). In addition to design complexities, the energy consumption of the high-speed I/Os can be significant mainly due to their need for performing complex equalization schemes. Depicted in Figure 4.3, the power consumption of the I/O circuitry is growing faster than the computing power. The required signal BW and the channel ISI are the two main design constraints for implementing a successful high-throughput low-power link. Therefore, developing signaling methods, which are inherently less sensitive to ISI, and consequently require less equalization overhead, are essential for relaxing the design tradeoffs and open the doors toward implementing data transceivers operating at much higher data transfer rates. This requirement is one of the main driving forces behind developing advanced signaling methods, such as OMWS, which exhibit low sensitivity to ISI.

4.2 WIDEBAND WIRELINE COMMUNICATIONS

As mentioned before, high-speed and energy-efficient serial data transceivers are key components in the modern HPC systems, and their importance grow more and more in the Von Neuman–based machines. Today MCM systems require links that can transfer data at more than 50 Gb/s/lane, while consuming much less than 1 pJ per transmitted bit (1 pJ/b). To achieve such a performance, both optical and electrical links can be considered. While optical links provide a very wideband communication media, complex interface circuitry and fabrication technologies are required to realize such systems. On the other side, wireline links carrying data over the conventional copper wires is more compatible with the IC fabrication and packing technology.

Figure 4.4 shows a typical serial wireline communication system transferring data over a differential pair of copper lanes. Differential signaling helps to suppress common-mode and crosstalk noises, and improve the signal integrity in a link. In such systems, a major part of the power is dissipated in the equalization and amplification to compensate for the frequency-dependent channel loss. Clocking is the second source of energy consumption, as precise timing is key to implement a low bit error rate (BER) link. The power dissipated to drive the channel also plays an important role in the overall system power consumption. With channel losses typically increasing with frequency, moving towards higher data rates requires more sophisticated equalization schemes, and correspondingly a higher power dissipation is expected. The rest of this article focuses on techniques that can be employed to reduce the sensitivity of the link to the frequency-dependent channel loss (or equivalently ISI), enabling better power-efficiency, and simultaneously supporting higher data rates over copper interconnects.

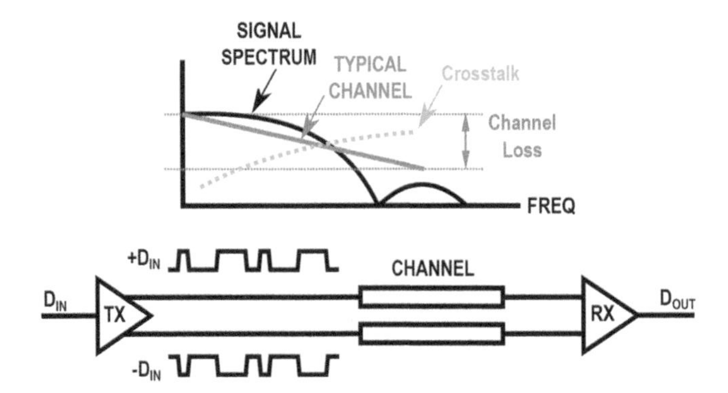

FIGURE 4.4 A generic serial data communication link, based on differential signaling (bottom). Channel loss and crosstalk (top) degrade signal integrity. Thus, proper signaling methods, as well as equalizers are required to implement a robust link achieving a very low bit error rate.

4.3 INTER-SYMBOL INTERFERENCE

The main bottlenecks in design of wideband wireline communications is inter-symbol interference (ISI). In wireless systems, channel attenuation and XTALK are the main limiting factors, determining the overall performance that can be measured in terms of signal-to-noise ratio (SNR). However, in wireline communications, most of the time, this is ISI which acts as the primary link quality and speed limiting factor. While signal power is fairly high in wireline systems, still the final BER can be low, as ISI degrades the quality of the link.

The effect of ISI on the quality of a link is analyzed in [14]. A simplified presentation of this analysis is brought here, in order to provide more insight into how ISI reduces the link margin. Consider a generic signaling, in which codewords are selected from a code book described by:

$$C = \{C_k\},\ k = 0,\ \ldots,\ K - 1 \tag{4.1}$$

There are K available codewords in this example. The number of coordinates of each codeword is equal to the number n of transmission wires. As such, codewords can be visualized as points in an n-dimensional Euclidean space. The detection of a codeword from other codewords is done using a circuit we call a "multi-input comparator" or MIC for short. For example, differential pair is the simplest form of a MIC that compares values over two wires and produces a bit after making a simple "subtraction" operation. Geometrically, a MIC can be viewed as a circuit which determines, for a given point, on which side of a given hyperplane the point lies. A code built upon OMWS is then a collection of points (codewords) and hyperplanes (MICs) such that each codeword is uniquely determined by which side of each hyperplane the point is. The distance between a codeword C_k, and a hyperplane MIC_m, plays an important role in quantifying the effect of ISI, and will be denoted by:

$$d\{C_k,\ \ MIC_m\} \tag{4.2}$$

As mentioned before, differential signaling is an example of a multi-wire system, in which a bit is placed over two wires. It can be shown that if different code words have different distances respect to the comparison level, then coding will be very sensitive to ISI (e.g., PAM4). Conversely, when all different code words show the same distance to the comparison level, then ISI sensitivity is very low (e.g., single-ended signaling and PAM2).

Based on this observation, ISI-ratio can be defined as [14]:

$$ISI - ratio\,(m) = \frac{max_{0 \leq k < K}\, d\,(C_k,\, MIC_m)}{min_{0 \leq k < K}\, d\,(C_k,\, MIC_m)} \tag{4.3}$$

The ISI-ratio is defined to be ratio between the maximum distance of the codewords to the reference level, respect to the minimum distance, both calculated at the output of MIC (i.e., input of the slicer). This ratio can be as low as one (such as in a simple NRZ differential system). In this case, ISI-ratio = 1, and signaling exhibits low sensitivity to channel ISI. However, in some other signaling methods such as in PAM-4, the ratio can be much higher, which translates into more sensitivity to ISI. For example, in PAM-4 signaling, the ISI-ratio is 3, which means the minimum possible number of levels at the output of MIC will be three. Conceptually, this means that having a multi-level signal in front of a slicer creates more eye-width closure compared to a binary system. As a conclusion, effort must be spent to minimize the ISI-ratio in order to relax the design and the required equalization, and consequently improve the inherent signal integrity of a link, and lower the consumption.

4.4 CHOICE OF SIGNALING METHOD

The analysis carried out here shows that ISI-ratio is a key parameter for designing reliable, low-power, and high-speed links. Common signaling methods used in the industry today are PAM-2 (two- level pulse amplitude modulation), and PAM-4 (four level amplitude modulation), which are based on very basic encoding schemes. There are various other modulation and signaling methods that can be used to exploit the available channel bandwidth more efficiently. However, still there is no fundamental study that show which scheme can be used to enhance the capacity of the copper wireline channels.

Indeed, ISI-ratio can provide a very strong basis to study and compare different signaling and modulation schemes for wireline communication. Comparing, PAM2 and PAM4 signaling, for example, it turns out that the ISI-ratio for PAM-2 and PAM-4 are 1 and 3, respectively. Based on (9), PAM-4 suffers much more from channel or circuit ISI compared to PAM-2. Therefore, the available link margin diminishes much faster in a PAM-4 system compared to a PAM-2 one. In other words, the horizontal opening of a PAM-2 link with a PAM-4 one reveals that the rate at which the horizontal opening degrades is much stronger in PAM-4. Moving to higher-order PAMs will exacerbate the situation even furthermore [15]. The ISI sensitivity of PAM_N signaling is $(N - 1)$, which translates into much faster loss on the horizontal opening, as indicated in (9).

Transferring more number of bits per wire, or more BW efficiency, is the primary motivation for higher-order PAM being utilized to increase the data rate. However, as the ISI sensitivity grows quickly with the number of levels, the link margin eventually does not improve at higher order PAM. Because of this issue, all the existing standard protocols support only PAM-2 and PAM-4, and still thre. OMWS, however, proposes a different approach to increase the data rate without sacrificing ISI sensitivity. While PAM_N uses amplitude domain modulation to increase the bandwidth, OMWS exploits spatial domain coding, or multi-wire signaling to enhance the link efficiency.

4.5 OMWS

Figure 4.5 shows a conventional differential link, in which two wires are utilized to transfer only one bit, hence exhibiting a wire or pin efficiency of 0.5 (i.e., 0.5 bit transferred over two wire). In differential signaling, the data is distributed over two wires, or on a two-dimensional space, to enhance resistivity against common-mode and crosstalk noises. As mentioned before, this signaling method is the simplest form of multi-wire signaling (1b over 2 wires). It will be shown later on that this signaling method also is based on an orthogonal transformation. The lower pin efficiency is the direct cost of reducing the link sensitivity to common-mode noise and cross-talk. The superior reliability of differential signaling over single-ended scheme is due to the increased

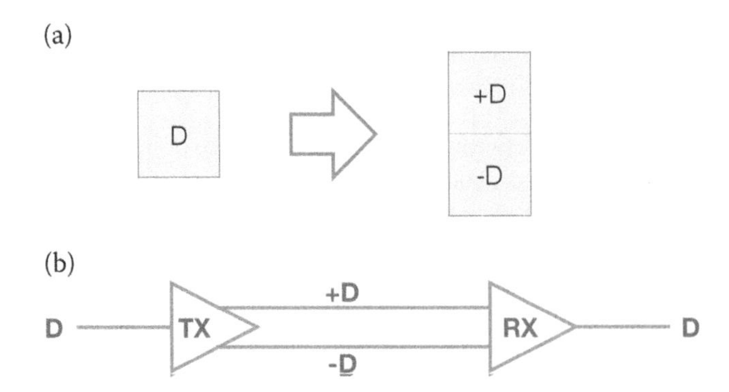

FIGURE 4.5 Differential link.

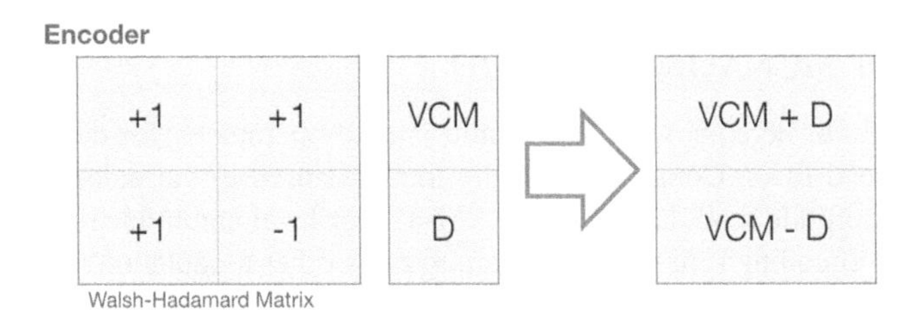

FIGURE 4.6 Single-ended to differential transformation.

noise immunity (common-mode noise, crosstalk noise, and supply noise), which can be explained using the transformation depicted in Figure 4.6 [16]. Indeed, transforming a single bit into a differential one can be explained using the Walsh-Hadamard (WH) matrix, which is an orthogonal one. There are two inputs fed into the transformer: the data bit (here denoted as D), and a common-mode value (e.g. V_{CM}, or common-mode voltage). The absolute value of the common-mode parameter does not really matter, as it will be rejected on the receiver side. As the WH transformation is orthogonal, the same matrix can be used in the receiver in order to decode the original data. The orthogonality of this transformation explains why the conventional differential transceivers can employ the same circuit topology to transmit (e.g., LVDS: low-voltage differential signaling), and receive (i.e., differential amplifier) the signal. Indeed, both LVDS driver and a differential pair can be described by WH transformation.

Based on this, differential signaling is part of OMWS family. The main difference between OMWS, and any other conventional multi-wire signaling methods is that the orthogonal transformations are exploited in OMWS to minimize the ISI-ratio (or, ISI sensitivity).

One key observation from Figure 4.6 is that the basic WH transformation puts one bit over two wires (1b2w, or 0.5 b/w). Interestingly, using the WH transformation order N, it is possible to put (N – 1)-bits over N-wires, which translates into a pin-efficiency equal to:

$$PE = \frac{N-1}{N} = 1 - \frac{1}{N} \tag{4.4}$$

Based on (4.4), pin-efficiency can approach one (i.e., as good as single-ended signaling), when large number of wires (N) are used for communication. For example, a fourth-order WH transformation provides a pin efficiency as high as 0.75 (3b over 4w), which is 50% higher than basic differential signaling. Shown in Figure 4.7a, the WH transformation of order four (3b4w) is called Ensemble NRZ (ENRZ) signaling in the context of wireline communication. Similarly, more complex transformations may be used to improve the pin-efficiency without trading ISI-ratio.

(a)

$$ENRZ = \begin{bmatrix} +1 & +1 & +1 & +1 \\ +1 & -1 & +1 & +1 \\ +1 & +1 & -1 & -1 \\ +1 & -1 & -1 & +1 \end{bmatrix}$$

(b)

$$CNRZ = \begin{bmatrix} +3 & +3 & +3 & +3 & +3 & +3 \\ -3 & -3 & -3 & +3 & +3 & +3 \\ +2 & +2 & -4 & 0 & 0 & 0 \\ +3 & -3 & 0 & 0 & 0 & 0 \\ 0 & 0 & 0 & -4 & +2 & +2 \\ 0 & 0 & 0 & 0 & -3 & +3 \end{bmatrix}$$

FIGURE 4.7 (a) ENRZ and (b) CNRZ transformations.

For example, correlated NRZ (CNRZ) signaling, shown in Figure 4.7b, puts 5 bits over 6 wires (i.e., 5b6w) [17,18].

4.6 TRANSCEIVER ARCHITECTURE

Design of high-speed, energy-efficient transceiver circuitry is a major challenge in high data rate serial communications. Any complexity in signaling method used to implement the system will directly affect performance and efficiency of the circuits. Thereby, any proposal for implementing signaling methods should be carefully studied in order to minimize the circuit level complexities. For example, PAM-4 signaling requires a complex receiver architecture in order to amplify, equalize, and slice the data. PAM signaling is based on coding data in the amplitude of the signal, hence, digital-to-analog converter (DAC) and analog-to-digital converter (ADC) circuits are ideal encoder and decoders for this type of signaling. To implement PAM_N, data converters with resolution of at least $\log_2(N - 1)$ are required to implement the transceiver. Therefore, ADC/DAC are matched circuits (matched hardware) for implementing PAM. The required resolution of the data converters has to be more than $\log_2(N - 1)$ to realize proper equalization. As OMWS is not based on amplitude modulation, the matched hardware topology to be used for receiver and transmitter is also different from DAC and ADC.

4.7 ANALOG ENCODER AND DECODER CIRCUITS

As mentioned before, OMWS is orthogonal, and designed to carefully keep the ISI-ratio as close as possible to one. Therefore, the matched circuit topologies, used for OMWS, may be described by the matrix transformation governing the signaling method. If a different circuit topology is used, it is possible to degrade the ISI-ratio and fail to provide the inherent signal integrity expected from OMWS. For example, a differential pair matches well to the transformation described in Figure 4.6. A differential pair is a linear combiner that can take V_{CM} and data and produce the desired differential output. Alternatively, a differential pair can receive $V_{CM} + D$ and $V_{CM} - D$ at its input and produce the decoded data at its output. This linear combination happens in both cases in the analog domain, and does not add any latency. As a generalized form of differential signaling, the same approach is valid in OMWS, whereas both encoder and decoder are combinational analog circuits. On the transmitter side, the digital bits are fed into an analog combiner (such as single-side terminated, SST, drivers), and produce the wire values. Those wire values, on the other end, are received by front-end linear combiners to produce binary values at their outputs. All the circuits before the final stage of the transmitter, and after the first stage of the receiver, remain the relatively simple, conventional circuits used for differential systems.

Figure 4.8 shows one example for ensemble NRZ (ENRZ) link. The three input bits on the transmitter side are fed into an analog encoder and launched onto the wires. The signals over the wires are all multi-level. On the receiver side, multi-input combiners or comparators (MICs) are used to convert the multi-wire, multi-value signals to binary voltages. The circuit topology of the following equalizer and slicer blocks will be very similar to a conventional differential system.

Figure 4.9 shows an exemplary implementation of ENRZ transmitter and receiver analog combiners. The encoder is a simple SST driver which receives three input bits (B_0, B_1, B_2). The output of each wire is a proper linear combination of these bits placed on the four output wires. The circuit used to decode the wire values inside the receiver is shown in Figure 4.9 (left). The combination of the four wire values ($W_{<3:0>}$) is calculated and a binary output (V_{OUT}) is produced. Both linear encoder and decoder circuits are implemented in the analog domain, with no additional latency and no digital domain encoding or decoding are required. Both of the circuit diagrams shown in Figure 4.9 match to the matrix transformation depicted in Figure 4.7a. Figure 4.10 illustrates the analog encoder (transmitter), and decoder (receiver) for 5b6w CNRZ signaling, which are very similar to the ENRZ encoder and decoder. The binary bits are fed into five linear encoders (Figure 4.10 left shows one example), while a multi-value signal is placed on the wire. On the receiver side, the multi-level signal is received at the input of a linear decoder, while the output is binary.

Figure 4.11a illustrates a transceiver fabricated in 28 nm that can transfer data at 20.83 Gb/s/wire (effective BW), while consuming 0.94 pJ/b, using CNRZ signaling, which sends 5 bits over 6 wires

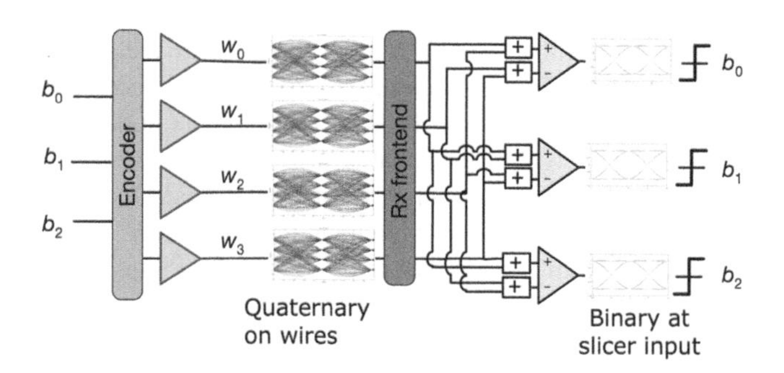

FIGURE 4.8 OMWS transceiver based on ENRZ signaling.

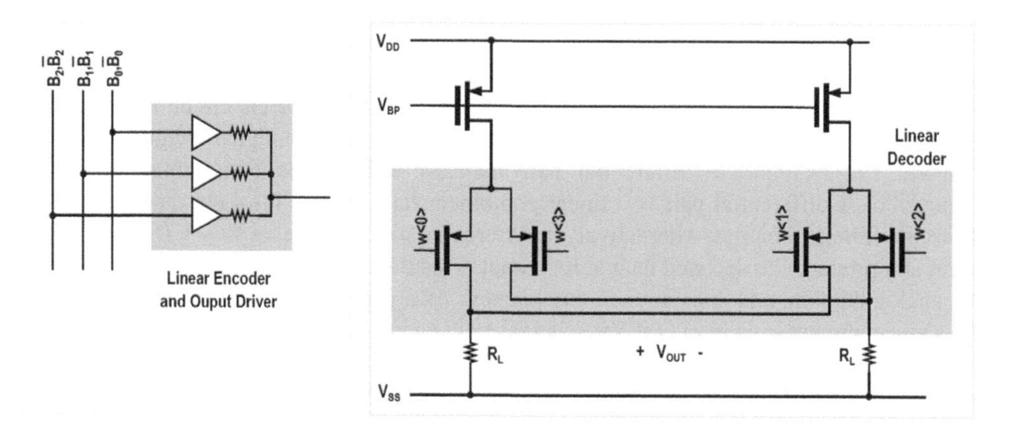

FIGURE 4.9 The ENRZ transmitter encoder (left), and multi-input comparator (MIC) circuit topology (right) [19,20]. The rest of circuitry on the transmitter or receiver side are similar to the conventional differential PAM2 systems. The logic combination in the transmitter (driver) and receiver (MIC) vary depending on wire or sub-channel.

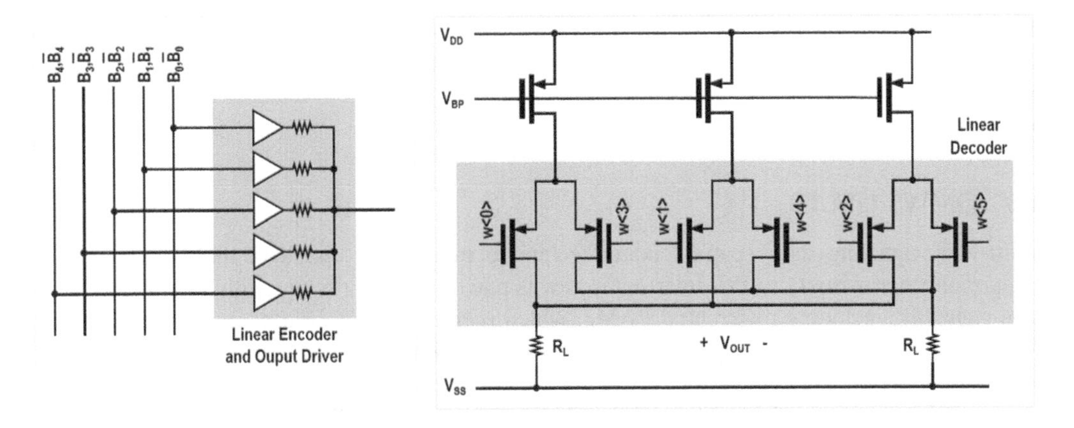

FIGURE 4.10 The CNRZ transmitter encoder (right), and multi-input comparator (MIC) circuit topology (left) [19,20]. The rest of circuit on the transmitter or receiver side are similar to the conventional differential PAM2 systems. The logic combination in the transmitter (driver) and receiver (MIC) vary depending on wire or sub-channel.

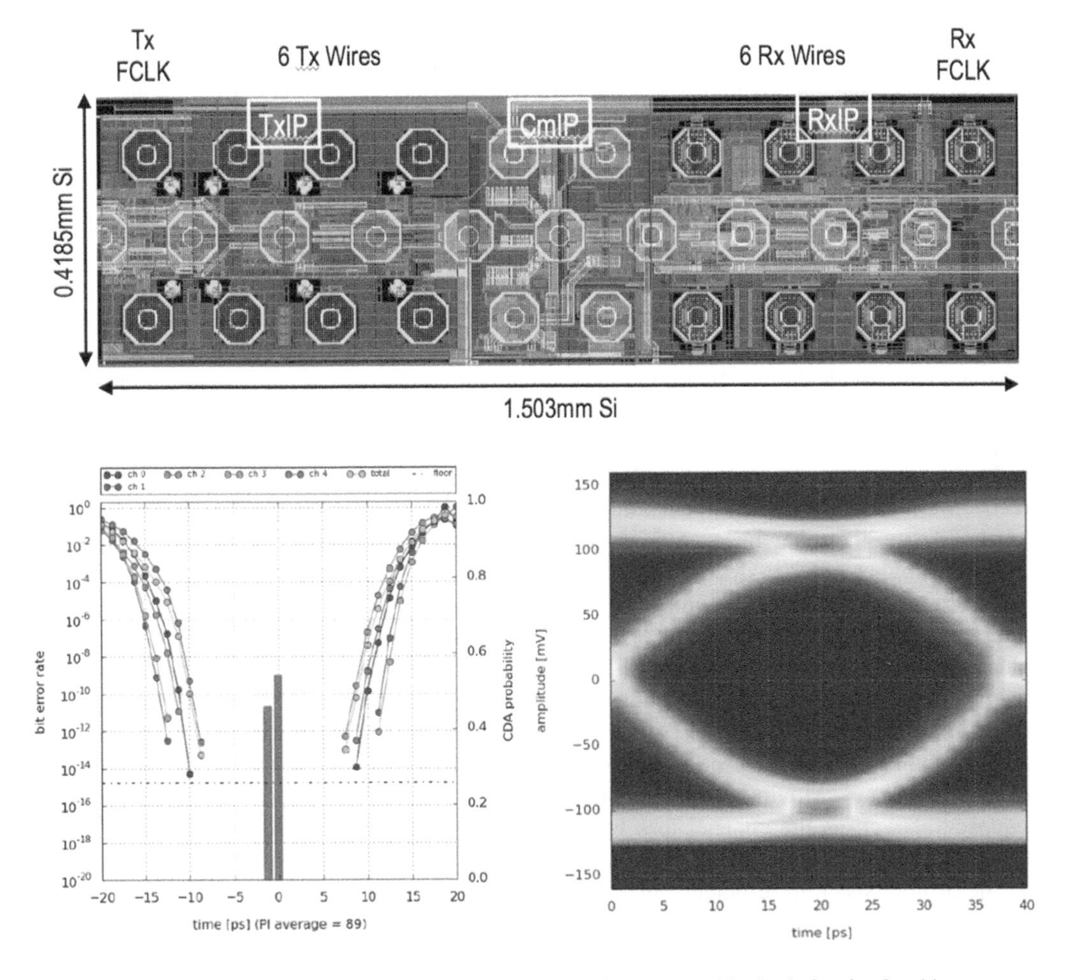

FIGURE 4.11 (a) A CNRZ transceiver reported [21], (b) the measured bathtub for the five binary output bits (sub-channels), and eye diagram of one of the sub-channels.

(PE = 5/6). Figure 4.11b shows the bathtub for the five output bits (called five sub-channels), and also a sample eye diagram, exhibiting more than 12 ps opening at 1E-15 BER (bit error rate). The proposed transceiver employs a clock forwarding scheme in order to enhance the jitter tolerance of the link, and reduce power consumption [22].

4.8 OMWS ISSUES

Multi-wire signaling uses spatial coding technique to enhance the data throughput, without compromising sensitivity to ISI. Inter-wire skew is one of the main issues with any kind of multi-wire signaling, including differential PAM2, which is the simplest form of multi-wire signaling. Any skew between wires will cause inter-sub-channel interference, which means part of information from one sub-channel penetrates into the other sub-channels, thus resulting in eye-closure and performance degradation. When inter-wire skew is small, for example 10% of the unit interval (UI), or lower, generally its effect can be neglected. If the inter-wire skew is larger, then skew correction circuits can be.

Based on Figure 4.7a and Figure 4.7b, the transmitted data is distributed over multiple wires. Therefore, still the transmitted data can be reconstructed accurately, if there is small skew between wires. When skew is high, however, the eye closure can be considerable. In this case, a proper skew correction circuit has to be employed. The skew correction can be employed either on the transmitter side, or on the receiver side. Implementing skew correction on the transmitter side is more convenient, however it requires a backchannel communication mechanism in order to estimate and correct the imperfection. Otherwise, the receiver side skew correction can be employed to adjust the signal skew before the receiver front-end.

4.9 SUMMARY AND THE FUTURE DIRECTIONS

There is a growing demand to increase the data transfer BW in the HPC industry. Applications such as Autonomous Driving, Machine Learning, and Artificial Intelligence, require a continuous flow of data moving between memory and processor at rates much faster than what is achievable today, with better energy-efficiency. Today's wireline industry is mainly based on PAM signaling, with PAM-2 being the most dominant modulation scheme below 56 Gb/s, and PAM-4 the basis for most links at higher data rates (Figure 4.12a). Transferring more number of bits per the same number of lines allows to enhance the data rate with the same available channel bandwidth. There are already reports demonstrating 112 Gb/s using PAM-4 implementing high-precision DACs and ADCs for encoding and decoding purposes, as well as for equalization [9,10]. To date, there has been no demonstration showing 224 Gb/s or higher data rates using PAM-4, or higher-order PAM.

OMWS technique, however, proposes a different roadmap. Based on the theory of orthogonal multi-wire signaling, the number of levels at the slicer input should be minimized in order to reduce the sensitivity of the system to ISI. Figure 4.12b shows the copper wireline communication roadmap based on the OMWS concept. At low data rates (such as 16 Gb/s and 28 Gb/s), differential binary (the simplest form of the OMWS) has proven to be a proper choice. The differential binary signaling (PAM2) can be implemented using reasonably simple transmitter and receiver circuit architectures. Moving to higher data rates (e.g., 56 Gb/s) and beyond, the sampling clock frequency as well as the channel loss becomes excessively large. Therefore, a more spectrum efficient signaling is required. If multi-wire signaling possible, then OMWS (e.g., ENRZ) can provide a more spectrum efficient link compared to PAM2, with much lower sensitivity to ISI compared to PAM4. As a conclusion, Figure 4.12b proposes using higher-order OMWS for 56 Gb/s and beyond. For example, ENRZ can be a proper choice for both 56 Gb/s and 112 Gb/s. If more

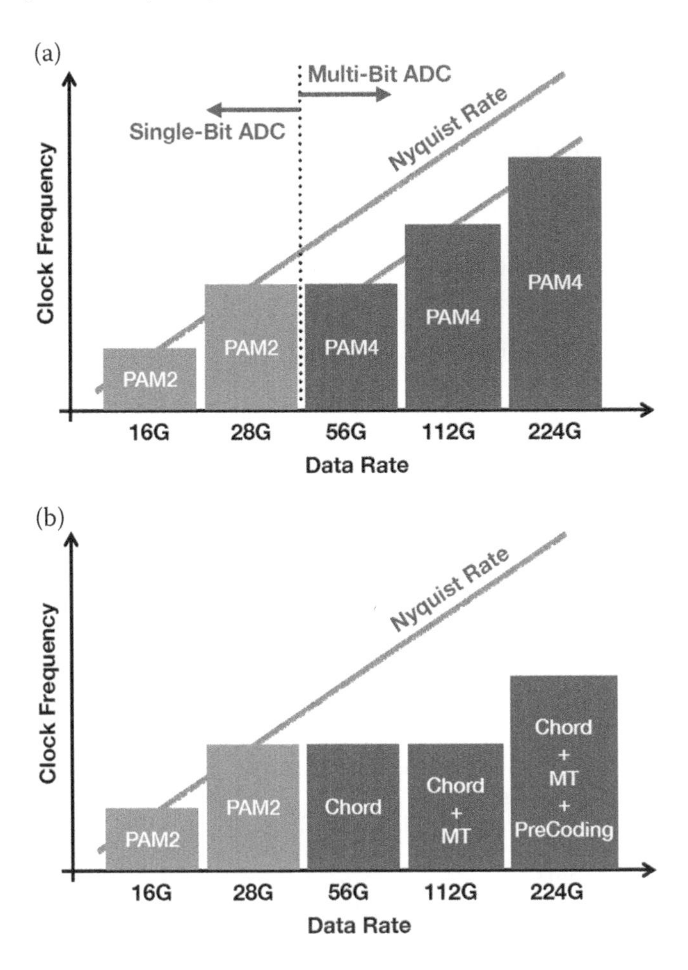

FIGURE 4.12 The evolution of the signaling methods used in the copper wireline communications. The required BW is compared with the Nyquist rate of the PAM2 signaling: (a) conventional technology trend for wireline communications, (b) the proposed wireline communication roadmap based on OMWS.

spectrum efficient signaling methods are required for 112 Gb/s and beyond, then multi-tone (MT) signaling can be employed [23,24]. Combined MT/OMWS allows to improve the spectrum efficiency and maintain a low ISI sensitivity, simultaneously. Each tone, can be considered as an independent OMWS, where the ISI ratio is kept to be as low as one. Using this methodology, the accumulation of multiple tones will provide the target throughput without compromising ISI ratio. As shown in [23], advanced pre-coding (pre-filtering) techniques can be used in addition to MT/OMWS in order to reach data rates as high as 224 Gb/s, while resilience against ISI has been kept to be very high. Based on this strategy, Figure 4.12b provides a clear roadmap to reach data rates as high as 224 Gb/s at a low level of energy consumption and circuit complexity.

ACKNOWLEDGMENT

The concept of "orthogonal multi-wire signaling" has been developed by scientists and engineers in Kandou Bus, Switzerland, between 2010 and 2019. The author would like to thank the Kandou team, especially Prof. Shokrollahi and Dr. Hormati, for their contribution in this work. This work is in part supported by Kandou Bus, under the University of Utah's Grant 50503434.

REFERENCES

[1] K. Rupp, "42 years of microprocessor trend data," on: www.karlrupp.net.
[2] G. Moore, "Cramming more components onto integrated circuits," Electronics, vol. 38, no. 8, 1965.
[3] D. Pham, et al., "The design and implementation of a first-generation CELL processor," IEEE ISSCC, 2001.
[4] N. Beck, et al., "Zeppelin: an SoC for multichip architecture," IEEE ISSCC, Feb. 2018.
[5] L. Ceze, et al., "Arch2030: a vision of computer architecture research over the next 15 years," CCC Computer Community Consortium.
[6] D. Brooks, "Computer architecture for deep learning applications," University of Harvard, Apr. 2017.
[7] C. Berry, et al., "IBM z14TM: 14nm microprocessor for the next-generation mainframe," IEEE ISSCCC, 2018.
[8] S. M. Tam, et al., "SkyLake-SP: a 14nm 28-core Xeon R Processor," IEEE ISSCC, 2018.
[9] J. Kim, et al., "A 112Gb/s PAM-4 transmitter 3-tap FFE in 10nm CMOS," in IEEE ISSCC, Feb. 2018.
[10]. P. Upadhayaya, et al., "A fully adaptive 19-to-56Gb/s PAM-4 wireline transceiver with a configurable ADC in 16nm FinFET," IEEE ISSCC, 2018.
[11] A. Ramachandran, et al., "A 0.5-to-0.9V, 3-to-16Gb/s, 1.6-to-3.1pJ/b wireline transceiver equalizing 27dB loss at 10Gb/s with clock-domain encoding using integrated pulse-width modulation (iPWM) in 65nm CMOS," IEEE ISSCC, 2018.
[12]. H. Lee, et al., "A 16 Gb/s/Link, 64 GB/s bidirectional asymmetric memory interface," IEEE JSSC, 2009.
[13] G. Keeler, "Photonics in package for extreme scalability," Proposers Day, Nov. 2018.
[14] A. Hormati, et al., "ISI tolerant signaling: a comparative study of PAM4 and ENRZ," *Designcon*, 2016.
[15] R. Bindiganavile and A. Tajalli, "Spectrum-efficient communication over copper using hybrid amplitude and spatial signaling," *IEEE MWSCAS*, Aug. 2019.
[16] H. Cronie, et al., "Orthogonal differential vector signaling," US Patent 20110268225A1, May 2011.
[17] A. Shokrollahi, "High speed and low power SerDes architectures using Chord signaling," IEEE ISSCC Forum, Feb. 2016.
[18] A. Tajalli, et al., "A 1.04pJ/b 417Gb/s/mm USR link in 16nm FinFET," *IEEE VLSI Symp.*, Tokyo, Japan, 2019.
[19] A. Tajalli, et al., "Efficient processing and detection of balanced codes," US Patent 8593305B1, 2013.
[20] H. Cronie, et al., "Methods and systems for noise resilient, pin-efficient, and low power communications with sparse signaling codes," US Patent 8649445B2, Feb. 2014.
[21] A. Shokrollahi, et al., "A pin-efficient 20.83Gb/s/wire 0.94pJ/bit forwarded clock CNRZ-5-coded SerDes up to 12mm for MCM packages in 28nm CMOS," *IEEE ISSCC*, Feb. 2016.
[22] A. Tajalli, "Matrix phase detector for high bandwidth and low jitter frequency synthesis," IEE Electronics Letters, 2017.
[23] A. Hormati, et al., "High speed communications," US Patent 2018/0083807A1, 2018.
[24] K. Gharibdoust, et al., "Hybrid NRZ/multi-tone serial data transceiver for multi-drop memory interface," IEEE J. Solid-State Circuits, vol. 50, no. 12, pp. 3133–3144, Dec. 2015.

Section II

Device- and System-Level Challenges and Improvements

5 System-Level Management of Silicon-Photonic Networks in 2.5D Systems

Aditya Narayan, Ajay Joshi, and Ayse K. Coskun

CONTENTS

5.1 INTRODUCTION

To support emerging graph processing, artificial intelligence, and cognitive applications, computing systems in supercomputers and data centers are rapidly moving towards large manycore chips. 2.5D integration is developing into a promising platform for integrating hundreds or thousands of logic cores in a single chip. In 2.5D systems, multiple smaller chiplets are integrated on a large interposer chip, which may be fabricated on a mature technology node. 2.5D systems provide higher manufacturing yield, thereby lower fabrication costs compared to 2D systems [1]. As 2.5D systems also enable heterogeneous integration of different IPs, several academic and industrial efforts have focused on developing energy-efficient 2.5D manycore systems [2–5].

In large 2.5D manycore chips, data processing of applications with high inherent parallelism results in increased data traffic in the on-chip communication network. Consequently, the bandwidth requirement of such applications goes up to 1–2TBps [6]. The throughput in such systems, therefore, is highly dependent on the provided bandwidth density of the on-chip communication network. Traditional electrical links fall short of providing the required bandwidth density on such large manycore chips [7].

With technological advances in CMOS integration of silicon-photonic technology [8–12], photonic links are developing as a high-bandwidth and low-latency alternative to electrical links for large manycore chips [13–18]. In photonic links, laser sources emit optical signals that are routed across the chip using on-chip waveguides. Owing to wavelength division multiplexing (WDM), multiple optical signals can be multiplexed in the same waveguide resulting in increased on-chip bandwidth density [19]. In a typical photonic link, microring resonators (MRRs) are used as the primary optical devices. MRRs are used for data modulation at the transmitting site and for data filtering at the receiving site. These MRRs are designed such that their resonant wavelength

DOI: 10.1201/9780429292033-5

matches that of the optical signal through which the data is sent. The physical dimensions and the material properties of an MRR determine its resonant wavelength.

There are two primary factors that introduce major shifts in the resonant wavelength of the MRR – thermal variations [20] and manufacturing process variations [21,22]. MRRs are typically made of silicon, which have a high thermo-optic coefficient, leading to high sensitivity to thermal variations [11,20]. Moreover, the manufacturing process introduces variations in MRR dimensions, further shifting their resonant wavelengths [21,22]. As a result, these MRRs no longer resonate at the same wavelength as that of the optical signal. Such deviations in the resonant wavelength of the MRR introduces undesired bit errors in data communication [14].

In large manycore chips that use multiple optical channels for communication among hundreds of cores, there are several thousands of MRRs that are operating at different optical signals. The compute activity across the chip introduces thermal hot spots and large thermal gradients on the chip. To ensure reliable on-chip communication in the photonic link, it is, therefore, essential to mitigate the impact of thermal and process variations on the resonant wavelength of the MRRs.

In this chapter, we first describe the impact of thermal and process variations on the MRR resonant wavelength. We discuss several device-level strategies that are currently employed in photonic links to mitigate and rectify the resonant shifts of MRRs. At the chip design level, we discuss placement and routing techniques that help reduce the impact of thermal variations on MRR resonant shifts. Though such device-level and design-level techniques are effective, the runtime application characteristics significantly impact the thermal distribution across the chip, thereby increasing thermal-variation-induced MRR resonant shifts. The primary focus of this chapter is, therefore, to understand the cross-layer implications of device and design strategies on MRR resonance shifts and explore system-level optimizations to reduce the overall photonic network power.

5.2 THERMAL AND PROCESS SENSITIVITIES OF OPTICAL DEVICES

Figure 5.1 shows an example of communication on a silicon-photonic link. MRRs are the basic building blocks used in silicon-photonic links as shown in Figure 5.1. At the transmitter site (Tx), MRRs are utilized to modulate the digital data onto optical signals. Depending on the input data (logic 0 or logic 1), a modulation driver applies the appropriate drive voltage across the MRR. At the receiver site (Rx), MRRs filter out the modulated optical signal. The filtered optical signal is converted back into electrical data by photodetectors, and amplified by transimpedance amplifiers (TIA). Such MRR modulation and filtering of optical signals is achieved when the MRR is designed to operate at the same resonant wavelength as that of the optical signal. This phenomenon diverts most of the optical power from waveguide to the MRR. Figure 5.2 shows the ideal scenario, when the MRR is in resonance with the optical signal with resonant wavelength λ. The MRRs at Tx and Rx are designed to operate at the same resonant wavelength λ_0 as that of the optical signal used

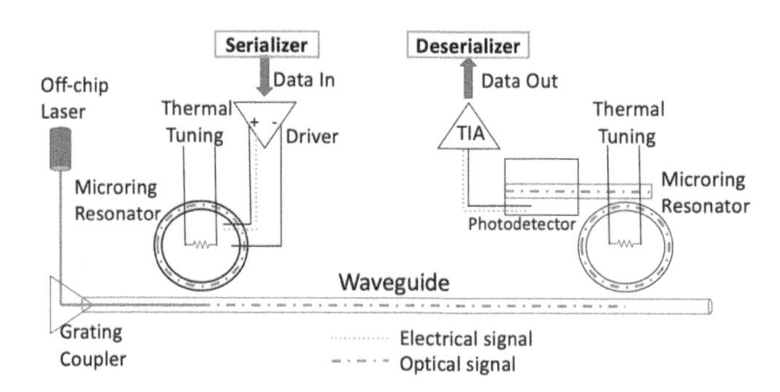

FIGURE 5.1 Communication on a silicon-photonic link.

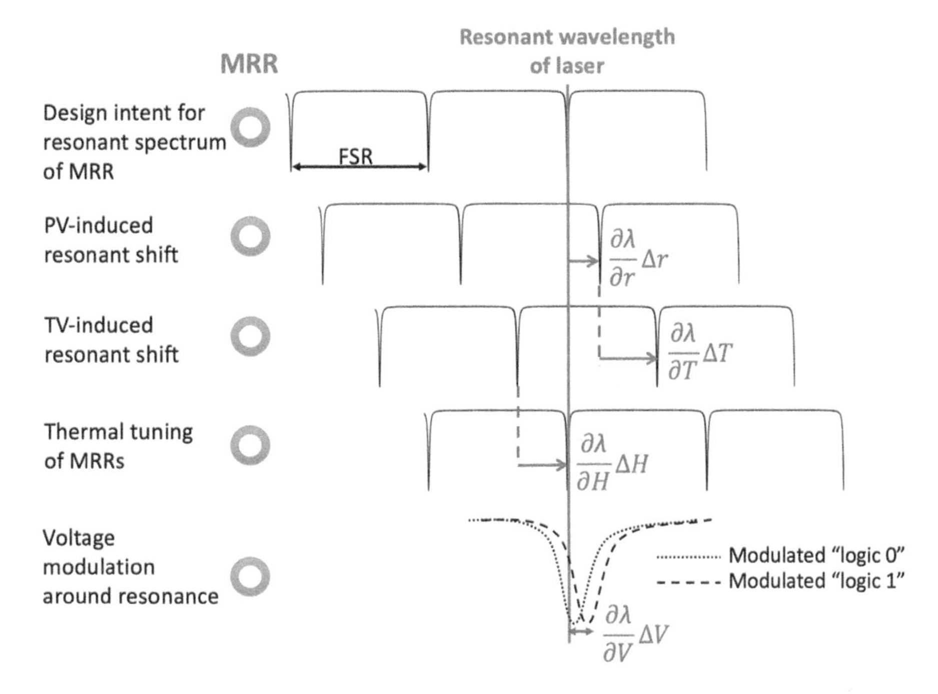

FIGURE 5.2 Resonant wavelength shift of MRRs due to thermal and process variations.

for communication. We now describe how thermal variations (TV) and process variations (PV) introduce shifts in the resonant wavelength of MRRs.

5.2.1 Thermal Variations (TV) Induced MRR Resonance Shifts

MRRs are typically fabricated using silicon, which has a high thermo-optic coefficient (1.86×10^{-4} K^{-1}) [23]. The high thermo-optic coefficient induces variations in the refractive index of the MRR, which in turn shifts its resonant wavelength to a higher value, as shown in Figure 5.2. As a result, the MRR moves out of resonance with the optical signal's resonant wavelength λ_0. Silicon MRRs have been demonstrated to have a high sensitivity to thermal variations, about 70–100 pm/K [20,24].

On-chip temperature gradients of about 20–25 K have been observed across manycore chips due to high compute activity of applications at runtime [25,26]. Therefore, the MRR resonant wavelength shift due to thermal variations are almost as high as the full spectral range (FSR). Moreover, the MRRs at Tx and Rx experience different shifts in their resonant wavelengths due to such thermal gradients. The resulting mismatches in the MRR resonant wavelengths at Tx and Rx may impact the link integrity during data transmission in the silicon-photonic link.

5.2.2 Process Variations (PV) Induced MRR Resonance Shifts

Challenges in CMOS fabrication process at lower technology nodes introduce variations in the thickness, width and radii of the MRRs. Krishnamoorthy *et al.* [21] quantify the intrawafer and interwafer variations on the resonance wavelengths of MRRs. Their study shows that absolute resonances of MRRs cannot be controlled across the wafers or even across reticles within a wafer. The different manufacturing steps such as lithography, etching, and chemo-mechanical polishing induce variations in the MRR dimensions by up to 10 nm. Due to variations in waveguide width, silicon thickness and etch-depth nonuniformities, the effective refractive index of silicon changes.

As a result, the resonant wavelength of MRR shifts significantly from the design intent. These PV have a small random component and are mostly geometric variations. Therefore, during the fabrication process of a die reticle, two distant MRRs in the same die experience completely different variations in their resonant wavelengths. These PV-induced resonance shifts further add to the TV-induced mismatches in Tx and Rx MRR resonant wavelengths. Inability to control TV and PV makes it impossible to fix an MRR's resonant wavelength during design; instead it needs to be determined post-fabrication.

5.3 DEVICE- AND DESIGN-LEVEL TECHNIQUES TO MITIGATE MRR RESONANCE SHIFTS

While designing energy-efficient silicon-photonic links, there are two key approaches to minimize the impact of thermal and process sensitivities of MRRs. At the device-level, it is possible to reduce the thermal sensitivity of Si MRRs by using negative thermo-optic coefficient materials, or using athermal materials as MRRs. At the chip-design level, placement of MRRs and on-chip waveguide routing can be performed to minimize the impact of high thermal gradients. In this section, we look at several such device and design-level techniques that enable reducing the overall photonic power.

5.3.1 Device-Level Techniques

The overall shift in the resonant wavelength of an MRR ($\Delta\lambda_{shift}$) due to TV and PV can be expressed as follows:

$$\Delta\lambda_{shift} = \frac{d\lambda}{dT} \cdot \Delta T + \Delta\lambda_{shift,\ PV} \tag{5.1}$$

where $\frac{d\lambda}{dT}$ is the thermal sensitivity of the MRR, ΔT is the change in temperature, $\Delta\lambda_{shift,\ PV}$ is the resonance shift due to manufacturing process variation of the MRR. The solutions at device-level to counteract the effect of TV- and PV-induced resonant shifts of MRRs mainly consist of techniques to actively control the temperature of MRRs or materials to thermally insulate the MRRs.

Active control of resonant wavelength of MRRs is typically carried out by thermally tuning the MRRs to a higher order of resonant wavelength of the optical signal in the wavelength spectrum [24,26,27]. The thermal tuning of MRRs is achieved by controlled local heat injection using resistive heaters inside the MRRs. These heaters supply energy to the MRRs using Joule effect, thereby increasing the MRR temperature and right-shifting the MRR resonant wavelength. The feasibility of MRR resonant wavelength shift using Joule effect is strongly dependent on MRR thermal sensitivity and the heater efficiency in terms of *K/W*. For a single optical signal in a waveguide, the theoretical maximum resonant wavelength shift required for an MRR is one *FSR*. Equation 5.2 calculates the tuning range required ($\Delta\lambda_{heat}$) for an MRR to lock the MRR to the nearest order of resonant wavelength of the optical signal:

$$\Delta\lambda_{heat} = FSR - \Delta\lambda_{shift} \tag{5.2}$$

With WDM, it is possible to multiplex multiple optical signals (say n) in a waveguide, with peak resonant wavelength of each signal evenly spaced in the FSR. Therefore, it is possible to tune the MRR to the nearest resonant peak within FSR. Therefore, due to WDM, the maximum theoretical resonant wavelength shift of an MRR is reduced to *FSR/n*. Equation 5.3 calculates the tuning range required ($\Delta\lambda_{heat}$) for an MRR with WDM, where n multiple optical signals multiplexed in a single waveguide:

$$\Delta\lambda_{heat} = \frac{FSR}{n} - \left(\Delta\lambda_{shift}\ mod\ \frac{FSR}{n}\right) \tag{5.3}$$

Equation 5.4 calculates the heating power (P_{heat}) required to shift the resonant wavelength of the MRR by a tuning range of $\Delta\lambda_{\text{heat}}$:

$$P_{heat} = \frac{\Delta\lambda_{heat}}{\frac{d\lambda}{dH}} \tag{5.4}$$

Here $\frac{d\lambda}{dH}$ is the heater efficiency in pm/mW. Controlled local heat injection requires a closed-loop feedback system. This feedback loop monitors the resonant wavelength shift of an MRR due to TV and PV, and the heating power required to thermally tune the MRR to the resonant wavelength of the nearest optical signal in the wavelength spectrum. These feedback loop techniques monitor the optical power and derive a heating level. The heater then maintains a fixed temperature within the MRR to lock the MRR at the resonant wavelength. There are analog as well as digital circuit mechanisms for closed-loop feedback monitoring [24,28,29]. More robust techniques add a second level of control in these closed-loop feedback systems to handle the large temporal TV at runtime. This level of control enables the dynamic remapping of the resonant wavelengths of MRR to different optical signals at runtime during an application execution. As remapping requires larger amounts of thermally-controlled shifts, it is a relatively slow process of about 100 μs, but occurs less than once per second due to the thermal inertia of chips [24].

It is also possible to reduce the thermal sensitivity of MRRs by designing athermal MRRs. As explained earlier, silicon exhibits high thermo-optic coefficient. A common approach to reduce the thermal sensitivity of silicon MRRs is to clad these MRRs with a material with negative thermo-optic coefficient. Overlaying a layer of polymer over a silicon MRR reduces the thermal dependence of MRR resonant wavelength to less than 5 pm/°C [30,31].

The thermal sensitivity of MRRs can also be addressed by embedding MRRs with Mach-Zehnder interferometers (MZI) [32]. By engineering the optical modes in each arm of the MZI, it is possible to make a silicon MZI athermal. The two arms of the MZI can be tailored in such a way to exhibit a strong negative temperature coefficient.

5.3.2 Design-Level Techniques to Mitigate MRR Resonance Shifts

Though device-level solutions help towards reducing the thermal sensitivity of silicon MRR or localized heater design, we can further employ design-level techniques for thermal management of silicon-photonic networks on manycore chips. Such design-level techniques involve placement of MRRs on-chip, floorplanning strategies and thermally-aware routing of waveguides to mitigate thermal effects on MRRs.

A design technique to reduce the impact of high-power cores on MRR is to thermally decouple the processor die from the photonic layer. This can be performed by inserting an insulator layer in between [33]. This insulator maintains roughly constant thermal gradient around the MRRs, and therefore, helps in reducing the heating power for thermal tuning. Figure 5.3 shows the cross-section

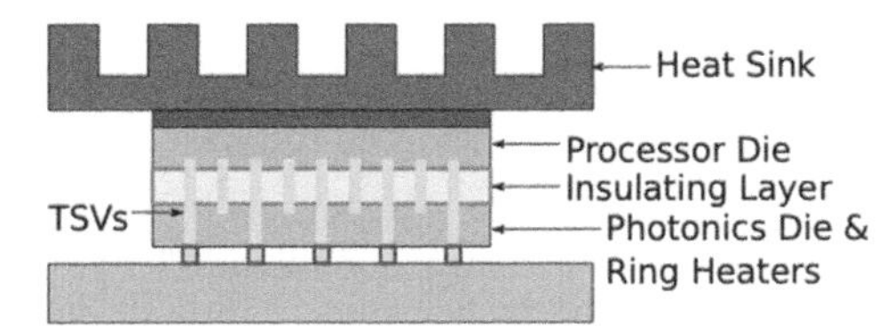

FIGURE 5.3 Cross-sectional view of a manycore system with insulation layer between the processor die and photonics die.

of a chip with an insulation layer between the processor die and photonic die. The experimental results on such a design reduces the heating power by about three to five times compared to a chip without the insulation layer.

Moreover, place and route (PnR) of photonic links on-chip also impacts the thermal coupling between the processing cores and MRRs. GLOW [34] proposes a global routing solution for optical-electrical interconnects by accounting for (1) thermal reliability and functionality, (2) minimal optical driving power, (3) signal integrity and data conversion quality, (4) timing constraints, (5) optical channel utilizations, and (6) legal DRC checks for optical-electrical interconnects. GLOW formulates the waveguide routing as an ILP problem and reduces the photonic power by 23–50% compared to a greedy routing approach.

Design-time PnR techniques can also preemptively reduce the thermal gradients between MRRs.

A cross-layer optimization approach based on integer linear programming (ILP) monitors the temperature around the MRRs and provides the placement of MRR groups, core clusters and routing pattern of waveguides. This ILP-based thermally aware optimizer the laser power consumption, power consumed in the electrical-optical conversion circuitry and thermal tuning power across different workloads [35]. The optimization results demonstrate that photonic power is sensitive to the power profiles of workloads, optical data rate and the system architecture. The ILP-based optimizer saves up to 15% photonic power compared to a thermally-agnostic PnR technique.

Another design approach integrates redundant MRRs at every MRR group [36]. The resonant wavelength of these redundant MRRs along with the other MRRs are equally spaced in the FSR. The basic idea with this approach is to minimize the tuning shift required for the MRRs. Without the redundant MRRs, there may arise situations where the tuning shift is significantly higher resulting in larger heating power (see Figure 5.4a). Using these extra MRRs provides the opportunity to minimize the thermal tuning shift by creating a sliding MRR window (Figure 5.4c). The simulation results demonstrate that adding additional MRRs per group increases the thermal gradient from 2.1°C to 5.6°C in a 64-bit crossbar network.

5.4 SYSTEM-LEVEL MANAGEMENT TECHNIQUES

Device-level and chip design-level techniques are effective towards thermal management of silicon-photonic links. However, these techniques do not account for the runtime characteristics of workloads. There is a strong diversity in workloads' runtime bandwidth and resource utilization that result in highly workload-specific power and thermal profiles. Therefore, the thermal effect on the MRR resonance shift may vary significantly across workloads, thereby resulting in different heating power for thermal tuning of MRRs. Moreover, the laser power and the power consumed in the electrical circuitry for electrical-optical (E-O) and optical-electrical (O-E) conversion increases proportionally to the provided bandwidth in the silicon-photonic link. Therefore, the overall photonic power is a strong function of the system architecture and runtime application behavior, which opens up opportunity for system-level policies and optimizations. This section focuses on cross-layer modeling methodologies that take a hierarchical approach to model device- and design-level thermal management strategies under different system-level constraints, and develop power-management policies at the system-level.

In a large manycore system, high compute activity in a processing core (e.g., Core1) may create a localized thermal hot spot. MRRs within an optical router for Core1, therefore, experience significantly high shifts in resonance wavelengths due to such hot spots. Another processing core (e.g., Core2), at a distance from the high activity core, may have low compute activity at the same time. The MRRs within an optical router for Core2 experience minimal to no resonance shift due to TV. When Core1 and Core2 communicate, there is a high thermal gradient between the MRRs used for this communication. It is, therefore, essential to lock the MRRs to the same laser wavelength by supplying different heating power to the MRRs in the optical routers of both the

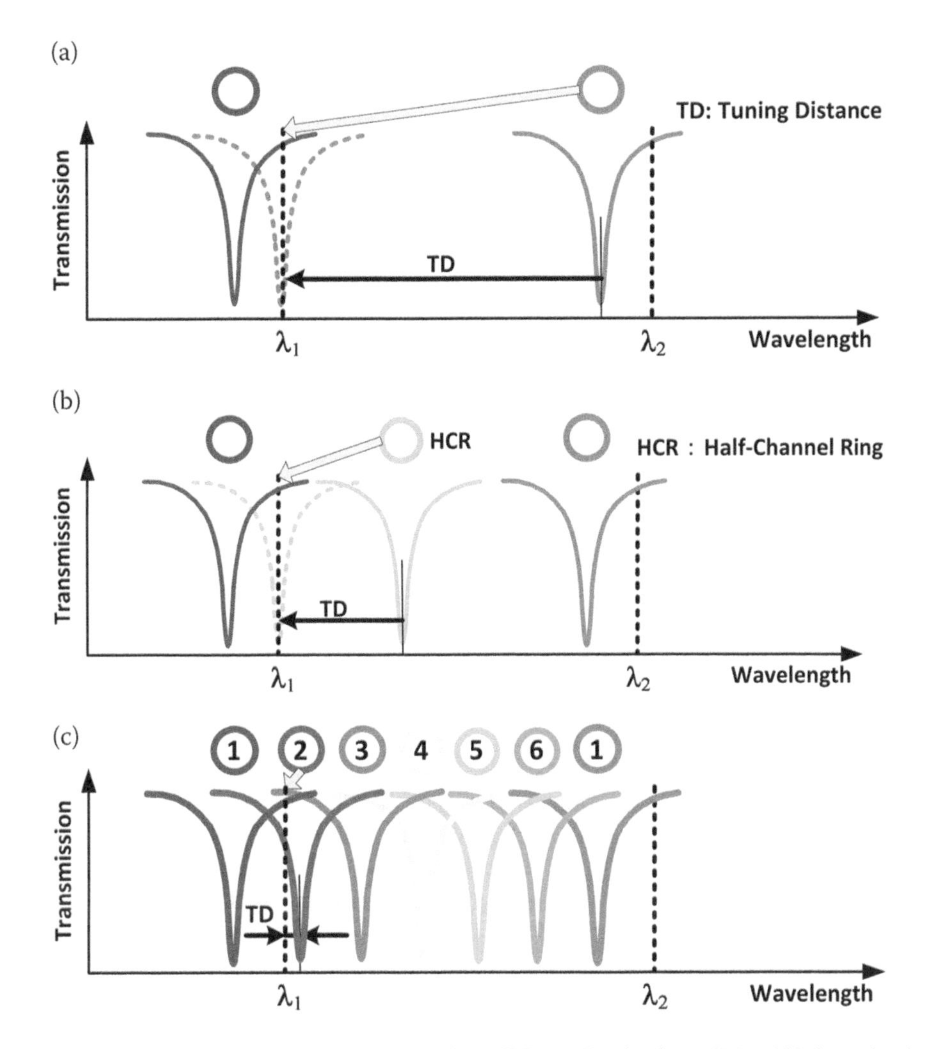

FIGURE 5.4 (a) No redundant MRRs, (b) 1 redundant MRR, (c) 5 redundant MRRs, 1/6 channel redundant.

cores. At the system-level, it is possible to perform workload allocation or migration accounting for the thermal activity around communicating MRRs to match the resonant wavelength of these MRRs. Such policies enable a stable thermal environment around the MRRs and thereby, reduce the heating power for MRR thermal tuning.

Workload allocation policies are effective in reducing the thermal gradients across the chip. *RingAware* [25] is a workload allocation policy that maintains similar power profiles around communicating MRRs to minimize the impact of TV-induced resonance shift. The temperature around the communicating MRRs are affected by the compute activity of the cores that are in close proximity. Consequently, *RingAware* accounts for the MRR locations during job allocation. *RingAware* classifies the cores with the physical distance from MRRs – RD#, where # is the core's relative distance to the MRR. The thermal impact of RD0 cores is the highest, followed by RD1, and so on. During workload allocation, *RingAware* first allocates threads in RD0 cores evenly across the MRR groups in the chip. The policy then greedily allocates the threads across the RD1 cores to minimize the thermal gradient across the chip and maintain similar thermal activity around MRRs. For workloads consisting of multiple programs, *RingAware* assumes that the programs' power levels are known a priori. *RingAware* starts by allocating high-power threads first, as their impact on temperature is the highest.

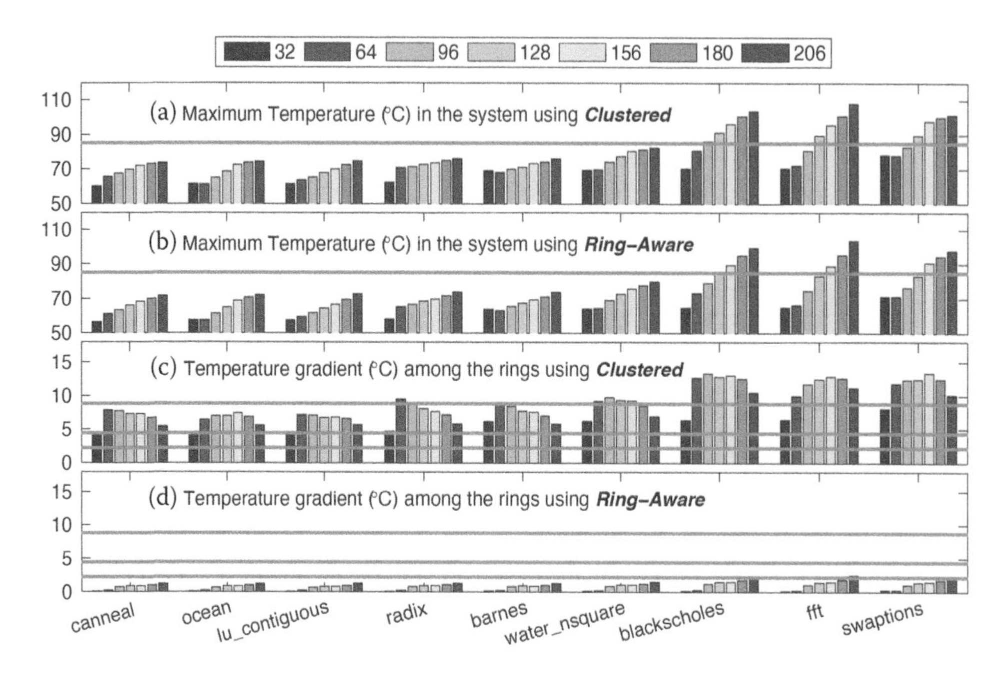

FIGURE 5.5 Maximum on-chip temperature and the thermal gradient using Clustered and RingAware policies.

Figure 5.5 shows the maximum temperature and the thermal gradient in the system with *RingAware* policy compared to *Clustered* policy, which allocates threads to cores from one corner of a chip to diagonal corner. *RingAware* reduces the maximum temperature in the system compared to Clustered. Moreover, *RingAware* is very effective in reducing the thermal gradient among the MRRs. The MRR thermal gradient is lower than 2.5°C with *RingAware*, whereas the gradient is more than 10°C with *Clustered*.

One limitation of *RingAware* arises from the fact that it only performs static workload allocation, and does not consider varied compute activity in different application phases. As the temperature of the MRRs change during the workload execution after the initial allocation, the resulting thermal gradients across the chip lead to high resonance shifts in MRRs due to TV. *Therma* [37] is runtime workload migration technique, where the threads are migrated across cores to maintain similar thermal activity around communicating MRRs during workload execution. *Therma* uses the thermal history of the cores and the nearby MRRs, which together form a cluster. Threads are moved from one cluster to another cluster, but within the same cluster. *Therma* assigns a thermal index to each core, which quantifies how prone a core is to thermal hot spots. In addition, Therma assigns an opto-thermal index to each MRR group within a cluster, which quantifies the thermal sensitivity of the group to resonance shifts. By monitoring the thermal activity during workload execution, *Therma* sets and updates these two indices. To make a migration decision at each interval in the workload execution, *Therma* calculates a score which indicates whether the core can receive a thread or send an executing thread. This score uses a weight factor that accounts for the temperature of the neighboring cores and MRR group temperatures. After updating the scores across all the cores, the cores with highest and lowest score swap the threads. *Therma* is effective in reducing the heating power for thermal tuning for single-threaded and multi-threaded workloads. *Therma* reduces the overall photonic power by up to 6.1% for single-threaded workloads and 20.7% for multi-threaded workloads compared to an interconnect-oblivious thread migration policy.

Aurora [38] uses a cross-layer approach, which encompasses circuit, architecture and operating system levels to mitigate the impact of TV-induced MRR resonance shifts. At the circuit level, *Aurora* adjusts the bias current that is flowing through the MRR to compensate for small thermal variations. This circuit level technique can only compensate for small thermal variations resulting in wavelength shift of <2 nm. For larger thermal variations, Aurora performs rerouting of packets away from MRRs around large thermal variations. These packets are routed through MRRs in cooler regions to their final destinations. Aurora uses shortest distance algorithms to select a path with lowest temperature, by accounting for the route length and utilization. The final level of thermal control of MRRs is performed at the operating system level, using a thermal/congestion-aware co-scheduling algorithm. This algorithm prioritizes thread allocation to the outer cores of the chip to minimize the impact of TV on MRR resonance shift. Adjusting the bias current at circuit level reduces the power consumption in the photonic link by 32% compared to using conventional metal strip heaters. The architectural and OS-level policies in Aurora further reduces this power consumption by 4% due to lower message retransmissions.

The above system-level policies are very effective in reducing the impact of on-chip thermal variations on MRR resonance shift. However, a major drawback with these policies stem from the fact that they ignore the impact of manufacturing process variations on MRR resonance shift. By maintaining the constant thermal activity across communicating MRRs, these works assume that the MRRs experience similar shifts in resonant wavelengths. As we observed earlier, fabrication processes introduce different variations in the dimensions and width of the MRRs across the die reticle. Therefore, no two MRRs have the exact resonance shift due to process variations. Therefore, even though the MRRs experience similar thermal activity, a high portion of heating power needs to be applied to still ensure that the process variation-induced resonance shifts are mitigated.

FreqAlign [26] is a workload allocation and migration policy that accounts for the resonance shifts of MRRs due to thermal and process variations. Instead of maintaining similar thermal activity around communicating MRRs, *FreqAlign* aims at matching the resonant wavelength of MRRs by accounting the combined effect of thermal and process variations. *FreqAlign* uses a weight matrix to capture the resonance shifts of all MRRs in a group. The weights in this matrix correspond to the resonant shifts of the MRRs. Initially, all the weights are maintained at 0. After the fabrication of the chip, *FreqAlign* updates the weight values in this matrix based on the deviation of the MRR resonant wavelength from the design intent. During thread allocation at runtime, *FreqAlign* estimates the resonant wavelength difference between MRRs in separate groups. This estimation is performed using the power consumption of the thread and the impact of the thread execution on resonant wavelength of the MRR. After *FreqAlign* estimates the impact of a thread, the thread is allocated to a core such that the resonant wavelength gradient between the MRR groups is minimized.

During communication, the resonant wavelength of all MRRs in the chip and the resonant wavelength of the laser source needs to be tuned to the same wavelength to ensure reliable communication. If all the MRRs are tuned to a fixed target wavelength (say the resonant wavelength of the laser), some MRRs may experience a larger resonant shift leading to increased heating power for MRR tuning. To address this issue, *FreqAlign* also implements an adaptive frequency tuning (*AFT*). With *AFT, FreqAlign* dynamically selects the target resonant wavelength as the lowest resonant wavelength among all the MRRs and the laser source. Once the target resonant wavelength is fixed, all the other MRRs and/or the laser source are, then, tuned to this target resonant wavelength. With such a runtime control at the system-level, *FreqAlign* is first able to reduce the resonant wavelength gradient across the MRRs by accounting for the impact of thermal and process variations. Second, by setting the target resonant wavelength adaptively at runtime, *FreqAlign* is able to further reduce the tuning range for MRRs, and thereby, the heating power required. Figure 5.6 shows the average differences in resonant frequency with different policies (Clustered, RingAware, and FreqAlign), when the system executes different utilization of

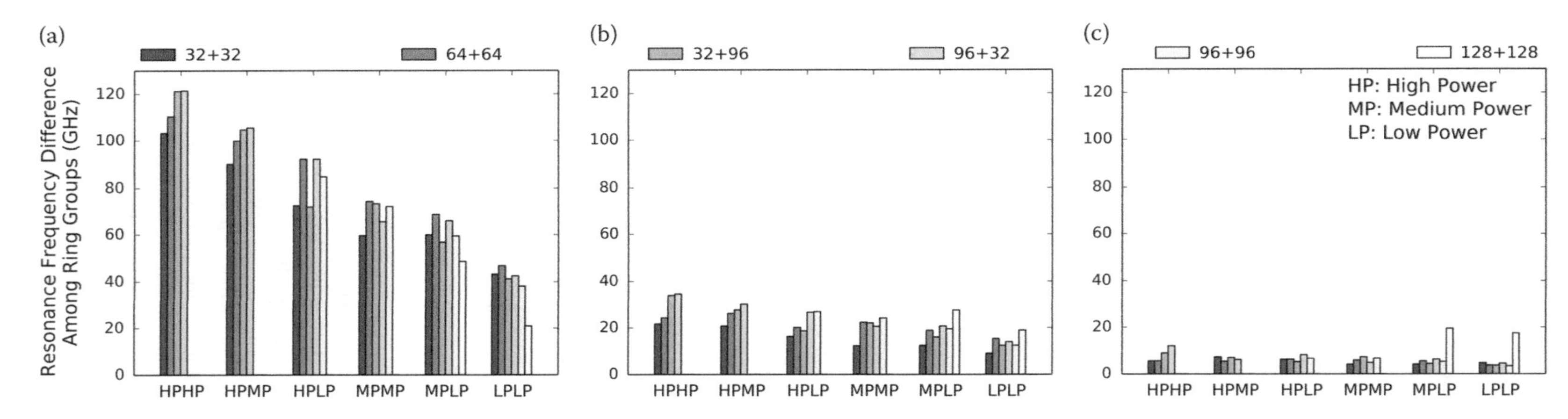

FIGURE 5.6 Average difference in resonant frequencies between MRR groups when using (a) Clustered, (b) RingAware, and (c) FreqAlign policies. Each bar corresponds to a workload + system utilization.

workloads. *FreqAlign* is able to reduce the resonant frequency difference between MRRs by 60.6% on average compared to the *RingAware*, which only accounted for thermal variations. Consequently, as seen from Figure 5.7, the overall heating power for thermal tuning is substantially lower with *FreqAlign* compared to *Clustered* or *RingAware*.

LIBRA [39] uses a reactive technique at device-level and a proactive thread migration policy at system-level to reduce the impact of thermal and process variations on MRR resonance shifts. *LIBRA* uses thermal trimming to reduce or thermal tuning to increase the resonant wavelength of an MRR to align with the nearest resonant peak. *LIBRA* chooses a boundary temperature where the trimming power and tuning power are equal. It then dynamically selects either to thermally trim the MRR if the temperature is below this boundary temperature or thermally tune the MRR otherwise. This reactive temperature-aware MRR assignment (*TMA*) enables adapting to the on-chip thermal fluctuations. *LIBRA* also accounts for the process variations (*TPMA*), by calibrating the boundary temperature between thermal trimming and thermal tuning. During initialization, the boundary temperature for each MRR is adjusted based on its process variations.

In addition, *LIBRA* incorporates system-level aware anti wavelength-shift dynamic thermal management (*VADTM*). It uses a support vector regression-based temperature predictor using performance metrics of the workload. When the predicted temperature goes above a threshold, *VADTM* migrates threads from hotter cores to cooler cores. Figure 5.8 shows the actual and predicted temperatures over the execution of two applications. The thread migration policy significantly reduces the high thermal gradients across the chip, and thereby the thermal trimming/tuning power. *LIBRA* is able to reliably satisfy thermal thresholds of the system and reduces the total power in the photonic network by 61.3% compared to other thermal and process variation aware solutions.

The above works target on reducing the heating power to thermally tune/trim the MRRs as the heating power is a major portion of the overall photonic power. However, in order to support the large bandwidth demands of emerging applications, a high density of multiplexed optical signals is used in the photonic link. The laser power and the power consumed in the digital/analog circuitry for E-O and O-E conversion increases linearly with increasing number of optical signals in the photonic link. To ensure energy-efficient computing systems, it is therefore essential to address the high-photonic power concerns at the system-level.

One approach to reduce the high photonic power is to scale the optical channels to provide the required bandwidth for an application. This can be achieved by dynamically powering the laser sources on or off. This scaling technique reduces the excess laser power and electrical-optical conversion when a high provided bandwidth is redundant. Figure 5.9 shows the framework of power scaling technique in PEARL. PEARL [40] performs such a power scaling technique on a heterogeneous system with CPUs and GPUs that uses a reservation-assist single-writer-multiple-reader (R-SWMR) photonic network topology. PEARL incorporates a coarse-grained laser scaling technique that uses buffer occupancy of a prior application phase to predict the bandwidth requirement for the next phase. Building on such a reactive policy, PEARL uses a proactive regression-based machine learning technique by predicting the number of packets injected in the network. The power scaling technique in PEARL consumes 40–65% lower photonic power with 0–14% loss in throughput performance.

WAVES [41] is another approach that addresses the high photonic power concerns by activating the required number of optical signals that satisfies the bandwidth requirements of an application. *WAVES* is a wavelength selection technique that first selects the minimum number of optical signals (laser wavelengths), required for an application. It then accounts for the thermal and process variation-induced resonance shifts of MRRs, and activates the best combination of optical signals that result in the lowest tuning range.

WAVES uses a cross-layer simulation framework that enables modeling the device-level MRR locking to optical signals under different bandwidth constraints and thermal environment. To perform the wavelength selection, the user first specifies the performance loss threshold that is tolerable for a workload. During offline characterization, *WAVES* determines the minimum

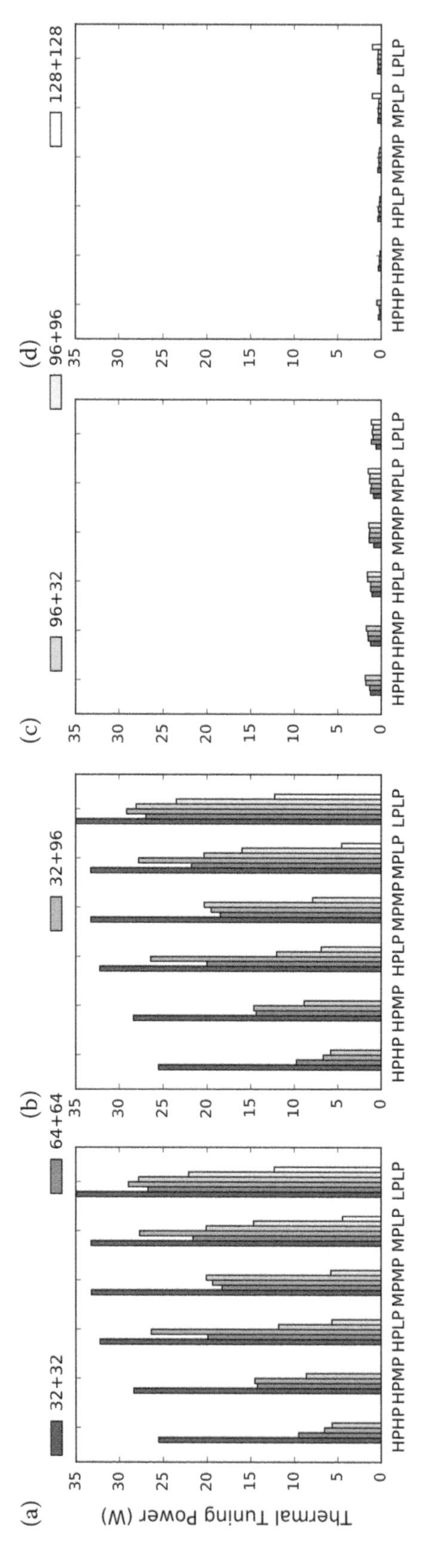

FIGURE 5.7 Heating power required for thermal tuning when using (a) Clustered, (b) RingAware, (c) FreqAlign + TFT, and (d) FreqAlign + AFT.

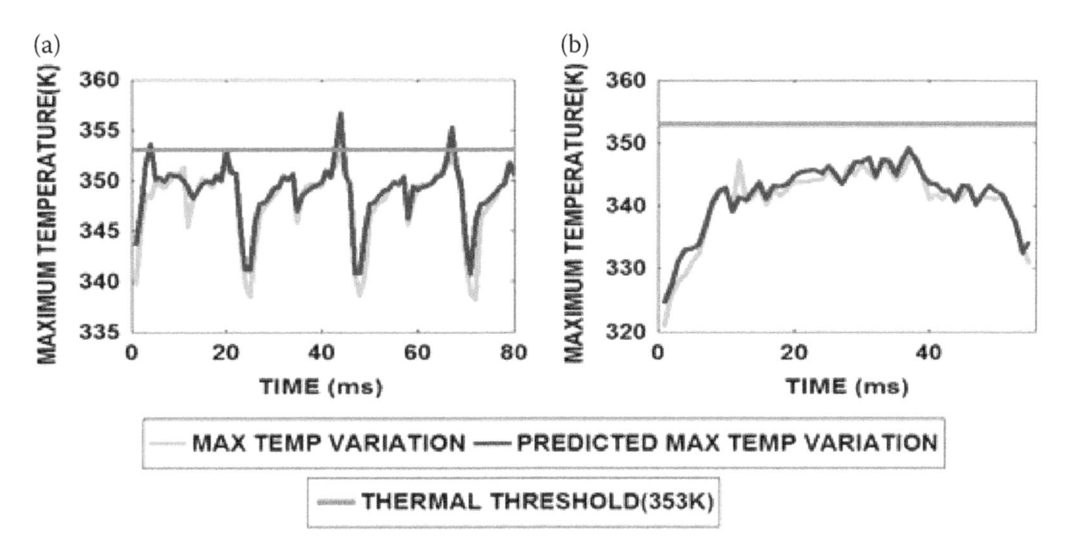

FIGURE 5.8 Actual and predicted temperature during application execution for (a) fluidanimate, and (b) radiosity application.

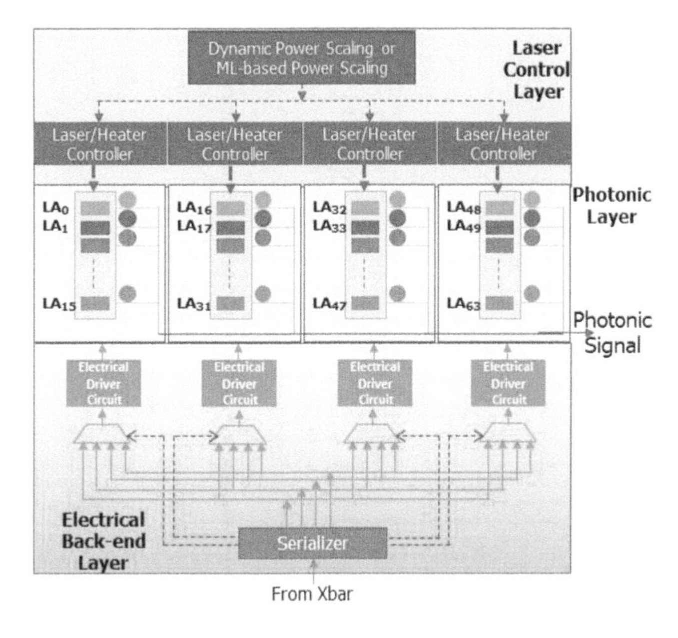

FIGURE 5.9 Framework of the dynamic power scaling in PEARL.

bandwidth requirement for the workload that satisfies the performance loss threshold. Based on the bandwidth requirement, the number of optical signals required is determined (say k among a total of n optical signals that are evenly spaced in the FSR). At the hardware level, *WAVES* maintains the process variation of each MRR in every group in lookup tables. During workload execution, *WAVES* determines the temperature of each MRR group. It then calculates the heating power required to thermally tune the MRRs for all possible combinations of activating k optical signals. Depending on the combination of k optical signals that result in the lowest thermal tuning range, *WAVES* activates these k optical signals. This additional control knob in *WAVES* enables mitigating the effect of process and thermal variations.

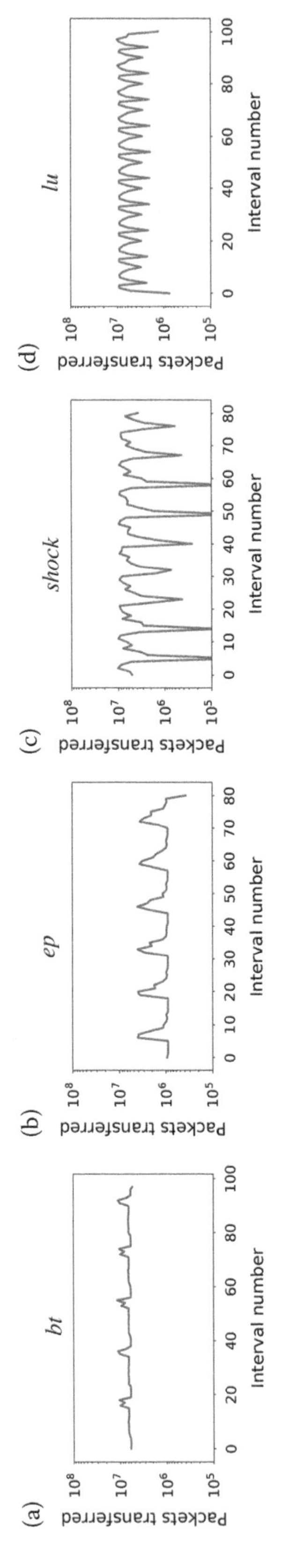

FIGURE 5.10 Number of packets transferred in the silicon-photonic link during application execution. Applications have phases of higher number of packet transfers and phases with lower number of packet transfers.

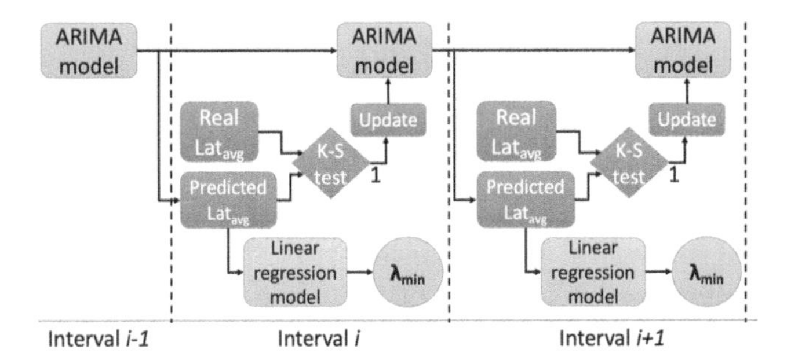

FIGURE 5.11 Flow of PROWAVES. Every interval, ARIMA predicts the network activity. The predicted value is compared to real value to tune the model in case of divergence. A linear regression model selects the optical signals using predicted network activity.

WAVES reduces the photonic power consumption in a 2.5D manycore system by 23% and 38% with 1% and 5% loss in performance, respectively. Thus, by running an application with a lower number of optical channels, the overall power consumption in the photonic link is lowered without a significant impact on performance. Moreover, the above thread migration and allocation policies such as *RingAware*, *Therma*, *Aurora*, and *FreqAlign* are orthogonal to *WAVES*, and therefore, can be applied in parallel to further lower the heating power.

WAVES and other system-level power scaling policies, however, do not account for the dynamic bandwidth needs during the application execution. *WAVES* assumes that the bandwidth needs during application is constant throughout the execution and selects the minimum optical channels to satisfy the average bandwidth needs. However, as an application progresses through different execution phases, the bandwidth needs vary significantly, as shown in Figure 5.10.

We observe from Figure 5.10 that the bandwidth needs during application execution are also periodic. *PROWAVES* [42] leverages this property to predict the bandwidth requirement for the next phase of an application execution using past trends, and proactively selecting the number of optical channels required for the next phase. PROWAVES uses ARIMA time-series forecasting to predict the network activity for the next interval. Figure 5.11 shows the flow of PROWAVES. As the thermal profile of the application varies during the application execution, the MRRs experience different shifts during the execution. Thus, there is an opportunity to lock the MRRs to different optical signals (laser wavelengths) at runtime, and reduce the tuning range for the MRRs in each group using the analog thermal control loop. Therefore, in addition to predicting the network activity and selecting the number of optical signals for the next interval, PROWAVES selects the optimal set of optical signals that reduces the thermal tuning range for MRRs.

Compared to a power-scaling policy that does not account for the thermal tuning and the modeling of the thermal control loop, PROWAVES reduces the heating power for MRRs by 26.3 W. Moreover, compared to WAVES, PROWAVES consumes 16.4% lower photonic power with a performance loss threshold of 5%.

The above runtime system-level policies work at the hardware-level and the OS-level to address the bandwidth-power tradeoffs in silicon-photonic networks. However, the on-chip communication traffic and chip temperature profile at runtime also depends on the software implementation of the application. Therefore, it is possible to instrument privileged information such as network-specific instructions or data structures that provide information to the system-policies. Figure 5.12 shows the framework of such an application-instrumentation approach [43]. An application-instrumentation approach on a widely used graph processing algorithm, PageRank, is able to reduce 35% higher photonic power with WAVES compared to a non-instrumented PageRank.

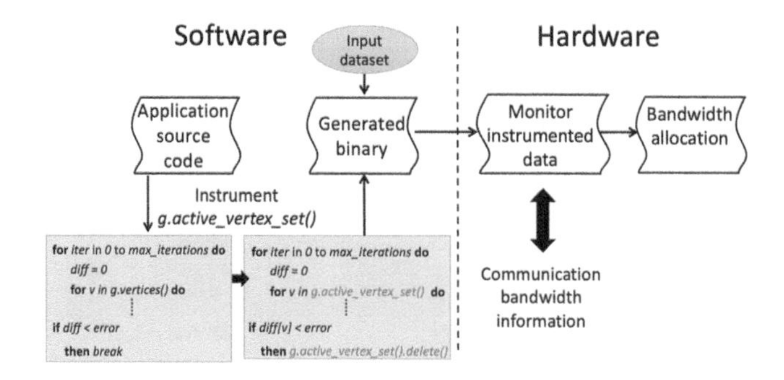

FIGURE 5.12 Framework of an application-instrumentation-assisted bandwidth allocation technique.

5.5 CONCLUSION

Silicon-photonic technology are promising candidates to provide high-bandwidth and low-latency on-chip communication in large manycore chips. A major roadblock towards wide-spread adoption of silicon-photonic technology is the high sensitivity towards thermal and process variations and the resulting increased power. This chapter reviews the device-level and chip design-level strategies to minimize the impact of the thermal and process sensitivities. This chapter then focuses on exploring the cross-layer simulation frameworks that can help bridge the gap between the device-level strategies and the system-level policies. Finally, a detailed in-depth review of the system-level runtime approaches to mitigate thermal tuning and power scaling is presented.

ACKNOWLEDGMENT

The authors thank Yvain Thonnart, Pascal Vivet and Cesar Fuguet Tortolero from CEA-Tech, Grenoble, France, and Tiansheng Zhang for their invaluable contributions in the project that formed the basis of this chapter. This work was partially funded by NSF CCF-1716352.

REFERENCES

[1] D. Stow, I. Akgun, R. Barnes, P. Gu, and Y. Xie, "Cost analysis and cost-driven IP reuse methodology for SoC design based on 2.5D/3D integration Invited Paper," in @@@*Proc.* Int. Conf. Comput.-Aided Des., 2016.
[2] J. Macri, "AMD's next generation GPU and high bandwidth memory architecture: FURY," in *Proc. Hot Chips Symp.*, 2015, August, pp. 1–26.
[3] A. Kannan, N. E. Jerger, and G. H. Loh, "Enabling interposer-based disintegration of multi-core processors," in *Proc. Int. Symp. Microarchitecture, Waikiki, HI*, 2015, pp. 546–558.
[4] A. Coskun *et al.*, "Cross-layer co-optimization of network design and chiplet placement in 2.5D systems," *IEEE Trans. Comput. Des. Integr. Circuits Syst.*, vol. X, no. X, pp. 1–14, 2020.
[5] I. Akgun, J. Zhan, Y. Wang, and Y. Xie, "Scalable memory fabric for silicon interposer-based multi-core systems," in *Proc. 34th IEEE Int. Conf. Comput. Des.ICCD 2016*, pp. 33–40, 2016.
[6] A. Narayan, Y. Thonnart, P. Vivet, A. Joshi, and A. K. Coskun, "System-level evaluation of chip-scale silicon photonic networks for emerging data-intensive applications," in *Proc. 2020 Des. Autom. Test Eur. Conf. Exhib. DATE 2020*, pp. 1444–1449, 2020.
[7] R. Kumar, V. Zyuban, and D. M. Tullsen, "Interconnections in multi-core architectures: Understanding mechanisms, overheads and scaling," in *Proc. Int. Symp. Comput. Archit.*, pp. 408–419, 2005.
[8] J. Zheng *et al.*, "Single-chip microprocessor that communicates directly using light," *Opt. Mater. Express*, vol. 4, no. 8, p. 1551, 2015.
[9] L. Virot *et al.*, "Germanium avalanche receiver for low power interconnects," *Nat. Commun.*, vol. 5, pp. 3–8, 2014.
[10] J. Cardenas, C. B. Poitras, J. T. Robinson, K. Preston, L. Chen, and M. Lipson, "Low loss etchless silicon photonic waveguides," *Opt. InfoBase Conf. Pap.*, vol. 17, no. 6, pp. 4752–4757, 2009.

[11] W. Bogaerts *et al.*, "Silicon microring resonators," *Laser Photonics Rev.*, vol. 6, no. 1, pp. 47–73, 2012.
[12] Y. Thonnart *et al.*, "POPSTAR: A robust modular optical NoC architecture for chiplet-based 3D integrated systems," in *Proc. 2020 Des. Autom. Test Eur. Conf. Exhib. DATE 2020*, Section II, pp. 1456–1461, 2020.
[13] M. J. Cianchetti, J. C. Kerekes, and D. H. Albonesi, "Phastlane: A rapid transit optical routing network," in *Proc. Int. Symp. Comput. Archit.*, pp. 441–450, 2009.
[14] A. Joshi *et al.*, "Silicon-photonic clos networks for global on-chip communication," in *Proc. 2009 3rd ACM/IEEE Int. Symp. Networks-on-Chip, NoCS 2009*, pp. 124–133, 2009.
[15] N. Kirman *et al.*, "Leveraging optical technology in future bus-based chip multiprocessors," in *Proc. Annu. Int. Symp. Microarchitecture, MICRO*, pp. 492–503, 2006.
[16] Y. Pan, P. Kumar, J. Kim, G. Memik, Y. Zhang, and A. Choudhary, "Firefly: Illuminating future network-on-chip with nanophotonics categories and subject descriptors," *Architecture*, vol. 37, no. 3, pp. 429–440, 2009.
[17] L. Ramini, D. Bertozzi, and L. P. Carloni, "Engineering a bandwidth-scalable optical layer for a 3D multi-core processor with awareness of layout constraints," in *Proc. 2012 6th IEEE/ACM Int. Symp. Networks-on-Chip, NoCS 2012*, pp. 185–192, 2012.
[18] D. Vantrease *et al.*, "Corona: System implications of emerging nanophotonic technology," in *Proc. Int. Symp. Comput. Archit.*, pp. 153–164, 2008.
[19] A. Alduino *et al.*, "Demonstration of a high speed 4-channel integrated silicon photonics WDM link with hybrid silicon lasers," in *Integrated Photonics Research, Silicon and Nanophotonics*, 2010, p. PDIWI5.
[20] K. Padmaraju and K. Bergman, "Resolving the thermal challenges for silicon microring resonator devices," *Nanophotonics*, vol. 3, no. 4–5, pp. 269–281, 2014.
[21] A. V. Krishnamoorthy *et al.*, "Exploiting CMOS manufacturing to reduce tuning requirements for resonant optical devices," *IEEE Photonics J.*, vol. 3, no. 3, pp. 567–579, 2012.
[22] X. Chen, M. Mohamed, Z. Li, L. Shang, and A. R. Mickelson, "Process variation in silicon photonic devices," *Appl. Opt.*, vol. 52, no. 31, pp. 7638–7647, 2013.
[23] A. Densmore *et al.*, "Compact and low power thermo-optic switch using folded silicon waveguides," *Opt. Express*, vol. 17, no. 13, p. 10457, 2009.
[24] Y. Thonnart *et al.*, "A 10Gb/s Si-photonic transceiver with 150μW 120μs-lock-time digitally supervised analog microring wavelength stabilization for," in *IEEE Int. Solid-State Circuits Conf.*, 2018, pp. 350–352.
[25] T. Zhang, J. L. Abellán, A. Joshi, and A. K. Coskun, "Thermal management of manycore systems with silicon-photonic networks," in *Proc. Design, Autom. Test Eur. DATE*, pp. 3–8, 2014.
[26] J. L. Abellan *et al.*, "Adaptive tuning of photonic devices in a photonic NoC through dynamic workload allocation," *IEEE Trans. Comput. Des. Integr. Circuits Syst.*, vol. 36, no. 5, pp. 801–814, 2017.
[27] C. Nitta, M. Farrens, and V. Akella, "Addressing system-level trimming issues in on-chip nanophotonic networks," in *Proc. Int. Symp. High-Performance Comput. Archit.*, pp. 122–131, 2011.
[28] M. Rakowski *et al.*, "A 4×20Gb/s WDM ring-based hybrid CMOS silicon photonics transceiver," *Dig. Tech. Pap. - IEEE Int. Solid-State Circuits Conf.*, vol. 58, pp. 408–409, 2015.
[29] C. Sun *et al.*, "A 45 nm CMOS-SOI monolithic photonics platform with bit-statistics-based resonant microring thermal tuning," *IEEE J. Solid-State Circuits*, vol. 51, no. 4, pp. 893–907, 2016.
[30] J. Teng *et al.*, "Athermal SOI ring resonators by overlaying a polymer cladding on narrowed waveguides," *IEEE Int. Conf. Gr. IV Photonics GFP*, vol. 17, no. 17, pp. 77–79, 2009.
[31] J. M. Lee, D. J. Kim, H. Ahn, S. H. Park, and G. Kim, "Temperature dependence of silicon nanophotonic ring resonator with a polymeric overlayer," *J. Light. Technol.*, vol. 25, no. 8, pp. 2236–2243, 2007.
[32] B. Guha, B. B. C. Kyotoku, and M. Lipson, "CMOS-compatible athermal silicon microring resonators," *Opt. Express*, vol. 18, no. 4, p. 3487, 2010.
[33] Y. Zhang, H. Oh, and M. S. Bakir, "Within-tier cooling and thermal isolation technologies for heterogeneous 3D ICs," *2013 IEEE Int. 3D Syst. Integr. Conf. 3DIC 2013*, pp. 1–6, 2013.
[34] D. Ding, B. Yu, and D. Z. Pan, "GLOW: A global router for low-power thermal-reliable interconnect synthesis using photonic wavelength multiplexing," in *Proc. Asia South Pacific Des. Autom. Conf. ASP-DAC*, pp. 621–626, 2012.
[35] A. K. Coskun *et al.*, "Cross-layer floorplan optimization for silicon photonic NoCs in many-core systems," in *Proc. 2016 Des. Autom. Test Eur. Conf. Exhib. DATE 2016*, pp. 1309–1314, 2016.

[36] Y. Zheng *et al.*, "Power-efficient calibration and reconfiguration for optical network-on-chip," *J. Opt. Commun. Netw.*, vol. 4, no. 12, pp. 955–966, 2012.
[37] M. V. Beigi and G. Memik, "Therma: Thermal-aware run-time thread migration for nanophotonic interconnects," *Proc. Int. Symp. Low Power Electron. Des.*, pp. 230–235, 2016.
[38] Z. Li, A. Qouneh, M. Joshi, W. Zhang, X. Fu, and T. Li, "Aurora: A cross-layer solution for thermally resilient photonic network-on-chip," *IEEE Trans. Very Large Scale Integr. Syst.*, vol. 23, no. 1, pp. 170–183, 2015.
[39] S. V. R. Chittamuru, I. G. Thakkar, and S. Pasricha, "LIBRA: Thermal and process variation aware reliability management in photonic networks-on-chip," *IEEE Trans. Multi-Scale Comput. Syst.*, vol. 4, no. 4, pp. 758–772, 2018.
[40] S. Van Winkle, A. K. Kodi, R. Bunescu, and A. Louri, "Extending the power-efficiency and performance of photonic interconnects for heterogeneous multicores with machine learning," in *Proc. Int. Symp. High-Perform. Comput. Archit.*, 2018, vol. 2018-February, pp. 480–491.
[41] A. Narayan, Y. Thonnart, P. Vivet, F. Tortolero, and A. K. Coskun, "WAVES: Wavelength selection for power-efficient photonic NoCs," in *Des. Autom. Test Eur. Conf. Exhib.*, 2019, pp. 516–521.
[42] A. Narayan, Y. Thonnart, P. Vivet, and A. K. Coskun, "PROWAVES: Proactive runtime wavelength selection for energy-efficient photonic NoCs," *IEEE Trans. Comput. Des. Integr. Circuits Syst.*, vol. X, no. X, pp. 1–14, 2020.
[43] A. Narayan, Aditya; Joshi, Ajay; Coskun, "Bandwidth allocation in silicon-photonic networks using application instrumentation," in *Proc. IEEE High Perform. Extreme Comput. Conf.*, 2020.

6 Thermal Reliability and Communication Performance Co-optimization for WDM-Based Optical Networks-on-Chip

Mengquan Li and Weichen Liu

CONTENTS

6.1 INTRODUCTION

By combining silicon photonics and NoC architecture, optical network-on-chip (ONoC) architecture [1,2] provides an innovative solution to satisfy the communication requirements on high-bandwidth, low-latency, and low-power dissipation. The employment of wavelength division multiplexing (WDM) technology further multiplies the bandwidth of optical communications. We take a 2D-mesh based ONoC as an example in Figure 6.1(a). Vertically on top of processing elements, a photonic network provides optical interconnects between optical routers for inter-processor bulk data transmission. An electronic network comprised of control routers and metallic interconnects is designed to perform logical control. The photonic and electronic networks together constitute an ONoC. In this chapter, we consider WDM-based ONoCs with 2D-mesh topology

DOI: 10.1201/9780429292033-6

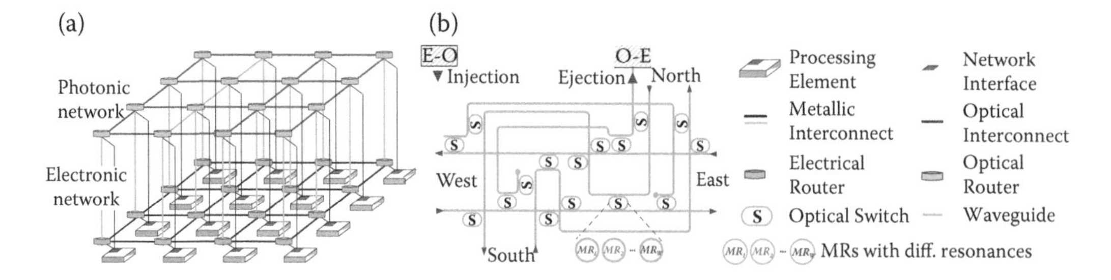

FIGURE 6.1 (a) Schematic of a mesh-based ONoC; (b) optical router *Cygnus* [3].

as the target platforms. Mesh topology has been widely used by ONoC designers thanks to its simplicity and good scalability.

As the core component of ONoCs, an optical router is comprised of a *transmitter*, a *receiver* and a *switching network*. The transmitter (i.e., E-O module) converts electrical signals into optical signals. Built-in microlaser sources, such as VCSELs [4], can be implemented in it. The optical receiver (i.e., O-E module) uses high-resolution photodetectors (PDs) [5] to convert optical signals into electrical signals. Before optical signals reach the PDs, a basic optical filtering element (BOFE) constituted by cascaded micro-ring resonators (MRs) is used in the receiver to demultiplex the WDM wavelengths. Between the transmitter and the receiver, MR-based optical switches and waveguides, respectively performing the switching and the transmission operations for optical signals, compose the optical switching network. The switching network of an optical router Cygnus is provided in Figure 6.1(b). For every specific transmission, either no or only one optical switch is activated in a router. Optical router designs follow this principle to minimize optical power loss.

However, there remain two critical challenges in ONoCs. The first one is the **thermal susceptibility** [3,6]. Chip temperature fluctuates temporally and spatially as a result of uneven chip power density and limited heat dissipation techniques. In general, it varies by more than 30°C across the chip under typical operating conditions [7]. Considering the typical 3D-stacked structure of ONoC-based chips, large thermal gradients often exist in ONoCs [8]. Optical switches, as the versatile components in ONoCs, are highly susceptible to ambient temperature fluctuations. On every routing path in ONoCs, multiple switches switch signals in stages to implement data transmission. However, the resonant wavelengths of the switches shift with temperature fluctuation. The undesired mismatch between the signal wavelength and the central wavelengths of the switches will result in additional optical power loss, seriously threatening the communication reliability of ONoCs [9,10]. This problem becomes more serious in the ONoCs employing dense WDM (DWDM).

To guarantee the thermal reliability of ONoCs, extensive thermal-aware management techniques have been proposed from the device level to the system level [9–13]. Among them, device-level wavelength tuning is widely employed for optical switches to compensate for their power loss [13]. However, this technique is energy-hungry, such that the energy consumed for thermal regulation may null the benefits of ONoCs in high energy efficiency [14,15]. The problem becomes more acute in the ONoCs who have large on-chip temperature variations and a large number of switching stages. To improve communication energy efficiency, thermal-aware routing mechanisms have been presented [11,12]. In the presence of on-chip temperature variations, the latest thermal-sensitive routing (TSR) algorithm [12] selects the route with the minimum thermal-induced optical loss out of all the alternative shortest paths for every communication to minimize the total energy overhead.

As the basis of these techniques, an effective thermal monitoring solution is of fundamental importance for ONoCs. CMOS-compatible sensors are the most popular ways of obtaining on-chip temperature information. Compared to traditional electronic counterparts, MR-based sensors have favorable properties of compact size, well-modeled temperature dependence, fast response, low power consumption, immunity to electromagnetic interference and robustness against mechanical shock and humidity [16,17]. The latest proposed MR-based sensors [8,18] provide a new way to realize accurate and efficient thermal monitoring on ONoCs. These designs operate by monitoring the temperature-dependent optical power losses of MRs and use the laser sources, MRs and PDs that are all readily available in ONoCs for power loss measurement with trivial hardware cost.

Nevertheless, the measuring accuracy of the MR-based temperature sensors are affected by fabrication-induced process variations (PVs) [19] and the wavelength tuning mechanism. PV in photolithography affects the cross-section, i.e., the width and height, of etched silicon channels, resulting in significant wavelength drifts for MRs (termed PV-drifts for short, and we denote the resonant wavelength shifts due to ambient temperature variations as TV-drifts). Recent laboratory measurements have reported a standard deviation of 1.3 nm chip-scale width variation [15,20]. The resulting PV-drift is equivalent to the TV-drift under 12.7°C temperature variation. As the PV-drifts would potentially be mistaken for TV-drifts, PVs greatly decrease the accuracy of the MR-based sensors. So does the device-level wavelength tuning technique, which is widely applied in ONoCs to guarantee communication reliability. As this technique will compensate for the wavelength shifts of MRs due to PVs and TVs simultaneously, it is also difficult to precisely obtain the spectral response merely resulted from TVs, but it is the basis for accurate and reliable temperature measurement. These two factors have been overlooked in prior arts but need to be considered when designing accurate sensors.

The second challenge lies in **communication conflicts**. An approach of *optical circuit switching* is employed on ONoCs for routing. Data transmission is preceded by a path reservation process, in which a "*path-setup*" control packet is routed in the electronic network to establish an optical path; once the path is acquired, optical messages are transmitted end-to-end in the photonic network without inflight buffering and processing. This approach provides excellent communication performance for unconflicted messages, in which dedicated optical circuits guarantee the full bandwidth of the channels and remain connected for the duration of the communication sessions. However, conflicted messages have to be blocked with extra delays. It significantly degrades the communication performance and energy efficiency of ONoCs, especially for communication-intensive applications.

To tackle this problem, conflict-free ONoC architectures using wavelength routing have been proposed in literature. Bypass links are also exploited for conflict circumvention [21]. They possess high efficiency but introduce additional hardware overheads. In contrast, conflict-aware routing is an attractive solution for ONoCs [22,23], which mitigates communication conflicts by utilizing the inherent flexibility of adaptive routing. It requires no extra hardware but performs lightweight computation over the existing generic network architecture for routing control. However, the routing techniques proposed for traditional electronic NoCs are usually unsuitable to be transplanted directly into ONoCs, for three reasons. First, the recently proposed techniques [22] usually make routing decisions for NoCs based on the buffer utilization information, whereas there is no photonic equivalent of a buffer in ONoCs. Second, typical adaptive routing approaches, such as DyXY [23], are simple enough to be used in ONoCs but the unique characteristics of optical interconnects have not been taken into account. Third, these techniques make routing decisions for every communication pair individually, which does not facilitate conflict avoidance. For large-scale manycore systems where thousands and millions of processor cores are executed in parallel, multiple inter-processor communications are often conducted in NoCs at any given time. In this chapter, we propose conflict-avoidance adaptive routing for ONoCs by considering multiple communications.

Separate performance improvement or thermal management is insufficient for deep optimization on ONoCs. However, there are apparent contradictions between minimizing conflicts and guaranteeing thermal reliability. Device-level wavelength tuning techniques largely enhance the thermal reliability of generic ONoCs but introduce heavy regulation energy overhead. In order to minimize the energy consumed for tuning, messages are required to be routed along the path that has the lowest thermal gradient, whereas communication pairs prefer to select different routing paths that do not conflict with one another for better communication performance. Therefore, strategies to collaboratively solve these contradictions are urgent.

In this chapter, we co-optimize the communication performance, energy efficiency and thermal reliability of ONoCs. We first propose a brand-new PV-tolerant optical temperature sensor design, *PV-OTS*, for accurate and efficient thermal monitoring on ONoCs, on top of that, further develop novel routing approaches to minimize communication conflicts and maximize energy efficiency with guaranteed thermal reliability. Our contributions are summarized as follows:

- We model the intrinsic thermal sensitivity of MRs with fine-grained consideration of the effect of the wavelength tuning mechanism on it.
- A novel PV-tolerant optical temperature sensor design, *PV-OTS*, is then presented. By exploiting the hidden "redundancy" in the WDM technique, it achieves accurate and efficient temperature monitoring on ONoCs with the capability of PV robustness.
- Based on real-time temperature monitoring, we analyze the thermal effects in ONoC and theoretically formulate a boundary condition for the number of switching stages in routing paths. A routing criterion at the network level is then put forward. Combined with device-level wavelength tuning, it can implement thermal-reliable ONoC.
- We further propose two conflict-avoidance adaptive routing approaches, including a mixed-integer linear programming (MILP) model and a heuristic algorithm (called *CAR*), to minimize communication conflict and maximize energy efficiency based on guaranteed thermal reliability. By applying the proposed criterion, our approaches achieve excellent performance with largely reduced complexity of design space exploration.
- Evaluation results on professional nanophotonic simulations show that, *PV-OTS* improves the measuring accuracy by 86.49% compared with state-of-the-art optical temperature sensors, with an untrimmed inaccuracy of only 0.8650°C over a large operating range from 25 to 105°C. Based on synthetic communication traces and realistic benchmarks, our routing approaches averagely improve communication performance by 159.64% and reduce energy overhead by 8.10% with guaranteed thermal reliability. *CAR* only introduces 7.20% performance difference compared to the optimal solutions obtained by the MILP model and is more scalable to large-size ONoCs.

6.2 CHAPTER OVERVIEW

Figure 6.2 shows an overview of our work. There are three main parts. First, we implement accurate and efficient thermal monitoring on ONoCs to obtain real-time chip temperature information, which is the basis of thermal-aware optimization techniques. With fine-grained consideration of the effect of the wavelength tuning mechanism on MRs, we build a thermal sensitivity model, and based on it, propose a PV-tolerant optical temperature sensor design, *PV-OTS*. They are respectively elaborated in Sections 6.3 and 6.4. Second, we put forward techniques to co-optimize communication performance and energy efficiency with guaranteed thermal reliability. Based on the analysis of the thermal effect in ONoCs, we theoretically formulate a boundary condition for the number of switching stages in routing paths. A network-level routing criterion is then presented. The criterion, combined with device-level wavelength tuning, is able to guarantee the thermal reliability of ONoCs (Section 6.5). On top of the thermal-reliable ONoC platforms, we then model the problem by MILP formulations to find the optimal routing solution for the given

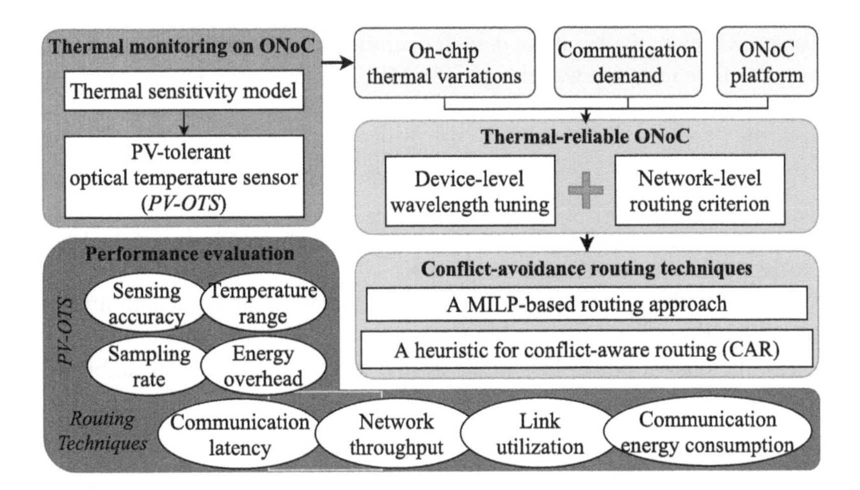

FIGURE 6.2 An overview of this chapter.

communication demand, which minimizes communication conflicts and maximizes communication energy efficiency under on-chip thermal variations (Section 6.6.2). A polynomial-time algorithm for conflict-aware routing, called *CAR*, is further proposed (Section 6.6.3). By applying the criterion, our approaches achieve excellent performance with largely reduced complexity of design space exploration. Last but not least, based on synthetic traffic pattern and realistic benchmarks, comprehensive evaluations using professional nanophotonic simulations are conducted to verify the performance of *PV-OTS* and the effectiveness of our routing approaches in network performance, communication energy consumption and thermal reliability, compared to the state-of-the-art techniques.

6.3 THERMAL SENSITIVITY OF MICRO-RING RESONATORS UNDER WAVELENGTH TUNING MECHANISM

In WDM-based ONoCs, MRs are widely used for wavelength filtering, multiplexing, switching, and modulating. As shown in Figure 6.3(a), a MR design consists of one ring and two straight waveguides. When a MR is configured to be switched *ON*, the optical signals whose wavelengths are within the MR's resonant wavelength range get resonated into the ring and delivered to the *Drop* port. Otherwise, they all will be delivered to the *Through* port. This wavelength selectivity lends MRs naturally for WDM operations. For example, as shown in Figure 6.3(b), cascaded MRs are employed as a BOFE in optical receivers to demultiplex the WDM wavelengths before optical signals reach the corresponding PDs.

The resonant wavelengths of MRs are sensitive to ambient temperature changes. A linear relationship between the wavelength shift and the temperature variation is well recognized. Given

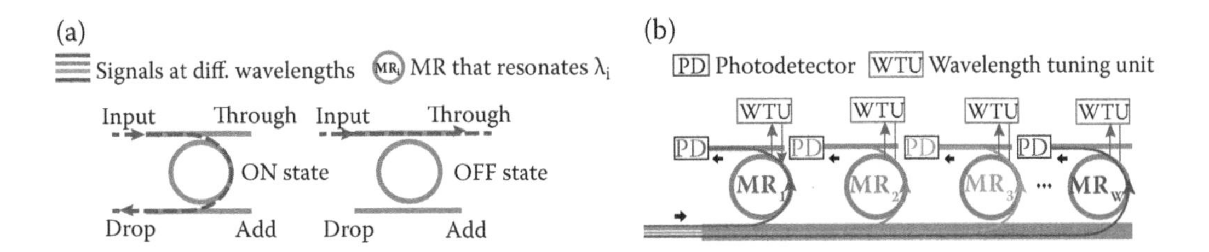

FIGURE 6.3 (a) Micro-ring resonator (MR) and (b) basic optical filtering element (BOFE) [15].

the initial resonant wavelength, λ_0, at the default operating temperature, T_0 (e.g., room temperature 25°C), the formula of the resonant wavelength of a MR, λ_{MR}, and its ambient temperature, T_{MR}, can be expressed as:

$$\lambda_{MR} = \lambda_0 + \rho \cdot (T_{MR} - T_0) \tag{6.1}$$

where ρ is the temperature-dependent resonant wavelength shift coefficient of the MR and is about 0.06 nm/°C at the 1,550 nm wavelength range [6].

Due to the TV-induced resonance peak deviation, the undesired wavelength mismatch between optical signals and MRs will result in additional optical power loss. According to the traveling wave theory [18], the optical power loss of a MR due to TV, L_{MR}^{T}, under its ambient temperature, T_{MR}, can be formulated as follows:

$$OL_{MR}^{T}=10log\left(\left(\frac{2\kappa^2+\kappa_p^2}{2\kappa^2}\right)^2\left(1+\frac{4(\lambda_{TX}-\lambda_0-\rho(T_{MR}-T_0))^2}{\theta^2}\right)\right) \tag{6.2}$$

where κ^2 is the fraction of optical power coupling between the waveguides and the ring. κ_p^2 is the fraction of intrinsic power losses (such as bending, absorption, and surface scattering due to roughness) per round-trip in the ring. λ_{TX} is the wavelength of the optical signal emitted by the transmitter and typically equals the initial resonant wavelength of the MR, i.e., $\lambda_{TX} = \lambda_0$. θ is the −3 dB bandwidth of the drop-port power transfer spectrum.

The formulas (6.1) and (6.2) well model the temperature dependence of MRs. However, the effect of the device-level wavelength tuning mechanism on MRs has been overlooked in these models. To ensure reliable data transmission, wavelength tuning is typically employed for each MR to compensate for their wavelength shifts. It enables wavelength stabilization for MRs, at the expense of extra power consumption. The regulation power consumed for a MR can be formulated as [13]:

$$P_{tuning} = \varepsilon \cdot (\lambda_{MR} - \lambda_0) \tag{6.3}$$

where ε is the tuning efficiency in mW/nm and $\lambda_{MR} - \lambda_0$ is the tuning distance.

As the resonant wavelength of a MR is realigned with the designated one through tuning, the optical power loss of the MR caused by the undesired TV-drifts would be largely reduced. From Eq. (6.2) and (6.3), the optical loss reduced by wavelength tuning can be expressed as follows:

$$OL_{tuning}=10\ log\left(\left(\frac{2\kappa^2+\kappa_p^2}{2\kappa^2}\right)^2\left(1+\frac{4P_{tuning}^2}{\varepsilon^2\theta^2}\right)\right) \tag{6.4}$$

We illustrate the scenario where a MR employs local wavelength tuning in Figure 6.4. Ideally, the tuning distance would equal the TV-drift and the total power loss of a tuned MR is zero. Given the tuning power, P_{tuning}, we can easily obtain the ambient temperature of the MR as $T_{MR} = T_0 + P_{tuning}/(\varepsilon \cdot \rho)$, where the rest parameters are constants for this MR. Nevertheless, wavelength tuning mechanism is not perfectly mature. The resonant wavelength of the MR after tuning is probably aligned not to the nominal wavelength (ideal value), but to a wavelength around the ideal one with a small tuning error; consequently, a small amount of optical power loss is caused. According to the law of energy conservation, for a MR under TVs, the optical

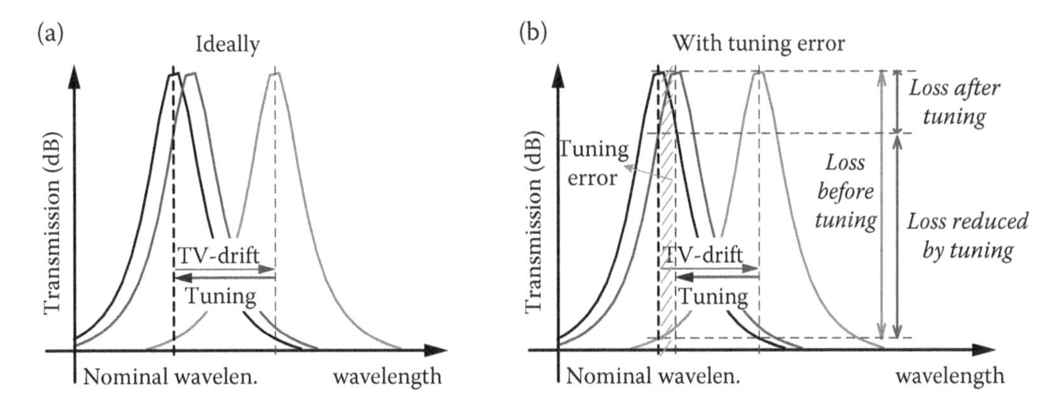

FIGURE 6.4 Effect of device-level wavelength tuning on MRs [15].

power loss before tuning, OL_{MR}^T, is the sum of the power loss after tuning and the power loss compensated by local wavelength tuning. It can be formulated as follows:

$$OL_{MR}^T = P_{TX}^o - P_{RX}^o + OL_{tuning} \tag{6.5}$$

where P_{TX}^o is the signal power emitted from the transmitter; P_{RX}^o is the optical power received by the PD at the receiver side; $P_{TX}^o - P_{RX}^o$ denotes the total optical power loss of the MR after tuning (i.e., the loss resulted from the tuning error) and OL_{tuning} is the power loss reduced by wavelength tuning shown in Eq. (6.4).

We formulate the thermal sensitivity of MRs in the presence of wavelength tuning as Eq. (6.6), which describes the relationship between the tuning power consumption (P_{tuning}), the optical power loss of a MR after tuning ($P_{TX}^o - P_{RX}^o$), and its ambient temperature (T_{MR}) [15].

$$T_{MR} = T_0 + \frac{\sqrt{\left(\frac{\theta^2}{4} + \frac{P_{tuning}^2}{\varepsilon^2}\right)\cdot 10^{\left(\frac{P_{TX}^o - P_{RX}^o}{10}\right)} - \frac{\theta^2}{4}} + (\lambda_{TX} - \lambda_0)}{\rho} \tag{6.6}$$

This physical model functions as the theoretical foundation of this work. With the single-wavelength incident signal whose wavelength equals to the initial resonant wavelength of a MR, Eq. (6.6) can be simplified to $T_{MR} = T_0 + f(P_{TX}^o, P_{RX}^o, P_{tuning})$ where the other parameters are constants for the MR. The temperature of the MR can be obtained once we know the optical power of the input signal, the measured optical power received by the PD and the power consumed by wavelength tuning, which are both easily known in ONoCs. With fine-grained analysis of the effect of the wavelength tuning mechanism on the thermal sensitivity of MRs, our model enhances the accuracy and applicability of MR-based temperature sensor designs.

6.4 *PV-OTS*: PV-TOLERANT OPTICAL TEMPERATURE SENSOR

Nevertheless, PVs complicate this problem. Under PVs, the wavelength tuning technique applied for MRs would correct PV-drifts and TV-drifts simultaneously. As the power loss caused by PV-drifts is mistaken for that caused by TV-drifts, an inaccurate temperature value will be derived from Eq. (6.6). Therefore, it is critical to precisely obtain the spectral response due to TVs for accurate temperature prediction, by removing the impact of PVs.

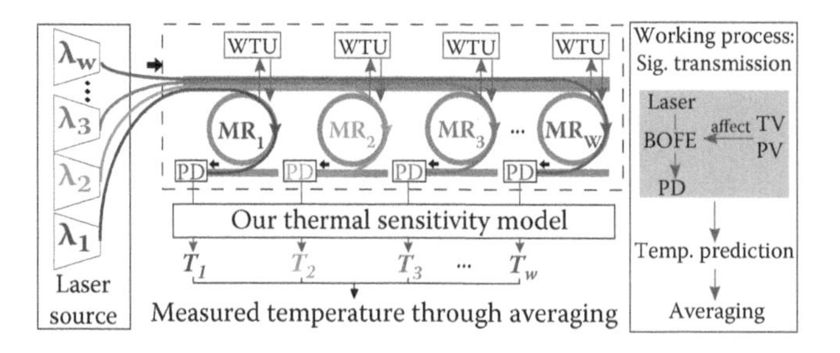

FIGURE 6.5 The PV-tolerant optical temperature sensor (*PV-OTS*) design [15].

A common solution for correcting PV-drifts is post-fabrication calibration at the device level, in which a variation map is determined via variation characterization during chip testing and then the MRs in the chip are calibrated directly based on it. No additional power is required if the MRs are successfully calibrated. Nevertheless, this technique is immature and less commercially practical [19]. Firstly, it is difficult to do post-fabrication calibration for ONoCs where thousands and millions of MRs exit. Every MR differs from one another and requires precise control. Secondly, the method of trimming the cladding material to implement calibration is unstable currently. Lastly, the post-fabrication trimming would decrease the quality factor of MRs, threatening their reliability. Therefore, current solutions for PV mitigation introduce huge overhead and lack scalability.

In this section, we achieve PV-tolerant thermal measurement by exploiting the "redundancy" in WDM technology at the system level. In WDM-based ONoCs, multiple optical signals at different wavelengths are concurrently transmitted on a single waveguide. To demultiplex the WDM wavelengths before the signals reach the PDs, a BOFE is used at every optical receiver. The cascaded MRs in a BOFE filter out their own resonant wavelengths correspondingly, which can be regarded as the duplication of the filtering operation using a single MR. It is an important observation as the hidden "redundancy" in WDM can be utilized for PV tolerance.

The PV-drift of every MR is determined by its physical dimensions. For a specific MR, its PV-drift is fixed based on the determined physical dimensions. If we implement temperature sensing based on the MR, a constant measurement error results from the fixed PV-drift. In contrast, the PV-drifts of the multiple MRs in a BOFE are random due to the fabrication non-uniformity, which result in various temperature measurement errors when employing them as thermal sensors separately. It is recognized in both academia and industry that the on-chip PVs are commonly characterized using Normal distribution [19,20]. Therefore, these temperature errors caused by PVs may potentially cancel each other out with an average value of zero. Taking a four-wavelength BOFE as an example, the ambient temperature of the four MRs is identical due to the compact structure and small footprint of BOFEs. When employing these MRs for thermal sensing separately, in an extreme case, the temperature errors caused by PVs potentially are +1°C, −3°C, +4°C, and −2°C, with an average error of zero. The impact of PVs on measurement accuracy can be significantly alleviated and even counteracted if we use their average temperature as the measured temperature. Evaluation results in Section 6.7 validate the feasibility of this idea.

As shown in Figure 6.5, we propose a PV-tolerant optical temperature sensor (*PV-OTS*) design. Each sensor is comprised of a multi-wavelength laser source, a BOFE and PDs, which are all readily available in typical WDM-based ONoCs. The MRs in the BOFE filter out their own designated resonant wavelengths from the input wavelengths, demultiplexing the W wavelengths before the optical signals reach the PDs. For every single MR, based on the known input optical power emitted from the laser, the received optical power measured by the PD and the power consumed by the corresponding wavelength tuning unit (WTU), we can derive a predicted temperature from Eq. (6.6) (e.g., T_1, T_2, …, T_w in Figure 6.5). Using the average value of the predicted temperatures as the measured temperature, *PV-OTS* is able to achieve accurate yet efficient temperature measurement with the ability to PV robustness.

To implement ONoC temperature monitoring based on the *PV-OTS* design, a basic scheme is to build and implement it as a stand-alone module, which can be placed in any area on ONoCs that requires temperature sensing. In this work, we consider WDM-based ONoCs with 2D-mesh typology as the target platforms, where each optical router is connected to a processing element and is evenly distributed across the networks (as shown in Figure 6.1(a)). In these platforms, optical routers are suitable to be used for ONoC temperature monitoring because it facilitates the run-time temperature monitoring of different network regions with a fine granularity of one sensor-per-router. Therefore, we can integrate a sensor for each optical router to perform ONoC temperature monitoring. It lays a good foundation for the implementation of thermal-reliable ONoCs.

6.5 THERMAL RELIABLE ONOCS

6.5.1 Methods for Reliable Optical Communication

Despite architecture diversity, in ONoCs, an optical connection for data transmission is generally composed of an optical transmitter (in the source router), an optical receiver (in the destination router), and a routing path connecting them. On every routing path, optical switches, combined with optical waveguides, transmit signals from the transmitter to the receiver. To ensure that ONoC functions properly, the signal power reaches the receiver on each routing path should be guaranteed not lower than the receiver's sensitivity. Otherwise, the fluctuations in received power may lead to dramatic reliability degradation and even communication failure [9,20]. We formulate the necessary condition as (6.7), where P_{TX}^{o} is the optical power output from the transmitter on a path; $\sum_{i=1}^{m} OL_{SW_i}$ is the total optical power loss of all the active switches in the path under temperature variations; m is the number of switching stages in the path (i.e., the number of active switches in this route); OL_{WG} includes the losses due to passive switches and waveguides and S_{RX} is the receiver's sensitivity.

$$P_{TX}^{o} - \sum_{i=1}^{m} OL_{SW_i} - OL_{WG} \geq S_{RX} \tag{6.7}$$

According to (6.7), there are two methods to achieve reliable data transmission on ONoCs subject to temperature fluctuations. One is to increase the output power of the transmitter to compensate for the power loss on the path such that enough optical power reaches the receiver. The other is to employ device-level wavelength tuning control for active switches whose power loss is the main contributor to the total loss of the routing path. Both methods enhance ONoCs' thermal reliability at the expense of extra energy consumption. To compare their energy costs, we conduct quantitative analysis with consideration of on-chip thermal variations.

Ideally, in every optical interconnection on ONoCs, the signal wavelength equals the initial resonant wavelengths of the switches; consequently, sufficient signal power reaches the receiver end after suffering a small amount of loss. Nevertheless, under thermal variations, more optical power is lost due to the resonance shift of active switches. Assumed the ambient temperature of the i-th switch in a routing path changes from the initial value of T_0 to T_{SW_i}, to compensate for the thermal-induced power loss for guaranteeing reliable communication, more input signal power is needed, which increases the transmitter energy consumption. From (6.2) and (6.7), we model the additional energy consumption required by a single-wavelength transmitter as (6.8) [24]. The impacts of thermal variations on waveguide propagation loss and receiver sensitivity are neglected because they are trivial compared to that on high-Q active switches [6].

$$\Delta E_{TX} \geq \frac{1}{\eta}\prod_{i=1}^{m}\left(1+\frac{4\rho^2\cdot(T_{SW_i}-T_0)^2}{\theta^2}\right)\cdot t_{TX} \tag{6.8}$$

where ΔE_{TX} is the difference between the electrical energy consumed by the single-wavelength transmitter with and without temperature variations, and in a W-wavelength ONoC, the total extra energy consumption of the transmitter will be $w \times \Delta E_{TX}$; η is the wall-plug efficiency of the transmitter, which is the conversion efficiency of electrical power to optical power; and t_{TX} is the duration time for which the transmitter remains turned on.

Formula (6.8) indicates that, it is very energy-inefficient to increase the transmitters' output power to compensate for the thermal-induced power losses in routing paths. There are three reasons. First, optical losses in routing paths are typically multiplicative in nature [14]. Second, the wall-plug efficiency of transmitters is also an obstacle. It is reported that VCSEL lasers can only achieve around 30% wall-plug efficiencies, whereas they are already much better than other competing off-chip lasers [4]. Third, the transmitter need be active most of the time because it requires a long time (hundreds of ns) to turn a laser on again if we turn it off for energy-saving [14]. To maintain the benefits of low latency for ONoCs, this delay should not be incurred. In addition to the heavy energy overhead, hardware cost also increases potentially. When employing VCSELs as the laser sources in transmitters, more VCSEL lasers are required to be integrated into a chip to generate strong optical signals, due to the limited produced optical power of a single VCSEL.

By contrast, control-based wavelength tuning is widely used for active switches to compensate for their power losses. Altering the free-carrier concentration in a MR core or tuning its local temperature using integrated microheaters, the tuning techniques dynamically maintain the local temperature of the switches and compensate for their thermal-induced power losses throughout the operation duration, via feedback control logic. We formulate the tuning energy consumption in a routing path as (6.9), where t_{dur} denotes the duration time of the switches' operation; T_{tgt} is their desired operating temperature and is usually equal to the initial temperature T_0.

$$E_{tuning} = \sum_{i=1}^{m} P_{tuning}\cdot t_{dur} = \sum_{i=1}^{m} \varepsilon\cdot\rho\cdot(T_{SW_i} - T_{tgt})\cdot t_{dur} \tag{6.9}$$

According to (6.8) and (6.9), unlike the former method whose energy consumption grows exponentially with the increasing number of active switches (m) and their temperatures (T_{SW_i}) in a routing path, the energy consumed by wavelength tuning only increases linearly. Besides, numerous techniques to improve the tuning efficiency and tuning speed have been explored [6,13], which further enhance the energy efficiency of the second method. Therefore, we employ the device-level wavelength tuning technique to implement reliable optical communication in this paper.

6.5.2 Routing Criterion: I/L/Z-Shaped Routing Path

Nevertheless, besides thermal variations, the communication reliability of ONoCs is also threatened by the large number of switching stages in routing paths. Even in the cases where every switch works stably and properly via wavelength tuning, a large number of active switches in a routing path would increase the total power loss on the path, resulting in a risk of unreliability. To ensure that the signal power received by the receiver on an optical path is not lower than its sensitivity, according to (6.7), we derive a boundary condition to limit the number of switching stages in routing paths as follow [24]:

$$m \leq (OP_{TX} - OL_{WG} - S_{RX})/OL_{SW_i}^{ideal} \tag{6.10}$$

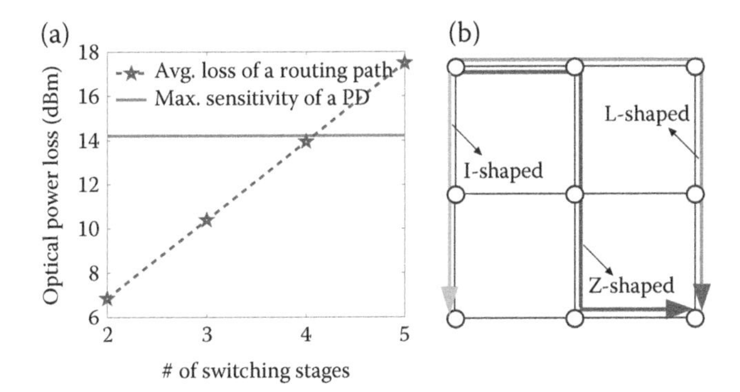

FIGURE 6.6 (a) The impact of the number of switching stages on the total loss of a route and (b) three types of routes in 2D-mesh ONoCs: I/L/Z-shaped routing paths [24].

where $OL_{SW_i}^{ideal}$ is the optical loss of the switches at their desired operating temperature. The power loss due to the switches under temperature fluctuations is always greater than the loss of ideal switches (i.e., $\sum_{i=1}^{m} OL_{SW_i} \geq m \cdot OL_{SW_i}^{ideal}$).

To show the impact of the number of switching stages on the power losses of routing paths more intuitively, we conduct a group of experiments to simulate the transmission processes of optical signals along routing paths that have different numbers of switching stages, based on current silicon photonics manufacturing level. To simulate the associated photonic components and integrated circuits, we used Lumerical FDTD and INTERCONNECT simulation infrastructure [25], which are widely used by the nanophotonics community for design and verification. A recognized optical router *Cygnus* (see Figure 6.1(b)) is employed in the simulations. Assumed that all the switches in routers work at their desired operating temperature by applying wavelength tuning, we summarize the average optical power loss of different input-output port pairs in a *Cygnus*. In every routing path, based on the network topology and the optical router structure, we can identify the input and output ports of each router, and then get the total optical power loss of this path.

According to the simulation results shown in Figure 6.6(a), the total power loss of a routing path is averagely proportional to the number of switching stages in it. State-of-the-art PDs achieve a sensitivity of -14.2 dBm for a BER of 10^{-12} [6]. Assumed the input power is 1 mW, we observe that if the number of switching stages in a route is more than 4, the power reached to the PD is lower than its sensitivity. This problem will be more serious under large temperature variations. Therefore, in this chapter, we restrict the number of switching stages to 4 for guaranteed system reliability and, as a side benefit, energy-saving. The less number of switching stages a route has, the less input energy the transmitter requires for reliable transmission. Besides, the energy consumed by wavelength tuning decreases with the reduced number of active switches. Note that with the development of silicon photonic technologies, this restriction can be relaxed by virtue of the low-loss optical device fabric.

For 2D-mesh ONoCs, there are three types of shortest routing paths for every communication pair according to their shape: I-shaped, L-shaped, and Z-shaped routes, as illustrated in Figure 6.6(b). We employ minimal routing in this work because it considers the route length and path utilization, which facilitates to mitigate link delay and conflict risk.

6.6 CONFLICT-AVOIDANCE ROUTING APPROACHES

Combined device-level wavelength tuning and the proposed network-level routing criterion, we achieve thermal-reliable ONoCs. However, it brings other issues. First, how to minimize the regulation energy consumed by wavelength tuning while avoiding conflicts for better communication performance? Second, we target to mitigate conflicts by utilizing the inherent flexibility of adaptive routing, whereas the flexibility would decrease when the number of switching stages is restricted, which is adverse to conflict reduction. To solve them, in this section, we formulate the problem and further propose two conflict-avoidance routing approaches to achieve communication optimization based on guaranteed thermal reliability.

6.6.1 Problem Formulation

6.6.1.1 Communication in ONoC

In ONoCs, the latency for inter-processor communication includes the path-setup latency in the electronic network, $L^e_{path-setup}$, and the data-transmission latency in the photonic network, $L^o_{payload}$. It can be formulated as follows:

$$L_{comm} = L^e_{path-setup} + L^o_{payload} \tag{6.11}$$

The control packet is routed in the electronic network in a packet-switching manner. It travels a number of electronic routers and undergoes several processes in each hop. After activating m associated switches, as well as their tuning control, in corresponding optical routers, an optical communication path is reserved. Local wavelength tuning works for the duration of the switches' operation with a tuning speed as fast as ~1 μs [13]. Negligible delay but extra tuning energy is incurred. We formulate the path-setup latency as follows:

$$L^e_{path-setup} = L^e_R \cdot (h + 1) + (W^e_{ctrl} - 1)/b^e + m \cdot L^e_{SW} + L^e_{conflict} \tag{6.12}$$

where L^e_R is defined as the number of stages in each control router (typically 2 or 3 stages) under the clock frequency of f^e; h denotes the number of hops from the source to the destination; W^e_{ctrl} is the size of the control packet in the electrical control network; b^e is the bandwidth of electronic channels; L^e_{SW} is the latency of activating a switch; $L^e_{conflict}$ is the additional delay caused by communication conflicts during the path-reservation phase, which is non-neglectable especially for communication-intensive applications.

Once the optical path is established, the photonic network can be perceived as a fast circuit-switched network. The data transmission latency can be formulated as:

$$L^o_{payload} = W^o_{payload}/R_{oeeo} + d \cdot n/c \tag{6.13}$$

where $W^o_{payload}$ refers to the volume of data transmitted in the photonic network; R_{oeeo} is the O-E and E-O interface data rate; d is the travel distance of optical data; n is the refractive index of silicon waveguide; and c is the light speed in vacuum.

Likewise, the total communication energy consumption has two portions shown in (6.14), where $E^o_{payload}$ is the energy consumed for data transmission and E^e_{ctrl} is the control overhead.

$$E_{comm} = E^e_{ctrl} + E^o_{payload} \tag{6.14}$$

E^e_{ctrl} can be calculated by Eq. (6.15), where E^e_{int} is the average energy required to transfer a single bit through an electrical interconnection; E^e_{cu} denotes the average energy required by the control unit to make a decision for a single packet; $E^e_{conflict}$ is the extra energy incurred by communication conflicts during path setup.

$$E^e_{ctrl} = E^e_{int} \cdot W^e_{ctrl} \cdot h + E^e_{cu} \cdot (h+1) + E^e_{conflict} \quad (6.15)$$

The energy overhead of data transmission includes the energy consumed for E-O and O-E conversions, switch activation and wavelength tuning, formulated as:

$$E^o_{payload} = E^o_{oeeo} \cdot W^o_{payload} + m \cdot P^o_{SW} \cdot t_{dur} + E_{tuning} \quad (6.16)$$

where E^o_{oeeo} is the energy required by a single bit for O-E and E-O conversions; P^o_{SW} is the average power consumed to activate an optical switch; $t_{dur} = W^o_{payload}/R_{oeeo} + d \cdot n/c$, which calculates the duration time of the switches' operation.

6.6.1.2 Problem Definition

These enlightening models reveal that, besides the number of switching stages in routing paths, communication conflicts and on-chip thermal variations are another two key factors influencing the communication performance and energy efficiency of ONoCs. Therefore, we define the problem addressed in this section as follow: *Given the floorplan of a thermal-reliable ONoC that is implemented using the solutions in Section 6.5, and a communication demand represented by a set of pairs that require inter-processor data transmission, determine a routing path for every pair to minimize communication conflicts and maximize energy efficiency in the presence of on-chip thermal variations.*

6.6.2 MILP-Based Routing Approach

We first model the problem by MILP formulations. Our MILP-based routing approach includes two phases: the main objective is to optimize communication performance by minimizing conflicts; based on it, we minimize the total communication energy consumption with consideration of on-chip thermal variations. The notations used in MILP formulations are summarized with detailed definition in Table 6.1. In the MILP model, by applying the routing criterion proposed in Section 6.5.2 for guaranteed thermal reliability, the alternative routing paths of every communication pair is a set of I/L/Z-shaped routes. It exponentially reduces the design search space and thus enhance the computation efficiency of our approaches.

In the first phase, we minimize communication conflicts. The main objective is shown as follows:

TABLE 6.1
Variables used in the MILP model

Notation	Definition
D	The communication demand.
rP_i	The set of candidate routing paths for the comm. pair i.
L	The set of optical links in ONoC.
$reql_{i,k}$	The set of required links to construct the routing path k for pair i.
T_r	The ambient temperature of the router r.
$SW_{i,k}$	The set of active switches in the path k of pair i.
$X_{i,k}$	Binary variable, equals to 1 if communication pair i routes along the path k.
N_{cx}	The number of communication conflicts.
E_i	The total communication energy consumption of pair i.

$$Objective1\colon\ minimize\,(N_{cx}) \tag{6.17}$$

The key constraints are described as follows:

1. Constraint for communication demand: Eq. (6.18) guarantees that at most one routing path is selected for every communication pair.

$$\sum_{k\in rP_i} X_{i,k} \le 1 \quad \forall\, i \in D \tag{6.18}$$

2. Constraint for route overlap: For each optical link $l \in L$, it can be occupied by zero or one routing path. If link l is occupied by a route, C_l=1. Then $\sum_{k\in rP_i} X_{i,k*}reql_{i,k} \le 1$, which guarantees no route overlap.

$$\sum_{k\in rP_i} X_{i,k*}reql_{i,k} \le C_l \quad \forall\, i \in D \tag{6.19}$$

3. Constraint for communication conflict: Eq. (6.20) calculate the number of conflicts for the given communication demand. $|D|$ is the size of the communication demand.

$$N_{cx} = |D| - \sum_{i\in D}\sum_{k\in rP_i} X_{i,k} \tag{6.20}$$

In the second phase, on the basis of the minimal communication conflicts obtained from the first phase, we minimize the total communication energy consumption with consideration of on-chip thermal variations, shown as follows:

$$Objective2{:}minimize\,(sum\,(E_i)) \quad i \in D \tag{6.21}$$

Besides the constraints (6.18) and (6.19), we add two key constraints in this phase as:

4. Constraint for minimum conflict: Eq. (6.22) guarantees that we optimize the communication energy efficiency based on the minimum conflicts.

$$\sum_{i\in D}\sum_{k\in rP_i} X_{i,k} = |D| - N_{cx} \tag{6.22}$$

5. Constraints for thermal-induced energy consumption: In this work, we consider a set of tasks running in a relatively stable power profile. The ONoC is in a steady temperature state during the time period we concern. Eq. (6.23) estimates the total communication energy consumption for each pair by taking on-chip thermal variations into account.

$$\begin{aligned} E_i = X_{i,k*}\ & (E_{int}^{e}\cdot W_{ctrl}^{e}\cdot h + E_{cu}^{e}\cdot(h+1) \\ & + E_{oeeo}^{o}\cdot W_{payload}^{o} + m\cdot P_{SW}^{o}\cdot t_{dur} \\ & + \textstyle\sum_{r\in SW_{i,k}} \varepsilon\rho\,(T_r - T_{tgt})\cdot t_{dur}) \end{aligned} \tag{6.23}$$

Our MILP model, as a formal method, can perform systematic exploration in the design space and provide a theoretically guaranteed optimal routing solution, which serves as the baseline solution for the heuristic algorithm proposed next.

6.6.3 Conflict-Aware Routing Heuristic Algorithm

As the problem is NP-hard, the running time of the MILP model grows exponentially with the increasing sizes of network and communication demand. To efficiently solve this complex issue, we further propose a polynomial-time algorithm, *CAR*.

As shown in Algorithm 6.1, we can firstly obtain the set of all the available candidate routing paths for every communication pair (i.e., a set of I/L/Z-shaped routes) by applying the proposed routing criterion, which constitutes its communication region (denoted as *cRegion*) (Lines 2–3). For example, in a 2D-mesh based ONoC, the communication region of a pair $< src, des>$ contains at most $|des.\ x - srx.\ x| + |des.\ y - src.\ y|$ paths in total. The routing path for each pair is selected from its communication region. For the pairs whose communication regions are independent of other regions, we are free to select a routing path based on the routes' temperature distribution, without the risk of contention (Lines 4–6). The function `GetRouteWithMaxEnergyEfficiency()` is designed to find the route that maximizes energy efficiency among the communication region. For the pairs that there are overlaps between their regions, we should consider the risk of link conflict, potentially resulting in communication contention. To handle the potential conflict regions, we sort pairs according to the sizes of their communication regions, from small to large (Line 8). The pairs with small communication regions are prior to the pairs with large regions because it has less alternatives. It is easier for the pairs with large communication regions to find feasible routes even if some of links are occupied by other communications. In the most extreme case, no feasible route can be obtained for the pairs that there is only one available routing path and it has been occupied by others. We make routing decisions for the pairs in the sorted order. For every communication pair, we select the route that achieves maximal energy efficiency on the basis of no communication conflict (Lines 9–14).

Figure 6.7 gives an example on a 4 × 4 ONoC to illustrate our *CAR* algorithm. There are three communication pairs ($Pair_1$: $< R_{13}, R_{10}>$, $Pair_2$: $< R_7, R_1>$ and $Pair_3$: $< R_{10}, R_4>$) and we can obtain

ALGORITHM 6.1 CONFLICT-AWARE ROUTING (*CAR*) ALGORITHM

Data: The *ONoC*, the communication demand *D*
Output: Routes for every communication pair *rPs*
1 *idleLs* = GetIdleLinks(*ONoC*);
2 **for** *cPair* ∈ *D* **do**
3   *cRegion* = GetCommPathCandidates(*cPair*, *idleLs*);
      # A set of I-shaped, L-shaped, and Z-shaped paths;
4 *IndePairs* = FindIndependPairs(*D.cRegion*);
5 **for** *cPair* ∈ *IndePairs* **do**
6   *cPair.rP* = GetRouteWithMaxEnergyEfficiency(*cRegion*);
7 Update(*D*);
8 $D = sort(D, D.cRegion, Ascend)$;
9 **for** *cPair* ∈ *D* **do**
10   *idleLs* = GetIdleLinks(*ONoC*);
11   **if** $src.x = des.x$ ***or*** $src.y = des.y$ **then**
12         $cPair.rP = (src \rightarrow des)$;   #I-shaped path;
13   **Continue**;
14   *cPair.rP* = GetRouteWithMaxEnergyEfficiency
       ($cRegion \cap idleLs$);
15 **return** (*rPs*);

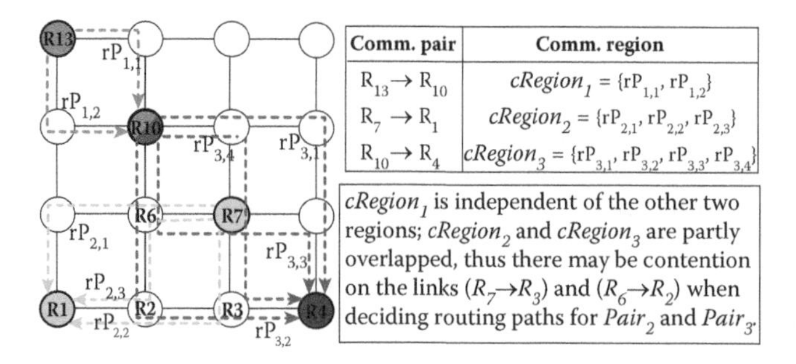

Comm. pair	Comm. region
$R_{13} \rightarrow R_{10}$	$cRegion_1 = \{rP_{1,1}, rP_{1,2}\}$
$R_7 \rightarrow R_1$	$cRegion_2 = \{rP_{2,1}, rP_{2,2}, rP_{2,3}\}$
$R_{10} \rightarrow R_4$	$cRegion_3 = \{rP_{3,1}, rP_{3,2}, rP_{3,3}, rP_{3,4}\}$

$cRegion_1$ is independent of the other two regions; $cRegion_2$ and $cRegion_3$ are partly overlapped, thus there may be contention on the links ($R_7 \rightarrow R_3$) and ($R_6 \rightarrow R_2$) when deciding routing paths for $Pair_2$ and $Pair_3$.

FIGURE 6.7 An example of the *CAR* algorithm [24].

their communication regions (i.e., $cRegion_1$, $cRegion_2$ and $cRegion_3$ respectively). As $cRegion_1$ is independent of the other two regions, we are free to select a route for $Pair_1$ according to the temperature distributions of the routes $rP_{1,1}$ and $rP_{1,2}$. However, as the regions $Pair_2$ and $Pair_3$ are partly overlapped, there may be contention on the links ($R_6 \rightarrow R_2$) and ($R_7 \rightarrow R_3$) when deciding routes for them. Therefore, we give priority to the pair that has smaller region and select a routing path for it first ($Pair_2$ in this example). Assume the routing path $rP_{2,2}$ achieves maximum energy efficiency for $Pair_2$, the link ($R_7 \rightarrow R_3$) is occupied; consequently, the available communication region of $Pair_3$ is reduced to $\{rP_{3,1}, rP_{3,2}, rP_{3,3}\}$ and we can select one of routing paths based on their temperature distributions without conflicts.

Complexity analysis: The complexity of Algorithm 6.1 is $O(M \times N)$, where M is the number of communication pairs in a $N \times N$ ONoC. The complexity of looping through the alternative paths to find a route with the maximal energy efficiency for a pair is $O(N)$ in the worst case, and there are M communication pairs in total.

6.7 PERFORMANCE EVALUATION

6.7.1 Accuracy of the Thermal Sensitivity Model

We use Lumerical FDTD and INTERCONNECT simulation infrastructure. A compact silicon-on-insulator (SOI) MR with a radius of 5 μm is used for simulation. The coupling gaps between the waveguides and ring are 100 nm with a cross-section of 400 nm × 180 nm. We simulate the MR and study its optical characteristics through FDTD, shown in Figure 6.8(a). The wavelength tuning mechanism applied for the MR is studied through INTERCONNECT. As shown in Figure 6.8(b), we import the MR's frequency responses obtained from FDTD into INTERCONNECT and analyze the optical characteristics of the MR under wavelength tuning through an optical analyzer. The wavelength tuning technique applied in the simulations enables wavelength-locking for the MR by reducing the resonance variations to below 50 pm, which has been demonstrated to be adequate for most silicon photonic applications [26]. As shown in Figure 6.8(c), simulation results show that this technique has a tuning efficiency of about 0.1692 nm/mW.

Setting the incident light as the fundamental TE mode for wavelength at 1,550 nm, with which the MR resonates at room temperature, we compare our thermal sensitivity model in Eq. (6.6) with the state-of-the-art model presented in [18] for prediction accuracy verification. Work [18] formally models the thermo-optic effect of MRs but does not consider the impact of wavelength tuning techniques on it. We conduct 80 groups of experiments, in which the ambient temperature of the MR increases from 25 to 105°C. The temperature of 105°C is an extremely high value for a silicon chip. As shown in Figure 6.9, the red curve is the true temperatures of the MR set in every group of simulations. The black dotted line and the green curvy are the measured temperature values obtained by Eq. (6.6) and the model in [18], respectively. With fine-grained consideration of the effect of tuning mechanism on wavelength locking, our model achieves high prediction

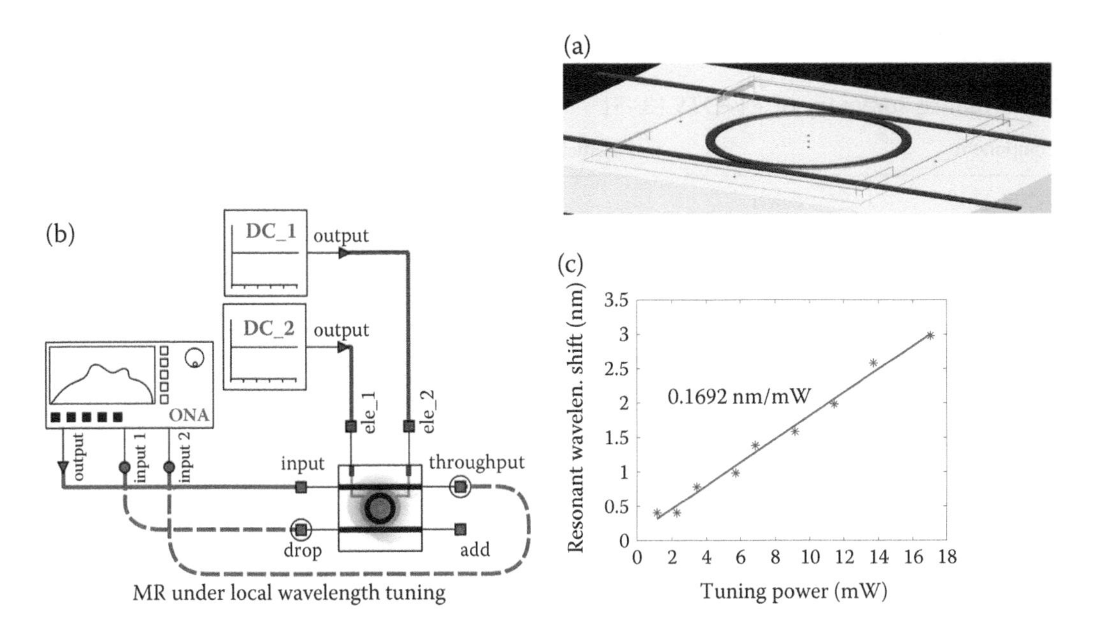

FIGURE 6.8 (a) The perspective view of a MR in FDTD; (b) the simulation of the MR under wavelength tuning mechanism in INTERCONNECT; (c) the tuning efficiency of the wavelength tuning technique applied in INTERCONNECT [15].

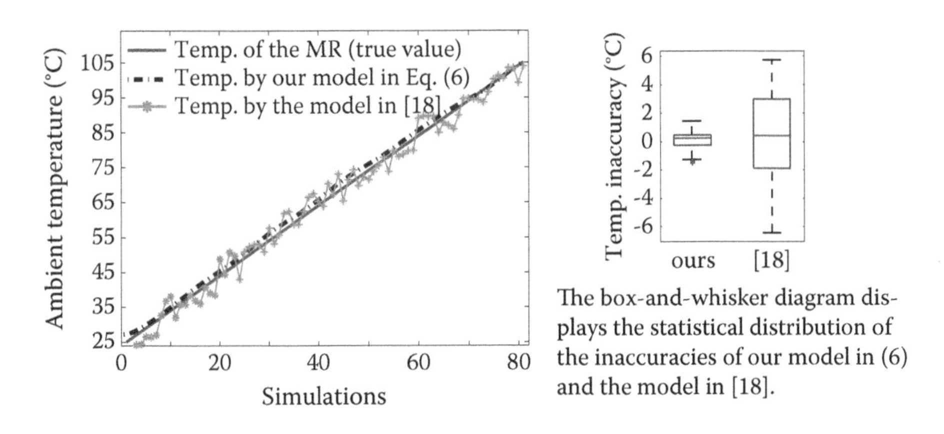

FIGURE 6.9 The accuracy of our thermal sensitivity model in Eq. (6.6) [15].

accuracy with an average difference of only 0.5521°C compared to the true values. It improves the accuracy by 79.5% on average over the model in [18]. We further present a box-and-whisker diagram to display the statistical distribution of their temperature inaccuracies at the right side of Figure 6.9. This accurate model provides a good theoretical foundation for the proposed *PV-OTS* design.

6.7.2 Performance of the *PV-OTS* Design

We build a *PV-OTS* module using an on-chip VCSEL array, an eight-wavelength BOFE and a PD array. Table 6.2 lists the conservative parameters of the associated components used in the sensor and summarizes its performance. As all the components are readily available in ONoCs, trivial additional hardware supports are needed.

TABLE 6.2
Performance summary of *PV-OTS* [3–5]

Component		Parameter	
VCSEL array	Power		25 mW
	Transm. rate		25 Gbps
PD array	Power		4.25 mW
	Transm. rate		25 Gbps
Eight-wavelength BOFE	MR	Radius	$5 + 0.008 \times (n - 1)$ μm ($1 \leq n \leq 8$, $n \in N^{+}$)
		Cross-section	400 nm × 180 nm
		Coupling gap	100 nm
		Adjacent MR spacing	5 μm
	Power		1.4 mW
	Transm. rate		25 Gbps
PV-OTS	Accuracy	0.8650°C	
	Temp. range	25–105°C	
	Sampling rate	~1 MSa/s	
	Energy	627.71 pJ/Sa	

Measuring accuracy under PVs. We conduct 20 groups of experiments based on the *PV-OTS*. We assume that the ambient temperature of the cascaded MRs in the BOFE is identical due to its compact structure and small footprint. In every group of experiments, the waveguide widths of the eight MRs are randomly generated and follow Normal distribution $N(400,1.3^2)$, similar to the models in [19,20]. Given a random ambient temperature within the range from 25 to 105°C, the optical characteristics of the eight MRs are obtained from FDTD, which are then imported into INTERCONNECT for tuning. Based on the simulation results and the thermal model in Eq. (6.6), we can obtain the measured temperature of our sensor.

As shown in Figure 6.10(a), the black dots and blue circles are the measurement errors obtained by sensing with every single MR in the BOFE, based on the thermal sensitivity model in Eq. (6.6) and the model in [18], respectively, and the red curve is the temperature inaccuracy obtained by

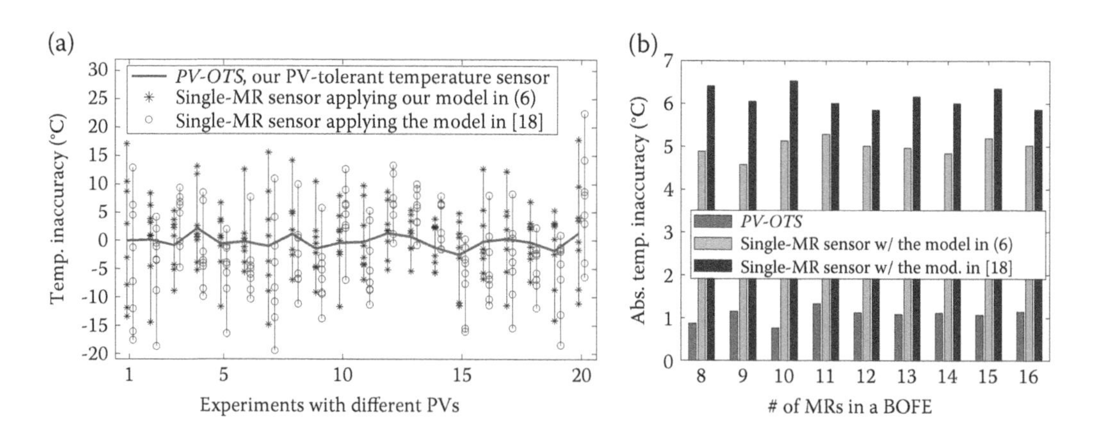

FIGURE 6.10 The effectiveness of the *PV-OTS* design. (a) Accuracy comparison between *PV-OTS* and the single-MR based sensors employing our model in Eq. (6.6) and the latest model in [18]. (b) Our sensor achieves stable performance in the ONoCs employing DWDM [15].

our sensor design. Under PVs, the predicted temperatures obtained by these MRs vary, with an error of up to >20°C. Compared to the solution that implements thermal sensing with a single MR, our sensor averagely improves the measuring accuracy by 86.49%. Integrated with the accurate thermal model in Eq. (6.6), the average inaccuracy of our thermal sensor is only 0.8650°C, further improving the measurement accuracy compared to the sensor using the latest model in [18]. Furthermore, similar experiments are conducted based on the BOFEs with different number of MRs (ranging from 8 to 16). The average absolute errors of the thermal sensors based on different sizes of BOFEs are illustrated in Figure 6.10(b). It can be observed that our sensor achieves stable performance in the ONoCs employing DWDM with an average of 82.47% accuracy improvement.

Temperature range. By utilizing the laser sources, BOFEs and PDs that are readily available in ONoCs, *PV-OTS* operates through monitoring the optical power losses of MRs. The temperature range is determined by the full width at half maximum (FWHM) bandwidth of the MRs. Generally, due to the high-Q, small-FWHM MRs are used in ONoCs to implement reliable communication. However, the smaller the FWHM, the sharper the resonance. When using a small-FWHM MR for thermal sensing, the power loss of the MR changes rapidly when the ambient temperature slightly changes; Consequently, high sensitivity can be achieved but with limited sensing range. To balance the temperature range, sensing sensitivity and communication reliability, in this chapter, we employ MRs with moderate FWHM based on typical dimensions, similar to the MR in [8,18]. Based on them, simulation results show a large sensing range of our *PV-OTS* from 25 to 105°C, which covers the typical temperature range for chip operating.

Sampling rate and energy overhead. *PV-OTS* achieves a high sampling rate with low energy consumption. For a sampling operation, an optical packet will be transmitted from the laser source to the PDs, passing through a BOFE. We assume a packet is 512 bits, which is the average size of payload data in ONoCs [27]. As all of the associated components achieve a transmission rate of higher than 25 Gbps, it costs only tens of nanoseconds to complete a sampling operation. This small delay enables an extremely high sampling rate. In fact, the sampling rate of our *PV-OTS* is limited by the intrinsic response time of the thermo-optic effect, which is on the order of μs. Therefore, the sampling rate of *PV-OTS* is about 1 MSa/s. Similarly, according to the power of the associated components, it costs 627.71 pJ per sampling operation.

6.7.3 Effectiveness of Our Routing Approaches

Our approaches are evaluated by comparing with two recognized techniques, DyXY [23] and TSR [12], which achieve excellent performance in conflict avoidance or thermal-induced energy overhead reduction, repectively.

We consider the 2D-mesh ONoCs with sizes ranging from 8 × 8 to 15 × 15 as the target platforms. A high-level simulator in MATLAB is built for technique evaluation. We use the out-of-order Alpha 21346 core array in 22 nm technology for computation. Vertically on top of the processing layer where the cores are located, 2D-mesh ONoCs are employed for inter-processor communication. We model the power consumption of the processor cores by McPAT v1.0 [28] and use HotSpot v5.02 [29] to obtain the steady-state chip thermal profiles. Inter-processor communications are conducted under this chip temperature profile. Based on extensive synthetic communication traces and a set of real-world applications, we compare the effectiveness of different routing techniques (including our approaches, DyXY and TSR). The routing criterion proposed in Section 6.5.2 is applied in the experiments. Gurobi optimization [30] with CVX v2.1 is employed as the MILP solver, which is designed to be the fastest, most powerful solver available for MILP problems. The values of the parameters used in the experiments are listed in Table 6.3.

We first evaluate the routing techniques based on *Uniform traffic pattern*. We conduct 100 groups of experiments for every size of ONoC. In every group of experiment, the source, destination and the volume of communications among processor cores are randomly generated and follows the traffic model. To focus on the effectiveness of these techniques in contention reduction,

TABLE 6.3
Parameters used in the experiments [27,28]

Parameter	Value	Parameter	Value
f^e	1 GHz	b^e	32-b bidirectional
L^e_{SW}	30 ps	R_{oeeo}	12.5 Gbit/s
n	3.48	c	3×10^8 m/s
W^e_{ctrl}	8 + 1 bits	$W^o_{payload}$	512 bits
E^e_{int}	0.52 pJ/bit	E^e_{cu}	1 pJ/pkt
P^o_{SW}	20 μW	E^o_{oeeo}	1 pJ/bit
ε	1.10 mW/nm	ρ	0.06 nm/K

we set a high load rate for the networks in the experiments. Therefore, the size of the generated communication demand increases with the growing size of ONoCs. The voltage and frequency levels of processor cores are randomly assigned, consequently resulting in a randomly-generated chip thermal distribution. Based on the generated chip thermal variation and communication demand, our approaches, TSR and DyXY algorithms are separately employed to perform routing control. Evaluation results on average communication latency, network throughput, link utilization, and communication energy consumption are compared. The link utilization is defined as the ratio of busy links to the total links in an ONoC during communications.

As shown in Figure 6.11, we set the results obtained by the MILP in 8 × 8 mesh-based ONoC as the baseline and normalize the other results accordingly. Owing to the minimum communication conflicts, the MILP-based approach achieves an average of 25.86% and 19.39% reduction in communication latency and an average of 159.64% and 118.25% improvement in network throughput, respectively, compared to the TSR and DyXY. The link utilization is increased by 61.20% and 43.76% as well. Results obtained by *CAR* are close to the optimal results obtained by the MILP approach, with a 13.49% difference on average. A stable performance gain is achieved and is scalable to large-scale ONoCs with heavy traffic. Based on the preserved communication performance in terms of communication latency and network throughput, the link utilization decreases with the increasing network size; consequently, more communication requests are potentially affordable. In addition, compared to the TSR that obtains minimal communication energy consumption, the MILP and the *CAR* only consume an average of 0.04 pJ/bit and 0.07 pJ/bit more energy, respectively, and they reduce 5.54% energy overhead than the DyXY.

TSR achieves excellent energy efficiency by selecting the routing paths with the minimum communication energy consumption for communication pairs. Significant delays are introduced to wait for the occupied optical channels to become available, thus increasing the communication latency and reducing the network throughput and link utilization. The adaptive DyXY is a contention-aware routing algorithm, which makes decisions by monitoring the optical channel occupancy status in the proximity. By selecting the next hop with more available optical outgoing links, this technique effectively reduces the possibility of communication conflicts for ONoCs. Therefore, compared to the TSR, DyXY achieves better communication performance with low latency, high throughput and link utilization, but is more energy-consuming because it does not take the thermal susceptibility of ONoCs into account. However, restricted by the routing criterion applied in ONoCs and using only local information to make decisions for communication pairs individually, the benefits of DyXY in communication performance optimization is potentially diminished. Compared with them, our approaches implement centralized control based on global information, which facilitates both conflict avoidance and energy efficiency optimization.

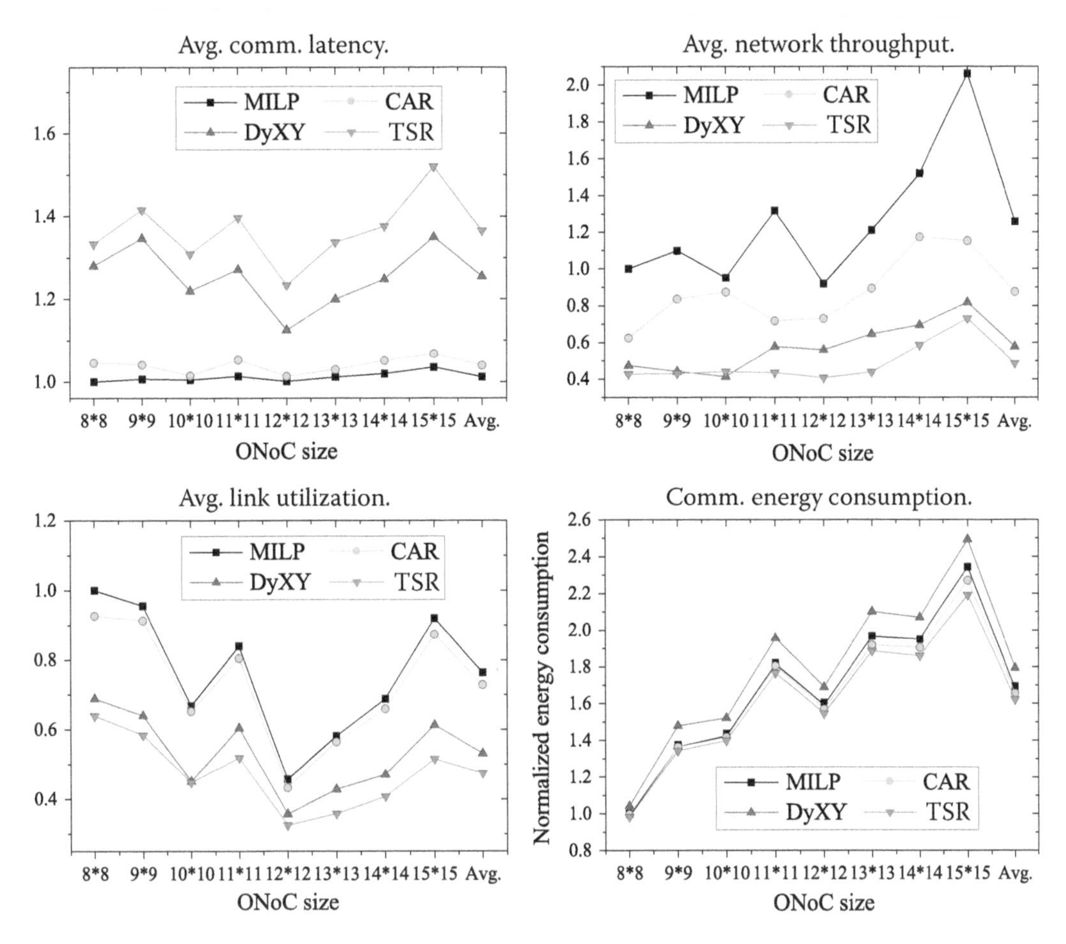

FIGURE 6.11 Effectiveness comparison of different techniques under Uniform traffic pattern.

To evaluate the feasibility of our approaches to real-world applications, we further conduct experiments based on a set of *realistic benchmarks*, including *autocor*, *audiobeam*, *tde_pp*, *fmradio*, *filterbank*, and *beamformer* in StreamIt and *IIRfilter* and *4_stagelattice* in DSP-stone benchmark [8,31]. The data communication traces are generated by the task mapping algorithm in [32] and the chip thermal distributions are generated by McPAT and HotSpot simulations, similar to the experiments that use synthetic communication traces. Figure 6.12 illustrates the results. Compared to TSR and DyXY, our MILP model achieves an average of 22.86% and 13.84% reduction in communication latency, an average of 123.47% and 85.75% improvement in network throughput, and an average of 50.93% and 31.63% improvement in link utilization, respectively. The heuristic *CAR* only has an average of 7.20% performance difference compared to the MILP. Regarding the communication energy, our approaches consume an average of 0.05 pJ/bit more energy than TSR but improve the energy efficiency by 8.10% compared with DyXY. The experimental results based on the realistic benchmarks are consistent with those based on the synthetic communication traces.

6.8 CONCLUSION

In this chapter, we have presented a series of novel techniques to collaboratively optimize the thermal reliability, communication performance and energy efficiency of WDM-based ONoCs. We first proposed a brand-new PV-tolerant optical temperature sensor design (*PV-OTS*) for

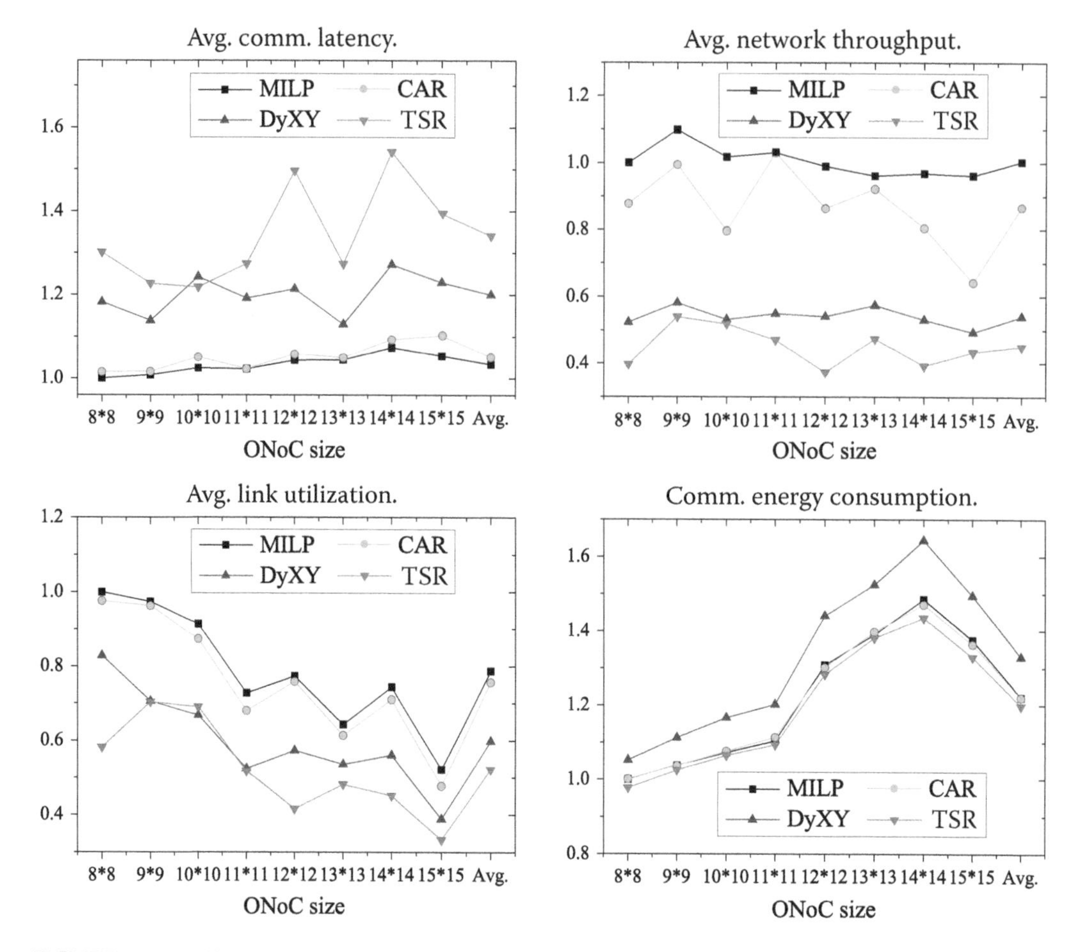

FIGURE 6.12 Effectiveness comparison of different techniques based on realistic benchmarks.

accurate and efficient thermal monitoring on ONoCs. Based on it, we further developed a routing criterion and two conflict-circumvented and thermal-aware routing approaches (including a MILP model and a *CAR* heuristic) to minimize communication conflicts and maximize energy efficiency with guaranteed thermal reliability. Extensive evaluation results have verified the excellent performance of *PV-OTS* and the effectiveness of the proposed routing approaches, compared with state-of-the-art techniques.

ACKNOWLEDGMENT

This work is partially supported by the Ministry of Education, Singapore, under its Academic Research Fund Tier 2 (MOE2019-T2-1-071) and Tier 1 (MOE2019-T1-001-072), and Nanyang Technological University, Singapore, under its NAP (M4082282) and SUG (M4082087).

REFERENCES

1. K. Bergman, L. P. Carloni, A. Biberman, J. Chan, and G. Hendry, *Photonic network-onchip design*. Springer, 2014.
2. S. Werner, J. Navaridas, and M. Luján, "A survey on optical network-on-chip architectures," *ACM Computing Surveys*, vol. 50, no. 6, p. 89, 2017.

3. M. Li, J. Zhou, and W. Liu, "Lightweight thermal monitoring in optical networks-on-chip via router reuse," in *IEEE Design, Automation & Test in Europe Conference & Exhibition*, 2020, pp. 406–411.
4. S. Spiga, W. Soenen, A. Andrejew, D. M. Schoke, X. Yin, J. Bauwelinck, G. Boehm, and M.-C. Amann, "Single-mode high-speed 1.5-μm VCSELs," *IEEE/OSA Journal of Lightwave Technology*, vol. 35, no. 4, pp. 727–733, 2016.
5. S. Saeedi, S. Menezo, G. Pares, and A. Emami, "A 25 Gb/s 3D-integrated CMOS/silicon-photonic receiver for low-power high-sensitivity optical communication," *IEEE/OSA Journal of Lightwave Technology*, vol. 34, no. 12, pp. 2924–2933, 2015.
6. Y. Ye, Z. Wang *et al.*, "System-level modeling and analysis of thermal effects in WDM-based optical networks-on-chip," *IEEE Transactions on Computer-Aided Design of Integrated Circuits and Systems*, vol. 33, no. 11, pp. 1718–1731, 2014.
7. M. Li, W. Liu, L. Yang, P. Chen, and C. Chen, "Chip temperature optimization for dark silicon many-core systems," *IEEE Transactions on Computer-Aided Design of Integrated Circuits and Systems*, vol. 37, no. 5, pp. 941–953, 2017.
8. M. Li, W. Liu, N. Guan, Y. Xie, and Y. Ye, "Hardware-software collaborative thermal sensing in optical network-on-chip–based manycore systems," *ACM Transactions on Embedded Computing Systems*, vol. 18, no. 6, pp. 1–24, 2019.
9. C. Sai Vineel Reddy, I. G. Thakkar, and S. Pasricha, "LIBRA: Thermal and process variation aware reliability management in photonic networks-on-chip," *IEEE Transactions on Multi-Scale Computing Systems*, vol. 4, no. 4, pp. 758–772, 2018.
10. M. Li, W. Liu, L. Yang, P. Chen, D. Liu, and N. Guan, "Routing in optical network-on-chip: Minimizing contention with guaranteed thermal reliability," in *ACM Asia and South Pacific Design Automation Conference*, 2019, pp. 364–369.
11. Z. Li, A. Qouneh, M. Joshi, W. Zhang, X. Fu, and T. Li, "Aurora: A cross-layer solution for thermally resilient photonic network-on-chip," *IEEE Transactions on Very Large Scale Integration (VLSI) Systems*, vol. 23, no. 1, pp. 170–183, 2014.
12. K. Yao, Y. Ye, S. Pasricha, and J. Xu, "Thermal-sensitive design and power optimization for a 3D torus-based optical NoC," in *IEEE/ACM International Conference on Computer-Aided Design*, 2017, pp. 827–834.
13. K. Padmaraju and K. Bergman, "Resolving the thermal challenges for silicon microring resonator devices," *Nanophotonics*, vol. 3, no. 4-5, pp. 269–281, 2014.
14. J. Bashir, E. Peter, and S. R. Sarangi, "A survey of on-chip optical interconnects," *ACM Computing Surveys*, vol. 51, no. 6, p. 115, 2019.
15. W. Liu, M. Li, W. Chang, C. Xiao, Y. Xie, N. Guan, and L. Jiang, "Thermal sensing using micro-ring resonators in optical network-on-chip," in *IEEE Design, Automation and Test in Europe*, 2019.
16. C.-T. Wang, C.-Y. Wang, J.-H. Yu, I.-T. Kuo, C.-W. Tseng, H.-C. Jau, Y.-J. Chen, and T.-H. Lin, "Highly sensitive optical temperature sensor based on a SiN micro-ring resonator with liquid crystal cladding," *Optics Express*, vol. 24, no. 2, pp. 1002–1007, 2016.
17. H.-T. Kim and M. Yu, "Cascaded ring resonator-based temperature sensor with simultaneously enhanced sensitivity and range," *Optics Express*, vol. 24, no. 9, pp. 9501–9510, 2016.
18. W. Liu, P. Wang, M. Li, Y. Xie, and N. Guan, "Quantitative modeling of thermo-optic effects in optical networks-on-chip," in *ACM Great Lakes Symposium on VLSI*, 2017, pp. 263–268.
19. Y. Xu, J. Yang, and R. Melhem, "Tolerating process variations in nanophotonic on-chip networks," in *IEEE/ACM International Symposium on Computer Architecture*, 2012, pp. 142–152.
20. Z. Li, M. Mohamed *et al.*, "Reliability modeling and management of nanophotonic on-chip networks," *IEEE Transactions on Very Large Scale Integration (VLSI) Systems*, vol. 20, no. 1, pp. 98–111, 2010.
21. S. Werner, J. Navaridas, and M. Luján, "Amon: An advanced mesh-like optical NoC," in *IEEE Symposium on High-Performance Interconnects*, 2015, pp. 52–59.
22. C. Li, D. Dong, X. Liao, F. Lei, and J. Wu, "CCAS: Contention and congestion aware switch allocation for network-on-chips," in *IEEE International Conference on Computer Design*, 2016, pp. 444–447.
23. M. Li, Q.-A. Zeng, and W.-B. Jone, "DyXY: A proximity congestion-aware deadlock-free dynamic routing method for network on chip," in *Design Automation Conference*, 2006, pp. 849–852.
24. M. Li, W. Liu, L. H. Duong, P. Chen, L. Yang, and C. Xiao, "Contention-aware routing for thermal-reliable optical networks-on-chip," *IEEE Transactions on Computer-Aided Design of Integrated Circuits and Systems*, 2020.

25. "Lumerical Inc.," https://www.lumerical.com/ [Online].
26. A. H. Atabaki, A. A. Eftekhar, M. Askari, and A. Adibi, "Accurate post-fabrication trimming of ultra-compact resonators on silicon," *Optics Express*, vol. 21, no. 12, pp. 14139–14145, 2013.
27. H. Gu, J. Xu, and W. Zhang, "A low-power fat tree-based optical network-on-chip for multiprocessor system-on-chip," in *IEEE Design, Automation and Test in Europe*, 2009, pp. 3–8.
28. S. Li, J. H. Ahn *et al.*, "McPAT: An integrated power, area, and timing modeling framework for multicore and manycore architectures," in *IEEE/ACM International Symposium on Microarchitecture*, 2009, pp. 469–480.
29. W. Huang, S. Ghosh *et al.*, "Hotspot: A compact thermal modeling methodology for early-stage VLSI design," *IEEE Transactions on Very Large Scale Integration (VLSI) Systems*, vol. 14, no. 5, pp. 501–513, 2006.
30. "Gurobi optimization," http://www.gurobi.com/ [Online].
31. W. Liu, J. Yi, M. Li, P. Chen, and L. Yang, "Energy-efficient application mapping and scheduling for lifetime guaranteed MPSoCs," *IEEE Transactions on Computer-Aided Design of Integrated Circuits and Systems*, vol. 38, no. 1, pp. 1–14, 2019.
32. L. Yang, W. Liu, P. Chen, N. Guan, and M. Li, "Task mapping on smart NoC: Contention matters, not the distance," in *IEEE Design Automation Conference*, 2017, pp. 1–6.

7 Exploring Aging Effects in Photonic Interconnects for High-Performance Manycore Architectures

Ishan G. Thakkar, Sudeep Pasricha,
Venkata Sai Praneeth Karempudi, and
Sai Vineel Reddy Chittamuru

CONTENTS

7.1 INTRODUCTION

To meet the inter-core communication demand of the state-of-the-art manycore chips, the use of electrical networks-on-chip (ENoCs) has become a norm [1]. However, with an ever-increasing core count, the performance and energy efficiency of such ENoCs are projected to scale poorly. Due to recent advances in silicon photonic interconnects technology, photonic NoCs (PNoCs) are being considered as a potential solution to overcome the drawbacks of traditional ENoCs [2]. PNoCs can provide several prolific advantages over traditional ENoCs, including near-light speed transfers, high bandwidth density, and low dynamic power dissipation [3]. These advantages of PNoCs have catalyzed research for their integration into future manycore systems.

DOI: 10.1201/9780429292033-7

The basic building blocks of PNoC architectures (e.g., [4–8]) are photonic links). Typically, a photonic link connects two or more nodes of the PNoC and employs multiple photonic waveguides and microring resonator (MR) devices as its key components. A photonic waveguide typically supports dense wavelength division multiplexing (DWDM) of a large number of wavelengths. Each of these wavelengths corresponds to a transmission channel used for data transfers. An MR is a compact and highly wavelength-selective device (i.e., a compact device with a narrow resonance passband), which incorporates a PN-junction in its silicon (Si) core and a microheater in its silicon dioxide (SiO_2) surroundings. The resonance wavelength of an MR can be adjusted by manipulating either the free-carrier concentration in the MR's core through voltage biasing of its PN-junction or the MR's local temperature through voltage biasing of its microheater. This resonance adjustive property of MRs renders them energy-efficient and promotes their use in PNoC architectures as modulators, receivers, and switches. The MR modulators (that are in resonance with the utilized wavelengths) at the source node modulate electrical signals (i.e., sequence of logical "1" and "0" voltage levels) onto the utilized wavelengths to convert them into photonic signals that travel through the waveguide. MR receivers at the destination node receive photonic signals from the waveguide and recover electrical signals. Moreover, MR switches can route the photonic signals in the PNoC. Thus, DWDM photonic waveguides and wavelength-selective MRs enable high-bandwidth parallel data transfers across the PNoC.

The application of voltage bias across an MR's PN-junction to adjust its resonance generates an electric field across the MR's Si (core) and SiO_2 (cladding) boundary. Similar to MOSFETs, this electric field generates voltage bias temperature-induced (VBTI) traps at the Si-SiO_2 boundary of the MR over time (i.e., VBTI aging) [9]. From [9], these VBTI aging-induced traps alter carrier concentration in the Si core of MRs, leading to redshifts in the MRs' resonance wavelengths and an increase in the widths of the MRs' resonance passbands (or decrease in the MRs' Q-factor). This VBTI aging-induced degradation in the MRs' resonance passbands increases photonic signal degradation and decreases energy efficiency in PNoCs that utilize these MRs. For example, over a span of 5 years, VBTI aging can increase the power penalty of signal degradation in PNoC architectures by up to 7.6 dB and energy-delay products by up to 26.8% [9]. A detailed analysis of VBTI aging in MRs and its impacts on PNoCs is presented in [9], *but no prior work to date has focused on mitigating the adverse impacts of VBTI aging in PNoC architectures.*

This chapter presents four pulse amplitude modulation (4-PAM) signaling as a means of proactively mitigating the adverse impacts of VBTI aging on the energy efficiency of PNoC architectures. The motivation of using 4-PAM signaling to counter the energy impacts of aging comes from the fact that 4-PAM signaling-based PNoC architectures can achieve significantly better energy efficiency compared to the traditional on-off keying (OOK) signaling based PNoC architectures [10]. This improved energy efficiency of 4-PAM-based PNoCs is utilized in this chapter to provide proactive guarding against the adverse impacts of VBTI aging.

The key contributions of this chapter are summarized below:

- We perform frequency-domain analysis of the photonic link-level impacts of VBTI aging. Our analysis shows that the energy impacts of VBTI aging are primarily caused by signal degradation due to aging-induced overlap of photonic signal and MR resonance spectra.
- In our frequency-domain analysis, we also show that the use of 4-PAM signaling instead of traditional OOK signaling can reduce signal degradation caused by aging-induced spectral effects.
- We analyze the system-level impacts of VBTI aging on two variants of the well-known CLOS PNoC architecture [4] and show that the 4-PAM signaling-based CLOS PNoC architecture yields better energy efficiency than the OOK signaling based CLOS PNoC after 5 years of aging.

7.2 VBTI AGING IN MICRORINGS (MRS)

To understand the mechanism of VBTI aging in MRs, consider Figure 7.1, which illustrates a typical MR structure with a PN-junction in its Si core that is cladded by SiO_2 surroundings. To impart high-speed and low-power resonance adjustive property to the MR, its PN-junction needs to be reversed (or negatively) biased [11], which is accomplished by applying higher voltage on the n side of the PN-junction (Figure 7.1). When a negative voltage is applied across the PN junction of the MR, an electric field "E" is generated from right to left across the Si-SiO_2 boundaries B1, B2, B3, and B4 (Figure 7.1). We used the Lumerical Solutions DEVICE [12] tool to construct and model the PN junction of an MR. For our preliminary analysis, we consider an MR waveguide similar to the one reported in [13] with a radius of 2 μm, fabricated using standard Si-SiO_2 material with a core cross-section of 450 nm × 250 nm. We simulated the MR using the charge transport solver in the DEVICE tool and then obtained the distribution of electric field across the MR waveguide with a bias voltage of –4 V. The results from the DEVICE tool demonstrate the presence of electric field E across all the Si-SiO_2 boundaries (i.e., B1, B2, B3, and B4). This electric field present across the Si-SiO_2 boundaries B2 (shown in Figure 7.2) and B4 attracts holes towards

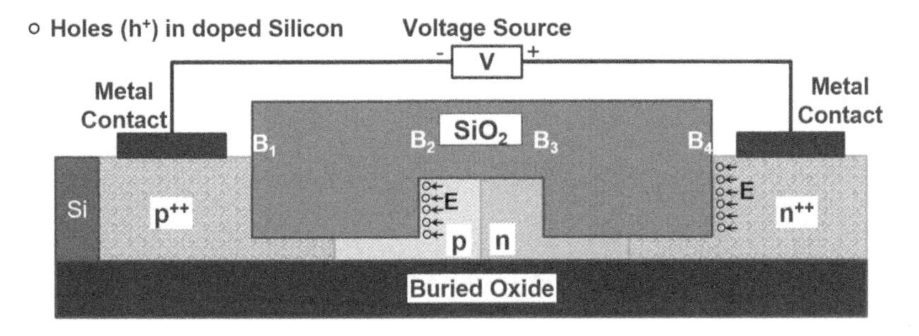

FIGURE 7.1 Cross-section of a tunable MR with PN junction in its core to facilitate carrier injection and depletion with voltage biasing.

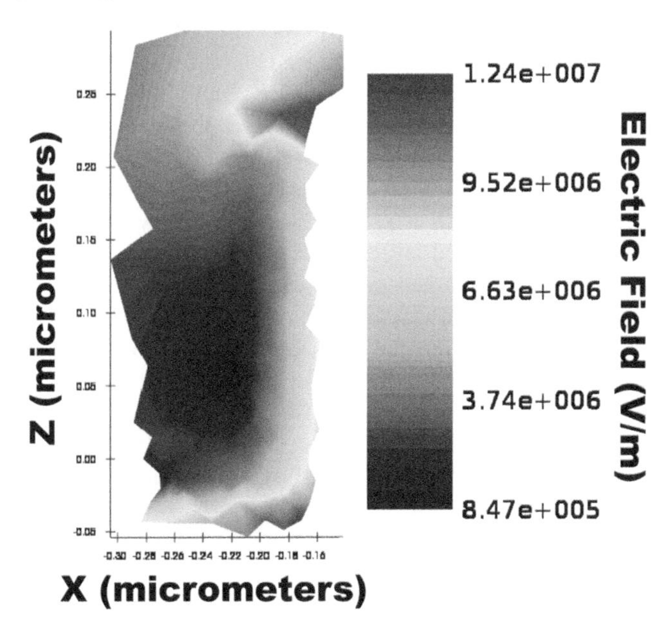

FIGURE 7.2 Distribution of electric field (E) across Si-SiO_2 boundary B2 when -4 V bias voltage is applied across PN junction.

them (Figure 7.1) and generates traps across these boundaries similar to pMOSFETs [14]. These voltage bias-induced traps on the B2 boundary change the electro-optic dynamics of the MR with time, which causes aging in the MR.

7.2.1 Analytical Models for VBTI Aging Mechanism

Several works in the literature (e.g., [14–16]) use reaction-diffusion (RD) models to characterize boundary trap generation at the MOSFET Si-SiO_2 boundary. As boundary traps in MRs are similar to boundary traps in MOSFETs, we use the same RD model to simulate the boundary trap generation at the MR's Si-SiO_2 boundary. This trap generation mechanism is represented as a chemical reaction in Eq. (7.1), where holes (h+) in the MR's Si core weaken a Si–H bond and hydrogen (H) is detached [15] in the presence of an electric field (E_B) and thermal variations (ΔT):

$$Si - H + h^{+} \overset{\Delta T,\ E_B}{\Leftrightarrow} Si* + H, \quad (7.1)$$

The generated Si dangling bond (Si*) acts as a donor-like boundary trap. The H ion released from the bond can diffuse away from the Si-SiO_2 boundary or anneal an existing trap. At any given time, the boundary trap density (N_{BT}) depends on the Si–H bond breaking rate k_F, Si–H bond annealing rate k_R, Si–H bond density available before stress (N_0), and the hydrogen (H) density at the MR's Si-SiO_2 boundary (N_H) through the following equations, the detailed description of which can be found in [9].

$$N_{BT} N_H = k_F N_0 / k_R, \quad (7.2)$$

$$k_F = B\sigma_0 E_{ox} e^{\frac{E_{ox}}{E_0}} e^{\frac{-E_F}{K_B T}}, \quad (7.3)$$

$$k_R = k_{R0} e^{\frac{-E_R}{K_B T}}, \quad (7.4)$$

In the above equations, E_F is the activation energy of forward bond dissociation, E_R is the activation energy of reverse bond annealing, K_B is the Boltzmann constant, E_{ox} is electric field strength across Si-SiO_2 boundary induced due to voltage biasing, $\exp(E_{ox}/E_0)$ is the field-dependent tunneling of holes into SiO_2 cladding, σ_0 is the capture cross-section of the Si–H bonds, T is operating temperature, and B determines field dependence of the Si–H bond dissociation. It is evident from these modeling equations that the rate of boundary trap generation and hence the boundary trap density N_{BT} depends on the electric field strength across the Si-SiO_2 boundary E_{ox} and operating temperature T. Because of this dependence, elevated levels of boundary electric field E_{ox} (or voltage bias) and temperature T accelerate boundary trap generation, which in turn accelerates MR aging.

Fabrication process and temperature variations cause drifts in an MR's resonance wavelengths, which must be rectified for the error-free and low-overhead operation of the MRs. From [17] and [18], to remedy these variation-induced resonance drifts in MRs, the use of elevated (or altered) levels of voltage bias and/or temperature with the MRs is inevitable. These elevated (or altered) levels of voltage bias and temperature are likely to accelerate boundary trap generation in the MRs. Therefore, MRs of a PNoC are likely to be inflicted by VBTI aging and its adverse impacts.

7.3 IMPACTS OF VBTI AGING

The VBTI aging-induced boundary traps change the electro-optic dynamics and resonance characteristics of MRs with time, which in turn deteriorates the energy-efficiency of PNoCs that utilize these MRs. In this section, we provide a detailed analysis of the impacts of VBTI aging.

7.3.1 Impacts of VBTI Aging on MRs' Resonance Characteristics

The VBTI aging-induced boundary traps in an MR ultimately affects the MR's resonance characteristics in two ways: (1) induce redshift in the MR's resonance, and (2) increase the width of the MR's resonance passband (or decrease the MR's Q-factor). With the onset of boundary trap generation, the concentration of holes in an aging-affected MR's core decreases, which increases the MR core's refractive index. However, the refractive index of the MR's SiO_2 surroundings remains relatively unaltered. Consequently, the refractive index contrast between the MR's core and surroundings, along with the MR's effective refractive index increases. The increase in the MR's effective refractive index induces redshift in its resonance passband. On the other hand, the increase in the refractive index contrast between the MR's core and surroundings increases optical scattering loss in the MR's cavity, which in turn decreases the MR's Q-factor (i.e., increases the MR's passband width).

Using the analytical models for VBTI aging and the design parameters of an example MR presented in [9], we evaluated the change in the example MR's resonance characteristics (redshift in resonance and Q-factor) with aging for different levels of bias voltage and operating temperature. For this evaluation, we adopted the initial resonance characteristic of the example MR from [9]. Figure 7.3 shows the variation in resonance wavelength redshift ($\Delta\lambda_{RWRS}$) and Q-factor (Q_A) with aging in MRs at different temperatures: 300 K, 350K, and 400 K. Moreover, Figure 7.4 shows $\Delta\lambda_{RWRS}$ and Q_A with aging in MRs at different levels of negative voltage bias: −2 V, −4 V, −6 V, and −8 V. From Figure 7.3 and Figure 7.4, it can be observed that at a particular temperature and bias level, with the increase in MR aging (i.e., increase in usage time) $\Delta\lambda_{RWRS}$ increases and Q_A decreases. Moreover, it can also be observed from Figure 7.3 and Figure 7.4 that the rates of change in $\Delta\lambda_{RWRS}$ and Q_A increase with the increase in temperature or voltage bias level. *Thus, it can be concluded from Figure 7.3 and Figure 7.4 that higher operating temperatures and voltage bias levels accelerate VBTI aging in MRs.*

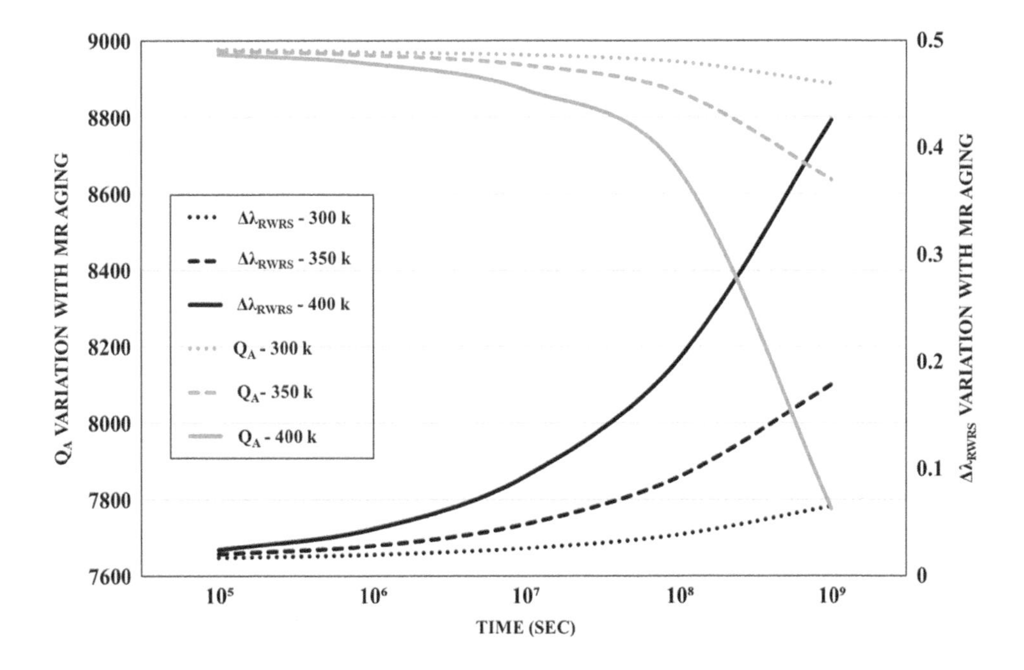

FIGURE 7.3 Variation of resonance wavelength red shift ($\Delta\lambda_{RWRS}$) and Q_A with time at three operating temperatures 300 K, 350 K, and 400 K.

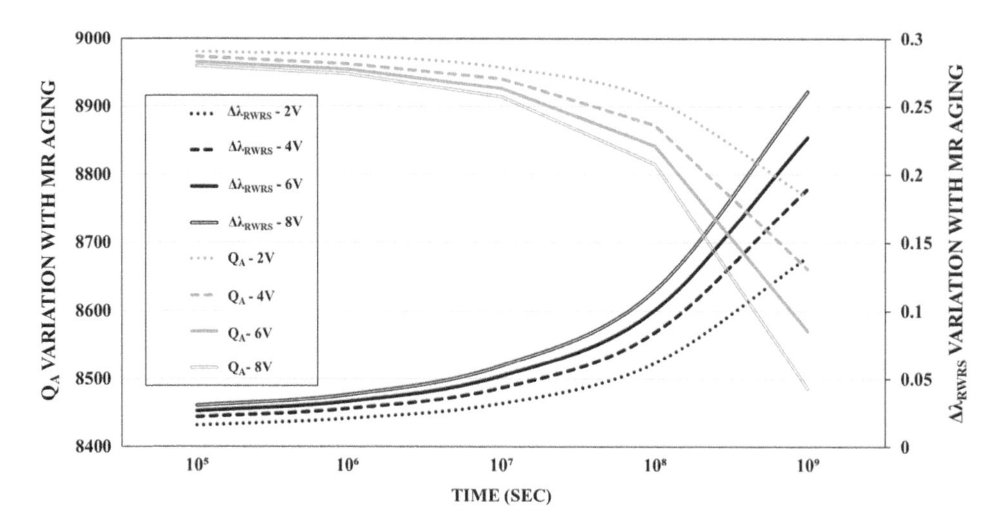

FIGURE 7.4 Variation of Q_A and resonance wavelength red shift ($\Delta\lambda_{RWRS}$) with operation time at four bias voltages -2 V, -4 V, -6 V, and -8 V.

7.3.2 Impacts of VBTI Aging on DWDM-Based OOK Links

A DWDM-based point-to-point on-off keying (OOK) link supports the parallel transfer of multiple OOK-modulated DWDM signals from a source node to a destination node through a photonic waveguide. The source node employs multiple modulator MRs to modulate the utilized DWDM wavelengths available in the waveguide, whereas the destination node employs multiple receiver MRs to receive the modulated DWDM signals from the waveguide. As VBTI aging primarily affects MRs' spectral (i.e., resonance) characteristics and MRs are primarily employed at the source and destination nodes of an OOK link, the impacts of VBTI aging on the link can be best explained using frequency domain (i.e., spectral) representation of the source and destination nodes.

7.3.2.1 VBTI Aging in Modulator MRs of Source Node

Consider Figure 7.5, which illustrates an example source node in the frequency domain before (Figure 7.5(a)) and after (Figure 7.5(b)) aging. From the figure, all four modulator MRs are represented in the frequency domain as Lorentzian frequency/wavelength passbands that are centered on the MRs' resonance frequencies/wavelengths. Similarly, unmodulated wavelength signals are represented as narrow-band frequency spectra that are identical to regularized Dirac-delta functions. From Figure 7.5(a), in an ideal scenario before aging, the unmodulated signals' spectra perfectly coincide with the modulator MRs' resonance wavelengths (i.e., centers of the MRs' resonance passbands). In this ideal case, the modulator MRs can modulate the utilized wavelength signals in OOK format efficiently. However, after aging, due to the induced redshifts in the modulator MRs' resonances and the increase in their passband widths, the signal spectra no longer perfectly coincide with the MRs' resonance wavelengths (Figure 7.5(b)). As a result, the modulator MRs cannot modulate the utilized wavelength signals efficiently. The inefficient modulation of the utilized wavelengths increases intermodulation crosstalk, which results in attenuated average spectral power for the modulated signals. A detailed explanation of and analytical models for the intermodulation crosstalk phenomenon can be found in [19].

7.3.2.2 VBTI Aging in Receiver MRs of Destination Node

Consider Figure 7.6 which illustrates an example destination node in the frequency domain before (Figure 7.6(a)) and after (Figure 7.6(b)) aging. From Figure 7.6(a), unlike the

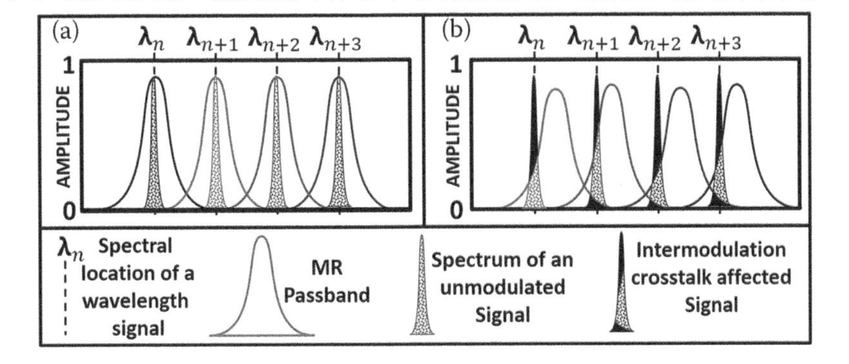

FIGURE 7.5 Illustration of an example source node in the frequency domain (a) before aging and (b) after aging.

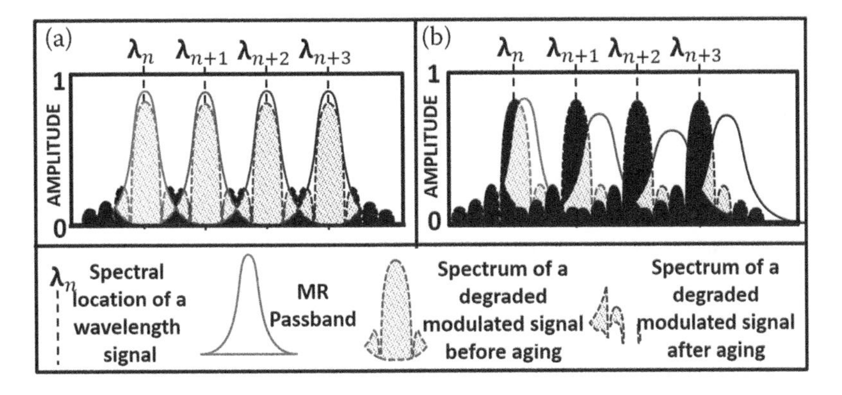

FIGURE 7.6 Illustration of an example destination node in the frequency domain (a) before aging and (b) after aging.

unmodulated signals' spectra at the source node, the modulated signals' spectra at the destination node have frequency side lobes, the widths of which depend on the modulation baud rate (i.e., number of level transitions in unit time). Ideally, at a destination node, the centers of the modulated signals' spectra coincide with the centers of the receiver MRs' passbands (Figure 7.6(a)). In spite of this ideal spectral alignment, the receiver MRs' passbands do not completely overlap with their respective signals' spectra as intended for efficient filtering and reception of the signals. This incomplete spectral overlap results in signal side lobe truncation [20]. Moreover, depending on the spacing between the adjacent wavelength channels, the receiver MRs' passbands partially overlap the neighboring non-resonant signals' spectra, which results in off-resonance filtering or heterodyne crosstalk [20]. The combined effect of the signal side lobe truncation and heterodyne crosstalk phenomena results in signal degradation and attenuation of average spectral power for the filtered/received photonic signals. Typically, while designing OOK links, the channel spacing and the receiver MRs' passband widths are optimized to trade-off signal truncation and crosstalk phenomena so that the effective signal degradation is minimized. But after aging, as shown in Figure 7.6(b), due to the induced redshifts in the modulator MRs' resonances and the increase in their passband widths, the effects of both signal truncation and crosstalk phenomena become more adverse, which further increases signal degradation. Detailed explanations and analytical models for the signal sidelobe truncation and heterodyne crosstalk phenomena can be found in [21].

Thus, VBTI aging in individual MRs of an OOK link produces three spectral effects: (1) intermodulation crosstalk at the source node, (2) signal side lobes truncation at the destination node, and (3) heterodyne crosstalk at the destination node. The ultimate impact of these three-fold spectral effects of VBTI aging is power attenuation of the utilized wavelength signals in the OOK link. This signal power attenuation needs to be compensated for, to achieve error-free link operation. The predominant way of doing this is to increase the input signal power. But increasing input signal power reduces the energy-efficiency of OOK links and PNoCs that utilize these links. *Hence, the end effect of VBTI aging in MRs is the degradation of energy efficiency in photonic links and PNoCs.*

7.4 MITIGATING THE IMPACTS OF VBTI AGING

As discussed in Section 7.2, we cannot avoid VBTI aging of MRs in PNoC architectures, as the causes (i.e., elevated levels of voltage bias and operating temperature) of VBTI aging in PNoC architectures are inevitable. Therefore, to resolve the problem of VBTI aging, the only alternative is to mitigate the adverse impacts of VBTI aging. As the end effect of VBTI aging is the degradation of energy efficiency in PNoCs, any technique or optimization that improves the energy efficiency of PNoCs can be used to remedy the adverse impacts of VBTI aging. These aging mitigative techniques and optimizations can be reactive or proactive in nature. In this section, we first discuss reactive mitigation techniques before presenting the four pulse amplitude modulation (4-PAM) signaling approach as a low-overhead proactive method of mitigating the impacts of VBTI aging.

7.4.1 Reactive Mitigation of VBTI Aging Impacts

Mitigation techniques or optimizations that are triggered upon the onset of VBTI aging are referred to as reactive mitigation techniques. Several techniques (e.g., [22–26]) presented in prior work can be used to reactively counter the following two device-level effects of VBTI aging on MRs' resonance characteristics: (1) redshift in MRs' resonances and (2) broadening of MRs' resonance passbands. In the remainder of this section, we discuss how to go about mitigating these two device-level effects using reactive mitigation techniques.

The VBTI aging-induced redshifts in MR resonances are similar to the redshifts induced by variations (i.e., fabrication process and temperature variations [32]) in MR resonances. Free-carrier injection aka localized trimming, which is a predominantly used technique to counter the variation-induced redshifts in MR resonances [17], can also be used to counter the VBTI aging-induced redshifts in MR resonances. Localized trimming induces blue shifts in MRs' resonances, which can counter the aging-induced resonance redshifts and bring back the MRs' resonances in perfect alignment with the corresponding signals' spectra. This spectral realignment caused by the localized trimming technique eliminates signal side lobe truncation as one of the link-level spectral effects of VBTI aging. However, from [22], localized trimming can cause additional broadening of the MRs' passbands, which in turn exacerbates the intermodulation crosstalk and heterodyne crosstalk effects of VBTI aging [25]. Nevertheless, dealing with two exacerbated link-level spectral effects is still more manageable than dealing with three link-level spectral effects of VBTI aging. Therefore, *we promote the use of localized trimming to counter the VBTI aging-induced redshifts in MR resonances.*

On the other hand, to counter the intermodulation crosstalk and heterodyne crosstalk effects caused by the VBTI aging and localized trimming induced broadening of MRs' resonance passbands, any of the crosstalk mitigation techniques proposed in prior work (e.g., [22–25]) can be utilized. However, the reactive nature of all of these crosstalk mitigation techniques from prior work incur significant performance and/or area overhead. For example, the data-encoding-based heterodyne crosstalk mitigation technique presented in [22] can incur a performance overhead of up to 20%. Moreover, the crosstalk mitigation technique presented in [25] can incur an area overhead of up to 15%. The high overheads of these reactive crosstalk mitigation techniques make

(a)

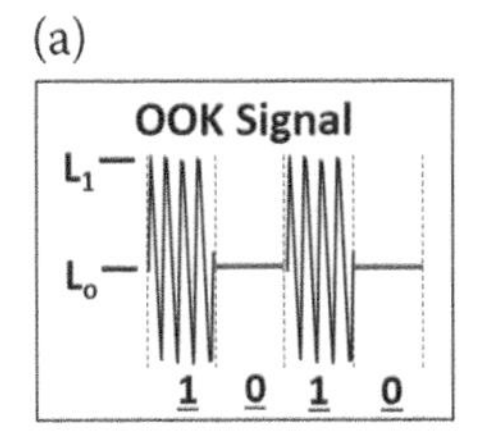

(b)

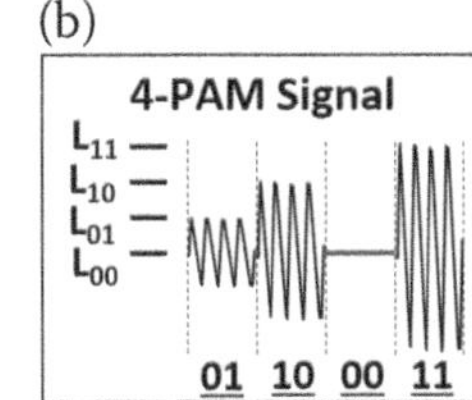

FIGURE 7.7 Illustration of the time-domain representation of (a) on-off-keying (OOK) signaling method and (b) four pulse amplitude modulation (4-PAM) signaling method.

them less preferable. *This motivates the need for low-overhead techniques that can proactively counter the intermodulation crosstalk and heterodyne crosstalk effects of the VBTI aging-induced broadening of MRs' resonance passbands.*

7.4.2 4-PAM Signaling: A Technique for Proactive Mitigation of VBTI Aging Impacts

In this section, we present the 4-PAM signaling method as a low-overhead technique that can proactively mitigate the intermodulation crosstalk and heterodyne crosstalk effects of VBTI aging. Figure 7.7 illustrates the time-domain representations of a traditional OOK signal and a 4-PAM signal. As evident from the figure, the OOK signal utilizes two optical transmission levels (i.e., L_0 and L_1) to represent one bit of information in one data symbol. On the other hand, the 4-PAM signal utilizes four optical transmission levels (i.e., L_{00}, L_{01}, L_{10}, L_{11}) to represent two bits of information in one data symbol. Thus, for a given signal baud-rate (i.e., the number of level transitions in unit time of a data signal), the 4-PAM signaling method achieves 2× bandwidth (or bit-rate) compared to the OOK method.

This ability to double the bandwidth without altering the baud-rate of data transmission allows the 4-PAM signaling method to reduce the number of DWDM wavelengths per photonic link by two times and still achieve the same aggregate link bandwidth as an OOK link. The ability to use two times fewer wavelengths allows for two times wider channel spacing between adjacent wavelength channels. This fact is evident from Figure 7.8, which shows the frequency-domain representations of the OOK and 4-PAM signaling-based destination nodes. From the figure, the 2× channel spacing and 2× bandwidth with unaltered signal sidelobe widths for 4-PAM signaling (i.e., signal sidelobe widths for OOK and 4-PAM signals of the same baud-rate are same) naturally minimize the heterodyne crosstalk effect, as the overlap of the MRs' passbands with the neighboring signals' spectra is reduced. Moreover, it also provides a proactive guard against the exacerbated crosstalk effects caused by VBTI aging-induced broadening of MRs' resonance passbands. As a result, 4-PAM-based PNoCs can undergo significant aging and still achieve better energy efficiency compared to OOK-based PNoCs with no aging. This fact is corroborated by our evaluation results presented in the next section.

7.5 EVALUATION

We analyze the adverse impacts of VBTI aging on two variants of the well-known CLOS PNoC architecture [4]. The OOK signaling-based variant is referred to as CLOS-OOK PNoC, whereas the 4-PAM signaling-based variant is referred to as CLOS-4PAM PNoC.

7.5.1 CLOS PNoC Architecture

We adopt the configuration of the CLOS PNoC from [9]. Each photonic link in the CLOS-OOK PNoC uses 64 DWDM wavelengths between 1525 nm and 1575 nm at 0.8 nm channel spacing, with 32 wavelengths for forward communication and the remaining 32 wavelengths for backward

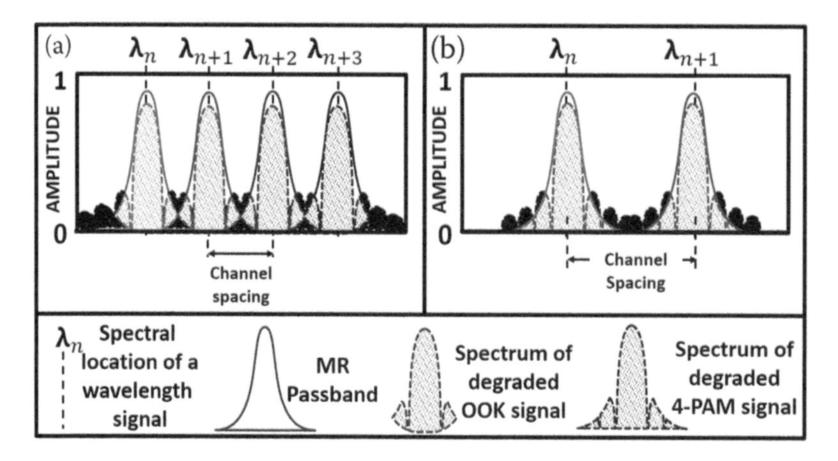

FIGURE 7.8 Illustration of (a) OOK-based and (b) 4-PAM-based destination nodes in the frequency domain.

communication. On the other hand, each link in the CLOS-4PAM PNoC uses 32 DWDM wavelengths at 1.6 nm channel spacing, with 16 wavelengths for forward communication and 16 wavelengths for backward communication. We use the signal power loss models presented in [9] to determine the worst-case power loss in the CLOS-OOK PNoC. For the CLOS-4PAM PNoC, we add the 4.8 dB power penalty related to 4-PAM signaling to the worst-case power loss obtained for the CLOS-OOK PNoC, similar to what is done in [10].

7.5.2 Evaluation Setup

We modeled and performed asimulation-based analysis of the CLOS-OOK and CLOS-4PAM PNoCs using a cycle-accurate NoC simulator, for a 256 core single-chip architecture at 22 nm. We generated 100 process variation (PV) maps to evaluate MR aging impact on these PNoCs for different PV profiles. We used real-world traffic from applications in the PARSEC benchmark suite [27]. GEM5 full-system simulation [28] of parallelized PARSEC applications was used to generate traces that were fed into our cycle-accurate NoC simulator. We set a ‘warm-up’ period of 100 million instructions and then captured traces for the subsequent 1 billion instructions. We performed geometric calculations for a 20 mm × 20 mm chip size, to determine lengths of photonic links in the CLOS-OOK and CLOS-4PAM PNoCs. We consider a 5 GHz clock frequency of operation for the cores. A 512-bit packet size is utilized for both CLOS PNoCs.

The static and dynamic energy consumption of electrical routers and concentrators in both CLOS PNoCs is based on results from the open-source DSENT tool [29]. We extended the photonic model of the DSENT tool before use to include the OOK and 4-PAM signaling methods, as done in [10]. For energy consumption of photonic devices, we use 0.42 pJ/bit for every modulation and detection event and 0.18 pJ/bit for the driver circuits of modulators and photodetectors. We used 15% laser wall-plug efficiency and –20 dBm OOK detector sensitivity to estimate the photonic laser power and correspondingly the electrical laser power from the worst-case signal power loss analysis described in the previous section. We considered the DAC-based implementation of 4-PAM signaling, as described in [10], for our CLOS-4PAM PNoC.

7.5.3 Modeling of Fabrication Process Variations in MRs

We adapt the VARIUS tool [30] to model die-to-die (D2D) as well as within-die (WID) process variations in MRs for the CLOS PNoC. We consider a 256-core chip with a die size of 400 mm^2 at a 22 nm process node. For the VARIUS tool, we use the parameters and procedures given in [23] and [25] to generate 100 process variation (PV) maps, each containing over 1 million points indicating the PV-induced resonance shift of MRs. The total number of points picked from these maps equals the number of MRs in the CLOS PNoC.

7.5.4 Modeling of Signal Degradation-Related Power Penalty

We adopt the VBTI aging models and initial parameters defining MR characteristics from [9] for our evaluation. We compensate for the VBTI aging-induced resonance redshifts by using the localized trimming method. To model the additional broadening of MR passbands caused by localized trimming, we use the models presented in [25]. Moreover, we adopt the equations from [19] to simulate the intermodulation crosstalk effect for both the CLOS-OOK and CLOS-4PAM PNoCs. Furthermore, we adopt the models presented in [21] to simulate the heterodyne crosstalk and signal side lobe truncation effects for the CLOS-OOK PNoC. We also extend these models from [21], to simulate the heterodyne crosstalk and signal side lobe truncation effects for the CLOS-4PAM PNoC, as done in [10]. Our simulations of the intermodulation crosstalk, heterodyne crosstalk, and signal side lobes truncation effects allows us to determine total signal power penalty, which we add to our worst-case signal power loss evaluation to consequently analyze the total power consumption and energy efficiency (in terms of energy-per-bit) of the CLOS-OOK and CLOS-4PAM PNoCs.

7.5.5 Evaluation Results

Our first set of evaluations compares the worst-case signal losses of the baseline CLOS-OOK and CLOS-4PAM PNoCs with their variants that undergo 1 year, 3 years, and 5 years of VBTI aging. We have performed this aging analysis across 100 PV maps as explained in Section 7.5.3, after the PV-induced and VBTI aging-induced resonance drifts are compensated for using localized trimming. The presented results are averaged across the PV maps. Furthermore, we performed this analysis at the highest possible on-chip temperature of 357 K [31].

Figure 7.9 compares the worst-case signal loss of the baseline CLOS-OOK and CLOS-4PAM PNoCs with three variants of these PNoCs that undergo 1 year, 3 years, and 5 years of VBTI aging. The confidence intervals represent the variation in signal loss across the 100 PV maps considered. From Figure 7.9, it can be observed that compared to their respective baselines, the CLOS-OOK PNoC with 1 year, 3 years, and 5 years of VBTI aging has 1.2 dB, 2.1 dB, and 2.7 dB higher signal

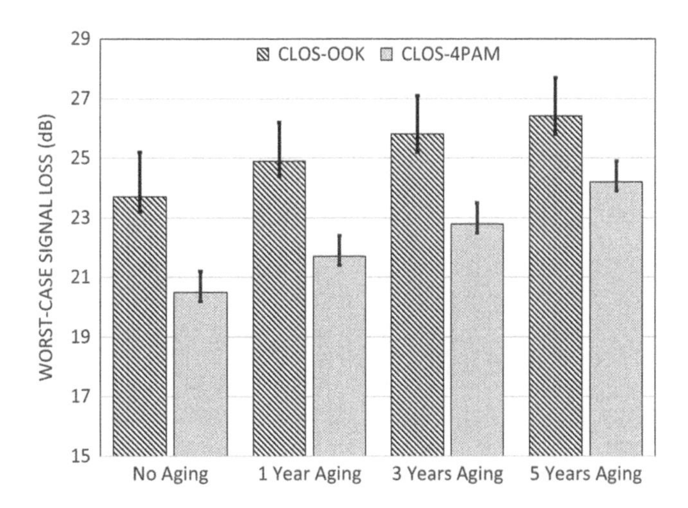

FIGURE 7.9 Worst-case signal power loss for CLOS-OOK and CLOS-4PAM PNoCs, with 1 year, 3 years, and 5 years of aging across 100 PV maps.

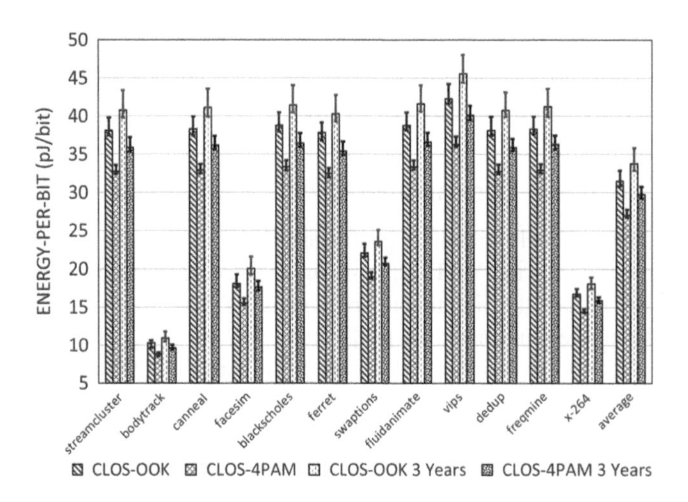

FIGURE 7.10 Energy-per-bit (EPB) comparison of the baseline CLOS-OOK and CLOS-4PAM PNoCs with their variants with 3 years of VBTI aging considering 100 PV maps across PARSEC benchmarks.

losses, and the CLOS-4PAM PNoC has 1.2 dB, 2.3 dB, and 3.7 dB higher signal losses. As explained in Section 7.3, the VBTI aging and use of localized trimming in MRs cause signal power attenuation due to the crosstalk effects, which in turn increases the worst-case signal power loss in the CLOS PNoCs. Moreover, it can also be observed that the CLOS-4PAM PNoC with 1 year, 3 years, and 5 years of VBTI aging has 3.2 dB, 3 dB, and 2.2 dB less signal losses compared to the CLOS-OOK PNoC with 1 year, 3 years, and 5 years of VBTI aging, respectively. The relaxed channel spacing for the CLOS-4PAM PNoC reduces the spectral effects of VBTI aging, which in turn results in less signal power loss for the CLOS-4PAM PNoC with aging than the CLOS-OOK PNoC.

Figure 7.10 presents energy-per-bit (EPB) results for the baseline CLOS-OOK and CLOS-4PAM PNoCs along with their variants with 3 years of VBTI aging. Results are shown for twelve multithreaded PARSEC benchmarks. It is evident that on average, CLOS-OOK PNoC with 3 years of VBTI aging has 7.5% and CLOS-4PAM PNoC with 3 years of VBTI aging has 10% higher EPB compared to their respective baselines with no aging. Similarly, Figure 7.11 presents EPB results for the baseline CLOS-OOK and CLOS-4PAM PNoCs along with their variants with 5 years of VBTI

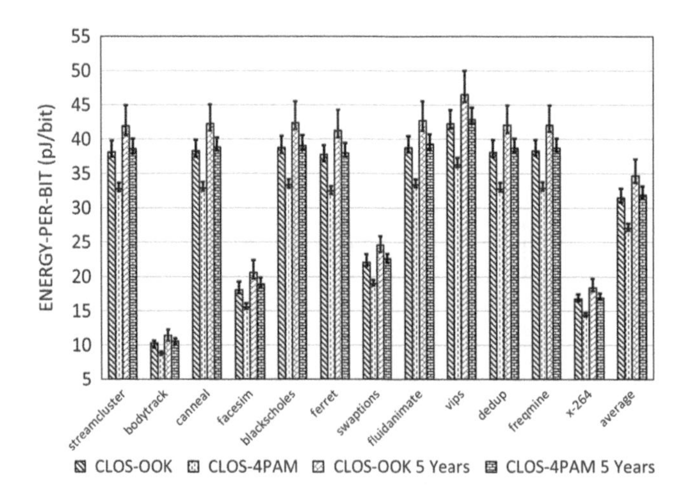

FIGURE 7.11 Energy-per-bit (EPB) comparison of the baseline CLOS-OOK and CLOS-4PAM PNoCs with their variants with 5 years of VBTI aging considering 100 PV maps across PARSEC benchmarks.

aging. It is evident that on average, CLOS-OOK PNoC with 5 years of VBTI aging has 10.3% and CLOS-4PAM PNoC with 5 years of VBTI aging has 18% higher EPB compared to their respective baselines with no aging. The increase in the signal power loss with VBTI aging increases total laser power consumption in PNoCs, which in turn increases the EPB values for PNoCs with aging.

Moreover, it can also be observed from Figure 7.9 and Figure 7.10 that the CLOS-4PAM PNoC with 3 years of VBTI aging has 0.9 dB less worst-case signal power loss and 5.5% less EPB compared to the baseline CLOS-OOK without aging. The relaxed channel spacing for 4-PAM signaling-based PNoCs provide proactive protection against the adverse spectral effects of VBTI aging. Therefore, 4-PAM-based PNoCs can undergo up to 3 years of aging and still achieve better energy efficiency compared to OOK-based PNoCs with no aging.

7.6 CONCLUSIONS

This chapter presents the frequency-domain analysis of the adverse impacts of VBTI aging on photonic links and PNoCs. Our analysis identifies three key spectral effects of VBTI aging at the link level, namely intermodulation crosstalk, heterodyne crosstalk, and signal side lobes truncation. These spectral effects cause signal degradation and can reduce the energy efficiency of PNoCs in terms of energy-per-bit by up to 10.3% over a span of 5 years. Moreover, our frequency-domain analysis shows that the use of 4-PAM signaling instead of traditional OOK signaling can proactively reduce signal degradation caused by aging-induced spectral effects. Our system-level evaluation results indicate that 4-PAM-based PNoCs can undergo 3 years of aging and still achieve 5.5% better energy efficiency compared to OOK-based PNoCs with no aging. Thus, these results corroborate the excellent capabilities of 4-PAM signaling-based PNoC architectures to proactively mitigate VBTI aging impacts.

REFERENCES

[1] W. J. Dally and B. Towles, "Route packets, not wires," in proceedings of IEEE/ACM Design Automation Conference (DAC), June, 2001.
[2] Y. A. Vlasov, "Silicon CMOS-integrated nanophotonics for computer and data communications beyond 100G," in IEEE Communications Magazine, vol. 50, no. 2, pp. 67–72, Feb. 2012.
[3] J. D. Owens, W. J. Dally, R. Ho, D. N. Jay, S. W. Keckler, and L.-S. Peh, "Research challenges for on-chip interconnection networks," in IEEE Micro, Sep.-Oct. 2007.
[4] A. Joshi, C. Batten, Y. J. Kwon, S. Beamer, I. Shamim, K. Asanovic, and V. Stojanovic, "Silicon-photonic clos networks for global on-chip communication," in proceedings of ACM/IEEE International Symposium on Networks-on-Chip (NOCS), 2009.
[5] S. V. R. Chittamuru, S. Desai, and S. Pasricha, "A reconfigurable silicon-photonic network with improved channel sharing for multicore architectures," in proceedings of ACM Great Lakes Symposium on VLSI (GLSVLSI), May 2015.
[6] S. Pasricha and S. Bahirat, "OPAL: A multi-layer hybrid photonic NoC for 3D ICs," in proceedings of IEEE/ACM Asia and South Pacific Design Automation Conference (ASPDAC), Jan. 2011.
[7] S. Bahirat and S. Pasricha, "METEOR: Hybrid photonic ring-mesh network-on-chip for multicore architectures," in ACM Journal of Emerging Technologies in Computing (JETC), vol. 13, no. 3, 2014.
[8] I. Thakkar, S. V. R. Chittamuru, and S. Pasricha, "A comparative analysis of front-end and back-end compatible silicon photonic on-chip interconnects," in proceedings of ACM/IEEE System Level Interconnects Predictions (SLIP) Workshop, June 2016.
[9] S. V. R. Chittamuru, I. Thakkar, and S. Pasricha, "Analyzing voltage bias and temperature induced aging effects in photonic interconnects for manycore computing," in proceedings of ACM/IEEE System Level Interconnects Predictions (SLIP) Workshop, June 2017.
[10] I. Thakkar, S. V. R. Chittamuru, and S. Pasricha, "Improving the reliability and energy-efficiency of high-bandwidth photonic NoC architectures with multilevel signaling," in proceedings of ACM/IEEE International Symposium on Networks-on-Chip (NOCS), Oct. 2017.
[11] P. Dong, S. Liao, D. Feng, H. Liang, D. Zheng, R. Shafiiha, C.-C. Kung, W. Qian, G. Li, X. Zheng, A. V. Krishnamoorthy, and M. Asghari, "Low Vpp, ultralow-energy, compact, high-speed silicon electro-optic modulator," in Optics Express, vol. 17, pp. 22484–22490, 2009.

[12] Lumerical Solutions Inc. – DEVICE toolkit. https://www.lumerical.com/tcad-products/device/
[13] K. Preston, N. Sherwood-Droz, J. S. Levy, and M. Lipson, "Performance guidelines for WDM interconnects based on silicon microring resonators," in proceedings of IEEE Conference on Lasers and Electro-Optics (CLEO), May 2011.
[14] M. A. Alam, H. Kufluoglu, D. Varghese, and S. Mahapatra, "A comprehensive model for PMOS NBTI degradation," in Microelectronics Reliability, vol. 45, pp. 71–81, 2005.
[15] H. Kufluoglu, "MOSFET degradation due to negative bias temperature instability (NBTI) and hot carrier injection (HCI) and its implications for reliability-aware VLSI design," PhD thesis, Purdue University, 2007.
[16] H. Kufluoglu and M. A. Alam, "Theory of interface-trap-induced NBTI degradation for reduced cross section MOSFETs," in IEEE Transactions on Electron Devices (TED), 2006.
[17] J. Ahn, M. Fiorentino, R. G. Beausoleil, N. Binkert, A. Davis, D. Fattal, N. P. Jouppi, M. McLaren, C. M. Santori, R. S. Schreiber, S. M. Spillane, D. Vantrease, and Q. Xu, "Devices and architectures for photonic chip-scale integration," in Applied Physics A: MSP, vol. 95, pp. 989–997, 2009.
[18] C. Nitta, M. Farrens, and V. Akella, "Addressing system-level trimming issues in on-chip nano-photonic networks," in proceedings of IEEE Symposium on High-Performance Computing Architecture (HPCA), April 2011.
[19] K. Padmaraju, X. Zhu, L. Chen, M. Lipson, and K. Bergman, "Intermodulation crosstalk characteristics of WDM silicon microring modulators," in IEEE Photonics Technology Letters (PTL), vol. 26, no. 14, 2014.
[20] M. Bahadori, D. Nikolova, S. Rumley, C. P. Chen, and K. Bergman, "Optimization of microring-based filters for dense WDM silicon photonic interconnects," in proceedings of IEEE Optical Interconnects (OI) Conference, April 2015.
[21] M. Bahadori, S. Rumley, H. Jayatilleka, K. Murray, N. A. F. Jaeger, L. Chrostowski, S. Shekhar, and K. Bergman, "Crosstalk penalty in microring-based silicon photonic interconnect systems," in IEEE Journal of Lightwave Technology (JLT), vol. 34, no. 17, pp. 4043–4052, 2016.
[22] S. V. R. Chittamuru, I. Thakkar, and S. Pasricha, "Process variation aware crosstalk mitigation for DWDM based photonic NoC architectures," in proceedings of IEEE International Symposium on Quality Electronic Design (ISQED), Mar. 2016.
[23] S. V. R. Chittamuru, I. Thakkar, and S. Pasricha, "PICO: Mitigating heterodyne crosstalk due to process variations and intermodulation effects in photonic NoCs," in proceedings of IEEE/ACM Design Automation Conference (DAC), June 2016.
[24] I. Thakkar, S. V. R. Chittamuru, and S. Pasricha, "Mitigation of homodyne crosstalk noise in silicon photonic NoC architectures with tunable decoupling," in proceedings of ACM/IEEE International Conference on Hardware/Software Codesign and System Synthesis (CODES+ISSS), Oct. 2016.
[25] S. V. R. Chittamuru, I. Thakkar, and S. Pasricha, "HYDRA: Heterodyne crosstalk mitigation with double microring resonators and data encoding for photonic NoCs," in IEEE Transactions on VLSI (TVLSI), vol. 26, no. 1, 2018.
[26] I. Thakkar, S. V. R. Chittamuru, and S. Pasricha, "Run-time laser power management in photonic NoCs with on-chip semiconductor optical amplifiers," in proceedings of ACM/IEEE International Symposium on Networks-on-Chip (NOCS), Oct. 2016.
[27] C. Bienia, S. Kumar, J. P. Singh, and K. Li, "The PARSEC Benchmark Suit: Characterization and Architectural Implications," in proceedings of International Conference on Parallel Architectures and Compilation Techniques (PACT), Oct. 2008.
[28] N. Binkert, B. Beckmann, G. Black, S. K. Reinhardt, A. Saidi, A. Basu et al., "The gem5 Simulator," in Computer Architecture News, May 2011.
[29] C. Sun, C.-H. O. Chen, G. Kurian, L. Wei, J. Miller, A. Agarwal, L.-S. Peh, and V. Stojanovic, "DSENT - A tool connecting emerging photonics with electronics for optoelectronic networks-on-chip modeling," in proceedings of ACM/IEEE International Symposium on Networks-on-Chip (NOCS), 2012.
[30] S. R. Sarangi, B. Greskamp, R. Teodorescu, J. Nakano, A. Tiwari, and J. Torrellas, "Varius: A model of process variation and resulting timing errors for microarchitects," in IEEE Transactions on Semiconductor Manufacturing (TSM), vol. 21, no. 1, 2008.
[31] M. Cho, C. Kersey, M. P. Gupta, N. Sathe, S. Kumar, S. Yalamanchili, and S. Mukhopadhyay, "Power multiplexing for thermal field management in many-core processors," in IEEE Transactions on Components, Packaging and Manufacturing Technology (TCPMT), vol. 3, no. 1, 2013.
[32] S. V. R. Chittamuru, I. Thakkar, and S. Pasricha, "LIBRA: Thermal and process variation aware reliability management in photonic networks-on-chip," in IEEE Transactions on Multi-Scale Computing Systems (TMSCS), 2018.

8 Improving Energy Efficiency in Silicon Photonic Networks-on-Chip with Approximation Techniques

Febin P. Sunny, Asif Mirza, Ishan Thakkar, Sudeep Pasricha, and Mahdi Nikdast

CONTENTS

8.1 INTRODUCTION

The continuous growth in data volumes consumed in emerging applications is causing an overall, rapid increase in energy consumption in computing systems. Ensuring fault-free computing for such large quantities of data is becoming difficult due to various reasons. One is the fact that the increasing resource demands for big data processing limit the resources available for traditional redundancy-based fault tolerance; another more fundamental problem is the ongoing scaling of semiconductor devices, which makes them increasingly sensitive to variations, e.g., due to imperfect fabrication processes. Approximate computing, which trades-off 'acceptable errors' during execution to reduce energy and runtime, is a promising solution to both these challenges [1]. Leveraging such aggressive techniques to achieve energy-efficiency is becoming increasingly important, with diminishing performance-per-watt gains from Dennard scaling.

To cope with the data processing needs of emerging applications, the core counts in manycore processors have also been rising. The increasing core counts, in response to increasing processing load, creates greater core-to-core and core-to-memory communication. Consequently, the traffic in the on-chip communication architecture fabric has been increasing to the point where today it costs more energy to retrieve and move data than to process it. Conventional electrical interconnects and electrical networks-on-chip (ENoCs) today dissipate very high power to support the high bandwidths and low latency requirements of data-driven parallel applications [2]. Fortunately, chip-scale silicon photonics has emerged in recent years as a very promising development to enhance

DOI: 10.1201/9780429292033-8

on-chip communication with light speed photonic links that can overcome the bottlenecks of slow and noise-prone conventional electrical links. Silicon photonics can enable photonic networks-on-chip (PNoCs) that can enable much higher bandwidths and lower latencies than ENoCs for on-chip communication [3].

Typical PNoC architectures employ several photonic devices such as photonic waveguides, couplers, splitters, and multi-wavelength laser sources, along with microring resonators (MRs) as modulators, detectors, and switches. A laser source (either off-chip or on-chip) generates light with one or more wavelengths, which is coupled by an optical coupler to an on-chip photonic waveguide. This waveguide guides the input optical power of potentially multiple wavelengths (referred to as wavelength-division-multiplexed (WDM) transmission), through a series of optical power splitters, to the individual nodes on the chip. Each wavelength serves as a carrier for a data signal. Typically, data signals are generated at a source node in the electrical domain as sequences of logical 1 and 0 voltage levels. These input electrical data signals are modulated onto the wavelengths using a bank of modulator MRs, using on-off keying (OOK) modulation. Once the data has been modulated on the wavelengths at the source node, it is routed over the PNoC to the destination node, where the wavelengths are coupled out of the waveguide by a bank of detector MRs, which drop the wavelengths of light onto photodetectors to recover the data in the electrical domain. Each node in the PNoC can communicate to multiple other nodes through such WDM-enabled photonic waveguides in the PNoC.

Unfortunately, light signals suffer various losses as they propagate through waveguides, requiring high laser power to compensate for such losses, so that the signal can be received at the destination with sufficient power to enable error-free recovery of the data. Power is also dissipated due to MR tuning at the source and destination MR banks, to ensure appropriate modulation and coupling of signals. Typically, the laser power requirement dominates overall power requirement of PNoCs. Novel solutions are therefore urgently needed to reduce this laser power, so that PNoCs can serve as a viable high-bandwidth and low-latency network in emerging and future manycores.

In this chapter, we explore how using data approximation can help reduce power and energy footprint of the laser power source in PNoCs.

8.2 RELATED WORK

By carefully relaxing the requirement for computational correctness, it has been shown that many applications can execute with a much lower energy consumption, without significantly hurting output quality. As an example, it is possible to approximate weights (e.g., from 32-bit floating point to 8-bit fixed point) in deep neural networks, with negligible changes in classification accuracy [4]. Many other approximation tolerant applications, beyond machine learning models, also exist, e.g., in the domains of video, image, and audio processing and big data analysis [5]. For such applications, approximation is an effective technique to improve energy efficiency.

Approximate computing approaches, proposed so far, can be broadly categorized into four types based on their scope [6]: hardware, storage, software, and systems. The approximation of hardware components allows a reduction in their complexity and thus a reduction in area and energy consumption [7] (e.g., using an approximate full adder that inexactly computes the least significant bits, compared to a conventional full adder). Storage approximation utilizes techniques, such as reduced refresh rates in DRAM [8], which results in a deterioration of stored data, but at the advantage of increased energy efficiency in memory units. Software approximation includes algorithmic approximation, which may leverage domain specific knowledge [9–11] or simplify the implementation [12]. It may also refer to approximating annotated data, variables, and high-level programming constructs (e.g., loop iterations), as specified by the software designer via annotations in the software program [13]. Approximation at the system level involves modification of architectures to support approximate operations. Attempts to create approximate NoC architectures to reduce the energy cost for communication at the system level (between processing cores and memories) would fall under this category.

Several efforts have attempted to approximate data transfers over electrical NoC architectures, by using strategies that reduce the number of bits or packets being transmitted, to reduce NoC utilization and thus reduce communication energy. An approximate NoC for GPUs was discussed in [12], where the authors proposed an approach for data approximation at the memory controller by coalescing packets with similar (but not necessarily the same) data, to reduce the packets that traverse over the reply network plane. A hardware data approximation framework with an online data error control mechanism for high performance NoCs was presented in [14]. The architecture facilitates approximate matching of data patterns, within a controllable value range, to compress them and thereby reducing the volume of data movement across the chip. A dual-voltage NoC is proposed in [15], where the lower priority bits in a packet are transferred at a lower voltage level, which may cause them to incur bit flips. The higher priority bits of the packet, including headers, are transmitted with higher voltage, ensuring a lower bit error rate (BER) for them. This approach allows a trade-off between errors introduced due to the low transmission voltage and the subsequent increase in the BER, with low-power consumption during transfers.

As for PNoCs, there have been various explorations on how to make them more energy efficient. Previously the explorations were based around how to save laser power, which is the largest contributor to PNoC power consumption, as mentioned earlier. The authors of [16] explored how to save runtime laser power by limiting the optical output to the bare minimum and then using semiconductor optical amplifiers (SOAs) to amplify the signals as needed. In [17] using other forms of optical signal modulation, other than OOK, was explored, as a way to increase energy efficiency of PNoCs while sustaining reliability needs. The authors of [17] used four-level pulse amplitude modulation (PAM4) as a substitute for OOK to achieve their goal. Methods to ensure energy-efficiency and reliability of PNoCs as the devices age due to usage and wear were explored in [18]. Many works have also explored how to ensure high reliability in PNoCs [19–22].

A recent paper [23] explored the use of approximate data communication on PNoCs for the first time. The authors explored different levels of laser power for transmission of bits across a single-writer-multiple-reader (SWMR) photonic waveguide, with a lower level of laser power used for bits which could be approximated, causing them to suffer higher BER. The work focused specifically on approximation of floating point data, which are known to be resilient to approximation compared to integer data. The least significant bits (LSBs) of the floating point data were subjected to lower laser power for transmission. However, the specific number of these bits to be transmitted as well as the laser power levels were decided in an application-independent manner, which ignores application-specific sensitivity to approximation. Moreover, the laser power is set statically, without considerations of varying loss that photonic signals encounter as they traverse through photonic waveguides.

The framework discussed in this chapter (called *LORAX*) overcomes the limitations of [23] by utilizing a novel loss-aware approach that adapts laser power at runtime to enable efficient approximate communication in PNoCs. We perform comprehensive analysis of the impact of adaptive approximation and laser power levels on application output quality, to enable an approach that can be tuned in an application-specific manner. The impact of discarding the conventional on-off keying photonic signaling approach in favor of a pulse amplitude modulation photonic signaling approach, on the energy savings achievable in PNoCs, was also additionally explored. To the best of our knowledge, this is the first work that considers loss-aware laser power management and multilevel signaling for approximation and energy-efficiency in PNoCs.

8.3 BACKGROUND: FLOATING POINT FORMAT

In many applications, floating point data is resilient to at least some level of approximation. The Least Significant Bits (LSBs) are considered for approximation in [23], as well as in this work.

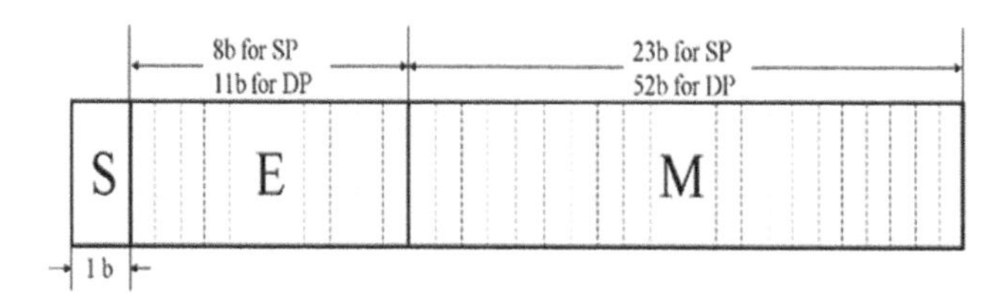

FIGURE 8.1 IEEE 754 floating point representation.

Most Significant Bits (MSBs) are avoided due to the unique data representation for floating point data as per the IEEE-754 standard, depicted in Figure 8.1 below.

The IEEE-754 standard defines a standardized floating point data representation with three parts: sign (*S*), exponent (*E*), and mantissa (*M*), as shown in Figure 8.1. The true value of the data stored is:

$$X = (-1)^S \times 2^{E-bias} \times (1 + M), \tag{8.1}$$

where *X* is the floating point value. The *bias* values are 127 and 1203 for single and double precision representation respectively, and are used to ensure that the exponent is always positive, thereby eliminating the need to store the exponent sign bit. The single precision (SP) and double precision (DP) representations vary in the number of bits allotted to the exponent and mantissa (Figure 8.1). *E* is 8 bits for SP and 11 bits for DP; while *M* is 23 bits for SP and 52 bits for DP. Also, *S* is 1 bit for both cases. From equation (8.1) we can observe how significant the *S* and *E* values are as they notably affect the value of *X*, but *M* is typically less sensitive to alterations in many cases, and it also takes up a significant portion of the floating point data representation. In accordance to this, [24] considered *S* and *E* as MSBs that should not be altered, whereas *M* makes up the LSBs that are more suitable for approximation to save energy during photonic transmission.

To establish how effective an approach that focuses on approximating floating point LSB data can be, we evaluated the breakdown of integer and floating point data usage across multiple applications. The ACCEPT benchmark suite [11], which consists of applications that have been shown to have a relatively strong potential for approximations was selected for this. The gem5 [25] system-level simulator was used to perform a benchmark characterization for this suite. The simulator was used to count the total number of integer and floating point packets in transit during the simulations. The breakdown of the float and integer packets across the applications for large input workloads is shown in Figure 8.2. The large input workloads were generated for applications such as *sobel* and *jpeg*, while for application from the PARSEC [26] suite, the large input workloads were selected from that suite.

From Figure 8.2, it is apparent that applications utilize varying number of floating point and integer data. To evaluate this proposed framework, we focused on five of these applications with notable and diverse floating point communication, while excluding *fluidanimate* and *x-264*, owing to their negligible floating point traffic. They also selected *jpeg* as a case study into the effects of approximation on low floating point traffic data.

8.4 LORAX FRAMEWORK: OVERVIEW

This section discusses the components of *LORAX* (LOss-awaRe ApproXimation) framework. Section 8.4.1 provides an overview of the loss-aware laser power management strategy. Section 8.4.2 discusses how integration of multilevel signaling can further enhance this approach.

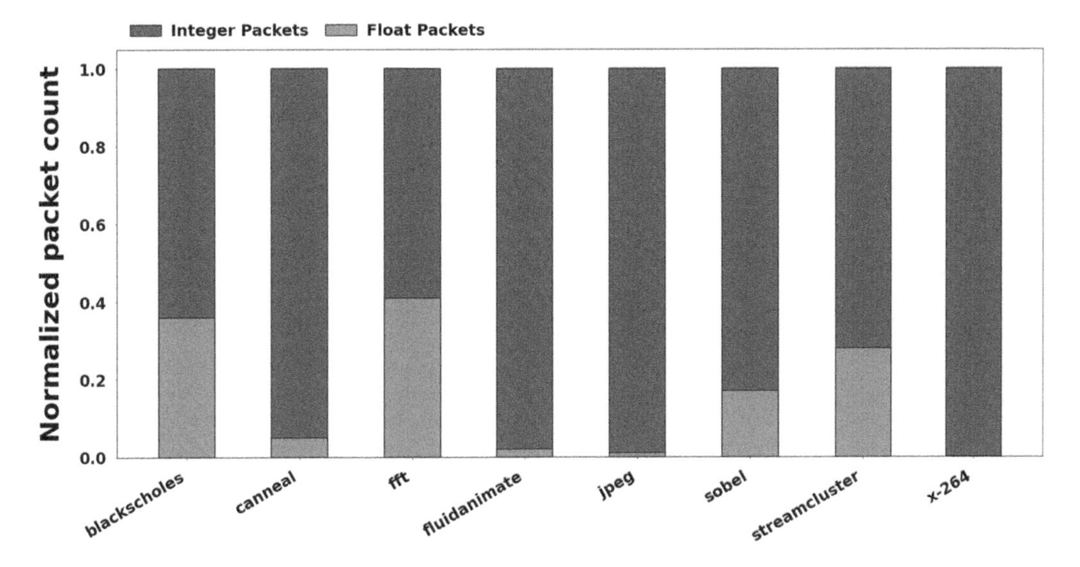

FIGURE 8.2 ACCEPT benchmark application characterization [24].

8.4.1 Loss-Aware Laser Power Management

The laser power required at a source node to transfer data on a WDM photonic waveguide (link) to a destination node is:

$$P_{laser} - S_{detector} \geq P_{phot\_loss} + 10 \times \log_{10} N_{\lambda} \tag{8.2}$$

where P_{laser} is the laser power in dBm, $S_{detector}$ is the MR detector sensitivity (e.g., -20 dBm [27]), and N_{λ} is the number of wavelength channels in the link. Also, P_{phot_loss} is the photonic loss incurred by the signal in its transmission, which includes propagation and bending losses in the waveguide, through losses in MR modulators and detectors, modulating losses in modulator MRs, and detection loss in detector MRs. P_{laser} thus depends on the link bandwidth in terms of N_{λ}, and the total loss P_{phot_loss} encountered by the photonic signals traversing the waveguide. The P_{phot_loss} encountered along the waveguide reduces the optical signal power, and the signal can only be accurately recovered at the destination node if the received signal power is higher than $S_{detector}$. Ensuring this requires a high enough P_{laser} to compensate for the losses.

To approximate data transmission for floating point data transfers, [23] used lower P_{laser} for transmitting LSBs (while keeping P_{laser} untouched for MSBs). However, if the destination node is relatively farther along a waveguide from a source node, the signals would encounter high losses and the signal intensity at the detector MRs would be less than $S_{detector}$, which would result in detecting logic "0" for all the LSB signals at the destination node. In the scenario where the destination is closer to the source, it may be possible to detect the LSB signals accurately, as long as the losses encountered are low enough that the signal power at the detector MRs would be higher than $S_{detector}$, even with the reduced P_{laser} for the LSBs.

We can make the following observation about the approach in [23]: for each communication on a waveguide, if we are aware of the distance of the destination from the source, it is possible to calculate the losses encountered for the signals. This calculated loss allows us to determine whether the signals can be recovered accurately, or if they will be detected as all "0"s. In such a scenario, it is more energy-efficient to simply truncate all the LSBs (i.e., reduce P_{laser} to 0 for LSB signals). Even when the destination is farther along the waveguide and there is no likelihood of the signal being recovered accurately [23] advocates for sending the LSB signals

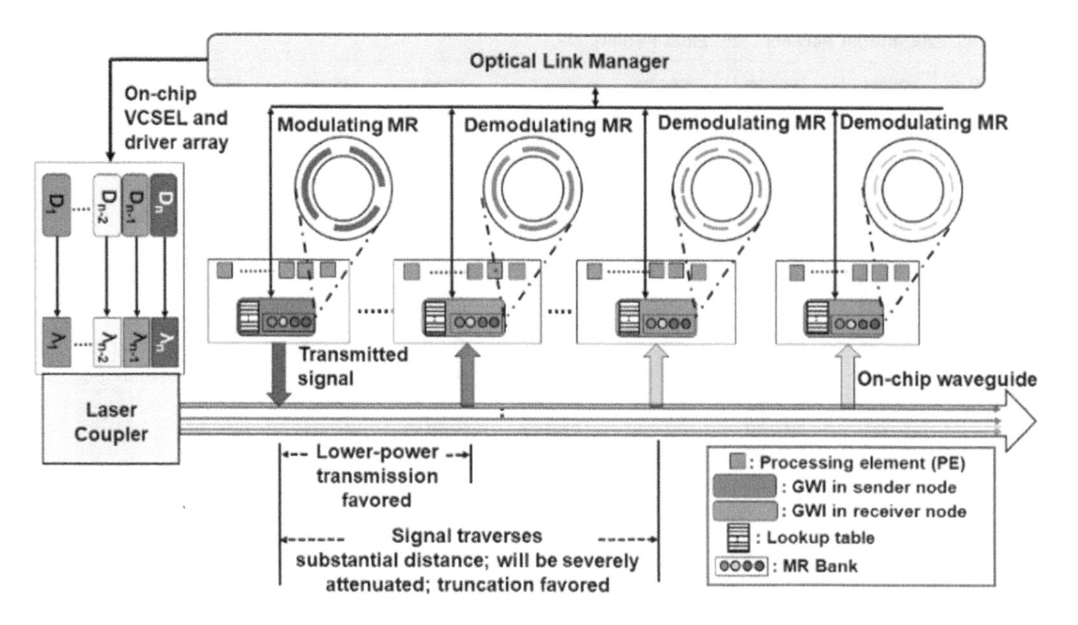

FIGURE 8.3 Overview of the proposed *LORAX* framework from [24].

at reduced P_{laser}, even if the signals cannot be recovered at the destination. In the cases where the destination is closer to the source, we can transmit the LSB signals with a lower P_{laser}, allowing some of the data be detected accurately at the destination, while approximating other data depending on its content and distance to the destination. Unlike [23] which reduces P_{laser} to a fixed value for a fixed subset of the LSB signals, irrespective of the application, we conjecture that it is important to tune the appropriate number of LSB signals and P_{laser} level in an application-specific manner. This is because the outputs for each application are sensitive to the LSB values in different ways, so a one-size-fits-all approach, as proposed in [23], may not make sense.

The proposal for the *LORAX* framework was motivated by the shortcomings in [23] and the observations discussed above. Figure 8.3 shows the operational details of this framework on a single writer multiple reader (SWMR) waveguide that is part of a PNoC architecture. Note that while the framework is illustrated with an SWMR waveguide, it is also applicable (with minimal changes) to multiple writer multiple reader (MWMR) and multiple writer single reader (MWSR) waveguides that are also used in many PNoCs. In the SWMR waveguide as shown in Figure 8.3, only one sender node is active per data transmission phase and one out of multiple (three in the figure) receiver nodes is the destination for the transmission. In a pre-transmission phase (called the receiver selection phase) the sender will notify the receivers about the destination for the upcoming data transmission, and only the destination node will activate its MR banks, whereas the other nodes will power down their MR banks to save power in the transmission phase. If the destination node is close to the sender node (e.g., the leftmost out of the three potential destination nodes in Figure 8.3), we can transmit the LSB signals with a lower P_{laser} as shown in Figure 8.4(b). Otherwise, if the destination node is farther away from the sender node (e.g., the second out of the three potential destination nodes shown in Figure 8.3), we determine that it would not be possible to detect the LSB signals at that destination due to the greater losses the signals will encounter. Considering this, *LORAX* dynamically turns off P_{laser}, essentially truncating the LSB bits, as shown in Figure 8.4(a).

To implement this framework, we require a laser control mechanism that can dynamically control the laser power injected into the on-chip waveguides. For this, an on-chip laser array with vertical-cavity surface-emitting lasers (VCSELs) [28], which can be directly controlled

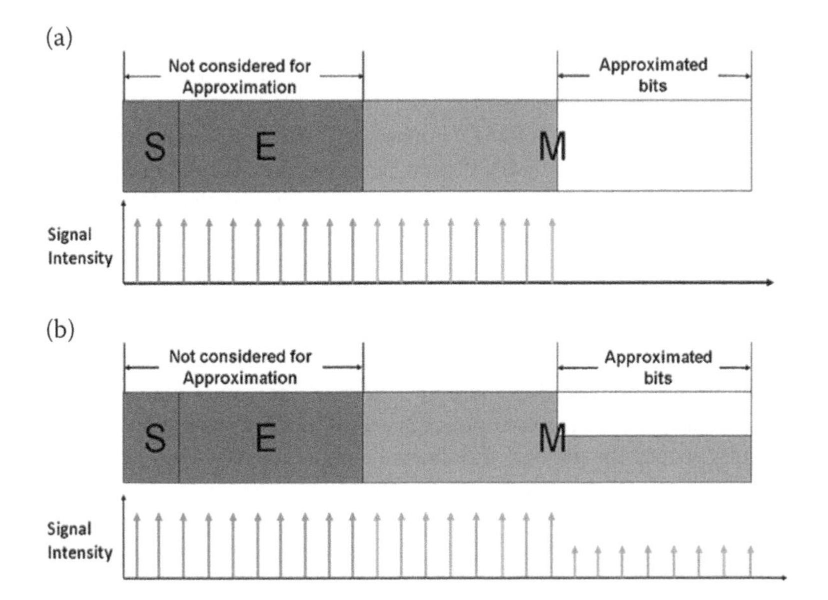

FIGURE 8.4 LSB signal: (a) truncation, (b) lower laser power [24].

using on-chip laser drivers, was utilized. With the laser drivers, we can control the power fed into each individual VCSEL, thus controlling the intensity of the laser output for a particular wavelength corresponding to that VCSEL. The Gateway Interface (GWI) that connects the electrical layer of the chip to the PNoC (Figure 8.3), communicates the desired P_{laser} intensity level (including 0 for truncation) to the drivers, via an optical link manager, similar in structure to the one proposed in [29].

The approach in *LORAX* also requires each source node to know when to switch between truncation and a lower P_{laser} level, and whether the packet contains approximable data or not. Identification of candidate packets to be approximated is done at the processing element level, via source code annotations, as in [13], to generate a flag for data (*e.g.,* floating point) that is approximable. This flag is inserted in the packet header. The GWI can then read the flag to determine if the packet is to be approximated. Then, whether the approximation is to be done via reduced power transmission or truncation is determined. This requires a lookup table at each GWI (Figure 8.3), with loss values to destinations from the source, with the IDs of the entire destination GWIs to which truncated transmission should be preferred. As the location of destination nodes as well as the cumulative loss to their GWI from the source does not change at runtime, these values can be calculated offline and used to populate the table. We discuss the overheads of the tables in Section 8.5.1. An application-specific P_{laser} for the LSB signals, discussed further in Section 8.5.2, can be used to determine if the signals can be detected at the given destination GWI, by consulting the loss value to that destination from the table, and then a decision can be made to either truncate or transmit the LSB bits. Once the decision to truncate or transmit at a lower laser power is made, the required intensity levels for the wavelengths are communicated to the VCSEL drivers via the optical link manager.

8.4.2 Integrating Multilevel Signaling for Approximation

The discussion in the previous section assumes the use of conventional on-off keying (OOK) signal modulation, where each photonic signal can have one of two power levels: high or on (when transmitting "1") and low or off (when transmitting "0"). In contrast, multilevel signaling is a signal modulation approach where more than two power levels of voltage are utilized to transmit multiple bits of data simultaneously in each photonic signal. Leveraging this technique in the photonic domain has, traditionally been a cumbersome process with high overheads, e.g., when using the signal superposition techniques from [30]. But with advances such as the introduction of Optical Digital to Analog Converter (ODAC) circuits [31] that are much more compact and faster than Mach-Zehnder Interferometers (MZIs) used in techniques involving superimposition [30], multilevel signaling has been shown to be more energy-efficient than OOK [17], making it a promising candidate for more aggressive energy savings in photonic links.

PAM4 is a multilevel signal modulation scheme where two extra levels of voltage (or photonic signal intensity) are added in between the 0 and 1 levels of OOK. This allows PAM4 to transmit 2 bits per modulation as opposed to 1 bit per modulation in OOK. This in turn increases the bandwidth when compared to OOK. While PAM4 promises better energy efficiency than OOK, it is prone to higher BER due to having multiple levels of the signal close to each other in the spectrum. Thus we cannot reduce the laser power level of the LSB bits to the level used in OOK, as it would significantly reduce the likelihood of accurate data recovery even when destination nodes are relatively close to the source. Therefore the reduced laser power level for PAM4 was kept at 1.5× that of OOK in [17]. This may seem like a backward step in conserving energy, but the reduced operational cost per modulation and the reduced wavelength count for achieving the same bandwidth as OOK, may reduce the overall laser power. The experimental results in the next section quantify the impact and trade-off of using PAM4 signaling with *LORAX* framework.

8.5 EXPERIMENTS

Here we describe the experimental setup (Section 8.5.1) adopted and the results (Section 8.5.2) obtained with our proposed framework.

8.5.1 Experimental Setup

To evaluate *LORAX*, the Clos PNoC architecture [32] was considered, with a baseline OOK signaling. The Clos PNoC (Figure 8.5) has an 8-ary 3-stage topology for a 64-core system with 8 clusters and 8 cores per cluster. Inter-cluster communication utilizes the photonic waveguides in the PNoC. Each cluster has two concentrators and a group of four cores connected to a concentrator, where the concentrators communicate with each other via an electrical router. The PNoC architecture was modeled and simulated using a SystemC based cycle-accurate simulator. The gem5 simulator was used for full system simulation, to generate traces for the entire application that were replayed on the PNoC simulator to determine energy savings in the PNoC. Then, details of the approximate data communication (i.e., whether a packet was truncated or transmitted at

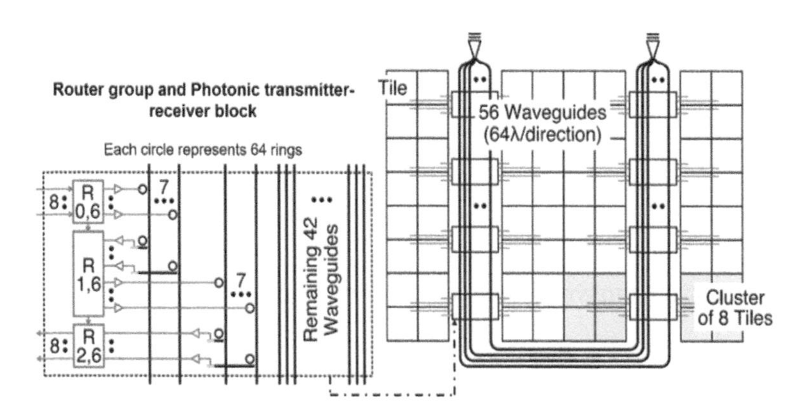

FIGURE 8.5 8-ary 3 stage Clos architecture with 64 cores [32].

lower power) were used to modify data in a subsequent gem5 simulation, to estimate the impact of the approximation on output quality for the application being considered.

Table 8.1 shows the gem5 architectural parameters considered for the platform used in experiments. As discussed earlier, six applications from the ACCEPT benchmark were used in evaluations. The performance was evaluated at the 22 nm CMOS node for a 400 mm^2 chip, with cores and routers operating at 5 GHz clock frequency. DSENT [33] was used to calculate the energy consumption by routers and the GWI at each node. CACTI [34] was used to evaluate the power and area for the lookup tables in the GWIs. These values were found to be: 0.105 mm^2 of area consumption for all tables, with a total power overhead of 0.06 mW. A single cycle latency overhead was considered for accessing the 64-entry table at 22 nm. We considered $N_\lambda = 64$ for OOK, which would enable 64 bit transmission across the waveguide per cycle. For PAM4, we only need to consider $N_\lambda = 32$ to achieve the same bandwidth as with OOK transmission.

Table 8.2 shows the energy values for losses and power dissipation in different photonic devices. These values were used to calculate laser power from (2) and total power after considering tuning and lookup table overheads.

An additional PAM4 induced signaling loss of 5.8 dB was considered in P_{phot_loss} for laser power calculations for PAM4. To compensate for the increased sensitivity of PAM4 to bit errors, laser power levels that are 1.5× than those used for OOK signaling was considered. Lastly, the output error incurred by the application due to an approximation approach can be calculated as:

$$Percentage\,(Output)\ Error = \frac{|approximate\ \ value - exact\ \ value|}{exact\ \ value} \times 100 \tag{8.3}$$

TABLE 8.1
64-core architecture configuration simulated using gem5 in [24]

Simulated Component	Specification
No. of cores, processor type	64, x86
DRAM	8 GB, DDR3
Memory controllers	8
L1 I/D cache, line size	128 KB each, direct mapped, 64 B
L2 cache, line size, coherence	2 MB, 2-way set associative, 64 B, MESI

TABLE 8.2
Loss and power values for photonic devices considered in [24]

Parameters Considered	Parameter Values
Detector sensitivity	–23.4 dBm [35]
MR through loss	0.02 dB [36]
MR drop loss	0.7 dB [37]
Waveguide propagation loss	0.25 dB/cm [38]
Waveguide bend loss	0.01 dB/90° [39]
Thermo-optic tuning	240 μW/nm [40]

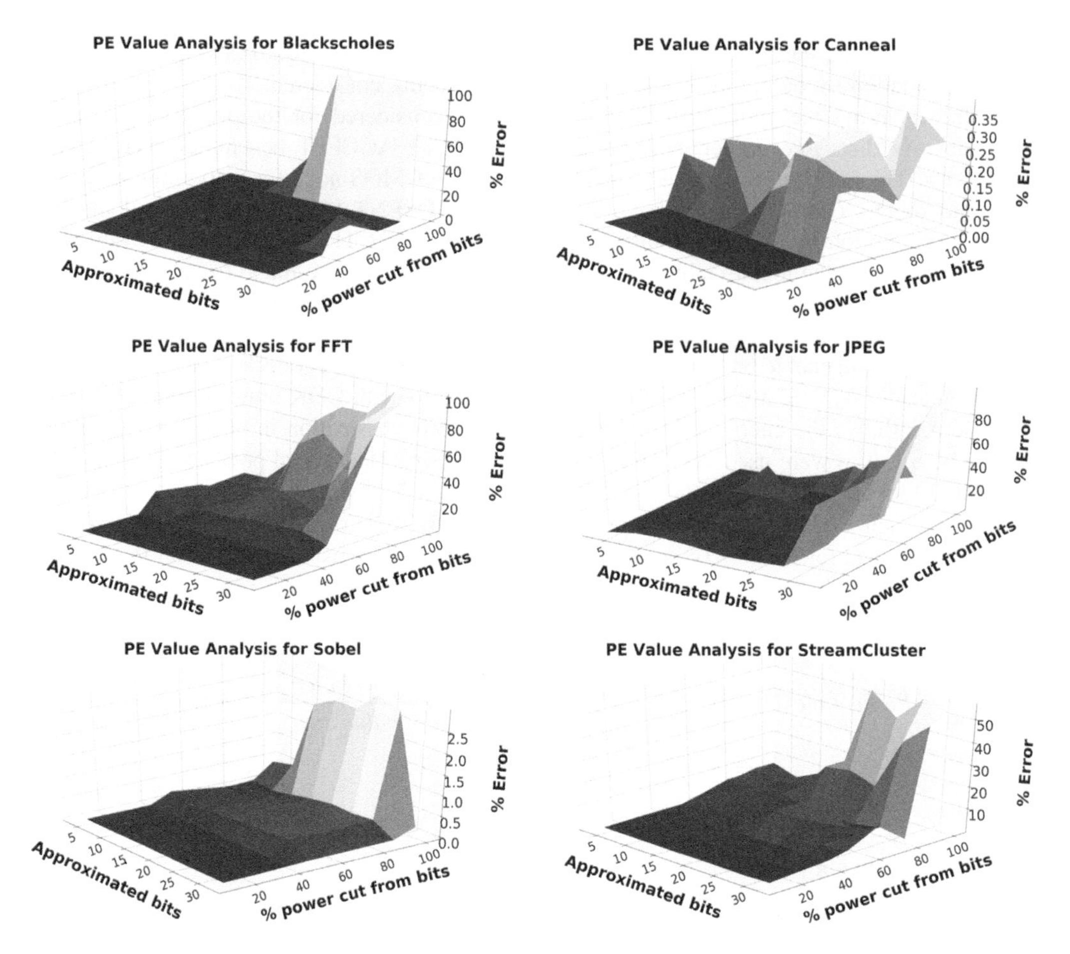

FIGURE 8.6 Percentage error (PE) in application output as a function of the number of approximated LSB signals and reduction in laser power for the LSB signals, for the blackscholes, canneal, fft, jpeg, sobel, and streamcluster benchmarks with large input workloads.

For analysis, an error threshold of 10% output error was assumed. This ensures that none of the approximation strategies degrade output quality by more than 10%.

8.5.2 Application-Specific Approximation Sensitivity Analysis

The first set of experiment involves analyzing the sensitivity of an application to varying degrees of approximation of their floating point data, to study the impact on output error of approximating a varying number of LSBs. Additionally, the impact on output error of varying levels of lowered laser power for the LSBs was studied. Figure 8.6 shows the results of this comprehensive study for the six benchmarks considered (see Figure 8.2). Each of the six surface plots presents insights into the behavior of the individual applications. The z-axis shows the percentage error (PE) in application output, as a function of the reduction in P_{laser} level for the photonic signals that carry the LSB bits (x-axis; varying from 0% to 100%, where 100% refers to truncation), and the number of LSBs that were considered for approximation (y-axis; with number of bits ranging from 4 to 32).

Table 8.3 summarizes the best combination of approximable bits (that are part of LSBs) and the laser power transmission levels for these bits, for each application, while ensuring that the application output error does not exceed 10% for the proposed framework (*LORAX*; rightmost two columns). In the next subsection, *LORAX* is compared with the framework from [23] and an

TABLE 8.3
Number of LSBs for approximation and laser transmission power level for LSB signals across benchmarks

Application Name	Truncation	[23]	LORAX	
	Truncated Bits		Approximated Bits	% Power Reduction
Blackscholes	12	16, with 20% power reduction	32	90
Canneal	32		32	100
FFT	8		32	50
JPEG	20		24	80
Sobel	32		32	100
Streamcluster	12		28	80

approach involving truncation. Table 8.3 also shows the number of bits that can be truncated, selected to meet the <10% PE constraint. For the approach in [23] we perform approximation on 16 LSBs transmitted at 20% laser power (advocated as an optimal choice in that work) which also satisfies the <10% PE constraint.

8.5.3 Comparative Results for Laser Power and EPB

The analysis from the previous subsection is used to determine the application-specific laser power intensity control in our framework. We compare the laser power and energy per bit (EPB) results for two variants of our framework: with OOK (*LORAX-OOK*) and with PAM4 (*LORAX-PAM4*). We compare our two framework variants with the framework from [23] and a truncation strategy that statically truncates a fixed number of bits, with the approximated LSBs and laser power levels for our *LORAX* frameworks chosen as discussed in the previous subsection.

Figure 8.7 shows the EPB and laser power comparison results for the various frameworks on the Clos PNoC architecture. Figure 8.7(a) shows that using *LORAX-OOK* results in lower EPB than [23] and the truncation approach. The truncation approach sometimes performs better than [23], as it avoids wasteful transmission at lower laser power when it is unlikely that the destination can recover the transmitted data due to high losses. But the lower number of truncated bits compared to approximated bits in [23] results in lower EPB for [23] in other cases. The *LORAX-OOK* framework improves upon both [23] and truncation, by adaptively switching between truncation and an application-specific laser power intensity level for LSBs. The *LORAX-PAM4* variant of our framework achieves the largest reduction in EPB, even though it uses higher power levels for the approximated bits. The use of fewer wavelengths in PAM4 allows for more energy savings, despite greater losses and the use of more laser power per wavelength than *LORAX-OOK*.

On average, *LORAX-PAM4* shows 13.01%, 12.16%, and 12.2% lower EPB compared to the baseline Clos, [23], and truncation approaches respectively. *LORAX-OOK* exhibits 2.5%, 1.9%, and 1% lower EPB on average compared to the same approaches. In the best-case scenarios for the *Blackscholes* and *FFT* applications, *LORAX-PAM4* has 13.7% and 13.5% lower EPB than the Clos baseline; and 12% and 12.2% lower EPB than [23], while against truncation it shows 12.45% and 12.4% lower EPB for these two applications.

Figure 8.7(b) specifically shows the laser power reduction. On average, *LORAX-PAM4* uses 34.17%, 30.1%, and 27.2% lower laser power compared to the baseline Clos, [23], and truncation approaches respectively, while *LORAX-OOK* exhibits 12.2%, 8.1%, and 7.8% lower average laser power consumption on average. For the best case *Blackscholes* and *FFT* applications laser power for *LORAX-PAM4* is 39.7% and 39.2% lower than the Clos baseline and 30.8% and 31.4% lower

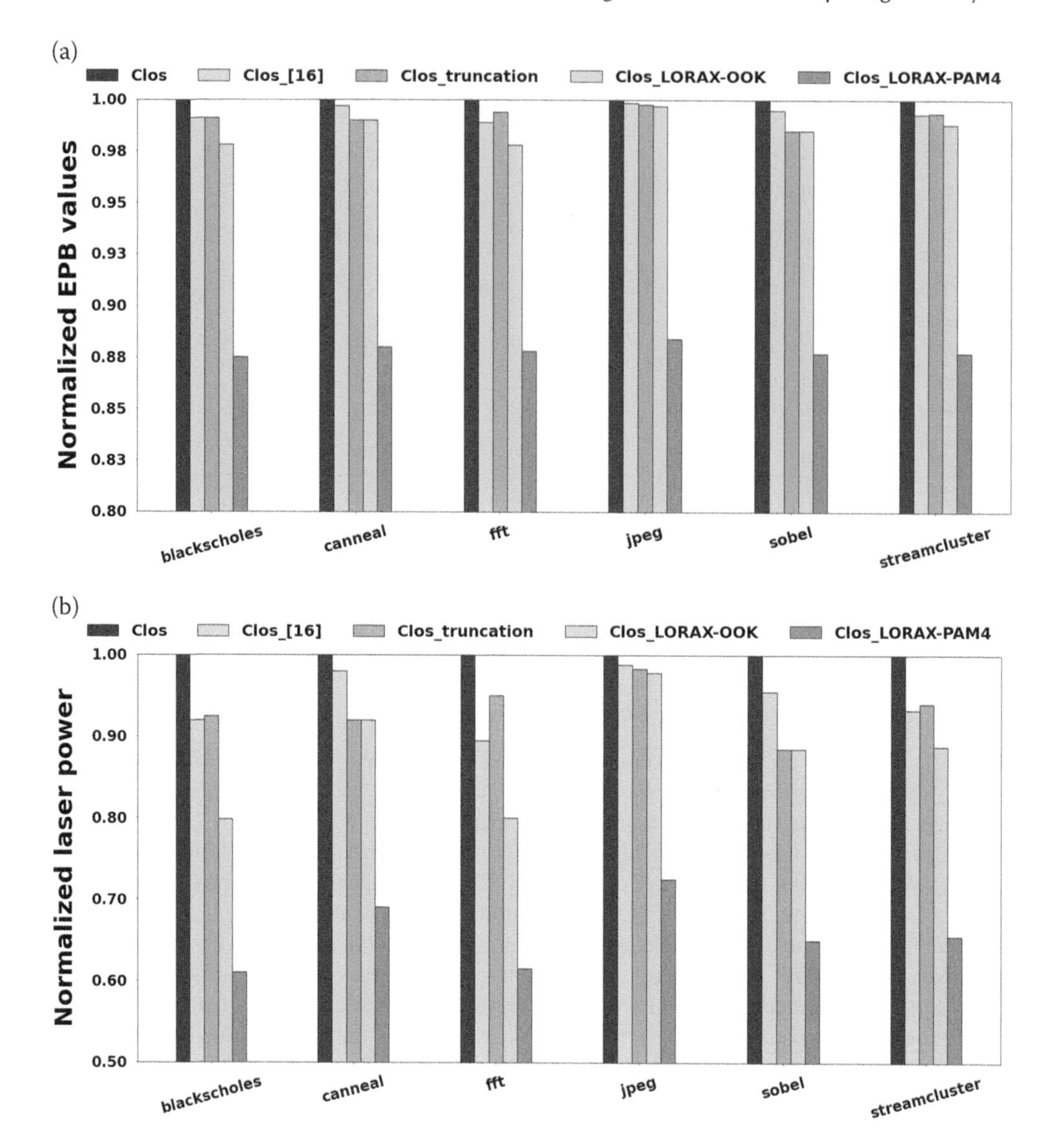

FIGURE 8.7 (a) Energy-per-bit (EPB) comparison across frameworks, (b) laser power comparison across frameworks [24].

than [23], while against truncation it is 32% and 33.6% lower. These results highlight the promise of our proposed *LORAX* framework, to trade off output correctness with energy efficiency and laser power savings in PNoC architectures executing selected applications.

8.6 CONCLUSIONS

In this chapter, we explored the *LORAX* framework, which used loss-aware approximation of floating point data communicated over PNoC architectures for reduction in energy utilization and power consumption. This exploration showcased how multilevel signaling can assist with the proposed approximation framework. The results indicate that utilizing multilevel signaling as part of our framework can reduce laser power consumption by up to 39.7% over a baseline PNoC architecture. The framework also shows up to 31.4% lower laser power and up to 12.2% better energy efficiency compared to the best known prior work on approximating communication in PNoCs. These results highlight the potential of using approximation strategies in PNoC architectures to reduce their energy footprint in emerging manycore platforms.

REFERENCES

[1] Q. Xu, T. Mytkowicz and N. S. Kim, "Approximate computing: A survey," in *IEEE D&T*, vol. 33, no. 1, 2016.
[2] J. D. Owens, W. J. Dally, R. Ho, D. N. Jayasimha, S. W. Keckler and L. Peh, "Research challenges for on-chip interconnection networks," in *Proc. IEEE Micro*, vol. 27, no. 5, 2007.
[3] T. Alexoudi, N. Terzenidis, S. Pitris, M. Moralis-Pegios, P. Maniotis, C. Vagionas, C. Mitsolidou, G. Mougias-Alexandris, G. T. Kanellos, A. Miliou, K. Vyrsokinos and N. Pleros, "Optics in computing: From photonic network-on-chip to chip-to-chip interconnects and disintegrated architectures," in *JLT*, 2019.
[4] D. T. Nguyen, H. Kim, H. Lee and I. Chang, "An approximate memory architecture for reduction of a reduction of refresh power consumption in deep learning applications," in *ISCAS*, 2018.
[5] F. Qiao, N. Zhou, Y. Chen and H. Yang, "Approximate computing in chrominance cache for image/video processing," in *IEEE ICMBD*, 2015.
[6] P. Yellu, N. Boskov, M. A. Kinsey and Q. Yu, "Security threats in approximate computing systems," in *GLVLSI*, 2019.
[7] J. Han and M. Orshansky, "Approximate computing: An emerging paradigm for energy-efficient design," in *IEEE ETS*, 2013.
[8] A. Raha, S. Sutar, H. Jayakumar and V. Raghunathan, "Quality configurable approximate DRAM," in *TC*, 2017.
[9] S. Venkataramani, V. K. Chippa, S. T. Chakradhar, K. Roy and A. Raghunathan, "Quality programmable vector processors for approximate computing," in *MICRO*, 2013.
[10] H. Esmaeilzadeh, A. Sampson, L. Ceze and D. Burger, "Neural acceleration for general purpose approximate programs," in *MICRO*, 2013.
[11] A. Sampson, A. Baixo, B. Ransford, T. Moreau, J. Yip, L. Ceze and M. Oskin, "ACCEPT: A programmer-guided compiler framework for practical approximate computing," white paper, University of Washington, 2014.
[12] Y. Raparti and S. Pasricha, "DAPPER: Data aware approximate NoC for GPGPU architectures," in *NOCS*, 2018.
[13] A. Sampson, W. Dietl, E. Fortuna, D. Gnanapragasam, L. Ceze and D. Grossman, "EnerJ: Approximate data types for safe and general low-power computation," in *PLD*, 2011.
[14] R. Boyapati, J. Huang, P. Majumder, K. H. Yum and E. J. Kim, "APPROX-NoC: A data approximation framework for network-on-chip architectures," in *ISCA*, 2017.
[15] A. B. Ahmed, D. Fujiki, H. Matsutani, M. Koibuchi and H. Amano, "AxNoC: Low-power approximate network-on-chips using critical-path isolation," in *NOCS*, 2018.
[16] I. Thakkar, S. V. R. Chittamuru, and S. Pasricha, "Run-time laser power management in photonic NoCs with on-chip semiconductor optical amplifiers," in *NOCS*, 2016.
[17] I. Thakkar, S. V. R. Chittamuru and S. Pasricha, "Improving the reliability and energy-efficiency of high-bandwidth photonic NoC architectures with multilevel signaling," in *NOCS*, 2017.
[18] I. Thakkar, S. V. R. Chittamuru and S. Pasricha, "Mitigating the energy impacts of VBTI aging in photonic networks-on-chip architectures with multilevel signaling," IEEE Workshop on Energy-efficient Networks of Computers (E2NC): From the Chip to the Cloud, Oct 2018.
[19] S. V. R. Chittamuru, I. Thakkar and S. Pasricha, "LIBRA: Thermal and process variation aware reliability management in photonic networks-on-chip," in *TMSCS*, vol. 4, no. 4, Oct-Dec 2018.
[20] S. V. R. Chittamuru, I. Thakkar and S. Pasricha, "HYDRA: Heterodyne crosstalk mitigation with double microring resonators and data encoding for photonic NoC," in *TVLSI*, vol. 26, no. 1, Jan 2018.
[21] S. Pasricha, S. V. R. Chittamuru and I. Thakkar, "Cross-layer thermal reliability management in photonic networks-on-chip," in *GLSVLSI*, 2018.
[22] S. V. R. Chittamuru, I. Thakkar and S. Pasricha, "PICO: Mitigating heterodyne crosstalk due to process variations and intermodulation effects in photonic NoCs," in *DAC*, 2016.

[23] J. Lee, C. Killian, S. L. Beux and D. Chillet, "Approximate nanophotonic interconnects," in *NOCS*, 2019.

[24] F. Sunny, A. Mirza, I. Thakkar, S. Pasricha and M. Nikdast, "LORAX: Loss aware approximations for silicon photonic networks-on-chip," in *GLSVLSI*, 2020, doi: 10.1145/3386263.3406919.

[25] N. Binkert, B. Beckmann, G. Black, S. K. Reinhardt, A. Saidi, A. Basu, J. Hestness, D. R. Hower, T. Krishna, S. Sardhashti, R. Sen, K. Sewell, M. Shoaib, N. Vaish, M. D. Hill and D. A. Wood, "The gem5 simulator," in *Comp. Arch. News*, 2011.

[26] C. Bienia, S. Kumar, J. P. Singh and K. Li, "The PARSEC Benchmark Suite: Characterization and architectural implications," in *PACT*, 2008.

[27] A. Biberman, K. Preston, G. Hendry, N. Sherwood-Droz, J. Chan, J. S. Levy, M. Lipson, and K. Bergman, "Photonic network-on-chip architectures using multilayer deposited silicon materials for high-performance chip multiprocessors," in *JETC*, 2011.

[28] H. Li, A. Fourmigue, S. Le Beux, X. Letartre, I. O'Connor and G. Nicolescu, "Thermal aware design method for VCSEL-based on-chip optical interconnect," in *DATE*, 2015.

[29] X. Wu, J. Xu, Y. Ye, Z. Wang, M. Nikdast and X. Wang, "SUOR: Sectioned unidirectional optical ring for chip multiprocessor," in *JETC*, vol. 10, no. 4, 2014.

[30] T. J. Kao and A. Louri, "Optical multilevel signaling for high bandwidth and power-efficient on-chip interconnects," in *PTL*, vol. 27, no. 19, 2015.

[31] A. Roshan-Zamir, B. Wang, S. Telaprolu, K. Yu, C. Li, M. A. Seyedi, M. Fiorentino, R. Beausoleil and S. Palermo, "A 40 Gb/s PAM4 silicon microring resonator modulator transmitter in 65nm CMOS," in *OIC*, 2016.

[32] A. Joshi, C. Batten, Y. Kwon, S. Beamer, I. Shamim, K. Asanovic and V. Stojanovic, "Silicon-photonic clos networks for global on-chip communication," in *NOCS*, 2009.

[33] C. Sun, C. O. Chen, G. Kurian, L. Wei, J. Miller, A. Agarwal, L. Peh and V. Stojanovic, "DSENT A tool connecting emerging photonics with electronics for opto-electronic networks-on-chip modeling," in *NOCS*, 2012.

[34] K. Chen, S. Li, N. Muralimanohar, J. H. Ahn, J. B. Brockman and N. P. Jouppi, "CACTI-3DD: Architecture-level modeling for 3D diestacked DRAM main memory," in *DATE*, 2012.

[35] H. T. Chen, J. Verbist, P. Verheyen, P. De Heyn, G. Lepage, J. De Coster, P. Absil, X. Yin, J. Bauwelinck, J. Van Campenhout and G. Roelkens, "High sensitivity 10Gb/s Si photonic receiver based on a low-voltage waveguide-coupled Ge avalanche photodetector," in *OE*, 2015.

[36] S. Bahirat and S. Pasricha, "OPAL: A multi-layer hybrid photonic NoC for 3D ICs," in *IEEE ASPDAC*, 2011.

[37] H. Jayatilleka, M. Caverley, N. A. F. Jaeger, S. Shekhar and L. Chrostowski, "Crosstalk limitations of microring-resonator based WDM demultiplexers on SOI," in *IEEE OIC*, 2015.

[38] http://www.aimphotonics.com/pdk.

[39] M. Bahadori, M. Nikdast, Q. Cheng and K. Bergman, "Universal design of waveguide bends in silicon-on-insulator photonics platform," in *JLT*, July 2019.

[40] C. Sun, M. T. Wade, Y. Lee, J. S. Orcutt, L. Alloatti, M. S. Georgas, A. S. Waterman, J. M. Shainline, R. R. Avizienis, S. Lin, B. R. Moss, R. Kumar, F. Pavanello, A. H. Atabaki, H. M. Cook, A. J. Ou, C. Leu, Y. Chen, K. Asanovic, R. J. Ram, M. A. Popovic and V. M. Stojanovic, "Single-chip microprocessor that communicates directly using light," in *Nature*, vol. 528, pp. 24–31, 2015.

Section III

Novel Design Solutions and Automation

9 Automated, Scalable Silicon Photonics Design and Verification

John Ferguson, Tom Daspit, Omar El-Sewefy, and Mohamed Youssef
Mentor, a Siemens business

CONTENTS

DOI: 10.1201/9780429292033-9

9.1 HISTORY/CONTEXT OF TRADITIONAL IC DESIGN AND FABLESS MODEL

Photonics technology has been around for decades, but in recent years, the incorporation of silicon photonics into integrated circuits (ICs) has gained traction in the semiconductor industry. Why? The historic electronic IC design paradigm faces increasing challenges in meeting the goals of faster signal processing, lower power consumption, and a smaller overall device footprint. Moving to smaller and smaller process nodes is becoming more difficult, and correspondingly takes longer and costs more. Photonics in general, and silicon photonics in particular, is one of several technologies that may help the industry overcome these challenges.

Still, while significant advancements in silicon photonic design have occurred in the course of the past decade, the semiconductor industry is only in the early stages of significant adoption. This positioning is not unlike the early days of the electronic design industry itself. In the early 1980s, ICs were designed primarily by engineers with PhDs and an advanced knowledge of electronic device physics, and typically manufactured at an on-site company fabrication facility (fab) by a similar set of engineers with advanced knowledge in materials processing. The focus was largely on who could create the best (fastest, lowest power, smallest) transistor. Designs consisting of dozens to maybe a few hundred transistors were laid out by hand. Each transistor in the design was characterized through time-consuming first principles simulation using technology computer-aided design (TCAD) tools. This kind of design was laborious, time-intensive, and often prone to errors and costly respins.

Today, the electronic IC industry designs and manufactures massive (in some cases, full reticle-sized) chips, containing billions of transistors and interconnect wires crossing dozens of metal layers, while keeping design to market times under control and avoiding costly manufacture and respin cycles. So, what changed? The short answer is the entire IC ecosystem experienced dramatic innovations in organizational structure and automation.

It began with the separation between IC design and manufacturing, which led to the emergence of independent manufacturing companies (foundries) and dedicated design companies (fabless companies). In 1970, all semiconductor companies owned their own fabs and controlled their IC production in-house. However, as with any manufacturing facility, there were times when the lines were idle. To keep production going, some semiconductor companies started selling their excess manufacturing capacity to other companies, and the model for "fabless" design companies was born. By the 1980s, the distinction between integrated device manufacturers (IDMs) and fabless design companies was firmly established. But more change was coming—in 1987, Taiwan Semiconductor Manufacturing Company (TSMC) was founded as the first "pure-play" foundry—a company that only manufactured ICs for other companies. With the arrival of pure-play foundries, the cost and risk of entering the semiconductor business was no longer prohibitive, and a veritable flood of fabless design companies joined the market [1].

9.2 PROCESS DESIGN KITS

Why did the cost and risk decrease so dramatically? In the foundry/fabless model, the foundries own the expertise on transistors and other key electronic components, along with the knowledge of the manufacturing process. To enable fabless designers to successfully design for a given process, there must be some standardized method for the foundry to communicate the requirements for a manufacturable design, and some method for the fabless design company to verify adherence to these design requirements. The foundry's process design kit (PDK) is the foundation that supports the foundry-fabless ecosystem. It not only enables design companies to produce designs that can be successfully manufactured, without requiring them to have in-depth expertise in the manufacturing process, but it also provides a foundation from which electronic design automation (EDA) companies can build tools to support automated design and verification processes. By minimizing manufacturing risk and providing the means for fast, automated, and standardized design and verification, the PDK made the "fabless revolution" possible, practical, and profitable.

9.3 ELECTRONIC PDKS

A PDK essentially represents an implied contract that, if companies create designs according to the rules and requirements of the PDK, they will produce a manufacturable, operable design in the target process. Every electronic PDK contains several key elements.

9.3.1 Layer Map

Design companies send designs to the foundry in the form of a digitized layout, typically in database formats such as GDSII or OASIS® [2]. In these formats, geometries are rendered on separately numbered layers. Each layer represents a different element, typically associated with a different manufacturing mask. To ensure the whole ecosystem works correctly, the foundry must define which layer numbers in an OASIS or GDSII database correspond to which mask layers. This information is known as the layer map.

9.3.2 Characterized Devices/Device Models and Symbols

From a device perspective, designers must know how devices will behave in their specific design environment. While descriptions of expected device behaviors can be included in the PDK, the fundamental building block for electrical behavior is delivered in the form of a device model, as a compact model in SPICE format. To enable early design and characterization, designers use schematic design or schematic capture tools to describe how various devices are connected electrically. The results of a schematic design are expressed in the form of a SPICE netlist that can be simulated by automated SPICE simulation tools. To enable this schematic capture, however, designers must have some way to differentiate one device type from another, and to distinguish the individual pins of each device. For this capability, the foundries typically supply device symbols along with their device models. Schematic capture tools allow designers to specify the physical parameters of an intended device that will drive the device model parameters during SPICE simulation.

9.3.3 Pre-characterized Cells

Once the design intent is adequately captured and verified in the schematic, designers create a programmatic layout representing the mask shapes that ultimately dictate the physical IC constructs. However, if designers had to focus on making sure every single transistor in the layout was assembled correctly, multi-billion transistor system-on-chips (SoCs) would not exist. Components that have the same function (e.g., a transistor), but which need to vary in size (length and width) throughout the layout can be replaced by pre-characterized parameterized cells (Pcells). Pcells enable designers to place multiple instances of different sizes that correspond to the varying device SPICE model parameters. Hierarchical processing and modeling enables these pre-characterized components to be placed repeatedly. Pcells are also used to create standard cells, as well as layout representations of intellectual property (IP).

9.3.4 Intellectual Property

As the industry expands designs from hundreds to potentially billions of transistors, compute requirements to design and simulate the entire system become overwhelming. Another level of abstraction is required. At this phase, the task is less about connecting individual devices, and more about connecting pre-characterized sets of devices in the form of hierarchical cells. Both foundries and 3rd-party IP providers can supply or sell pre-generated and validated cells known to work for a given process that can be easily inserted into a design. Typical examples of IP are standard cells and memory cell structures. Place and route (P&R) tools and specialized design tools (such as SRAM memory compilers) can automate the insertion of these cells to meet specified electrical parameters like timing or power, simplifying the layout process. However, these cells do require additional PDK elements, including the corresponding GDSII database, an abstraction layer (typically in the form of LEF files), corresponding timing files for each cell or block, and Verilog behavioral models.

9.3.5 Design Rules and Technology Files

For a given process, there are limitations on how well each mask layer can be effectively rendered and transferred to silicon. These limitations dictate what can or cannot be drawn into the layout, and take the form of design rules. While the design rule manual textually defines these rules for a given process, the PDK also provides design rule files, which can be read by automated design rule checking (DRC) EDA tools used by the design companies to verify design rule compliance prior to tape-out. P&R tools that automate the layout process must also understand these design rule constraints, but for these tools, the PDK provides tool-specific technology files.

These five elements are the basic components of a typical electronic PDK. Through these fundamental approaches, designers now have a way to capture their design intent (the schematic), and a method to create a physical layout.

9.4 PHYSICAL AND ELECTRICAL VERIFICATION

What comes next is the assurance that the physical layout not only correctly implements the schematic design, but also that it complies with all manufacturing restrictions (design rules). In this phase, designers verify that the design has been physically implemented in a layout that correctly executes the desired electrical operation, and that all physical constraints in the design rules have been met. Physical and electrical verification assures both the performance and manufacturability of the designed IC, and is achieved using a wide variety of automated checks incorporated into EDA verification tools. Physical and electrical verification of PDK requirements must be completed before a design can be taped out (sent to the foundry for manufacturing).

While design verification in its most basic form consists of DRC and layout vs. schematic (LVS) verification, more and more verification functionality has been added over the years. The primary categories of physical and electrical verification now include:

- Physical verification
  - DRC
  - LVS
  - Antenna checks
- Circuit verification
  - Parasitic extraction (PEX)
  - Electrical rule checking (ERC)
  - Reliability checking
- Design for manufacturability
  - Dummy fill insertion
  - Litho-friendly design (LFD) checking
  - Chemical mechanical polishing (CMP) modeling
  - Multi-patterning decomposition checking

Not all of these verification processes are mandatory for every design; their use depends on the design, the foundry, the process node, and the design company's internal requirements. DRC and LVS are the only two processes that are absolutely required by every foundry for every design and every process, while parasitic extraction is essential for ensuring accurate circuit behavior.

9.4.1 Design Rule Checking

DRC starts with geometrical rules defined by the foundry to ensure the physical design layout complies with their manufacturing requirements for a given process node technology. Because these rules are process-dependent, they vary layer to layer, node to node, and foundry to foundry. These geometric constraints ensure designs can be fabricated at or above a desired yield, while functioning properly and reliably. Design rules for PDKs are developed by foundry process engineers based on the capability of their manufacturing processes to realize design intent.

EDA DRC tools automate design rule verification by implementing the design rules as checks on specific constraints, such as the spacing between polygons, the width of a polygon, minimum sizing for enclosures, etc. Examples of typical DRC are shown in Figure 9.1.

As designs became smaller and more complex, new multi-dimensional physical requirements emerged that could not be resolved with basic DRC measurements. EDA companies responded by introducing equation-based DRC (eqDRC), in which measurements are performed and filtered based on custom equations. eqDRC allows more complex spacing and relational measurements to be implemented in DRC tools (Figure 9.2).

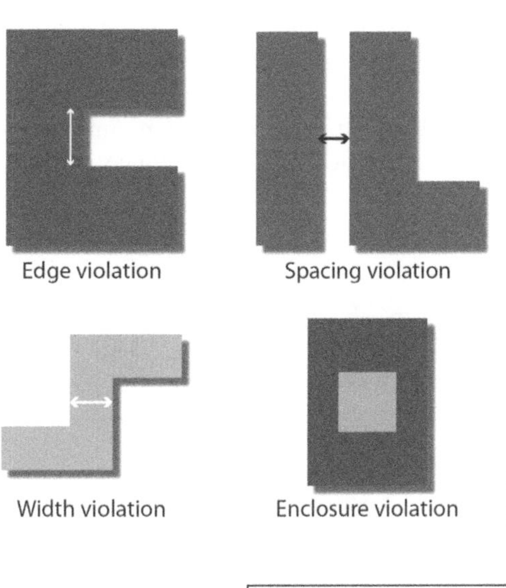

FIGURE 9.1 Automated DRC verifies compliance with physical configuration constraints.

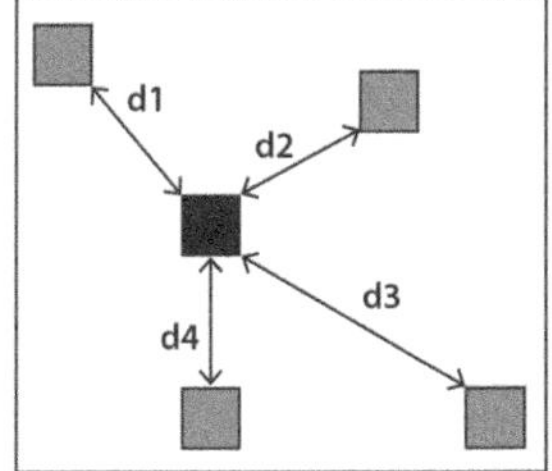

$$sensitivity = \frac{RR - \sum_{i=1}^{n} \frac{1}{d_i}}{RR - DR}$$

FIGURE 9.2 Example custom equation for complex DRC requirements.

9.4.2 Layout vs. Schematic Verification

LVS verification confirms that the physical layout correctly implements the schematic design. In its simplest form, LVS verification creates a SPICE netlist from the layout and compares it with the SPICE netlist created from the schematic. This comparison primarily ensures proper connectivity between the device elements, and validates the intended physical parameters. However, while LVS is used to ensure that the implemented layout matches the designers' intent exactly, it also ensures that no errors occurred during the transformation from schematic design to physical layout.

Automated LVS verification tools use PDK rule files to discern which layer combinations in a given layout represent device structures, and which represent interconnects. From this information, they extract a netlist representing the physical layout. This extracted netlist is then compared to the originally-validated schematic netlist representing the design intent. Figure 9.3 illustrates the LVS process.

9.4.3 Antenna Checks

Antenna effects refer to the potential (charge) acquired during manufacturing. This charge may build up, causing the transistor to be more susceptible to failure, either immediate or long term. Antenna problems can be avoided through careful adherence to design rules and the use of protective diodes, which otherwise serve no electrical function in the circuit.

9.4.4 Parasitic Extraction

LVS verification can be combined with PEX tools to go one step further, passing along physical parameters from the layout that cannot be known a priori during the schematic capture. Resistivity of metals, capacitive couplings between interconnects, inductive effects, and even stress impacts on device components can be captured and passed back into SPICE simulations for validation. PEX results can also be used to modify a physical layout to reduce parasitic effects and enhance IC reliability.

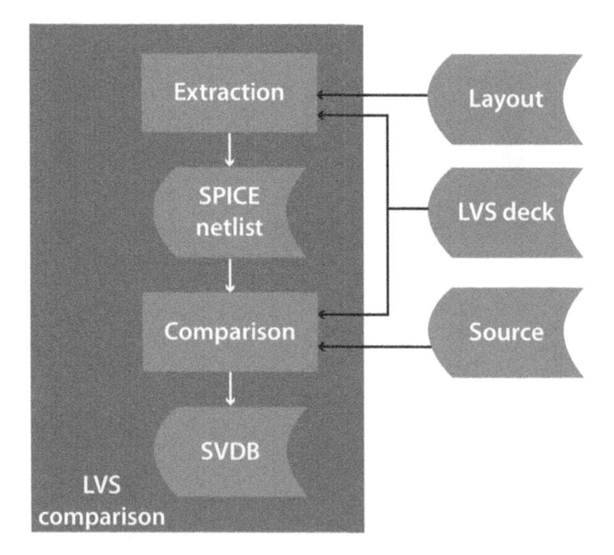

FIGURE 9.3 LVS verification compares the physical layout to the original schematic design to ensure the layout correctly implements the design intent.

9.4.5 Electrical Rule and Reliability Checking

ERC and advanced reliability checks are closely tied to LVS. Their goal is to ensure that the design layout not only represents the intended design, but also that the design and its layout representation follow best practices to ensure the device will operate correctly, not just on the testbench, but during repeated and continuous use in the consumer's hand. These checks detect susceptibility to operational issues like electrostatic discharge (ESD), which can cause a device to fail due to interaction with an outside source of static charge, or electromigration (EM), where copper wires physically change shape over time due to repeated use.

9.4.6 Design for Manufacturing

Even when layouts comply with all design rules, manufacturing processes can create flaws in the silicon. To mitigate the effects of various manufacturing processes on the final silicon, most design houses apply an additional set of DFM checks that help predict how a given design will react to the manufacturing process.

Historically, IC design processes, particularly at older nodes, assume that what is drawn is what will be delivered (mask to silicon). At advanced nodes, lithography can become an uncertain process, due to the finite size of the lens used (which does not capture all the mask diffraction order). LFD simulation is used to discover any areas in the layout that may cause resolution issues during the lithography process. Techniques such as optical proximity correction (OPC) can then be used to modify the layout before manufacturing to ensure the printed IC accurately represents the intended layout.

Fill consists of metal shapes or devices that serve no electrical purpose, and are not typically connected to any power source. During manufacturing, after each layer is created, the layer must be polished using CMP to remove excess materials and ensure a flat, uniform surface. Fill is often added to open areas in the layout to minimize the chance of peaks or valleys being created during CMP. CMP simulation can help designers determine optimum fill insertion.

Multi-patterning is a technique used at very advanced nodes, when the layout contains so many devices so closely spaced that they cannot be accurately reproduced with a single mask. The layout is split into two or more groupings, with each group representing a pattern with fewer elements spaced further apart. Multi-patterning rules are used to ensure the original layout can be properly split and recombined.

Over the years, PDKs and the automated design and verification flows that use them have been perfected for the IC industry, while they are still in the development phase for the silicon photonics industry. Foundries, EDA providers, and design houses are all working in conjunction to create a similar environment for silicon photonics designs.

9.5 A PDK FOR INTEGRATED SILICON PHOTONICS

The challenge to the industry is to replicate a similar PDK infrastructure in the silicon photonics world that enables designers to move from detailed engineering to a more automated domain, and support scaling similar to that achieved for IC designs. Ideally, we could just leverage the work already done in the electronics space. Of course, it's not that easy. There are a number of unique challenges associated with optical design in general, as well as the specific physical construction and manufacture of optical components. To create a silicon photonics PDK, we must understand the similarities and the differences between electronic and photonic design, verification, and manufacturing.

The requirements for creating a photonic design are simple. First, components must be placed, and then they must be interconnected. Sounds just like ICs, right? Except electronic ICs use a Manhattan geometry; circuits are designed in an orthogonal fashion, on a rectangular grid that only

FIGURE 9.4 Typical silicon photonics components.

allows for 0°, 45°, and 90° angles. In contrast, silicon photonics designs include a wide variety of curvilinear structures, such as delay lines, ring resonators, waveguides, grating couplers, etc. (Figure 9.4).

Special care must be taken when arbitrary curvilinear shapes are discretized to maintain a well-defined, smooth sidewall. This is true for the components themselves and for waveguide routing, which requires a curved format instead of rectilinear/Manhattan-style routing to ensure that light stays in the waveguide [3].

So, designers must now manage photons and curves instead of electrons and straight lines. What does that mean for silicon photonics? It means that tools and components essential to integrating photonics ICs (PICs) into the traditional IC design and verification processes must take into account these fundamental differences, starting with the photonics PDK.

The elements of a photonics PDK sound very similar to those of an electronic PDK: library symbols, Pcells, simulation models, and a physical verification rule deck. However, the maturity and availability of photonic PDKs is still an area of ongoing effort. While all foundries are transistor experts, most are far from being optical experts at this point in time.

9.5.1 Pre-characterized Devices

We are not yet at a point of having true pre-characterized photonics devices complete with a Pcell definition, but we are getting close. The ability to create such Pcells can be achieved through the use of Python™-based Pcells (Pycells), or by using tools like the PhoeniX OptoDesigner design platform or Luceda IPKISS.eda design framework [4–6].

EDA companies have stepped in to help both foundries and design companies define and implement modeling and characterization processes for photonics design and verification. For example, with an appropriate process model, which can be generated based on post-silicon measurements, engineers can predict how the drawn PIC layout will render through the manufacturing steps. EDA tools enable them to automatically capture the differences between the intended layout shapes and the final form of the manufactured shapes. In this way, foundries and design teams can characterize a device by generating layouts across multiple versions of possible physical parameters to determine how the differences will impact optical behavior. From this characterization, a better understanding of allowable combinations of allowable parameters will eventually enable the generation of certified/qualified Pcells for PICs.

9.5.2 Layout Implementation

Photons cannot turn at a 90° angle like an electron can: they require a gentle bend to change direction. In today's designs, in which photonics are limited to a single layer of optical interconnect, this restriction can make routing difficult. However, unlike the electrical interconnect on an IC, the optical interconnect (or waveguides) can cross each other without creating a short in the circuit. Then there are the required connections to the electrical circuitry—bond pads must be placed and their connections routed to the photonic component that they will control.

Existing silicon photonics design tools were built for the expert designers. These tools are based on a scripted custom layout methodology, in which the designer creates a script in a programming language (Python or C-like), and executes the script to create a photonic layout.

Using these tools, there are two ways to implement a photonics design.

The components can be described within the script, along with the design. The designer then executes the script with the design tool to create the layout. Alternatively, the designer places devices provided by a photonics PDK, and describes their interconnection in the script. With this second methodology, the first photonic component is placed at a set of coordinates in the layout. The next component is then placed at the next set of layout coordinates, and the designer defines the waveguide connection between these two components in the script. The designer continues to add to the script until all of the components are placed at their corresponding coordinates and all of the connecting waveguides are defined. The designer then executes the script in the design tool to create the layout.

In either approach, the resulting layout can be visualized in a layout viewer. If the photonics expert decides the photonic portion of the design is correct, the corresponding GDS database can be imported to an IC layout editor. The designer then places the corresponding electrical bond pads, and creates the routing from the bond pads to corresponding modulators and other photonic components as needed. At this point, the rest of the data needed for manufacturing (like alignment marks) are added.

One of the drawbacks of using the scripted custom layout methodology is that when a design rule violation is found, it may not be easy to understand which optical component or waveguide to move to fix that DRC violation. Not only that, but when fixing a violation, it is not uncommon to create a ripple effect in which the fix causes new violations to occur.

9.6 AUTOMATED PHOTONICS DESIGN METHODOLOGIES

As the optical component count increases, the scripted methodology becomes more and more limited, opening the door for a more EDA-like automated methodology. There are now multiple EDA-centric design methodologies that can support photonics: layout-centric design, schematic-based design, and automated implementation.

9.6.1 Layout-Centric Design

Layout-centric design is a “drag and drop” methodology, in which designers select optical components from the PDK component listing and place them into the layout view using an IC layout layout editor. They then connect multiple optical components with a waveguide. The waveguide can be edited to change the length or radius of the curve. When the design is complete, a netlist is generated to simulate the design in a photonic simulator. If the simulation matches the design intent, the design is ready for manufacturing finalization and physical verification.

9.6.2 Schematic-Based Design

As the design size increases, a schematic-based methodology can improve overall design productivity. This flow enables one engineer to create the design intent (captured in a schematic), while another engineer simultaneously creates the layout. A netlist is created from the schematic to verify that the optical portion of the design performs the required function. LVS verification confirms that the intent matches the physical implementation.

9.6.3 Schematic-Driven Layout

The next step up in productivity improvement is a schematic-driven layout flow. This flow uses the relative locations in the schematic to place both the photonic and electrical components, and connectivity information from the schematic is then used to verify LVS-clean routing.

9.6.4 Automated Layout

The highest level of productivity can be obtained by using layout automation to place the photonic components and their interconnections with curved waveguides. The corresponding electrical bond pads must be placed and their interconnection made using traditional IC multi-layer routing. There are several automation use models: scripted, schematic-driven layout, and test chip.

9.6.4.1 Scripted

The scripted use model is targeted to the expert photonic designers. All of the information needed to describe the design, as well as the instructions on how to implement that design, are contained in the script. The designer can choose from multiple routing engines for different parts of the design, enabling the designer to have complete control over the implemented layout.

9.6.4.2 Schematic-Driven

The schematic-driven design flow is targeted to the typical IC layout engineer. The designer does not have to be a photonic expert to quickly implement a design, because the schematic drives the layout implementation.

9.6.4.3 Test Chip

The test chip model is targeted to the PDK developer. The PDK developer needs to have multiple instances of the same component on a test chip, but each instance has a parameter with a different value.

In all three use models, the electrical portions of the design must be added to the layout by placing the associated bond pads and then routing them to their corresponding optical components.

9.7 SILICON PHOTONICS VERIFICATION

The design is now ready for physical verification. Many foundries now require silicon photonics designs to be DRC-clean before they accept the design for manufacturing. Figure 9.5 shows the types of verification required at each stage of the design flow.

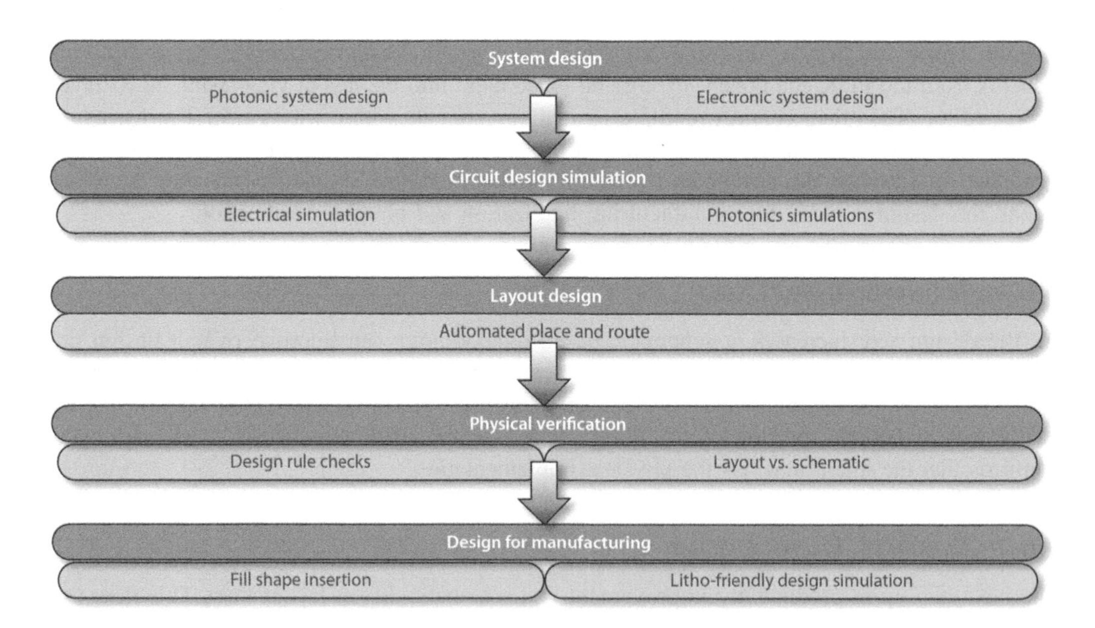

FIGURE 9.5 Design verification processes during the design flow.

9.8 PIC PHYSICAL LAYOUT VERIFICATION

In PICs, placing curvilinear structures on a rectangular grid presents a challenge for existing electronic IC verification tools and processes (Figure 9.6).

Precise measurement is problematic due to edge and vertex snapping, which can occur when the vertices of the curved shapes must adapt to the precision of piecewise linear approximation (Figure 9.7).

This effect must be compensated for during the geometrical parameter measurements. Extraction and careful validation of those non-traditional shapes requires unfamiliar parameters, such as bend curvature and curvilinear path length. Reconstructing or reinventing an entire PIC toolset and verification flow to fit such structures is unrealistic, given the time and resources that would be required. Alternatively, the EDA industry has developed new PIC verification techniques that can achieve the required degree of accuracy with modest modifications to existing electronic IC toolsets.

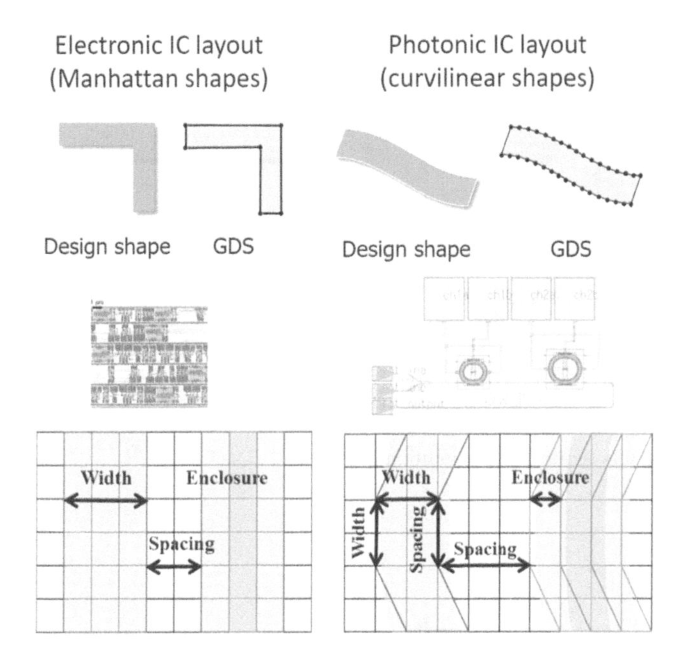

FIGURE 9.6 Physical verification challenges over curvilinear structures.

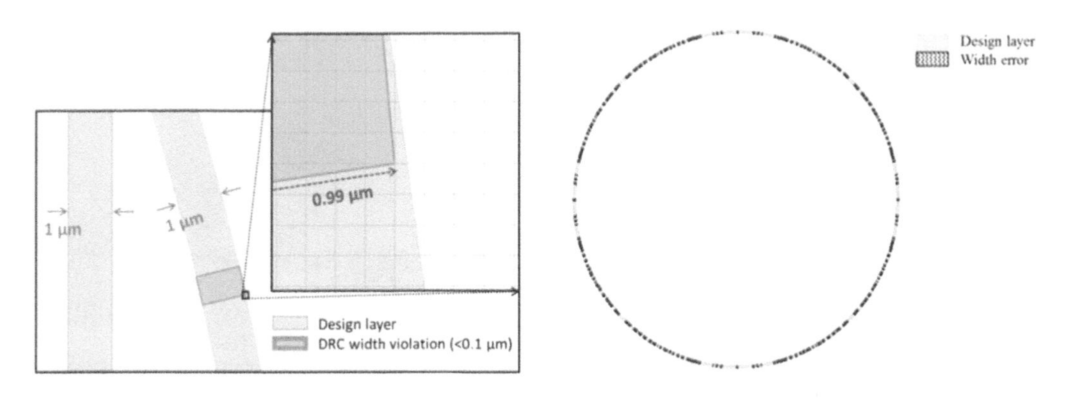

FIGURE 9.7 Snapping effect of curvilinear structure on a rectangular grid.

9.9 BEST PRACTICES FOR PHOTONICS DRC

By applying complex conditional eqDRC with multi-dimensional tolerance values in place of traditional DRC arithmetic calculations, designers can obtain a much more accurate analysis of a physical photonics structure. Without the use of such equation-based methods, physical verification generates many false errors, due primarily to edge snapping or rounding errors during measurements (Figure 9.8).

To filter out these false violations, equation-based DRC is added to traditional DRC to detect the curved segments of the design and apply the necessary tolerance factors to eliminate the false errors (Figure 9.9). The introduction of equation-based filtering and checking enables a whole new range of DRC capabilities for silicon photonics, where multi-dimensional equations can be evaluated to check the geometric validity of photonics designs.

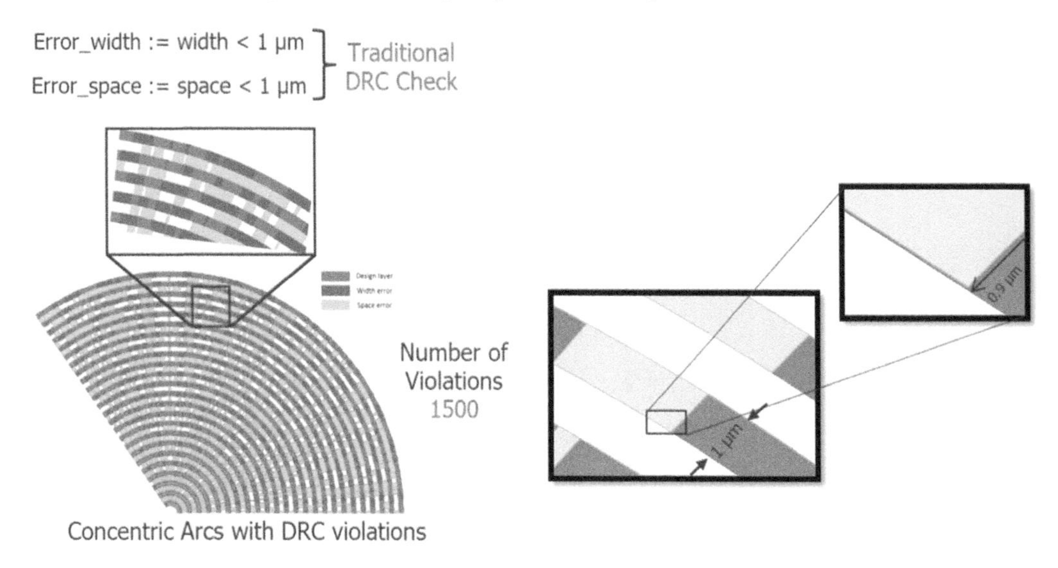

FIGURE 9.8 False errors generated by traditional DRC.

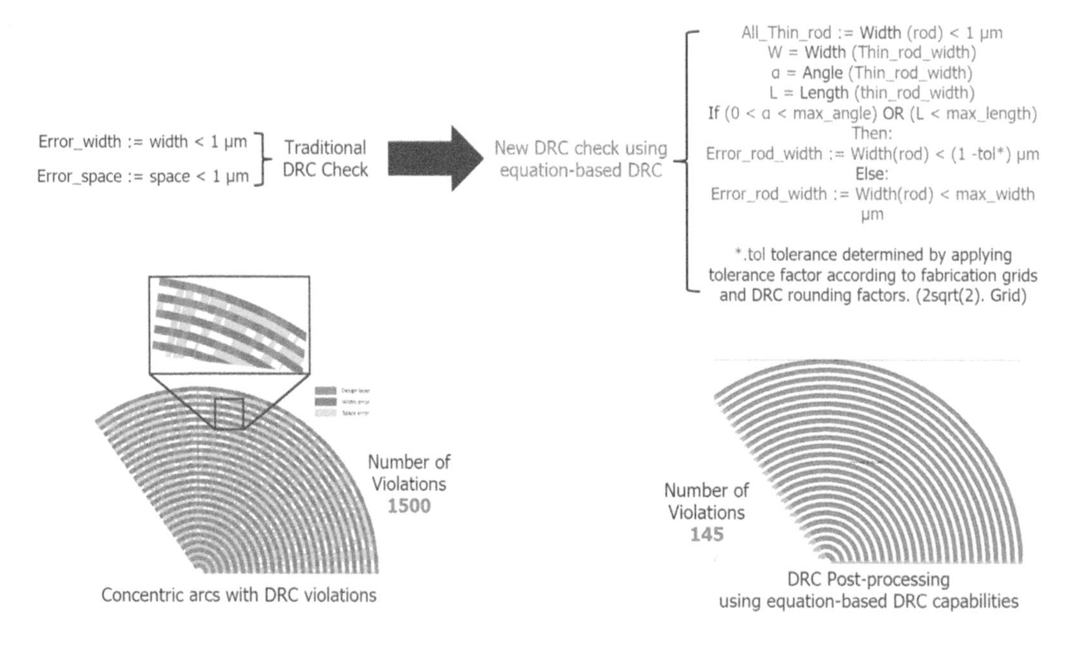

FIGURE 9.9 Eliminating false DRC errors using eqDRC post-processing.

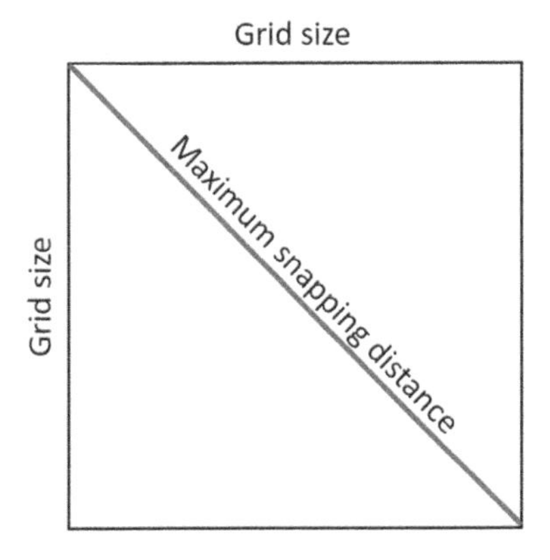

$\because$ *Diagonal length* $= \sqrt{2} * Grid\ size$
$\because$ *Maximum snapping distance lies on diagonal*
$\because$ *Square side* $-$ *Grid size*
$\because$ *Grid size* $= 5\ nm$
$\because$ *Tolerance per error side* $= 7.07\ nm$
$\because$ *Total tolerance* $= 14.14\ nm \approx 15\ nm$

FIGURE 9.10 Derivation of tolerance factor.

The tolerance is calculated based on the diagonal possible shift across a grid square, meaning tolerance is a function of grid size and must be calculated as the layout grid changes (Figure 9.10).

9.10 PIC LVS VERIFICATION

The waveguides in PICs act as an optical interconnect between various circuit components, but are also the building blocks of most PIC devices. Unlike interconnect in electronic ICs, waveguides must be treated as devices instead of ideal interconnect, due to the differences in the concept of connectivity in photonics. The parameters of a waveguide play a pivotal role in its operation, owing to their impact on the modes propagating the waveguides. Also, simple electronic concepts such as shorts and opens are different in photonics design. For example, two waveguides might overlap, creating a four-port network, without resulting in a shorted interconnect.

Comparing a classic electronic LVS flow to photonics LVS requirements can help define the missing LVS components, which are optical connectivity and the validation of curved design shapes (Table 9.1). Optical connectivity and photonic device functionality are validated through parameter extraction and comparison: width, curvilinear path length, and bend curvature, with the limitation being that we must assume a curve type for said curve (e.g., circular arc, Bessier, adiabatic, etc.).

9.11 BEST PRACTICES FOR PHOTONICS LVS

Traditional LVS extracts the assumed curvature and matches it to a source. Shape-matching LVS, a newer method of validating curvilinear design, starts with the source and validates curvature (Figure 9.11).

Table 9.2 describes the difference in parameter extraction between traditional LVS and shape-matching LVS.

TABLE 9.1
Classic LVS vs. photonics LVS

Classic Electronic LVS	Photonics LVS
• Device type and count • Connectivity • Device parameters	• Device type and count • Connectivity (optical) • Curved shape validation (photonic devices and waveguide interconnections)

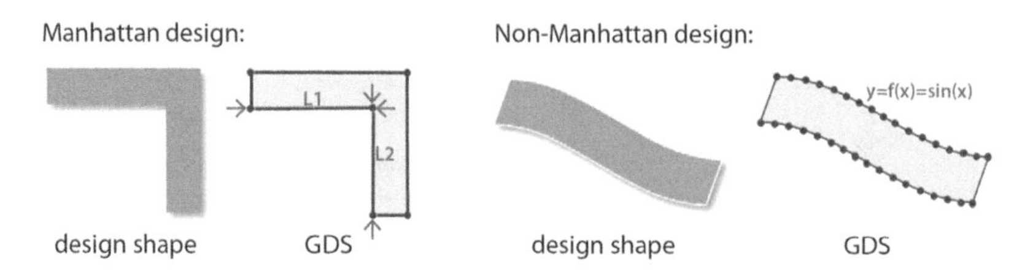

FIGURE 9.11 Shape-matching LVS.

TABLE 9.2
Parameter extraction comparison between traditional LVS and shape-matching LVS

Design Example	Critical Metal Line	Critical Waveguide Interconnect
Features to be extracted and/or validated	Width, Length	Design shape
Netlist	Instance pinA pinB Width Length	waveguide Instance pinA pinB + side-file

9.12 LITHO-FRIENDLY DESIGN SIMULATION

For silicon photonics, it is crucial that designers properly model the final shapes of the circuit, due to their direct impact on circuit performance. For multi-project wafer (MPW) runs, designers typically require multiple iterations of physical device manufacturing to understand and improve the circuit behavior. However, physical iteration is very time-consuming and extremely expensive.

Alternatively, designers can take advantage of LFD PDKs supplied by the foundry. Foundry lithographers and technology access groups (TAGs) use LFD software from EDA companies to develop these PDKs.

Using an LFD tool in conjunction with an LFD PDK enables designers to perform a variety of process simulation checks that can identify potential lithographic resolution issues prior to tapeout (Figure 9.12). Design teams can then apply the necessary design modifications or OPC techniques to ensure manufacturability and performance. With access to an automated virtual lithographic process, designers can shave months from their schedules while avoiding spending money on silicon that does not meet design intent.

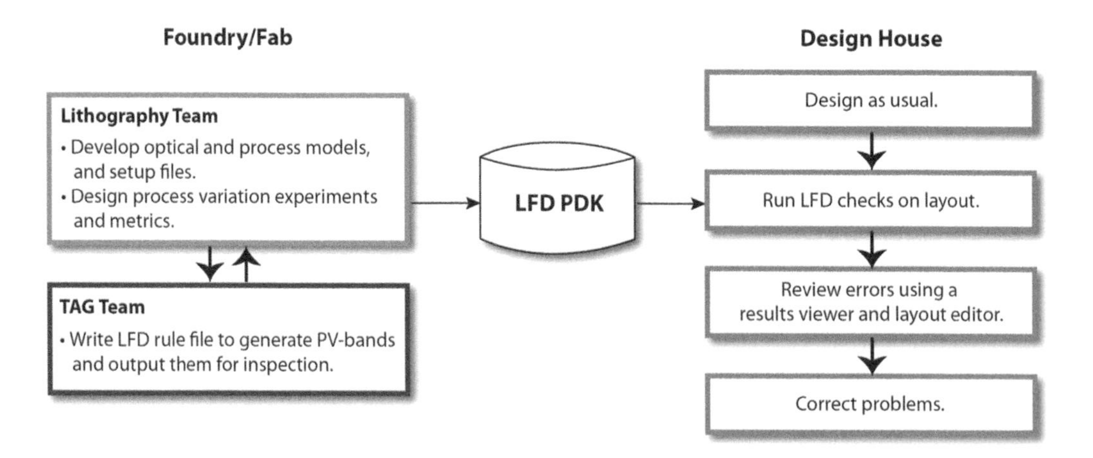

FIGURE 9.12 LFD workflow.

9.13 CONCLUSION

Silicon photonics offers up the promise of blazing fast data transmission and high bandwidth with low power consumption—essential for today's high-performance computing, telecommunications, military, defense, aerospace, medical, and research applications. However, to fully realize that promise, design companies must have the same level of support provided by foundries and EDA suppliers for the design and verification of ICs in the form of complete and proven PDKs. Fortunately, the industry is actively engaging foundries, designers, and EDA companies, with commitments to continue and expand on the progress achieved thus far. Advances in integrated circuit technology and fabrication have made it possible to leverage traditional CMOS fabrication processes and materials and apply them to the design of PICs. The ultimate goal remains the availability of an inexpensive and scalable platform that can transition silicon photonics to a true production design offering.

REFERENCES

[1] John Ferguson, "The process design kit: Protecting design know-how," Mentor, a Siemens Business. Sept. 2018. Link is now: https://resources.sw.siemens.com/en-US/white-paper-the-process-design-kit-protecting-design-know-how
[2] SEMI Standard P39-0304, OASIS—Open Artwork System Interchange Standard, 2004. OASIS specification. OASIS is a registered trademark of Thomas Grebinski and licensed for use to SEMI, San Jose, California. http://ams.semi.org/ebusiness/standards/SEMIStandardDetail.aspx?ProductID=1948&DownloadID=3748
[3] Jonas Flueckiger, "Layout driven design with L-Edit photonics," SiDx, Inc. Sept 2018. https://resources.sw.siemens.com/en-US/white-paper-layout-driven-design-with-l-edit-photonics
[4] Python Software Foundation. Python programming language. https://www.python.org/
[5] PhoeniX Software. OptoDesigner platform for integrated optics and photonic chip design. https://www.phoenixbv.com/product.php?submenu=dfa&subsubmenu=3&prdgrpID=3
[6] Luceda Photonics. IPKISS.eda framework for the design and the design management of integrated photonics chips. https://www.lucedaphotonics.com/en/product/ipkiss-eda

10 Inverse-Design for High-Performance Computing Photonics

Jinhie Skarda, Geun Ho Ahn, Rahul Trivedi, Tony Wu, Subhasish Mitra, and Jelena Vučković

CONTENTS

10.1 SYSTEM-LEVEL MOTIVATION

The increasing amount of data center traffic is driving significant interest in the development of new data transfer systems. Given that much of this traffic is intra-center data, it is now becoming important to optimize chip-to-chip, and even on-chip, data transmission and computation. As process technology has scaled down, the processing time for various compute blocks has decreased. However, since the RC time constant of the metal electrical interconnects is material-defined, the interconnect latency has not scaled down commensurately. This has led to the so-called interconnect bottleneck, where the information transfer time is now a major factor [1]. Through fiber-optics, transmitting data optically has helped increase bandwidth and speed for long-haul communications. Now, chip-scale photonics is being seriously explored as a potential means of increasing the bandwidth and decreasing the delay and operating power for information transfer within data centers at the chip-to-chip or even on-chip level. Given that developing a new platform for short-range data transfer is immensely challenging, it is worthwhile to estimate the system-level improvement one could actually expect if the short-range interconnects in a large computing system were improved. To this end, we provide an order of magnitude-scale estimate through a system-level analysis that considers the effect of improving the short-range interconnects that connect computing units/processors to memory. For these processor-memory interconnects, there are different system architectures to consider. In the 2D and 2.5D architectures, the processor and memory are laterally connected via either a PCB (2D) or a silicon interposer (2.5D) [2]. In these architectures, the high latency and energy of the electrical processor-memory interconnects are a major challenge for applications that are dominated by memory access. This problem is illustrated in Figure 10.1, which shows the fraction of time and energy consumed by the processor and by the memory

DOI: 10.1201/9780429292033-10

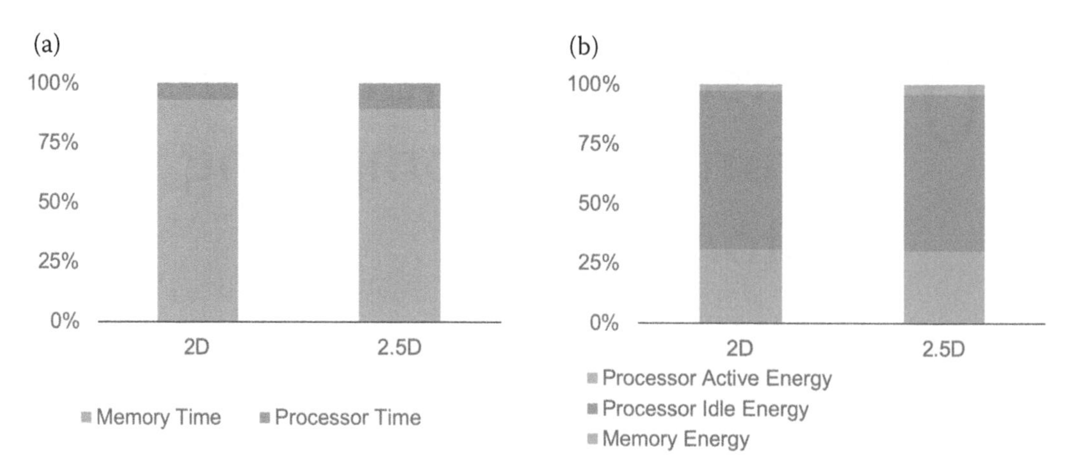

FIGURE 10.1 Breakdown of the total time (a) and total energy (b) consumed while training a Long Short Term Memory (LSTM) language model on a 64-core compute system with either 2D or 2.5D processor-memory chip architecture.

access system when a Long Short Term Memory (LSTM) language model [3,4] is trained on the compute system (as simulated using the simulator ZSim [5]). We expect this application to highlight the problem of the memory access bottleneck because training recurrent neural networks requires extensive memory access. Indeed, for both the 2D and 2.5D architectures, about 90% of the time is spent accessing memory (Figure 10.1a). Since the processor is still consuming energy while waiting for the memory access results, the processor idle energy accounts for about 65% of the total energy consumption (Figure 10.1b). Thus, improving memory accesses (latency, bandwidth, and energy) is crucial to improving the efficiency of the 2D and 2.5D architectures for memory-intensive applications. Given the typical low-latency and high-bandwidth of photonic systems, photonics may provide an attractive solution to this problem.

Another approach to overcoming the memory access challenge is improving the architecture structure such that the distance between the processor and the memory is minimized. This is the idea behind 3D architectures, where the memory sits directly above the processor on the same chip [6–9]. In this architecture, the interconnect lengths can be shorter and the density of vertical interconnects can be high (especially for dense 3D integration such as monolithic 3D). For high-density vertical interconnects in large-scale 3D architectures, it is conceivable that optical technologies can play an important role; a very interesting future research direction. The dense set of vertical through-layer connections in a 3D architecture, however, will be extremely difficult to implement in a scalable way using optics without significant developments in current state-of-the-art photonic fabrication techniques. Given that the lateral interconnect configurations in the 2D and 2.5D architectures are more suitable for implementation using current photonic fabrication techniques, we focus our analysis here on these architectures. To estimate the benefits of replacing electrical interconnects with arbitrarily efficient optical interconnects, we consider a system with 64 compute cores. We assume that the information transfer from a compute core to memory happens via an optical interconnect, and all other elements involved in the memory access are implemented with standard electronics. For this analysis, we model the optical interconnect only by its power consumption, in pJ/bit, and latency, in ns. The system-level figure of merit we compare is the system-level Energy Delay Product (EDP), defined as *EDP = energy × execution time*. The EDP gives an estimate of the system's efficiency, with lower values indicating more efficient systems. We can then provide a quantitative estimate of the optical interconnect system-level

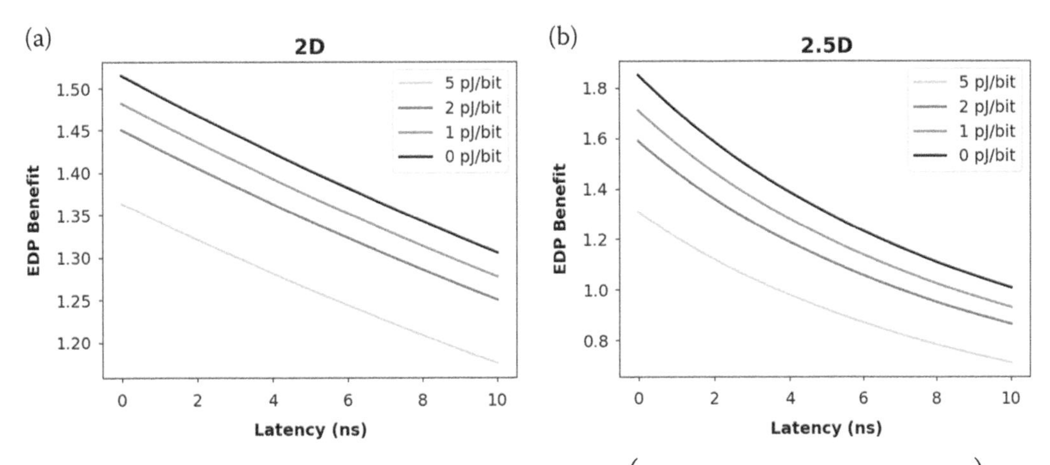

FIGURE 10.2 Average Energy Delay Product (EDP) benefit $\left(EDP_{benefit} = \frac{energy_{electronics} \times latency_{electronics}}{energy_{optics} \times latency_{optics}}\right)$ for a single memory access event that would result if the processor-memory interconnects had the latency values between 0 and 10 ns (x-axes) and energy/bit values of 0, 1, 2, or 5 pJ/bit (black, green, blue, and red lines, respectively). For both the memory read and memory write events, the analysis is performed for the 2D (a) and 2.5D (b) system architectures.

benefit by dividing the EDP of the system fully implemented with standard electronics by the EDP of the system implemented with optical interconnects. The details of how these quantitative estimates are calculated from the simulation data are provided in Appendix 10.1. We present the results of this analysis in Figure 10.2 by plotting the EDP benefit as a function of the optical interconnect latency for a selection of optical interconnect energies. We first study the EDP benefit for the memory access system alone, i.e. only the average energy and latency of a single memory access event are considered (Figure 10.2). When comparing the 2D and 2.5D architectures, we find that 2.5D tends to see higher EDP benefit than 2D. This is likely because the silicon interposer in the 2.5D architecture allows for more densely-packed wires than the PCB in the 2D architecture and, thus, improvements in the performance of these wires has a larger effect. In this case, if the interconnect latency is below about 2 ns and the interconnect energy is below about 1 pJ/bit, we estimate almost a 2x EDP benefit for a memory access event. These energy/bit and latency targets are within reach of optical interconnect architectures, if some of their photonic components can be improved and well-integrated with electronics.

However, for a given improved memory access system, the overall system-level EDP benefit will depend on the application run on the system. Hence, we also study the EDP benefit when an application is run on a 64-core computing system. In order to highlight the overall benefit to improving the processor-memory interconnects, we again use the memory-intensive LSTM word model application studied in Figure 10.1. For applications like these that require the processor to access memory frequently during computation, one can expect improvements to the EDP of the memory access system to translate well to system-level EDP improvements. Indeed, as shown in Figure 10.3 we see a similar EDP benefit for both of these cases, implying that the memory accesses are a dominant bottleneck when the LSTM application is run on this system. If the interconnect latency is around 2 ns and the energy/bit is below 1 pJ/bit, the 2.5D architecture sees about a 2x overall system-level EDP benefit for this LSTM application (Figure 10.3b). Repeating this type of study with a wider variety of applications would be useful in scoping out problems that will see large benefits when run on this type of improved memory access system. Additionally, we expect that the system-level improvements resulting from improvements to the processor-memory interconnects will grow as the number of cores in the compute system increases and as the

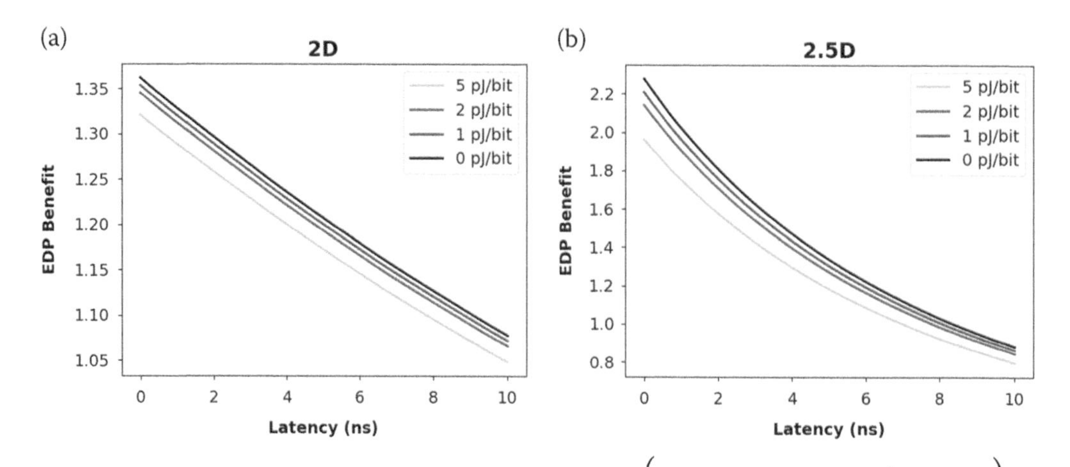

FIGURE 10.3 Average Energy Delay Product (EDP) benefit $\left(EDP_{benefit} = \frac{energy_{electronics} \times latency_{electronics}}{energy_{optics} \times latency_{optics}} \right)$ that would be expected when training a Long Short Term Memory (LSTM) language model on a 64-core compute system implemented with either 2D (a) or 2.5D (b) processor-memory chip architecture, if the processor-memory interconnects had the latency values between 0 and 10 ns (x-axes) and energy/bit values of 0, 1, 2, or 5 pJ/bit (black, green, blue, and red lines, respectively).

architecture is co-optimized with the interconnects. Though these calculations provide only a rough estimate, they give a relatively simple framework for estimating the system-level benefits of improving interconnects and the trends they show suggest that optical interconnects could help access large system-level EDP benefits.

10.2 COMPONENTS OF AN OPTICAL INTERCONNECT

This conclusion that optics could provide an attractive solution to the short-range data transfer problem for large-compute systems has been reached in a number of other more detailed studies [10–16]. There are several basic photonic components that form the building blocks for such chip-scale optical interconnects. First, there must be a coherent light-source either on or off the chip. If the light source is off the chip, a coupler to couple light into the chip's waveguides is necessary. This is most commonly implemented using either edge couplers, which tend to have larger bandwidths and higher efficiencies, or grating couplers, which are far easier to fabricate and allow access to optical circuits anywhere on the chips surface [17]. An optical modulator is then used to encode the electrical signals that are being communicated onto the light in the waveguide. This is usually done by shifting the resonances of a ring resonator or Mach-Zehnder interferometer structure through refractive index change [18]. Then, a series of linear optical elements handle multiplexing, taking advantage of optics' ability to transmit many different frequencies or modes through the same waveguide, to increase the bandwidth of the link. This is typically done using arrayed waveguide gratings, cascaded ring resonators, or multimode interference [19]. Finally, photodetectors convert the optical signal back to an electrical signal. These are typically implemented by integrating a smaller bandwidth material like germanium on the chip [20]. However, the existing photonic devices previously described have several limitations that present a large challenge in implementing a practical on-chip optical interconnect system. One challenge is the large footprint. For example, the Arrayed-Waveguide Grating multiplexing elements have areas on the order of 1,000 square microns [21], and the cascaded resonator-based multiplexing and modulator elements have areas on the order of 100 square microns per wavelength channel [22]. Furthermore, devices whose physical operational principle is resonance-based (e.g., ring resonator based multiplexers or modulators) are fundamentally sensitive to fabrication errors or temperature changes, making them challenging to integrate in large systems and requiring extra tuning elements (typically heater based) which consume additional energy. Thus, the active photonic devices pose another large design challenge, as they must be sensitive to their switching mechanism to maintain speed, while not being sensitive to their environment for robustness. Ring modulators tend to have low dynamic power consumption but require high static power consumption for stabilization [23], while MZI modulators tend to require high dynamic power consumption but less static stabilization power [24]. In the remainder of this chapter, we discuss a physics-guided optimization method called inverse-design and its potential to improve the design of these optical interconnect components.

10.3 INVERSE-DESIGN METHOD FORMULATION

Transitioning chip-scale interconnects from electronics to photonics requires devices that are mass-producible, compact, low-loss, and robust to fabrication errors and environmental factors like large temperature fluctuations. Designing current state-of-the-art photonic devices has conventionally relied on using brute-force parameter sweeps to fine-tune semi-analytical designs, which limits the number of free parameters that can be tuned. One approach that may help to discover device designs that can overcome the challenges faced by conventional photonic devices is using physics-guided optimization to explore the full parameter space afforded by a given footprint [25–28]. In this section, we first provide a brief description of the inverse design method that can be used to automate the design-by-specification process for photonic devices via a gradient based optimization method. Considerations of general optical devices, general optimization problem formulation, optimization process, and some practical implementations are presented. A waveguide is a fundamental building block of on-chip optical systems. A waveguide can be thought of as a wire for electromagnetic waves, and is schematically shown as gray wires in Figure 10.4. In general, a large number of waveguide-based optical devices can be considered as a linear system between a set of input ports and output ports. Schematically, it can be considered as the black design area in Figure 10.4.

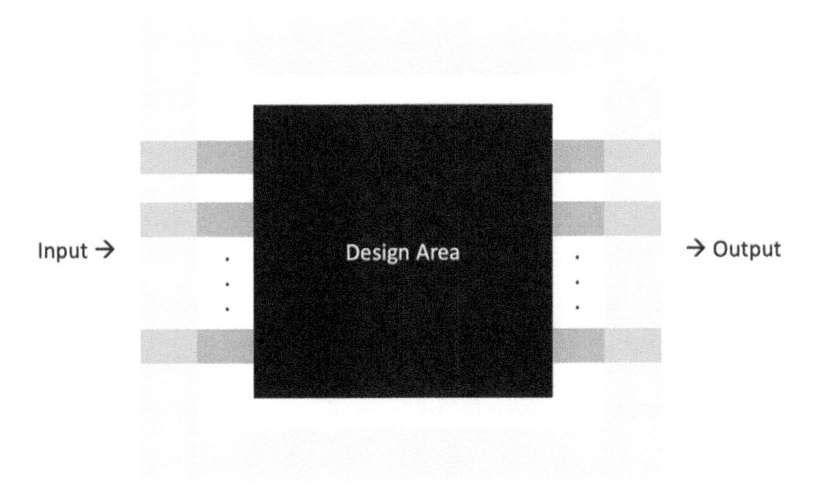

FIGURE 10.4 A schematic of a general optical device with inputs and outputs. The device can have as many inputs and outputs as necessary, as all of those can be accounted for in the scattering matrix. Design Area represents the area that connects inputs and the outputs. Yellow shaded boundary represents a simulation boundary, which sets a simulation size for the electromagnetic simulation to be performed.

Now one can consider exciting the nth input waveguide with a forward propagating complex source J_n. It is a complex numbered source, because it must contain amplitude and phase information. The propagation of its energy will scatter from the device and will propagate through the output waveguides. Accordingly, the overall relation of the propagating amplitudes between input waveguides and output waveguides can be related through a scattering matrix, $S(\omega)$, where ω is the frequency of the propagating electromagnetic wave. Notably, if there are n input and n output waveguide ports, then $S(\omega) \in \mathbb{C}^{n \times n}$. The structure of the device area will dictate the scattering behavior of the amplitudes from the input waveguides to the output waveguides. Hence the scattering matrix is a complicated function of the permittivity distribution of the device, governed by Maxwell's equation. Photonic designers have been utilizing this scattering matrix to analyze many conventional waveguide-based devices. Many of the conventional optical devices employ a library of analytically solvable designs, and the photonic designers further engineer the known device structures to meet their specifications. In recent years, there have been tremendous improvements in the design methodology of optical devices. One of the notable progress is to utilize a physics-guided optimization method to arrive at an optical device design, known as an Inverse Design Method. The Inverse Design method is an optimization method which finds a permittivity distribution in the design area of a photonic device to achieve a desired output performance, given the input source(s). The utilization of the optimization is crucial, as it allows for tractable exploration of the full design space, while the evaluation of the optimization is guided by Maxwell's equation. This means that the Inverse Design method attains the optimal permittivity distribution of the device that fulfills the desired scattering matrix based on the input and output specifications. Unlike the conventional optical devices, the resulting permittivity distribution may look highly non-intuitive with many aperiodically placed holes. In order to set up the optimization process, one first needs to specify the input and output waveguides and the design region. There are many aspects to consider in the specification stage, such as design area dimensions, number of waveguide ports, and their locations. An example of the specification for a 1 × 2 power splitter design is shown in Figure 10.5.

In Figure 10.5, one can see that there is an input defined with the current source, and two output ports to send the split power. The power splitting ratio can be dictated strictly by the scattering matrix. In Figure 10.6a, we illustrate the schematics of propagating amplitudes of the device specified in Figure 10.5. For a ratio of 50:50 power splitter, one can see that the

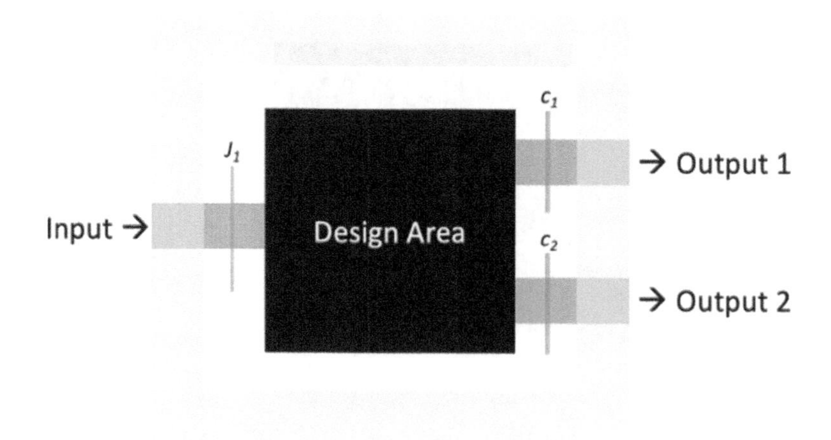

FIGURE 10.5 Schematic of the initial specification for the Inverse Design problem setup for 1 × 2 power splitter optimization. Here J_1, c_1, and c_2 represent the current source, locations of overlap integral 1 and overlap integral 2, respectively.

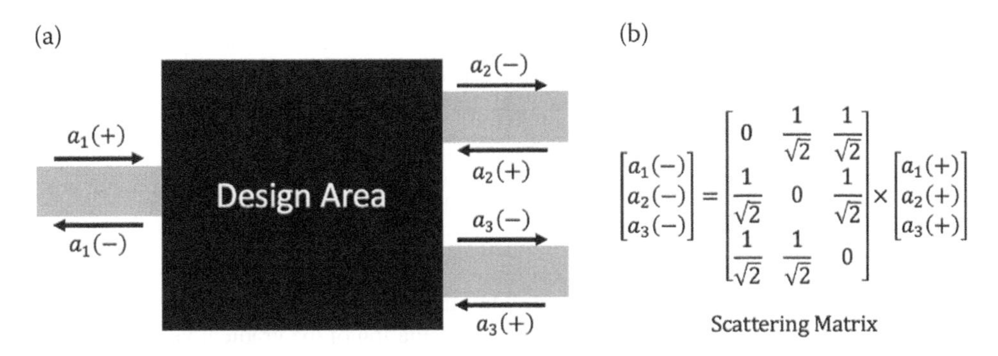

FIGURE 10.6 (a) Schematic of all amplitudes associated with the three port system waveguide based 1×2 power splitter, which is specified in Figure 10.5b. (b) The scattering matrix which would represent the 50:50 power splitting ratio amongst the waveguide ports.

scattering matrix would look like Figure 10.6b. With the scattering matrix in mind, the optimization by this specification will lead to some permittivity distribution in the design area.

In order to achieve this particular scattering matrix behavior, we need to formulate the optimization problem to match this performance. This can be done by defining an appropriate objective function. The formulation required for the inverse design, more generally an optimization of the photonic device design, can be a general optimization problem with an objective function as follows:

$$\min_{\varepsilon, \mathbf{E}} \quad f_{obj}(\mathbf{E}(\varepsilon))$$
$$\text{subject to} \quad \nabla \times \nabla \times \mathbf{E} - \frac{\omega^2}{\mathbf{c}^2}\varepsilon\mathbf{E} = -\mathbf{iw}\mu_0\mathbf{J}$$
$$\varepsilon \in \mathbf{C}(\mathit{Set\ of\ fabricable\ devices}).$$

where f_{obj} is the objective function capturing the optimization goal, E is the electric field distribution, ε is the permittivity distribution, representing the structure of the device, ω is the frequency, J is the input source and C is a set of fabricable epsilon distributions. By allowing permittivity distribution of the design area to be only from the set of fabricable permittivity distributions [27], one can enforce the corresponding result of optimization to be fabricable. Within this general optimization form, photonic device designers have the freedom to define the objective function to meet the required specifications. Therefore, it is important to choose an appropriate objective function. For instance, a suitable objective function for achieving a target transmission, t, and target phase, ϕ, at frequency ω would take the form of

$$f_{obj}(\omega, \varepsilon) = (|\mathbf{c}^{\dagger}\mathbf{E}(\omega, \varepsilon)| - te^{i\phi})^2$$

where $c^{\dagger}E$ computes the modal overlap of the target mode c with the electric field E at the output of the device [29]. By minimizing the objective function, which represents the difference between the simulated modal overlap and the target transmission and phase, a permittivity distribution that closely meets the target values can be obtained, producing a device with the specified performance and functionality. The optimization method can be extended further by utilizing multiple subobjective functions. One can employ multiple objective functions in one optimization, for instance, to optimize over multiple frequencies. This way one can trade off between multiple target parameters at different frequencies. In order to perform the gradient descent optimization on the desired objective function, the gradient $\partial f_{obi}/\partial\varepsilon$ must be computed. Considering the large size of the EM simulation, especially in the case of three dimensional simulation, the computational cost of gradient calculation can be very expensive. In this regard, inverse design technique utilizes the adjoint method calculation, which is shown in the equation below [28]. Its derivation is also shown in the Appendix.

$$\frac{\partial f_{obj}}{\partial E}\frac{\partial E}{\partial \varepsilon} = \left([\Delta - \omega^2 diag(\varepsilon)]^{-T} \frac{\partial f_{obj}^T}{\partial E} \right)^T \omega^2 diagE$$

With this equation, we only need to perform one additional electromagnetic simulation, which significantly shortens the computation. The efficient calculation of the gradient value with only two full-field simulations regardless of the degrees of freedom is very critical. as this allows for the optimization to handle a large number of elements in, and allow for a large number of iterations, which is necessary for gradient descent based inverse design approach. In order to implement this optimization scheme, one may find breaking down the optimization into sub-optimization stages practical. Sub-optimization stages can include continuous optimization, where permittivity distribution is relaxed to vary continuously between cladding and the device, and discrete optimization, where permittivity distribution is strictly binary, representing more realistic device configuration. Since the continuous optimization stage allows less constraints on the permittivity distribution, it is often faster to find a locally optimal device. At the end of the continuous optimization, the device is then converted into discrete device structure for further optimization in the discrete optimization stage. In this regard, one can utilize levelset parameterization in which the binary permittivity distribution is defined by the levelset function [30]. Depending on the value of the levelset function being greater or smaller than 0, the permittivity has one value or another value. This way we can ensure the resulting devices would be binary devices. In addition, the set of fabricable permittivity distributions can be determined by the designers input, such as minimum feature size of the fabrication facility. Another important consideration for the fabrication constraint is that realistic photonic devices will most likely be fabricated using top-down lithography. This means that along the vertical direction, the permittivity distribution must be the same. Given those practical considerations are met, one can implement a software system which provides an optimal photonic design given the specification [28]. With this basic understanding of the inverse design formulations, ranging from objective function, different optimization stages, fabrication constraints to calculation of the gradient using adjoint method, we can now explore some applications of the inverse design method in the following section.

10.4 PRACTICAL PHOTONIC DEVICES WITH INVERSE-DESIGN

In this section, we revisit some of the essential elements for chip-scale photonic links – grating coupler, splitter/multiplexer, and modulator—this time using inverse-design, to demonstrate how this method produces devices that overcome some of the challenges of the current conventionally-designed devices. In particular, we demonstrate that the inverse-designed devices tend to be more compact and broadband than their existing conventionally-designed counterparts. We discuss experimental demonstrations both for inverse-designed devices produced at academic fabrication facilities and inverse-designed devices produced at commercial foundry facilities, illustrating the potential practical scalability of this technique.

As previously discussed, grating-couplers are a key element of on-chip photonic systems because they provide a means of coupling light into waveguides anywhere on the chip. However, the conventionally-designed structures tend to suffer from low coupling efficiencies, small bandwidths, and limited functionality. Here, we discuss using the inverse-design method to design broadband, multifunctional, and very high efficiency grating couplers. To apply the inverse-design method to the grating coupler problem, the grating structure is parametrized by its edges and the

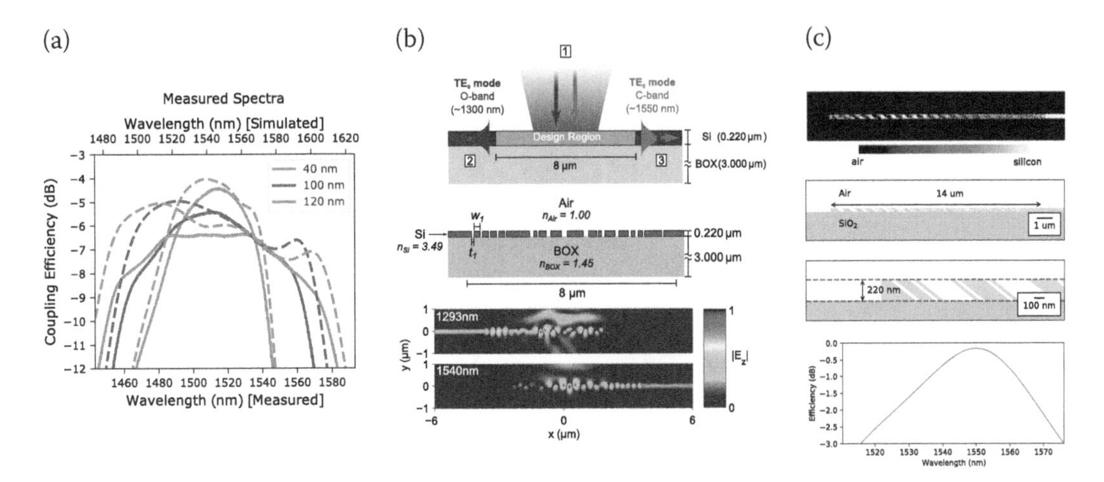

FIGURE 10.7 Inverse-designed grating couplers. (a) Comparison of simulated (dashed lines) and experimentally measured (solid lines) coupling efficiency spectra for fully etched SOI 220 nm grating couplers inverse-designed to have coupling bandwidths of 40 nm (blue), 100 nm (red), and 120 nm (purple) centered at 1,550 nm. Copyright 2019 IEEE. Reprinted, with permission, from [32]. (b) Inverse-design problem formulation (top), grating design (middle), and simulated performance (bottom) of a wavelength-demultiplexing grating coupler. Reprinted from [33]. (c) Continuous variable grating coupler design (top) that inspired the inverse-designed angled-etch gratings (middle) with a coupling efficiency over 99% at 1,550 nm. Reprinted from [31].

power coupled from a free-space Gaussian beam into a desired waveguide mode is optimized for a given material stack [31]. By specifying this high-coupling efficiency optimization objective over a range of wavelengths, grating couplers can be inverse-designed for a range of desired bandwidths. Experimental characterization of devices fabricated on a standard 220 nm Silicon on Insulator stack using Electron-beam Lithography at Stanford University shows good agreement with the simulated performance for the whole range of specified bandwidths [32]. When different functionality is specified for different wavelengths, the inverse-design method can also be used to design multi-function grating couplers, e.g., a vertical etch grating coupler that splits the O-band and S-band wavelengths and sends them in opposite directions [33]. Combining multiple functions in a single design footprint, which is quite challenging with conventional design, can help reduce the footprint for complex optical circuits. Finally, to improve the traditionally low efficiencies of conventional grating couplers, the inverse-design method can also be applied beyond vertically etched grating couplers to angled etches. When inverse-design is applied using an angled etch parametrization, a grating coupler design in the same SOI 220 nm material stack with nearly 100% simulated efficiency is discovered [31] (Figure 10.7).

The ability to significantly increase bandwidth via wavelength and mode multiplexing is one of the most attractive aspects of using photonics for high-performance computing, but current splitter/multiplexer photonic devices suffer from very large footprints and sensitivity to fabrication errors and environment. Using the inverse-design method for splitter/multiplexer elements, the device area can be decreased by a factor of 10 and the robustness can be increased by optimizing the device performance over a large wavelength range with fabrication constraints applied. These inverse-designed splitter and multiplexing devices have been fabricated at the AIM commercial photonics foundry using the standard oxide-cladded 220 nm SOI, and the experimental characterization showed excellent agreement with simulation and between multiple copies of the same device [34]. Thus, the inverse-design method can be compatible with commercial foundry processes, allowing the mass-production of new compact, efficient, and broadband devices (Figure 10.8).

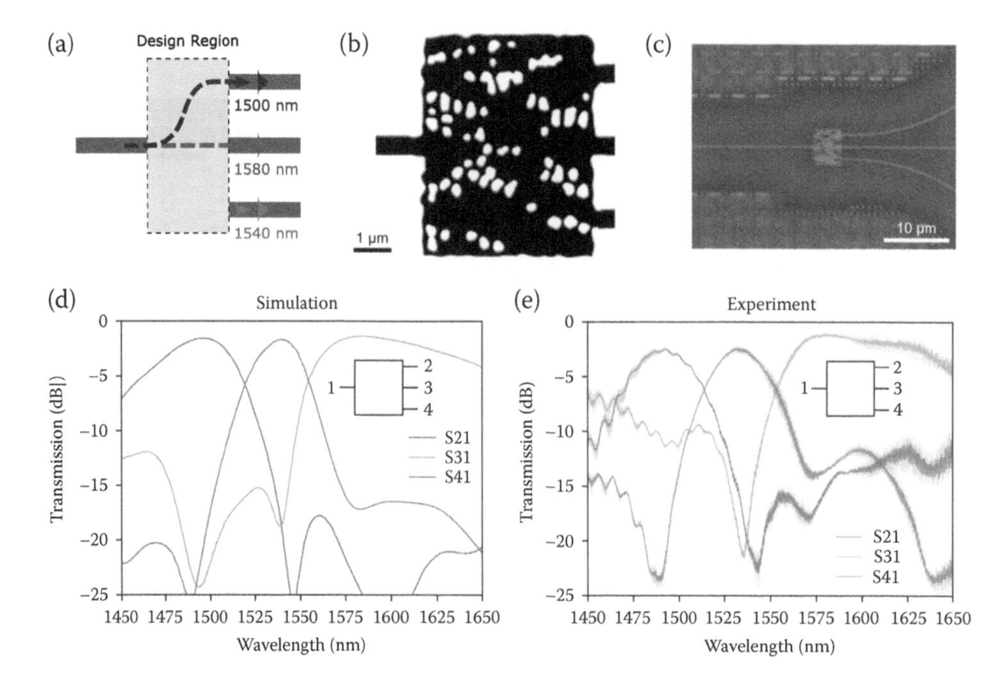

FIGURE 10.8 Foundry-fabrication of inverse-designed devices (reprinted with permission from [34]. Copyright 2020 American Chemical Society). (a) Wavelength De-Multiplexer (WDM) problem formulation. (b) Inverse-designed WDM device. (c) Optical microscopy image of the inverse-designed WDM device fabricated at AIM Photonics Foundry. (d) Simulated performance of WDM device. (e) Experimentally measured performance of the foundry-fabricated inverse-designed WDM.

Finally, modulators are a critical element of an optical interconnect but the current photonic modulators consume too much energy. Slow-light waveguide systems are of interest for designing more energy-efficient modulators because they produce large phase velocity change for small refractive index change, potentially allowing for more efficient optical switching and modulation in silicon [35]. The inverse-design approach can be applied to dispersion-engineer photonic devices like slow-light waveguides by posing an objective function based on the group velocity difference for a set of wave vectors of interest [36]. This approach has been well-demonstrated in 1D and 2D [37,38] but is too computationally expensive for use with full 3D simulations. However, if the objective is posed instead as the frequency-difference over a set of k-vectors subject to Maxwell's equations [39], the defect region of a photonic crystal waveguide can then be optimized with full 3D simulations to obtain a fabricable suspended-silicon structure that has the desired low-slope linear dispersion relation for a slow-light waveguide. If used in a resonator device, this type of dispersion-engineered device could then potentially achieve a larger resonance shift for a smaller applied power giving a more efficient but still robust photonic modulator. This type of device could also be useful to design more sensitive switching systems, more compact delay-line systems, or even better photonic structures for on-chip nonlinear optics processes (Figure 10.9).

Thus, the inverse-design method has been demonstrated to produce compact, reliable linear photonic devices that are key components of optical data transfer and computing platforms. Devices designed with this method can be fabricated with electron-beam lithography in university facilities or with photolithography in commercial CMOS foundries. Using this method to significantly shrink the footprint, and potentially the energy, necessary for coupling and routing light on a chip could be a big step toward more practical implementations of chip-scale photonic interconnects.

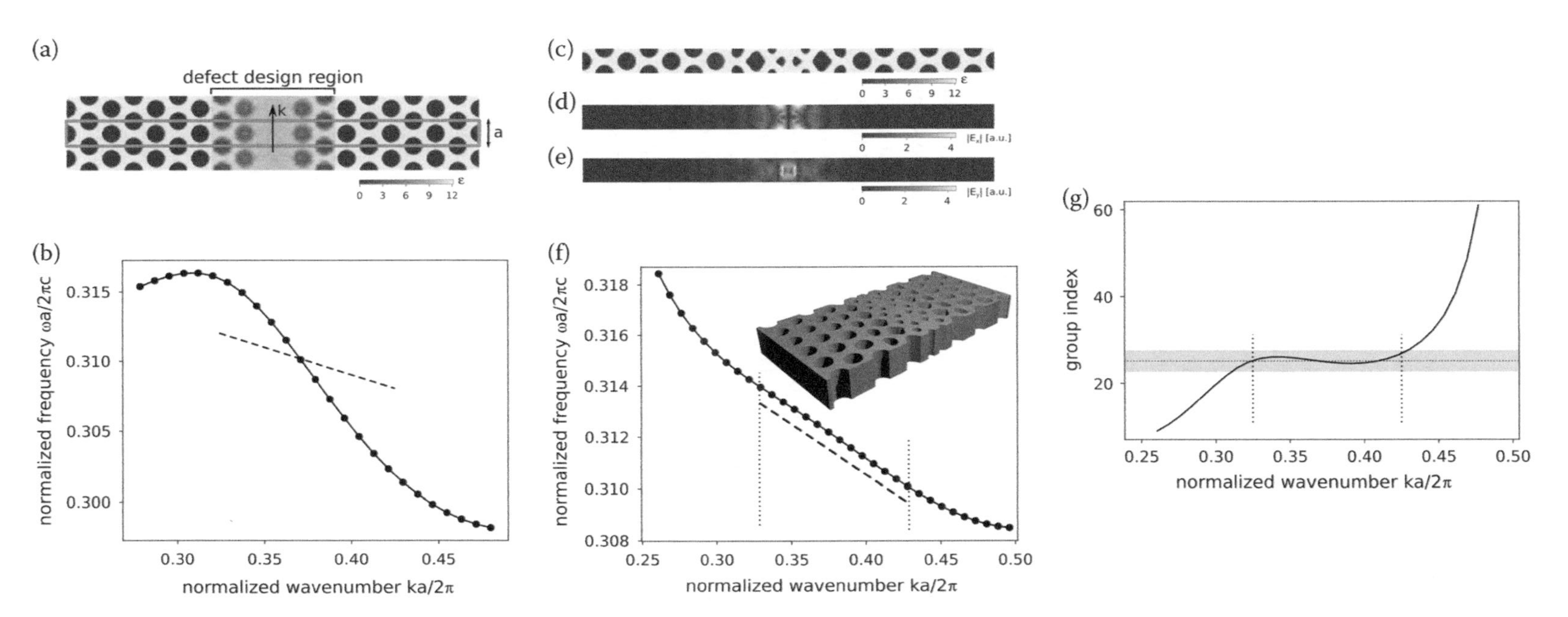

FIGURE 10.9 Inverse-design for dispersion engineering (Copyright 2019 IEEE. Reprinted, with permission, from [39]). The inverse-design approach is applied to a photonic crystal waveguide with permittivity as shown in (a) and dispersion as shown in (b) with the goal of producing a slow-light waveguide with low-slope linear dispersion (the dashed line in (b) shows the target dispersion relation for the optimization). The final inverse-designed device is shown in (c), the x and y components of the electric field in the final inverse-designed device is shown in (d) and (e) respectively, and the final dispersion relation and group index is shown in (f) and (g).

10.5 OUTLOOK OF INVERSE-DESIGN FOR HIGH-PERFORMANCE COMPUTING

The inverse-design method, with its ability to design devices that are compact, robust, multifunctional, and non-intuitive, shows promise as a tool enabling the implementation of practical chip-scale photonic systems for high-performance computing. However, there are several key future research directions, including integrating photonic devices with electronics and extending the inverse-design method to the challenging problem of photonic active device design. In order to implement practical chip-scale photonic interconnect links, the photonic components must be massively co-produced and integrated with electronics on the same chip. The electronics would be used for information processing in tandem with the photonics, as well as for the control circuitry used for stabilizing the performance of the photonic components. There has been impressive and promising work integrating conventional photonic devices with electronics [40,41]. Moreover, the recent demonstration of foundry-fabricated inverse-designed multiplexing and splitting devices [34] is a promising step towards a scalable production of inverse-designed devices, but their integration with electronics has yet to be demonstrated. Given the small footprint and large operating wavelength range achievable with inverse-designed devices, they are expected to be significantly more robust to chip-scale environmental fluctuations and the power required for the stabilizing control circuitry is likely to be far smaller. As previously discussed, active photonic devices such as electrooptic modulators remain a large limiting factor in the power consumption of photonic links. The recent work with applying inverse-design to slow-light waveguide systems [39] may be a promising step towards using this design tool for active devices that require less dynamic switching power and have smaller footprint than the conventional MZI-based modulators, and less static stabilization power than the conventional microring resonator-based modulators. However, applying the inverse-design method to active devices is challenging because a large design area is typically necessary to achieve a large phase change for a realistic small refractive index change. The 3D simulation time then becomes a large, and often prohibitive, bottleneck for these large design areas. In order to take full advantage of the inverse-design method for designing optoelectronics, the simulation efficiency and capabilities for larger-area devices must be improved. If these simulations can be performed in a reasonable time, potentially by using time-domain methods incorporating nonlinearities [42] or data-driven machine learning models to accelerate the simulations [43], the inverse-design method applied to a larger design area could generate exciting improvements. If improvements can be made to lower the energy/bit and footprint of the photonic components, such as the modulator and detector devices, chip-scale photonic interconnect systems could unlock significant system-level EDP benefits for large-compute systems. Given inverse-design's demonstrated ability to produce foundry-compatible compact, broadband, and multifunctional passive photonic devices, it shows much promise as a tool for realizing efficient and compact low-latency photonic links.

REFERENCES

[1] Kirchain, R., & Kimerling, L. (2007). A roadmap for nanophotonics. *Nature Photonics*, 1(6), 303–305.

[2] Deng, Y., & Maly, W. P. (2001). Interconnect characteristics of 2.5-D system integration scheme. *Proceedings of the 2001 International Symposium on Physical Design*, 171–175.

[3] Sundermeyer, M., Schlüter, R., & Ney, H. (2012). LSTM neural networks for language modeling. *Thirteenth Annual Conference of the International Speech Communication Association.*

[4] Jozefowicz, R., Vinyals, O., Schuster, M., Shazeer, N., & Wu, Y. (2016). Exploring the limits of language modeling. *arXiv preprint arXiv:1602.02410.*

[5] Sanchez, D., & Kozyrakis, C. (2013). ZSim: Fast and accurate microarchitectural simulation of thousand-core systems. *ACM SIGARCH Computer Architecture News*, 41(3), 475–486.
[6] Xie, Y., Cong, J., & Sapatnekar, S. (2010). *Three-dimensional integrated circuit design: EDA, design and microarchitectures*. New York: Springer, 20, 194–196.
[7] Aly, M. M. S., Gao, M., Hills, G., Lee, C. S., Pitner, G., Shulaker, M. M., ... & Mitra, S. (2015). Energy-efficient abundant-data computing: The N3XT 1,000 x. *Computer*, 48(12), 24–33.
[8] Loh, G. H. (2008). 3D-stacked memory architectures for multi-core processors. *ACM SIGARCH Computer Architecture News*, 36(3), 453–464.
[9] Woo, D. H., Seong, N. H., Lewis, D. L., & Lee, H. H. S. (2010, January). An optimized 3D-stacked memory architecture by exploiting excessive, high-density TSV bandwidth. In *HPCA-16 2010 The Sixteenth International Symposium on High-Performance Computer Architecture* (pp. 1–12). IEEE.
[10] Miller, D. A. B. (2000). Rationale and challenges for optical interconnects to electronic chips. *Proceedings of the IEEE*, 88(6), 728–749.
[11] Taubenblatt, M. A. (2011). Optical interconnects for high-performance computing. *Journal of Lightwave Technology*, 30(4), 448–457.
[12] Kachris, C., & Tomkos, I. (2012). A survey on optical interconnects for data centers. *IEEE Communications Surveys Tutorials*, 14(4), 1021–1036.
[13] Nitta, C. J., Farrens, M., & Akella, V. (2013). On-chip photonic interconnects: A computer architect's perspective. *Synthesis Lectures on Computer Architecture*, 8(5), 1–111.
[14] Koka, P., McCracken, M. O., Schwetman, H., Zheng, X., Ho, R., & Krishnamoorthy, A. V. (2010). Silicon-photonic network architectures for scalable, power-efficient multi-chip systems. *ACM SIGARCH Computer Architecture News*, 38(3), 117–128.
[15] Miller, D. A. B. (2009). Device requirements for optical interconnects to silicon chips. *Proceedings of the IEEE*, 97(7), 1166–1185.
[16] Meindl, J. D., Davis, J. A., Zarkesh-Ha, P., Patel, C. S., Martin, K. P., & Kohl, P. A. (2002). Interconnect opportunities for gigascale integration. *IBM Journal of Research and Development* 46(2.3), 245–263.
[17] Taillaert, D., Bienstman, P., & Baets, R. (2004). Compact efficient broadband grating coupler for silicon-on-insulator waveguides. *Optics Letters*, 29(23), 2749–2751.
[18] Reed, G. T., Mashanovich, G., Gardes, F. Y., & Thomson, D. J. (2010). Silicon optical modulators. *Nature Photonics*, 4(8), 518.
[19] Chrostowski, L., & Hochberg, M. (2015). *Silicon photonics design: From devices to systems.* Cambridge University Press.
[20] Long, C., & Lipson, M. (2009). Ultra-low capacitance and high speed germanium photodetectors on silicon. *Optics Express*, 17(10), 7901–7906.
[21] Fu, X., & Dai, D. (2011). Ultra-small Si-nanowire-based 400 GHz-spacing 15 × 15 arrayed-waveguide grating router with microbends. *Electronics Letters*, 47(4), 266–268.
[22] Xu, Q., Schmidt, B., Shakya, J., & Lipson, M. (2006). Cascaded silicon micro-ring modulators for WDM optical interconnection. *Optics Express*, 20, 9431–9436.
[23] Atabaki, A. H., Moazeni, S., Pavanello, F., Gevorgyan, H., Notaros, J., Alloatti, L., Wade, M. T., Sun, C., Kruger, S. A., Meng, H., Qubaisi, K. Al, Wang, I., Zhang, B., Khilo, A., Baiocco, C. V., Popović, M. A., Stojanović, V. M., & Ram, R. J. (2018). Integrating photonics with silicon nanoelectronics for the next generation of systems on a chip. *Nature*, 556(7701), 349–354.
[24] Green, W. M. J., Rooks, M. J., Sekaric, L., & Vlasov Y. A. (2007). Ultra-compact, low RF power, 10 Gb/s silicon Mach-Zehnder modulator. *Optics Express*, 15(25), 17106–17113.
[25] Jensen, J. S., & Sigmund, O. (2011). Topology optimization for nanophotonics. *Laser Photonics Reviews*, 5(2), 308–321.
[26] Piggott, A. Y., Lu, J., Lagoudakis, K. G., Petykiewicz, J. A., Babinec, T. M., & Vučković, J. (2015). Inverse design and demonstration of a compact and broadband on-chip wavelength demultiplexer. *Nature Photonics*, 9(6), 374–377.
[27] Piggott, A. Y., Petykiewicz, J. A., Su, L., & Vučković, J. (2000). Rationale and challenges for optical interconnects to electronic chips. *Proceedings of the IEEE*, 88(6), 728–749.
[28] Su, L., Vercruysse, D., Skarda, J., Sapra, N. V., Petykiewicz, J. A., & Vučković, J. (2020). Nanophotonic inverse design with SPINS: Software architecture and practical considerations. *Applied Physics Reviews*, 7, 011407.
[29] McIsaac, P. R. (1991). Mode orthogonality in reciprocal and nonreciprocal waveguides. *IEEE Transactions on Microwave Theory and Techniques*, 39(11), 1808–1816.

[30] Vercruysse, D., Sapra, N. V., Su, L., Trivedi, R., & Vučković, J. (2019). Analytical level set fabrication constraints for inverse design. *Scientific Reports*, 9, 8999.
[31] Su, L., Trivedi, R., Sapra, N. V., Piggott, A. Y., Vercruysse, D., & Vučković, J. (2018). Fully-automated optimization of grating couplers. *Optics Express*, 26(4), 4023–4034.
[32] Sapra, N. V., Vercruysse, D., Su, L., Yang, K. Y., Skarda, J., Piggott, A. Y., & Vučković, J. (2019). Inverse design and demonstration of broadband grating couplers. *IEEE Journal of Selected Topics in Quantum Electronics*, 25(3), 1–7.
[33] Piggott, A. Y., Lu, J., Babinec, T. M., Lagoudakis, K. G., Petykiewicz, J., & Vučković, J. (2014). Inverse design and implementation of a wavelength demultiplexing grating coupler. *Scientific Reports*, 4(1), 1–5.
[34] Piggott, A. Y., Ma, E. Y., Su, L., Ahn, G. H., Sapra, N. V., Vercruysse, D., ... & Vučković, J. (2020). Inverse-designed photonics for semiconductor foundries. *ACS Photonics*, 7(3), 569–575.
[35] Baba, T., Nguyen, H. C., Yazawa, N., Terada, Y., Hashimoto, S., & Watanabe, T. (2014). Slow-light Mach-Zehnder modulators based on Si photonic crystals. *Science and Technology of Advanced Materials*, 15(2), 024602.
[36] Wang, F., Jensen, J. S., & Sigmund, O. (2012). High-performance slow light photonic crystal waveguides with topology optimized or circular-hole based material layouts. *Photonics and Nanostructures-Fundamentals and Applications*, 10(4), 378–388.
[37] Elesin, Y., Lazarov, B. S., Jensen, J. S., & Sigmund, O. (2012). Design of robust and efficient photonic switches using topology optimization. *Photonics and Nanostructures-Fundamentals and Applications*, 10(1), 153–165.
[38] Kao, C. Y., Osher, S., & Yablonovitch, E. (2005). Maximizing band gaps in two-dimensional photonic crystals by using level set methods. *Applied Physics B*, 81(2-3), 235–244.
[39] Vercruysse, D., Sapra, N. V., Su, L., & Vučković, J. (2019). Dispersion engineering with photonic inverse design. *IEEE Journal of Selected Topics in Quantum Electronics*, 26(2), 1–6.
[40] Sun, C., Wade, M. T., Lee, Y., Orcutt, J. S., Alloatti, L., Georgas, M. S., ... & Stojanović, V. M. (2015). Single-chip microprocessor that communicates directly using light. *Nature*, 528(7583), 534–538.
[41] Atabaki, A. H., Moazeni, S., Pavanello, F., Gevorgyan, H., Notaros, J., Alloatti, L., ... & Ram, R. J. (2018). Integrating photonics with silicon nanoelectronics for the next generation of systems on a chip. *Nature*, 556(7701), 349–354.
[42] Hughes, T. W., Minkov, M., Williamson, I. A., & Fan, S. (2018). Adjoint method and inverse design for nonlinear nanophotonic devices. *ACS Photonics*, 5(12), 4781–4787.
[43] Trivedi, R., Su, L., Lu, J., Schubert, M. F., & Vučković, J. (2019). Data-driven acceleration of photonic simulations. *Scientific Reports*, 9(1), 1–7.

APPENDICES

A10.1 ESTIMATION OF OPTICAL INTERCONNECT SYSTEM-LEVEL BENEFIT

To arrive at our system-level benefit estimates, we take the system parameters computed through a ZSim [5] system simulation of a 64 core compute system with 2D or 2.5D processor-memory architecture and electrical interconnects, and recalculate these values given improved hypothetical interconnect parameters. We specify the interconnect by its power consumption (pJ/bit) and latency (ns). The overall system performance metric we calculate is the energy-delay product (EDP), defined as energy/bit x latency. To measure the system-level benefit we see for these improved hypothetical interconnect parameters, we divide the original EDP from the ZSim data by the EDP computed with the new interconnect parameters to give EDP benefit. Below, we detail how we rescale the original ZSim system parameters to arrive at the EDP benefits for the memory system we showed in Figure 10.2 and the EDP benefits for the deep learning application we showed in Figure 10.3.

A10.1.1 Memory System Analysis

For this analysis, our goal is to estimate the EDP benefit of a single memory read and a single memory write, as defined below.

$$EDP_{benefit} = \frac{1}{2}\left(\frac{read\ energy_{elec} \times read\ latency_{elec}}{read\ energy_{opt} \times read\ latency_{opt}} + \frac{write\ energy_{elec} \times write\ latency_{elec}}{write\ energy_{opt} \times write\ latency_{opt}}\right)$$

We consider the average of the read and write events because, in the 2D and 2.5D architectures, the energy and latency for the two are almost identical. The read energy and latency assuming electronic interconnects and assuming improved optical interconnects are defined as:

$$read\ energy_{elec} = read\ energy + interconnect\ energy_{elec}$$
$$read\ latency_{elec} = read\ latency + interconnect\ latency_{elec}$$

$$read\ energy_{opt} = read\ energy + interconnect\ energy_{opt}$$
$$read\ latency_{opt} = read\ latency + interconnect\ latency_{opt}$$

The write energy and latency assuming electronic interconnects and assuming improved optical interconnects are defined as:

$$write\ energy_{elec} = write\ energy + interconnect\ energy_{elec}$$
$$write\ latency_{elec} = write\ latency + interconnect\ latency_{elec}$$

$$write\ energy_{opt} = write\ energy + interconnect\ energy_{opt}$$
$$write\ latency_{opt} = write\ latency + interconnect\ latency_{opt}$$

The values *read energy*, *read latency*, *write energy*, *write latency*, $interconnect\ energy_{elec}$, and $interconnect\ latency_{elec}$ are obtained from the original ZSim simulation of the 2D and 2.5D architectures. To arrive at the plots in Figure 10.2, we sweep the $interconnect\ latency_{opt}$ from 0 to 10 ns for $interconnect\ energy_{opt}$ values of 0, 1, 2, and 5 pJ/bit.

A10.1.2 LSTM Application System Analysis

Here, our goal is to estimate the $EDP_{benefit}$ seen when a Long Short Term Memory (LSTM) language model [3,4] is trained on the system. The procedure we follow here to compute the $EDP_{benefit}$ is similar to the one for the memory system analysis, but we scale the read and write energy and latency by the read-to-write ratio for the LSTM application (denoted in all following equations as LRW) and use the resulting values to scale the run-time and energy consumption of the application.

The total system-level EDP benefit is given by:

$$EDP_{benefit} = \frac{total\ energy_{elec} \times total\ latency_{elec}}{total\ energy_{opt} \times total\ latency_{opt}}$$

For electronics, we compute the total memory latency and energy as:

$$total\ memory\ latency_{elec} = LRW \times (read\ latency + interconnect\ latency_{elec}) + (1 - LRW) \times (write\ latency + interconnect\ latency_{elec})$$

$$total\ memory\ energy_{elec} = LRW \times (read\ energy + interconnect\ energy_{elec}) + (1 - LRW) \times (write\ energy + interconnect\ energy_{elec})$$

For optics, we compute the total memory latency and energy as:

$$total\ memory\ latency_{opt} = LRW \times (read\ latency + interconnect\ latency_{opt}) + (1 - LRW) \times (write\ latency + interconnect\ latency_{opt})$$

$$total\ memory\ energy_{opt} = LRW \times (read\ energy + interconnect\ energy_{opt}) + (1 - LRW) \times (write\ energy + interconnect\ energy_{opt})$$

Once again, the *read energy*, *read latency*, *write energy*, *write latency*, $interconnect\ energy_{elec}$, and $interconnect\ latency_{elec}$ are obtained from the original ZSim simulation of the 2D and 2.5D architectures. The $interconnect\ latency_{opt}$ and $interconnect\ energy_{opt}$ are values we sweep for the analysis.

Then, we use the ratios of the total memory latency and energy in optics to the total memory latency and energy in electronics to rescale the amount of time spent for memory access (*memory time*), the energy consumed by memory access (*memory energy*), and the idle energy consumed by the processor while waiting for the results of memory accesses (*processor idle energy*). These original values are all obtained from the original ZSim simulation as percentages of the total time and energy, and the rescaled values estimate the new time and energy breakdowns for the optical processor-memory interconnect parameters.

$$memory\ time_{opt} = memory\ time \times \frac{total\ mem\ latency_{opt}}{total\ mem\ latency_{elec}}$$

$$memory\ energy_{opt} = memory\ energy \times \frac{total\ mem\ energy_{opt}}{total\ mem\ energy_{elec}}$$

$$processor\ idle\ energy_{opt} = processor\ idle\ energy \times \frac{memory\ time_{opt}}{memory\ time}$$

Finally, we use the rescaled time and energy percentages for the optical processor-memory interconnect parameters to recalculate the total time and energy for the LSTM application. Since these total time and energy values are percentages, the total system-level EDP benefit for the LSTM application is calculated as follows:

$$time\ total_{opt} = memory\ time_{opt} + processor\ time$$

$$energy\ total_{opt} = memory\ energy_{opt} + processor\ idle\ energyopt$$

$$EDP_{benefit} = \frac{1}{time\ total_{opt} \times energy\ total_{opt}}$$

The values *processor time* and *processor active energy* are obtained from the original ZSim simulation of the 2D and 2.5D architectures. To arrive at the plots in Figure 10.3, we sweep the $interconnect\ latency_{opt}$ from 0 to 10 ns for $interconnect\ energy_{opt}$ values of 0, 1, 2, and 5 pJ/bit.

A10.2 ADJOINT METHOD DERIVATION

In order to perform the gradient descent optimization on the desired objective function, the gradient $\partial f_{obi}/\partial\varepsilon$ must be calculated. There are many important considerations to make regarding the calculation of the gradient, such as caution with complex valued ε or E, and the computational cost of the calculation. The derivative generally would take the following form:

$$\frac{df_{obj}}{d\varepsilon} = \frac{\partial f_{obj}}{\partial\varepsilon} + \left(\frac{\partial f_{obj}}{\partial E}\frac{\partial E}{\partial\varepsilon} + frac\partial f_{obj}\partial E_{*}\frac{\partial E_{*}}{\partial\varepsilon}\right)$$

Where the partial derivatives are defined as the Wirtinger derivatives. Since f_{obj} is real, $\partial f_{obj}/\partial E = (\partial f_{obj}/\partial E_{*})_{*}$. Therefore, the equation above can be simplified to

$$\frac{df_{obj}}{d\varepsilon} = \frac{\partial f_{obj}}{\partial\varepsilon} + 2Re\left[\frac{\partial f_{obj}}{\partial E}\frac{\partial E}{\partial\varepsilon}\right]$$

Where Re denotes taking the real part. $\partial E/\partial\varepsilon$ is computationally expensive because it requires a number of electromagnetic simulations equal to the number of elements in ε. Instead, we calculate mathnormal $\partial f_{obj}/\partial E$ first, and then the calculation of quantity $(\partial f_{obj}/\partial E)(\partial E/\partial\varepsilon)$ is evaluated using the differentiated FDFD equation. Derivation is as following

$$[\Delta - \omega^2 diag(\varepsilon)]E = -i\omega J$$

where Δ is the discretized version of the $\nabla \times (1/\mu)\nabla\times$ operator. We can then take the derivative with respect to.

$$[\Delta - \omega^2 diag(\varepsilon)]\frac{\partial E}{\partial\varepsilon} = \omega^2 diag(E)$$

We can rearrange the equation further.

$$\frac{\partial E}{\partial\varepsilon} = [\Delta - \omega^2 diag(\varepsilon)]^{-1}\omega^2 diag(E)$$

This results in the gradient of E. However, the computation of this gradient requires as many simulations as the number of elements in the vector ε, which quickly becomes very expensive. Instead, we can compute $(\partial f_{obj}/\partial E)(\partial E/\partial \varepsilon)$ first, and backpropagate to achieve the value of $(\partial E/\partial \varepsilon)$. Such calculation of the can be evaluated using the following equations.

$$\frac{\partial f_{obj}}{\partial E}\frac{\partial E}{\partial \varepsilon} = frac\partial f_{obj}\,\partial E\,[\Delta - \omega^2 diag(\varepsilon)]^{-1}\omega^2 diag(E)$$

$$\frac{\partial f_{obj}}{\partial E}\frac{\partial E}{\partial \varepsilon} = \left([\Delta - \omega^2 diag(\varepsilon)]^{-T}\frac{\partial f_{obj}^T}{\partial E}\right)\omega^2 diag(E)$$

11 Efficiency-Oriented Design Automation Methods for Wavelength-Routed Optical Network-on-Chip

Tsun-Ming Tseng, Mengchu Li, Zhidan Zheng, Alexandre Truppel, and Ulf Schlichtmann

CONTENTS

11.1 INTRODUCTION

Nowadays, networks-on-chip (NoC) have been widely recognized as a scalable and power-efficient solution for the on-chip communication in multiprocessor systems-on-chip (MPSoC) (Jiang, et al. 2013). Conventional NoC systems use electrical interconnects. However, as the feature size continuously shrinks, the number of cores in NoC systems increases, and electrical interconnects gradually become insufficient to satisfy the bandwidth and latency demands (Ye, et al. 2013). Enabled by breakthroughs in silicon photonics, optical networks-on-chip (ONoC) emerge as an appealing next-generation architecture. As the name suggests, ONoCs convert electrical signals into optical signals on different wavelengths and transmit them through optical waveguides, which are the dual to electrical wires, from initiators (senders) to targets (receivers) in the network. Since a waveguide can accommodate multiple wavelengths (Preston, et al. 2011), which is known as wavelength-division multiplexing (WDM), ONoCs offer much higher bandwidth than offered by their electrical counterparts (Ramini, Bertozzi, and Carloni 2012).

Based on the signal routing mechanisms, we can classify current ONoC architectures into two categories (Werner et al. 2017): *control-network-based ONoCs* that dynamically set up/tear down signal paths on request; and *wavelength-routed ONoCs (WRONoCs)* that statically reserve all signal paths at design time. Compared to ONoCs in the other category, WRONoCs avoid the latency and energy overhead for arbitration and deliver the on-chip communication in a more predictable manner. These advantages come however at the expense of more extensive optical

DOI: 10.1201/9780429292033-11

resource usage, which is considered the major challenge for WRONoCs to be applied to large-scale networks.

This chapter will address the WRONoC design problem focusing on improving the efficiency of the network as well as the efficiency of the design process. Following a discussion of the WRONoC design criteria, we will cover two design automation strategies: *subtraction from fully connected router* and *design-template-based synthesis*.

11.2 WRONOC DESIGN CRITERIA

11.2.1 Common Setting and General Design Rules

Current research commonly considers WRONoCs in a 3-D setting composed of vertically stacked photonic and electronic layers (Li, et al. 2017). Figure 11.1 shows an example of a 3-D processor-memory network implemented with WRONoCs (Ramini, Grani, et al. 2013). The electronic layer consists of clusters of IP-cores, which are connected to hubs on the optical layer by through-silicon vias (TSVs). The optical layer also contains memory controllers (MCs), which are connected to off-chip memories. With laser sources providing wavelengths, electrical signals are converted to optical signals and distributed to hubs and memory controllers on the optical layer by a WRONoC router.

WRONoCs route optical signals based on their wavelengths. To avoid data collision, passive microring resonators (MRRs) are used to guide signals along their designated paths. An MRR consists of a looped waveguide that resonates with optical signals of particular wavelengths (Bogaerts 2012). As shown in Figure 11.2, on-resonance signals that approach an MRR through nearby waveguides will be coupled to the MRR and then switched to another direction, while off-resonance signals will just ignore the MRR and keep their original directions. Thus, by using a set of different wavelengths that are on-resonance with different MRRs, optical signals can be sent from the same initiator to different targets simultaneously without conflict.

In general, WRONoC design is to construct conflict-free signal paths between all required initiator-target pairs. The design tasks are threefold (Tseng, et al. 2019):

- Route waveguides to connect initiators to targets.
- Place MRRs of different resonant wavelengths in the waveguide topology.
- Decide the wavelengths used by each initiator-target pair for signal transmission.

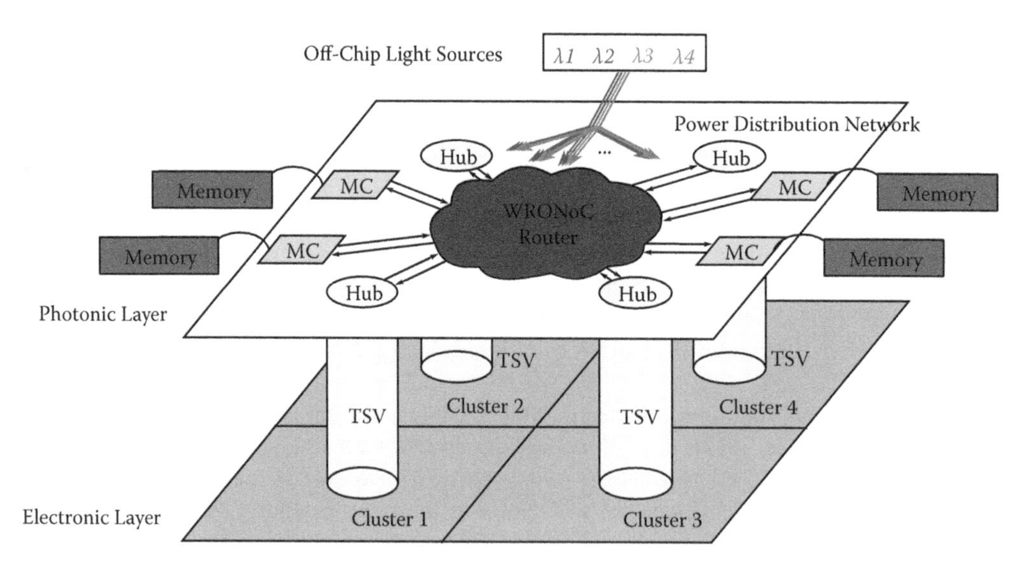

FIGURE 11.1 A common 3-D setting of WRONoCs.

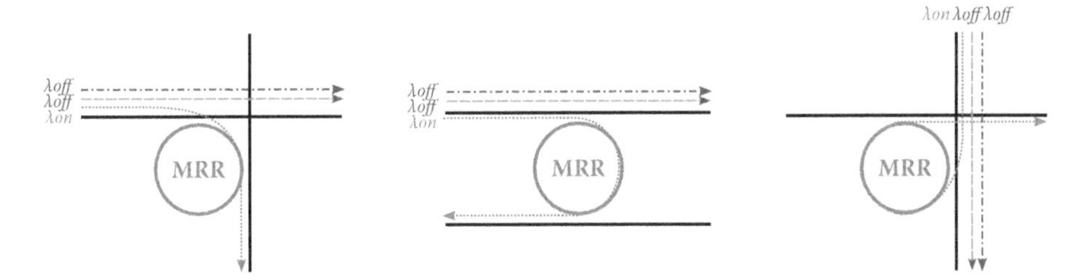

FIGURE 11.2 The working mechanism of microring resonators (MRRs). On-resonance signals are labeled as λ_{on} and off-resonance signals are labeled as λ_{off}. Depending on the arrangement of the MRR and the nearby waveguides, the directions of on-resonance signals can be altered by 90, 180, or 270 degrees.

As each initiator-target pair requires a dedicated signal path, WRONoCs require a relatively larger number of waveguides, MRRs and wavelengths compared to other ONoC architectures. In order to reduce the resulting energy overhead, efficient WRONoC topologies allow different signal paths to share some of their optical resources, as long as the following design rules are fulfilled (Peano, et al. 2016):

- [*Rule 1*] Optical signals sent from different initiators and tuned to have the same wavelength should never pass through the same waveguides.
- [*Rule 2*] Each target should receive signals from different initiators on different wavelengths to avoid any conflict at the receiver side.

Under the prerequisite that the above rules are satisfied, WRONoC designers try to construct the network with as few optical resources (wavelengths and MRRs in particular) as possible. The rest of this section addresses the wavelength and MRR usage in current WRONoC topologies.

11.2.2 Wavelength Usage

The maximum number of available wavelengths in WRONoCs is constrained by the WDM bandwidth and the required signal quality. Current technology supports up to 62 different wavelengths in 50nm WDM bandwidth (Preston, et al. 2011). However, since more wavelengths generally imply more crosstalk among optical signals and thus lower signal-to-noise ratio (SNR) at the receiver side, it is preferable to minimize the number of wavelengths in a WRONoC design.

Under the prerequisite that each initiator/target is connected by a single waveguide to the network, we can generalize the second design rule (Rule 2) and derive the following constraints on wavelength usage (Li, Tseng, and Bertozzi, et al. 2018):

- Signals sent from the same initiator to different targets must have different wavelengths.
- Signals sent from different initiators to the same target must have different wavelengths.

If we consider each initiator/target as a vertex and each signal path as an edge between a pair of initiator- and target-vertices, we can model the network as a bipartite graph $G = (I \cup T, E)$, where I and T are two disjoint and independent sets containing all initiator- and all target-vertices, respectively, and E is the set of edges. Based on this graph, we can model the wavelength minimization problem as an edge-coloring problem, i.e. coloring the edges with the smallest number of colors so that incident edges have different colors, where the color of an edge stands for the wavelength used by an initiator-target pair. By Vizing's theorem (Vizing 1964), the chromatic number for edge coloring a bipartite graph is the maximum degree of the graph. Thus, the minimum number of wavelengths required by a WRONoC topology is the maximum number of signal paths starting from (ending at) the same initiator (target). For a fully connected network where all initiators communicate with all targets, the minimum number of wavelengths is $max\{|I|, |T|\}$.

It is worth mentioning that for networks that require high communication parallelism, it makes sense to assign multiple wavelengths to a single signal path to transmit multiple signals at the same time, as long as the required signal quality can be achieved. The number of available wavelengths in each signal path depends on the resonant wavelengths of the MRRs along the path (Li, Tseng, and Tala, et al. 2020).

11.2.3 MRR Usage

MRRs are essential components of WRONoCs for wavelength-based signal routing. However, each time an optical signal passes through or resonates with an MRR, a small portion of energy will go to directions other than the designated direction, causing loss of the signal power and crosstalk noise in other signal paths (Nikdast, et al. 2014). To reduce the energy overhead and maintain the signal quality, it is desirable to avoid using redundant MRRs in a WRONoC topology.

Figure 11.3 shows a simple WRONoC router to transmit data from one initiator to four different targets. Three MRRs are used to implement four signal paths between different initiator-target pairs. Specifically, signals sent to T_1, T_2, and T_3 are tuned to have the wavelengths that resonate with the first, the second and the third MRR, respectively, and the signal sent to T_4 is tuned to have a wavelength that is off-resonance with all three MRRs. We can generalize the MRR usage in this simple router to a crossbar router where each initiator is directly connected to a target by a waveguide. Except for the directly connected initiator-target pairs, each initiator-target pair requires at least one MRR to resonate with the corresponding optical signals.

Depending on the coupling mechanism, an MRR can be used by either one or two signal paths. Figure 11.4 shows three different implementations of an optical switching element (OSE) with two input and two output ports that support different MRR coupling mechanisms:

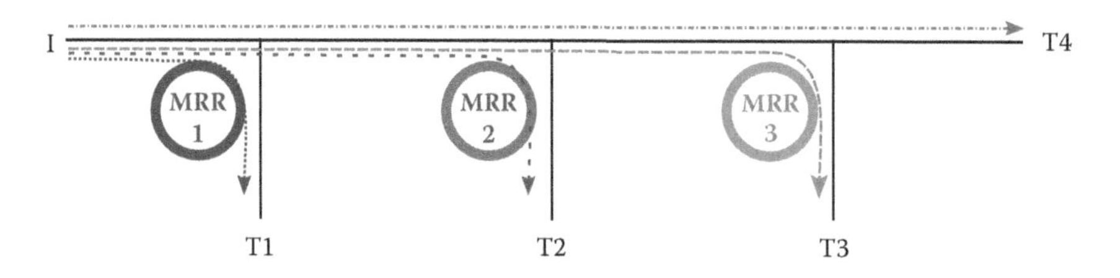

FIGURE 11.3 A simple WRONoC router consisting of three MRRs to route signals from one initiator to four different targets.

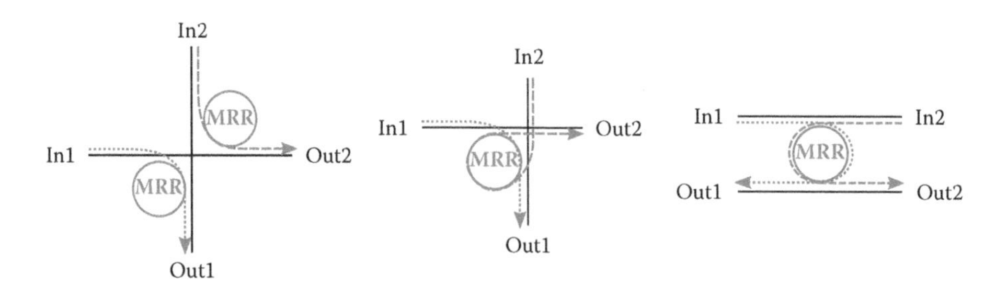

FIGURE 11.4 Different implementations of optical switching elements.

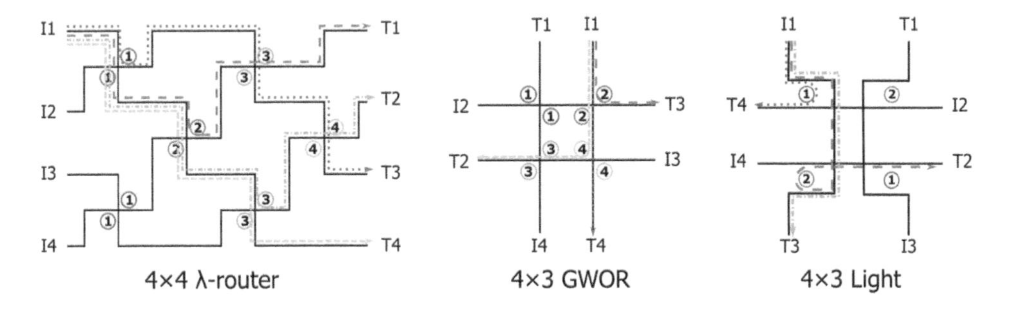

FIGURE 11.5 Three representative WRONoC routers.

- The left OSE consists of two MRRs placed in a diagonal direction so that on-resonance signals that enter the OSE will be coupled to the first MRR they meet and directed to the corresponding output-ports. One advantage of this OSE is its immunity to incoherent crosstalk. For example, when a signal enters the OSE from In1, most of its power will be directed by the bottom left MRR to Out1, while a small portion of its power will escape to the direction of Out2 and become crosstalk noise. However, before the crosstalk signal can arrive at Out2, it will be coupled to the upper-right MRR and experience a 270-degree change of the propagation direction. Eventually, it will still go to the same direction as the main signal and become its coherent crosstalk, which does not cause noise but only causes fluctuation of signal power (Shen, Lu and Gu 1999).
- The middle and the right OSEs consist of only one MRR. On-resonance signals that enter the OSE from different input ports will be directed to different output ports. By using one fewer MRR, these OSEs generally achieve better performance in energy efficiency and signal quality. However, these two OSEs also have their limitations: in the middle OSE, on-resonance signals that enter the OSE from In2 will experience a 270-degree change of the propagation direction, which results in much higher insertion loss and crosstalk noise than the usual 90-degree direction change; the OSE on the right is composed of two parallel waveguides, which require adaptations to fit the crossbar architecture.

In general, the minimum number of MRRs in a WRONoC consisting of n signal paths is $n - \omega$ if each MRR is used in only one signal path or $\left\lceil \frac{n-\omega}{2} \right\rceil$ if each MRR is used in two signal paths, where ω is the number of waveguides that directly connect an initiator to one of its target. For example, Figure 11.5 shows three manually designed WRONoC routers for a network consisting of four initiators and four targets. The λ-router (Briere, et al. 2007) supports 16 signal paths between all initiator-target pairs; the GWOR (Tan, et al. 2011) and the Light (Zheng, et al. 2021) support 12 signal paths under the assumption that an initiator does not communicate with a target of the same index. This assumption is realistic because an initiator and a target with the same index usually represent the in- and output ports of the same IP-core in the topology. All three routers directly connect each initiator to one of its targets by a waveguide, i.e., $\omega = 4$. In λ-router and GWOR, each MRR is used in only one signal path. Thus, λ-router uses 12 (i.e., 16 − 4) MRRs in total, and GWOR uses 8 (i.e., 12 − 4) MRRs in total. On the other hand, Light uses each MRR in two signal paths and thus uses only 4 (i.e., $\lceil \frac{(12-4)}{2} \rceil$) MRRs in total.

11.3 DESIGN AUTOMATION FOR WRONOCS

11.3.1 Necessity of Efficient Design Automation Methods for WRONoCs

Exhaustive exploration of the WRONoC design space is computational unaffordable. For example, it has been demonstrated that the number of $n \times n$ fully connected WRONoC topologies built upon a crossbar architecture and 2 × 2 OSEs amounts to: $[(n - 1)\cdot(n - 2)\cdots\cdot(n - n + 2)]^n$ (Tala, et al. 2016). Without systematic methods for design space exploration, manual design of WRONoC topologies strongly relies on the intuition of the designers.

In order to make the topologies generally applicable, WRONoC designers mostly assume a complete communication graph in which each initiator communicates with each target. This assumption is however often unnecessary. For example, in a 3-D stacked processor-memory network as shown in Figure 11.1, a memory controller typically only communicates with IP-core clusters but does not communicate with another memory controller, thus, it is not necessary to construct signal paths between two memory controllers. Furthermore, for short distances, electrical links are more energy-efficient than optical links as they do not require electrical-to-optical (E/O) and optical-to-electrical (O/E) conversions (Werner, Navaridas, and Luján 2017). Thus, it makes sense to combine a WRONoC with an electrical NoC so that signal paths between neighboring initiators and targets can be moved from the WRONoC to the electrical NoC.

For WRONoCs that do not require all-to-all communication, applying a fully connected router implies redundant usage of MRRs and wavelengths, which further results in higher power losses and worse signal quality. To design a customized WRONoC router with little redundancy for given communication requirements, computational efforts must be made to explore the design space in a systematic and yet efficient manner. Design automation for WRONoCs thus arises to solve this problem.

Current design automation works have demonstrated their capability of customizing WRONoC routers with minimum resource usage for small networks or sparse communication graphs (Li, Tseng, and Bertozzi, et al. 2018). However, available design automation tools hardly maintain their performances for large networks, as the design space increases exponentially with the number of the required signal paths. Thus, developing efficient design automation methods for WRONoCs is still a necessity. This section presents two design automation strategies for efficient WRONoC design.

11.3.2 Subtraction from Fully Connected Router

So far, WRONoC designers have proposed quite a few fully connected routers to support all-to-all communication between initiators and targets. Most of these routers, such as λ-router (Briere, et al. 2007), GWOR (Tan, et al. 2011), Snake (Ramini, Grani, et al. 2013), and Light (Zheng, et al. 2021), are well designed in terms of both scalability and resource-efficiency. In particular, the router structures can easily be generalized to support arbitrarily large networks, and the usage of wavelengths and MRRs in these routers is equal or very close to the theoretical minimum. These fully connected routers provide a good starting point for customization.

Specifically, given a communication graph of a target network consisting of n IP-cores, we can obtain a customized WRONoC router in three steps:

- *Select a fully connected router consisting of* n*initiators and* n*targets;*
- *Assign an initiator-target pair to the input- and output ports of each IP-core;*
- *Locate all redundant signal paths in the fully connected router and remove the wavelengths, MRRs and waveguide connections that are only used by the redundant signal paths.*

The target of the customization is to maximize the network performance with fewest possible optical resources. To this end, we need to choose the most appropriate fully connected router as the starting point and then select the most efficient signal paths to be retained in the router for data transmission.

In order to select the most appropriate fully connected router, we need to analyze the communication density and bandwidth requirements of the target application, and compare them to the SNR profiles of the fully connected routers. Figure 11.6 shows an analysis of the SNR distribution

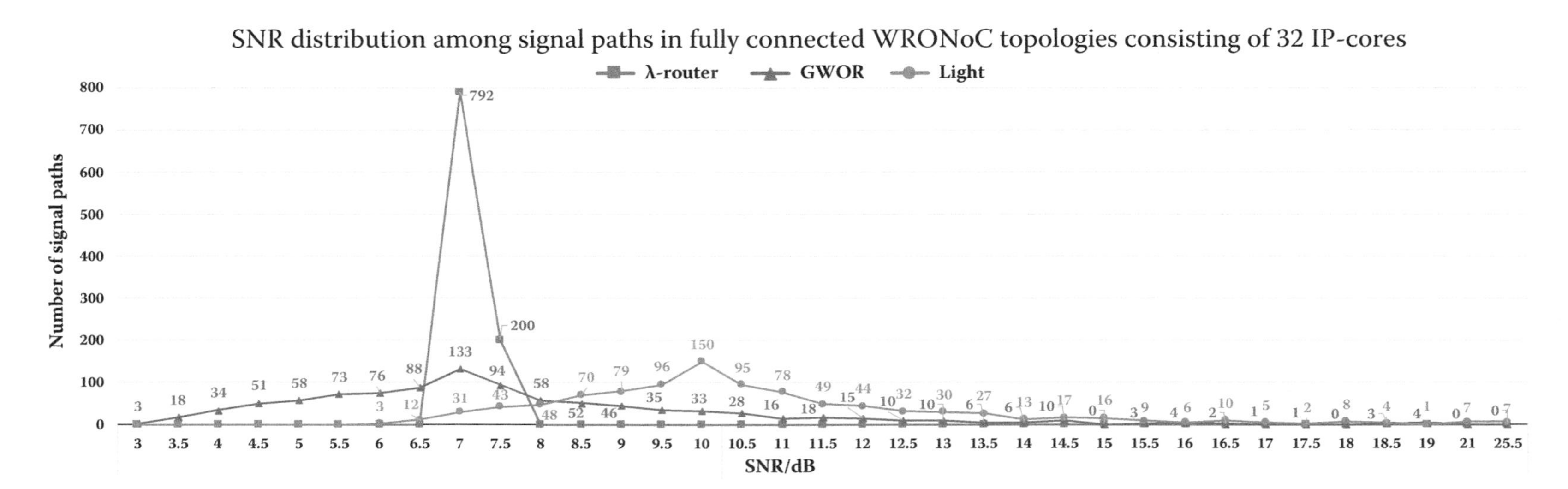

FIGURE 11.6 SNR distribution among different signal paths in three fully connected WRONoC topologies consisting of 32 IP-cores. The analysis only considers through loss, drop loss, crossing loss and first-order crosstalk noise. The SNR values are rounded to the nearest half. For example, 7.24 is rounded to 7 and 7.25 is rounded to 7.5.

among the signal paths in λ-router, GWOR and Light topologies consisting of 32 IP-cores. It is noteworthy that in λ-router, the SNR values of different signal paths are comparable, while in GWOR and in Light, there is a large variation of the SNR values among different signal paths. In general, if the communication graph is dense and the network performance of the target WRONoC application is dominated by the worst-case SNR, λ-router may be more appropriate for the customization. But if the communication graph of the target WRONoC is rather sparse, or the network has various bandwidth requirements among different signal paths, GWOR and Light may be more appropriate.

Once the initial router is determined, the next step is to map the input- and output- port of each IP-core to a pair of initiator and target, e.g., IP-core 1 may be mapped to initiator I_1 and target T_1. In a WRONoC router, the signal paths between all initiator-target pairs are fixed. Thus, the port assignment results decide the signal paths between the IP-cores. To achieve good network performance, one should try to assign the most efficient signals paths to the most critical communicating IP-cores, such as those that frequently exchange data or those that require high bandwidth. To note is that once the port assignment of two communicating IP-cores is determined, it does not only affect the signal paths between these two IP-cores, but also affects the signal paths that are used by other IP-cores to communicate with these IP-cores. For example, if we map the input-port of IP-core 1 to I_1, and the output-port of IP-core 2 to T_2, it does not only assign the signal path (I_1, T_2) to IP-core 1 and IP-core 2, but also confines that IP-core 1 can only use the signal paths starting from I_1 to send signals to other IP-cores and IP-core 2 can use only use signal paths ending at T_2 to receive signals from other IP-cores. Thus, the port assignment for different IP-cores should not be decided independently but should be optimized jointly.

Besides, the port assignment should also consider the physical proximity of the IP-cores on the optical plane. For example, in GWOR and in Light, initiators and targets with the same indices are usually placed next to each other in the logic topology. Thus, it is preferred to assign them to the input- and output-ports of the same IP-cores to reduce the waveguide routing efforts during the physical implementation. If two input- and output-ports are close to each other in the logic topology but far from each other on the physical plane, it may result in long detours of the waveguides or inevitable waveguide crossings.

At last, we should evaluate the consequent reduction of optical resources caused by removing each signal path. In a fully connected router, it is common to allow OSEs and waveguides to be shared among multiple signal paths. Thus, an OSE or a waveguide can only be removed from the fully connected router when all signal paths using it are not required in the customized router. In order to construct the network with fewest possible optical resources, we need to find out signal paths that share the same optical resources, and try to remove them together by assigning them to IP-cores that do not communicate.

The biggest advantage of this design strategy is that it can achieve a feasible customization solution in very short time. However, since the signal paths can only be selected from a fixed set of options provided by the fully connected router, the degree of customization is limited and the network performance greatly relies on the compatibility of the target network with the initial router (Le Beux, et al. 2013).

11.3.3 Template-Based Synthesis

Another strategy to improve the design efficiency for WRONoCs is to remove insignificant design options. In WRONoC design, most of the important performance factors such as insertion loss, crosstalk noise and SNR are dominated by a few crucial design features including MRRs, waveguide crossings, and wavelengths. In contrast, the shape and the lengths of waveguides contribute less significantly to the overall network performance. Thus, it is reasonable to confine the design space by fixing the insignificant design features and focusing on the potential variations of the significant design features.

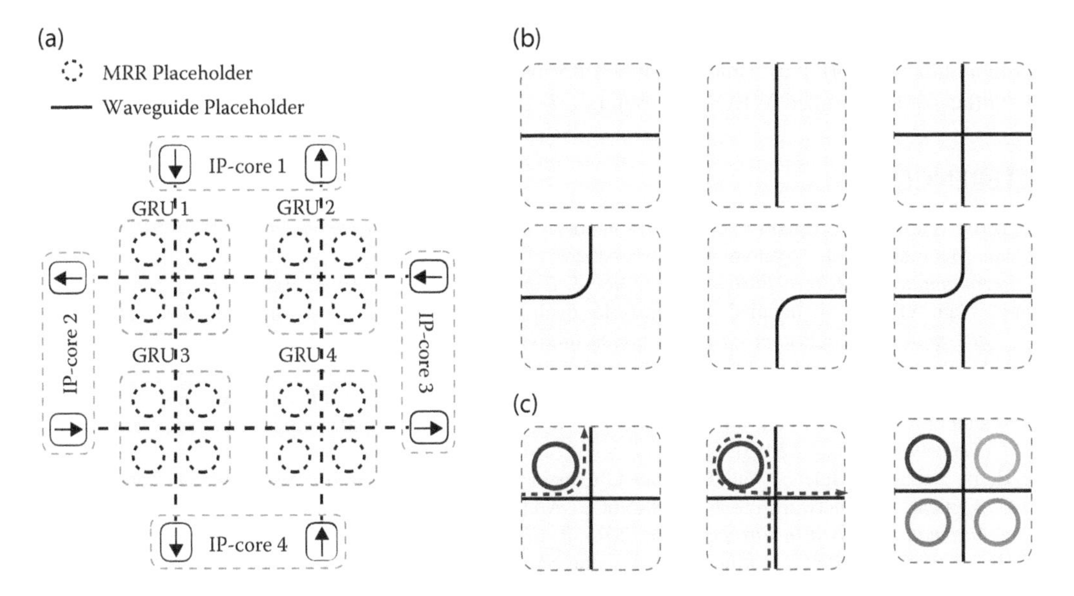

FIGURE 11.7 (a) A crossbar-based design template to interconnect 4 IP-cores. (b) Possible waveguide routing options. (c) Possible MRR configuration options.

We can model the significant design options with a design template, which defines a collection of waveguide- and MRR-placeholders to form a WRONoC router. Instead of exhaustively exploring the whole design space, we only explore the placeholders provided by the templates, and can thus greatly reduce the complexity of the problem.

For example, Figure 11.7(a) shows a crossbar-based design template (Truppel, Tseng, et al. 2019) (Truppel 2019Truppel, Tseng, et al. 2020) that interconnects four IP-cores. The template consists of four basic building blocks referred as *general routing units (GRUs)*. Each GRU further consists of four waveguide placeholders and four MRR placeholders. Figure 11.7(b) shows some waveguide connection options enabled with different combinations of the waveguide placeholders, and Figure 11.7(c) shows some MRR coupling options enabled with the MRR placeholders. Using this template, the WRONoC design problem is transformed into a discrete selection problem with a finite number of design options.

Besides the crossbar-based topology, a design template can also be built upon ring-based topologies (Le Beux, et al. 2011) or introduce MRR placeholders between parallel waveguide placeholders to cover the 180-degree MRR coupling options. The efficiency and the solution quality of the template-based synthesis hangs on the richness and the diversity of the design options defined in the template. In order to approximate the vast WRONoC design space with a limited set of options, we need to select the most significant design features and model their most representative variations. To note is that increasing the number of design options decreases the efficiency of the synthesis, but does not necessarily increase the solution quality if the additional design options are similar to the existing ones.

11.4 CONCLUSION

WRONoCs are promising optical interconnect architectures as they enable simultaneous and conflict-free communications among network components. However, in order to reserve all signal paths at design time, WRONoCs require a large number of optical resources, which further imply high power losses and crosstalk noise along the signal transmission paths. Thus, improving the network efficiency is crucial to the design of WRONoCs. Current research demonstrates that by customizing the networks to support application-specific connectivity, optical resources can be saved and the network performance can be improved. We address two design automation strategies to synthesize customized WRONoC routers efficiently.

REFERENCES

Bogaerts, Wim, et al. "Silicon microring resonators." *Laser & Photonics Reviews* 6, 2012: 47–73.

Briere, Matthieu, et al. "System level assessment of an optical NoC in an MPSoC platform." *Design, Automation & Test in Europe Conference & Exhibition (DATE)*. IEEE, 2007: 1–6.

Jiang, Nan, et al. "A detailed and flexible cycle-accurate network-on-chip simulator." *2013 IEEE international symposium on performance analysis of systems and software (ISPASS)*. IEEE, 2013: 86–96.

Le Beux, Sébastien, Ian O'Connor, Gabriela Nicolescu, Guy Bois, and Pierre Paulin. "Reduction methods for adapting optical network on chip topologies to 3D architectures." *Microprocessors and Microsystems (Elsevier)* 37, no. 1 (2013): 87–98.

Le Beux, Sébastien, Jelena Trajkovic, Ian O'Connor, Gabriela Nicolescu, Guy Bois, and Pierre Paulin. "Optical ring network-on-chip (ORNoC): Architecture and design methodology." *2011 Design, Automation & Test in Europe*. IEEE, 2011: 1–6.

Li, Hui, Sébastien Le Beux, Martha Johanna Sepulveda, and Ian O'connor. "Energy-efficiency comparison of multi-layer deposited nanophotonic crossbar interconnects." *ACM Journal on Emerging Technologies in Computing Systems (JETC)* 13, 2017: 1–25.

Li, Mengchu, Tsun-Ming Tseng, Davide Bertozzi, Mahdi Tala, and Ulf Schlichtmann. "CustomTopo: A topology generation method for application-specific wavelength-routed optical NoCs." *Proceedings of the International Conference on Computer-Aided Design (ICCAD)*. IEEE, 2018: 1–8.

Li, Mengchu, Tsun-Ming Tseng, Mahdi Tala, and Ulf Schlichtmann. "Maximizing the communication parallelism for wavelength-routed optical networks-on-chips." *25th Asia and South Pacific Design Automation Conference (ASP-DAC)*. Beijing: IEEE, 2020: 109–114.

Nikdast, Mahdi, et al. "Crosstalk noise in WDM-based optical networks-on-chip: A formal study and comparison." *IEEE Transactions on Very Large Scale Integration (VLSI) Systems* 23, 2014: 2552–2565.

Peano, Andrea, Luca Ramini, Marco Gavanelli, Maddalena Nonato, and Davide Bertozzi. "Design technology for fault-free and maximally-parallel wavelength-routed optical networks-on-chip." *IEEE/ACM International Conference on Computer-Aided Design (ICCAD)*. IEEE, 2016: 1–8.

Preston, Kyle, Nicolas Sherwood-Droz, Jacob S. Levy, and Michal Lipson. "Performance guidelines for WDM interconnects based on silicon microring resonators." *CLEO: 2011-Laser Science to Photonic Applications*. IEEE, 2011: 1–2.

Ramini, Luca, Davide Bertozzi, and Luca P. Carloni. "Engineering a bandwidth-scalable optical layer for a 3d multi-core processor with awareness of layout constraints." *2012 IEEE/ACM Sixth International Symposium on Networks-on-Chip*. IEEE, 2012: 185–192.

Ramini, Luca, Paolo Grani, Sandro Bartolini, and Davide Bertozzi. "Contrasting wavelength-routed optical NoC topologies for power-efficient 3D-stacked multicore processors using physical-layer analysis." *2013 Design, Automation & Test in Europe Conference & Exhibition (DATE)*. IEEE, 2013: 1589–1594.

Shen, Yunfeng, Kejie Lu, and Wanyi Gu. "Coherent and incoherent crosstalk in WDM optical networks." *Journal of lightwave technology* (IEEE) 17, no. 5 (1999): 759.

Tala, Mahdi, Marco Castellari, Marco Balboni, and Davide Bertozzi. "Populating and exploring the design space of wavelength-routed optical network-on-chip topologies by leveraging the add-drop filtering primitive." *Tenth IEEE/ACM International Symposium on Networks-on-Chip (NOCS)*. IEEE, 2016: 1–8.

Tan, Xianfang, Mei Yang, Lei Zhang, Yingtao Jiang, and Jianyi Yang. "On a scalable, non-blocking optical router for photonic networks-on-chip designs." *Symposium on Photonics and Optoelectronics (SOPO)*. IEEE, 2011: 1–4.

Truppel, Alexandre, Tsun-Ming Tseng, Davide Bertozzi, José Carlos Alves, and Ulf Schlichtmann. "PSION: Combining logical topology and physical layout optimization for Wavelength-Routed ONoCs." *ACM/SIGDA International Symposium on Physical Design (ISPD)*. ACM, 2019: 1–6.

Truppel, Alexandre, Tsun-Ming Tseng, Davide Bertozzi, José Carlos Alves, and Ulf Schlichtmann. "PSION+: Combining logical topology and physical layout optimization for Wavelength-Routed ONoCs." *IEEE Transactions on Computer-Aided Design of Integrated Circuits and Systems* 39, 2020.

Tseng, Tsun-Ming, Alexandre Truppel, Mengchu Li, Mahdi Nikdast, and Ulf Schlichtmann. "Wavelength-Routed Optical NoCs: Design and EDA-State of the Art and Future Directions." *IEEE/ACM International Conference on Computer-Aided Design (ICCAD)*. IEEE, 2019: 1–6.

Vizing, Vadim G. "On an estimate of the chromatic class of a p-graph." *Discret Analiz* 3, 1964: 25–30.

Werner, Sebastian, Javier Navaridas, and Mikel Luján. "A survey on optical network-on-chip architectures." *ACM Computing Surveys (CSUR)* 50, 2017: 1–37.

Ye, Yaoyao, et al. "3-D mesh-based optical network-on-chip for multiprocessor system-on-chip." *IEEE Transactions on Computer-Aided Design of Integrated Circuits and Systems* 32, 2013: 584–596.

Zheng, Zhidan, Mengchu Li, Tsun-Ming Tseng, and Ulf Schlichtmann. "Light: A scalable and efficient wavelength-routed optical networks-on-chip topology." *IEEE/ACM Asia and South Pacific Design Automation Conference (ASP-DAC)*. IEEE, 2021: 1–6.

Section IV

Novel Materials, Devices, and Photonic Integrated Circuits

12 Innovative DWDM Silicon Photonics for High-Performance Computing

G. Kurczveil, Y. Yuan, J. Youn, B. Tossoun, Y. Hu, S. Mathai, P. Sun, J. Hulme, and D. Liang

CONTENTS

12.1 INTRODUCTION

In the current big-data era, zettabytes of data are generated every year and the amount nearly doubles every two years. The conventional microprocessor and its I/O architecture are reaching physical limitations, leading to increased demands on memory systems due to frequent access patterns between processor and memory. As the global top commercial vendor of high-performance computing (HPC) solutions, Hewlett Packard Enterprise has dedicated enormous R&D resources for years to develop novel computing architectures, disruptive hardware and efficient software systems. High-speed, energy-efficient, and low-cost optical interconnects play a pivotal role in overall computing system innovation regardless for current dragonfly architecture, the revolutionary memory-driven computing concept, or next-gen all-to-all connected architectures.

In this chapter, we overview recent research and product development progress on key building blocks, an integration platform, wafer-scale testing, and packaging towards a silicon-based DWDM optical link solution. Starting from architecture, we discuss our rationale and strategies to realize scalable bandwidth and intrinsically better energy efficiency on low-cost silicon photonics and its large-scale production.

12.2 INTEGRATED DENSE WAVELENGTH DIVISION MULTIPLEXING ARCHITECTURE

Wavelength division multiplexing is one of the major technological differentiators between electrical and optical communications. While DWDM systems have been routinely deployed in

DOI: 10.1201/9780429292033-12

modern long-haul optical communications for decades, datacom applications in datacenters and HPCs are reluctant to adopt it due to control complexity in a temperature-fluctuating environment and much higher cost in each segment. The current coarse wavelength division multiplexing (CWDM) 4-λ solution is a main-stream choice for 100, 200, and 400 Gb/s pluggable transceivers. However, >100 Gbaud/s/λ modulation is extremely challenging to implement both in optical and electrical domains for post-400 Gb/s applications. Since coherent technology will inevitably cause more latency and power consumption, DWDM becomes a more viable solution for HPCs. The high integration density available through silicon photonics is a natural solution to the resultant increase in the number of components required.

Figure 12.1 is a schematic of our DWDM optical link architecture to provide >Tb/s bandwidth between high-radix switches. Compared with a traditional multi-wavelength source format of bundling up several single-wavelength lasers, comb lasers are the better choice here to generate tens or even over one hundred high-quality continuous-wave (CW) wavelengths. Fixed channel spacing among all adjacent wavelengths, with little change when temperature varies, is the primary operational advantage. An array of microring resonators cascaded along the same bus waveguide provides energy-efficient E/O data conversion and (de)multiplexing functionality simultaneously. In the receiver side, a similar array of microrings drops the signal to corresponding sensitive photodetectors (PD), such as avalanche photodetectors (APD), to finish O/E data conversion. To maximize the number of wavelengths and reach optimal comb laser operation, the free spectral range (FSR), i.e., channel spacing, can be 50 GHz or smaller and each comb line can contain a few mW of power. Narrow channel spacing inevitably increases the probability of crosstalk. A comb line power in the sub-mW range may be preferred due to a combination of low insertion loss in microring resonators, high receiver sensitivity, and nonlinear effects in Si nanowire waveguides. We, therefore, can add a wavelength deinterleaver or power splitter, or a combination of both between the comb source and modulator bank to effectively split one set of comb lines to N streams. The resultant equivalent larger channel spacing and/or lower individual comb line power will reduce risk of crosstalk and nonlinear effects, and increase bandwidth per laser by an integer time, largely reducing the energy consumption per bit and cost. Next we discuss recent progress in key building blocks to enable this DWDM architecture.

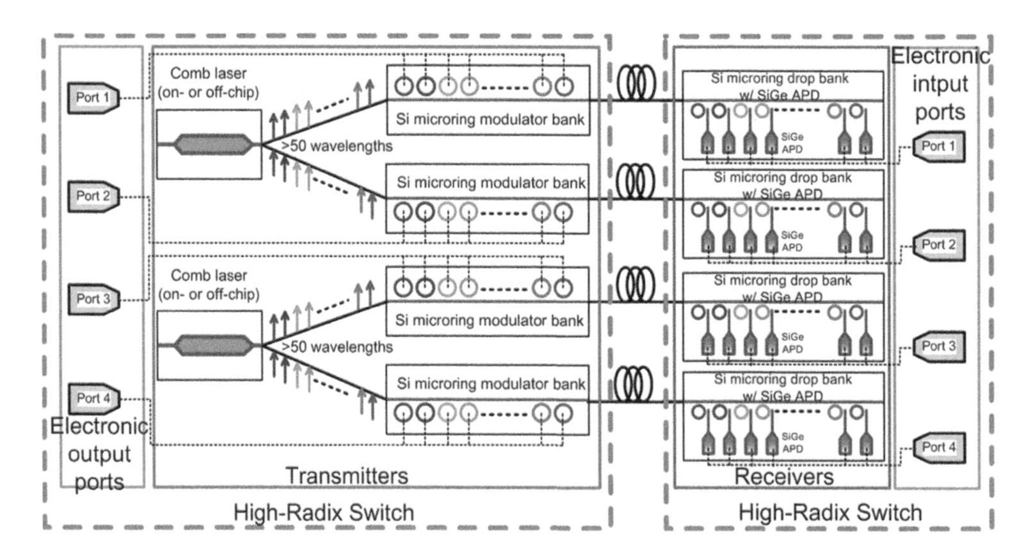

FIGURE 12.1 Illustration of DWDM optical transceiver links to enable large-bandwidth communications between high-radix switches.

12.3 ADVANCED SILICON PHOTONIC BUILDING BLOCKS

12.3.1 Multi-Wavelength Lasers

The light source is considered one of the most critical single optical components in the whole system. It is required to generate an optical signal with a required power intensity, good spectral purity and low noise. It can easily dominate the power budget and link cost, heavily affect the link data transmission architecture and packaging scheme, and finally but not least is likely to be a major limiting factor for overall system reliability. All those factors are amplified in datacom applications where the environment is harsh, space is limited and system cost margin is low, compared with the advanced but expensive long-haul fiber-optic systems.

A multi-wavelength source is the starting point for a DWDM-based optical communication architecture. As discussed above, comb lasers are particularly suitable as light sources where a high number of channels are needed. Quantum dot-based lasers are especially attractive comb lasers because the size distribution of the individual dots results in a wider spectral bandwidth than quantum well-based lasers [1]. In addition, due to their three-dimensional confinement of electrical carriers, quantum dot (QD) lasers also have an inherent benefit of efficient, high-temperature operation [2] as demonstrated by a CW laser that operated up to 220°C [3]. High-temperature operation is especially important in supercomputers, where room temperature operation is not an option.

The main requirement for comb lasers to be attractive for multi-channel links is that *each* individual comb line has to have low enough noise to enable error-free data transmission. For an amplitude-modulated signal, e.g., NRZ or PAM4, this means that the amplitude noise of each comb line must be low while the phase noise (optical linewidth) is less important as long as the signal fits inside the pass-band of the demultiplexer and modulator. While quantum well lasers can be used as comb lasers, they typically suffer from mode partition noise (amplitude noise in individual comb lines, while total power is constant), which greatly limits their use. On the other hand, QD-based comb lasers have shown low amplitude noise in each comb line [4].

The use of quantum dots has additional advantages. Because the dots are isolated islands they have a smaller overlap with the etched mesa sidewall. This results in a smaller current from non-radiative recombination (due to dangling bonds at the etched mesa) and therefore a smaller threshold current [5]. Finally, QD lasers were also shown to be less susceptible to optical feedback [6], which reduces the requirements for integrating isolators, thus reducing device complexity and costs. While this tolerance to feedback is the result of a low linewidth enhancement factor [7], it is more typically observed in *single-wavelength* QD lasers, while QD *comb* lasers have shown a high linewidth enhancement factor [8,9], and as a result they will perhaps not enjoy the advantage of isolator-free operation.

We are currently developing a comb laser chip on pure III-V substrate with a partner. Since existing III-V foundries are limited to relatively small wafer sizes and large processing nodes, it will hinder large-scale integration and low-cost manufacturing. Therefore, integrating QD comb lasers on silicon is a necessary step for future volume or full photonic integration needs. Direct wafer bonding of III-V material to silicon-on-insulator (SOI) wafers is the most promising short-term solution as it requires little alignment between the two materials. Wafer bonding also takes full advantage of the high-quality passive device properties (gratings, ring resonators, waveguides, etc.) that SOI wafers offer [10] as the optical mode can be efficiently transferred between the silicon waveguide and the QD-containing gain region [11].

Initial proof of concept devices consisting of simple Fabry-Perot lasers using wafer-bonding were demonstrated in [12]. A cross-sectional diagram and scanning electron micrograph of the device are shown in Figures 12.2(a-b), while a top-down diagram along with modal simulations are shown in Figures 12.3(c-e). The device consists of a 3 mm-long cavity formed by polished silicon mirrors and a 2 mm-long electrically pumped QD-based GaAs mesa. The cw light-current

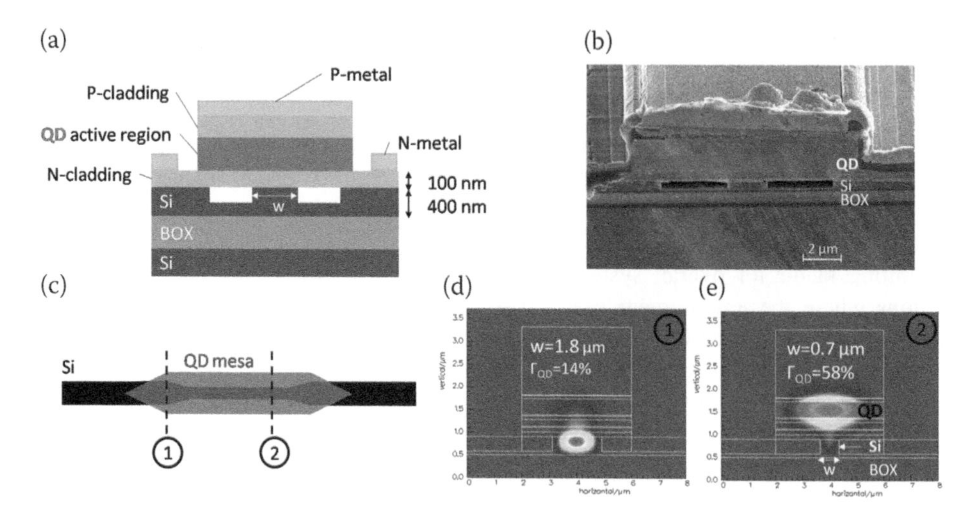

FIGURE 12.2 (a) Cross-sectional diagram of the wafer bonded QD laser. (b) Scanning electron microscope image of the fabricated device. (c) Top-down diagram of the device. Electrical contacts were omitted for clarity. Modal simulations for a wide silicon waveguide (d) and a narrow silicon waveguide (e).

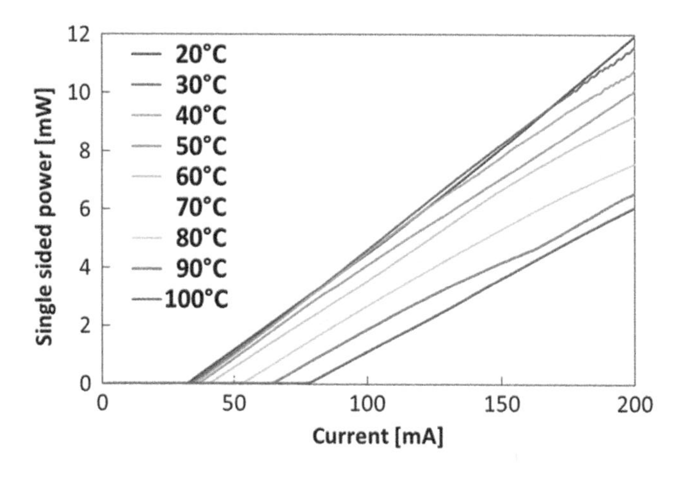

FIGURE 12.3 Light-current data as a function of stage temperature in continuous-wave operation.

characterization as a function of substrate temperature are plotted in Figure 12.3, showing operation up to 100°C. We note that the laser efficiency doesn't degrade much up to 100°C, and threshold current is relatively constant up to 60°C.

A more complex device was demonstrated in [13]. It is shown schematically in Figure 12.4(a), and it consists of a 2.3 mm-long cavity formed by a 100% back mirror and a 50% front mirror. The mirrors are based on multimode interferometers and loop mirrors. The gain region is 1.2 mm long, a 120 μm-long saturable absorber (SA) was placed at the center in the cavity, and they were electrically isolated using proton implantation. Light is coupled out of the device using a grating coupler to enable rapid wafer-level testing without having to dice and polish the wafer. A microscope image of the device is shown in Figure 12.4(b). The optical spectrum at gain and absorber biases of 395 mA and -6.2 V is shown in Figure 12.4(c), showing a relatively flat comb with a 3 dB bandwidth of 1.2 THz. We suspect that the spectral width of the comb is limited by the group velocity dispersion of the laser cavity. The channel spacing was measured to be 101 GHz, and it is the result of a coupled cavity that is formed between the main laser cavity (formed by the mirrors)

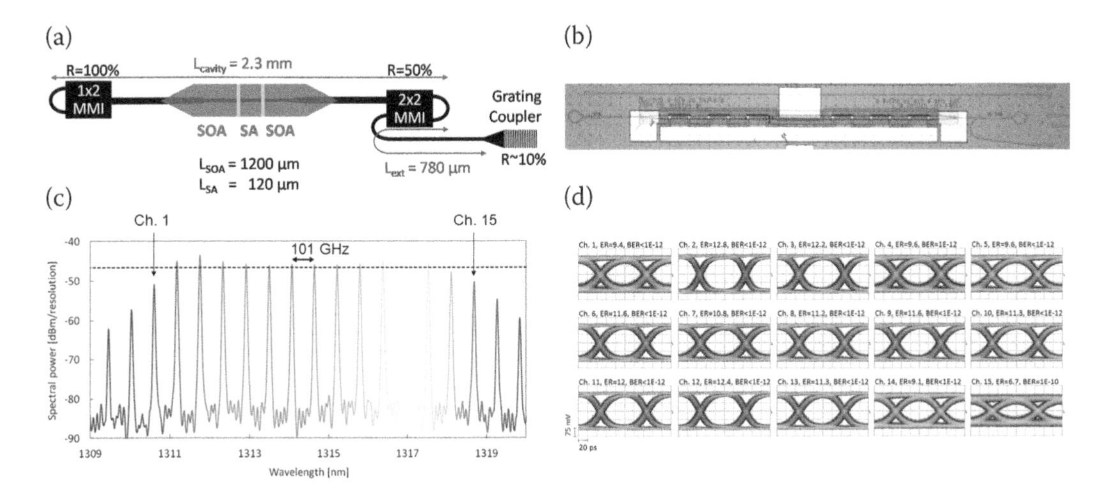

FIGURE 12.4 (a) Top-down diagram of the QD comb laser on silicon. On-chip mirrors and a grating coupler allow for rapid wafer-level testing without having to dice and polish out individual laser bars. (b) Optical microscope image of the fabricated device. (c) Optical spectrum at an SOA current of 395 mA and a SA voltage of −6.2 V. A coupled cavity determines the channel spacing. (d) Eye diagrams and BER for each comb line when using an external modulator.

and an external cavity formed by the front mirror and the grating coupler. The flat comb in this laser is a major advantage when compared to comb lasers that are based on Kerr non-linearities [14], which typically have a $sech^2$ amplitude distribution.

Data transmission experiments were carried out by filtering out each comb line using a tunable band-pass filter and modulating it with an external lithium niobate modulator at 10 Gb/s due to limit of high-speed instrument at the moment of experiment. The resulting eye-diagrams and bit error ratios (BER) are shown in Figure 12.4(d), and we note that 14 of the 15 measured channels show a FEC-free BER of $1{\cdot}10^{-12}$ or better. BERs higher than $1{\cdot}10^{-12}$ would require higher-latency and power consumption FEC which is not attractive for supercomputers.

While the layout of this comb laser looks similar to that of a colliding pulse mode-locked laser, this device does not emit optical pulses in the time domain for the bias conditions reported here. This is likely the result of a short gain recovery time. Having a CW output has two potential advantages. First, it results in a constant optical power (it lacks high instantaneous peak powers) and it therefore has a lower risk of triggering unwanted nonlinearities further down-link. Second, we hypothesize that the lack of high instantaneous powers also increases the device reliability.

In order to fully take advantage of the wide gain bandwidth of the QD gain material, the width of the comb needs to be extended. Initial simulations suggest that a careful balance between spatial hole burning, group velocity dispersion, and four-wave mixing are required for the widest combs [15], and experimental results show that comb lasers with a 3 dB bandwidth of 2.1 THz are possible as shown in a channel spacing of 15.4 GHz comb laser in Figure 12.5. Efforts are ongoing to integrate the on-chip comb lasers with modulators, photodetectors and other advanced silicon photonic circuits.

12.3.2 Efficient Phase Tuner and Modulators

Phase tuning is a critical function in a variety of photonic applications, and a phase tuner is a key element in many components like lasers, modulator, tunable add/drop filters, etc. For DWDM applications, individual device tuning is generally necessary to correct resonance drift from fabrication imperfection and external temperature variation, particularly for resonator-based devices

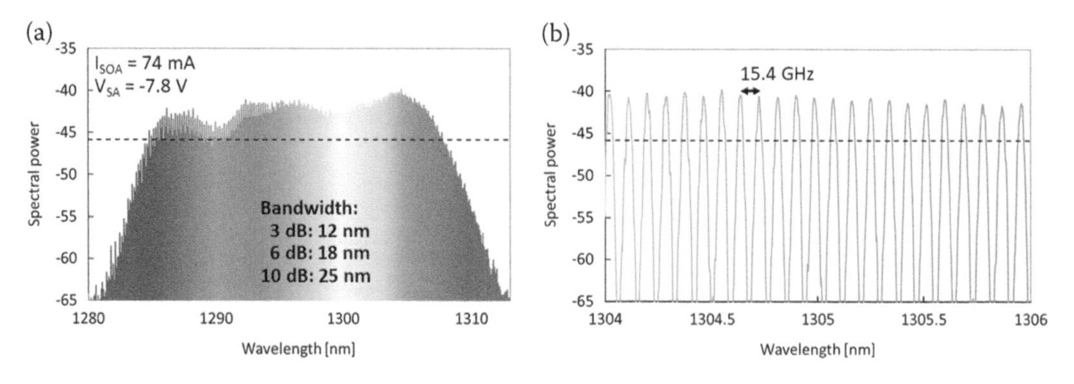

FIGURE 12.5 (a) QD comb laser with 3, 6, and 10 dB widths of 2.1, 3.1, and 4.4 THz, respectively. (b) The comb laser has a channel spacing of 15.4 GHz.

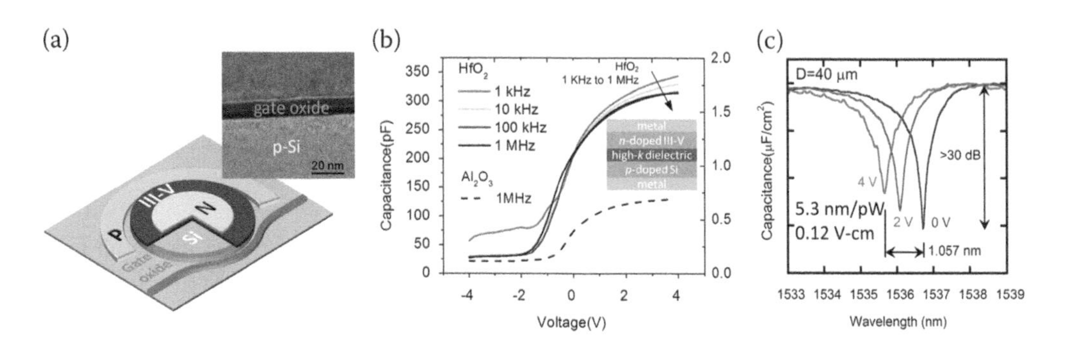

FIGURE 12.6 (a) TEM image of a heterogeneous MOSCAP and schematic of a microring resonator/modulator with MOSCAP integrated, (b) capacitance-voltage (CV) characteristics of heterogeneous MOSCAPs. Inset is a schematic of the MOSCAP structure, (c) measured spectra showing microring resonance shift was a function of MOSCAP bias.

like lasers and microring resonators, in order to avoid crosstalk between adjacent channels. It adds significant complexity in control circuitry and potentially more power consumption due to traditionally power-hungry tuning methods: slow thermal tuning or carrier injection which leads to self-heating effect as well [16].

We recently developed a novel heterogeneous metal-oxide-semiconductor capacitor (MOSCAP) structure which can provide a fine tuning mechanism with essentially zero power consumption [17], and is fully compatible with the entire heterogeneous silicon photonic platform [18]. The heterogeneous capacitor can be conveniently formed in this heterogeneous device platform by sandwiching a layer of high-quality thin dielectric, i.e., gate oxide like SiO_2 or Al_2O_3, at the bonding interface between III-V and Si during the wafer bonding step. When this capacitor is charged or discharged, the carrier concentration around the gate oxide can change several orders of magnitude to exceed 10^{20} cm^2. So it can enable a fast plasma dispersion effect for phase and power tuning and high-speed modulation. Figure 12.6(a) shows the cross-sectional transmission electronic microscopy (TEM) image of such a heterogeneous n-InP/Al_2O_3/p-Si MOSCAP, and a schematic of a microring resonator/modulator where the entire microring cavity overlaps with the MOSCAP [19].

The MOSCAP capacitance is a function of capacitor dimension and gate oxide property. By sandwiching high-k dielectric (e.g., HfO_2), we could easily obtain larger capacitance with the same gate oxide thickness. Circular-shaped test structures were made for MOS capacitor test, with the schematic shown in the inset of Figure 12.6(b). The measured capacitance-voltage (CV) characteristics from 1 kHz to 1 MHz are shown in Figure 12.6(b). For comparison, Al_2O_3- and HfO_2-based MOS structures with the same ~9.6 nm thickness were fabricated in parallel. The measured capacitance of the HfO_2-based MOS capacitor at 1 MHz is 1.7 $\mu F/cm^2$, which is 2.5 times that of the Al_2O_3-based MOS. The dispersion at 4 V is about 8.7% from 1 kHz to 100 kHz, but ~0 from 100 kHz to 1 MHz [19].

Wavelength blue shift over 1 nm under 4 V bias was demonstrated in this microring resonator device (Figure 12.6(c)) with larger capacitance design by using thinner gate oxide (~10 nm) with a high dielectric constant. It is equivalent to a very small $V_{\pi}L$ of 0.12 V/cm, approximately 10X smaller than pure silicon MOS-type modulators. It shows sufficient tuning capacity for correcting wavelength deviation from fabrication imperfection, locking and ~10°C temperature change (based on wavelength shift ~0.1 nm/°C in typical telecom diode lasers). Extremely small leakage current in the fA level through the gate oxide translates into a wavelength tuning power consumption as good as 5.3 nm/pW, over one billion times better than traditional thermal and carrier injection tuning. Since the heterogeneous MOS structure is generic, it can be readily integrated in heterogeneous lasers as well to form a novel three-terminal structure [18]. The integrated MOSCAP is an independent control knob to fine tune laser wavelength and output power. Such structure has been demonstrated in our heterogeneous InP-based MQW microring lasers, and can be applied to heterogeneous GaAs-based QD comb lasers to lock wavelengths in grid (typically channel spacing <1 nm) with essentially zero energy consumption. It is worth note that the plasma dispersion effect governed by Drude model is related to the effective mass and carrier mobility in semiconductor materials. So III-V materials like InP and GaAs have more efficient phase turning and simultaneously less free carrier absorption loss [20].

High-speed Mach–Zehnder interferometer (MZI) modulators based on this heterogeneous MOSCAP have been demonstrated as well. Up to 32 Gb/s NRZ modulation with help of a pre-emphasis driving signal [21] shows promising potential to quickly catch up or surpass the best result of 40 Gb/s demonstrated in pure silicon MOS counterparts [22]. At the time of writing this chapter, we just measured excellent performance up to 28 Gb/s in our most recently fabricated microring-type heterogeneous MOSCAP modulator without pre-emphasis driving signal. It was fabricated on the same chip containing high-performance lasers, amplifiers and photodetectors, which clearly paves the way to debut soon a fully integrated heterogeneous DWDM transceiver with ultra-high energy efficiency and scalable bandwidth.

12.3.3 Robust Si-Ge Avalanche Photodetectors

A photodetector is the final key building block to fulfill the mission of E-O-E conversion in any optical communication. As a widely used receiver on the silicon photonics platform, silicon-germanium (Si-Ge) avalanche photodiodes (APDs) have been a critical research element in our silicon photonics effort. In order to satisfy the high-operating temperature demand for data centers and high-performance computers, robust and high-efficiency Si-Ge APDs are explored.

As shown in Figure 12.7, a waveguide structure Si-Ge APD with thin Si and Ge layers was developed to enable high speed, high quantum efficiency (QE) and high sensitivity simultaneously. It consists of a 400 nm p-type Ge absorption layer, 50 nm p-type Si charge layer, 100 nm unintentionally doped (UID) Si multiplication layer, and 220 nm n + -type Si contact layer, which is known as a separate absorption and charge multiplication (SACM) structure [23]. The Si-Ge APDs dimension is determined by the Ge layer size, 4 μm wide and 10 μm long in this case. Thanks to the p-doped Ge absorption layer, the electric field in the Ge layer is low enough to reduce tunneling probability. In addition, the p-type doped absorption layer can reduce the relaxation time of majority holes generated by photons. This leads to a higher bandwidth like with uni-traveling-carrier (UTC) photodiodes [24]. The 100 nm thin Si multiplication layer results in a low breakdown voltage, high speed, low excess noise, and robust thermal stability.

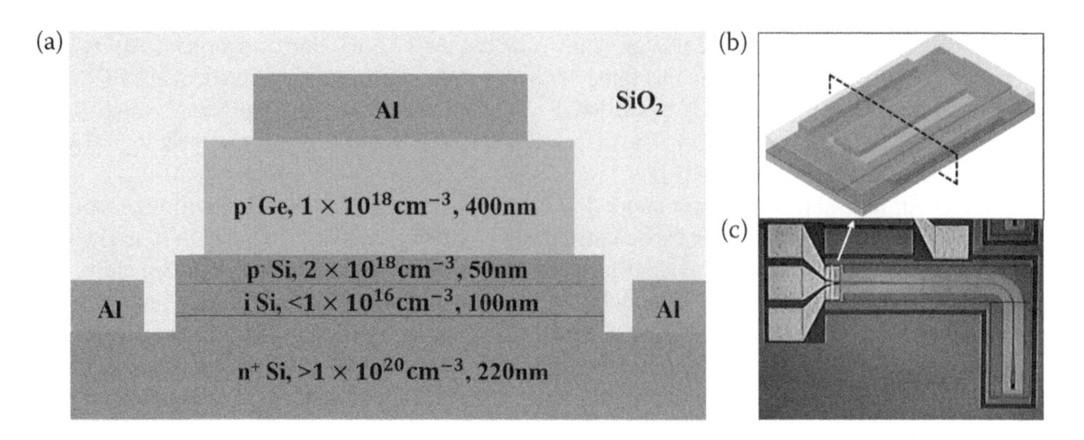

FIGURE 12.7 (a) Cross-sectional and (b) bird's eye-view schematic diagram of the waveguide Si-Ge SACM APD, and (c) top-view of a fabricated device.

For applications with dynamic operating temperature, APDs with robust thermal stability are desired. One significant figure of merit is the gain-temperature stability. Typically, it is characterized by the temperature coefficient of breakdown voltage, i.e., $\Delta V_{bd}/\Delta T$. At high temperature, increased phonon scattering will result in carrier energy loss, hence a higher electric field is needed to provide ionization threshold energy. The $\Delta V_{bd}/\Delta T$ not only depends on the multiplication layer material, but also depends on the device thickness. For SACM APDs, the device thickness contributing factor includes two parts: depletion region width, w_d, and multiplication region width, w_m. The temperature coefficient relationship between SACM and p-i-n APDs with the same multiplication region can be expressed by Equation 12.1 [25].

$$\Delta V_{bd}/\Delta T(\mathrm{SACM}) = \Delta V_{bd}/\Delta T(p - i - n) \times w_d/w_m \quad (12.1)$$

Figure 12.8 illustrates the gain curves of the Si-Ge SACM APD from 23°C to 90°C. As expected, the gain curve deviates to a higher bias as the temperature increases. The extracted breakdown voltages versus temperature are shown in the upper right corner, $\Delta V_{bd}/\Delta T$ value only around 4.2 mV/°C. It is much lower than other InAlAs-InGaAs and InP-InGaAs SACM APDs at telecommunication wavelength [26–33], as shown in Figure 12.8(b). The remarkably low $\Delta V_{bd}/\Delta T$ of the Si-Ge APD is due to its ultra-thin w_m and w_d. For the ultra-thin APDs, a higher electric field is needed to achieve the same gain, thus carriers acquire the ionization threshold energy faster. It leads to less phonon scattering during carrier acceleration, so that impact ionization events are less sensitive to temperature. Thanks to the p-doped Ge, the absorption layer is not depleted at breakdown voltage, and w_d is only around 0.15 μm. This contributes to the excellent $\Delta V_{bd}/\Delta T$.

In addition, the Si-Ge APD exhibits a stable bandwidth and gain-bandwidth product (GBP). The bandwidth versus gain characteristic at 30°C and 90°C is demonstrated in Figures 12.9(a-b), respectively. The 3 dB bandwidth decreases from 26.0 GHz to 24.6 GHz, and GBP reduces from 282.4 GHz to 241.1 GHz. As illustrated in Figure 12.9(c), the temperature related degradation is about 22 MHz/°C (~ 0.09%/°C) for bandwidth and 0.695 GHz/°C (~0.24%/°C) for GBP, which is a small decay percentage.

The dark current of the Si-Ge APD shows a uniform increase with temperature, results in higher thermal noise. Fortunately, the Si-Ge APD exhibits an elevated internal QE at high temperatures, which can compensate for the APD's sensitivity. The photocurrent at unity gain region versus 1,550 nm injected laser power is represented by Figure 12.10(a), the internal QE is the slope of the

(a)

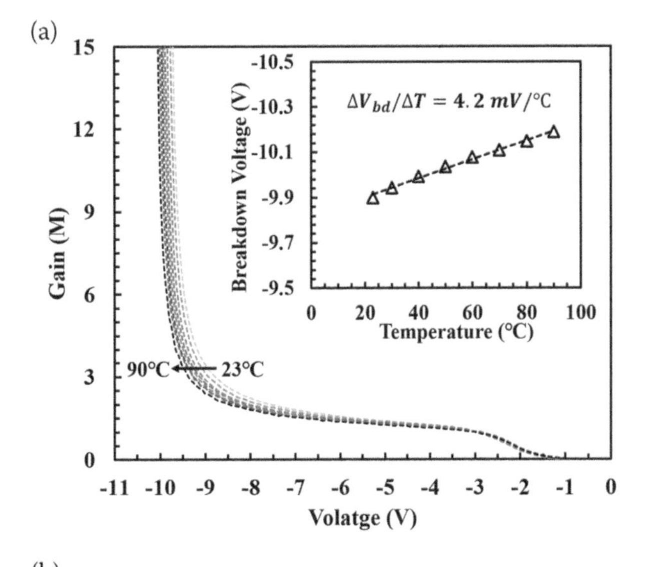

(b)

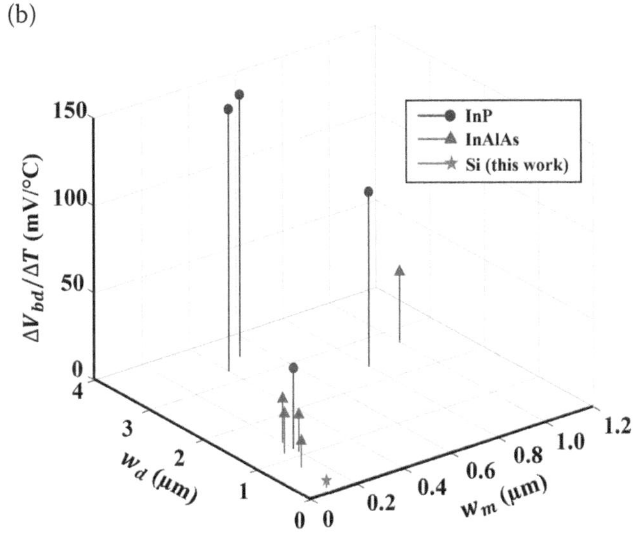

FIGURE 12.8 (a) Multiplication gain and breakdown voltages at different temperatures of the Si-Ge SACM APD. (b) Experimental $\Delta V_{bd}/\Delta T$ of different telecommunication SACM APDs.

dash linear lines. As displayed in Figure 12.10(b), the measured internal QE increases from 56% at 23°C to nearly 100% when temperature higher than 80°C. The increased temperature expands the Ge lattice constant, and thus introduces a smaller bandgap. The square root of the increased Ge absorption coefficient is proportional to temperature [34], $(\Delta\alpha)1/2 \propto T$, hence the absorption coefficient and internal QE yield as below.

$$\alpha(T) = [C \times (T - T_0)]2 + \alpha(T_0) \tag{12.2}$$

$$QE(T) = 1 - \exp[-\alpha(T) \times L] \tag{12.3}$$

In here, C is a fitting constant, T_0 is the initial temperature (23°C), and L is the absorption layer length (10 µm). The fitted QE(T) is shown as the dash line in Figure 12.10(b), it behaves a great agreement with measured QE dots.

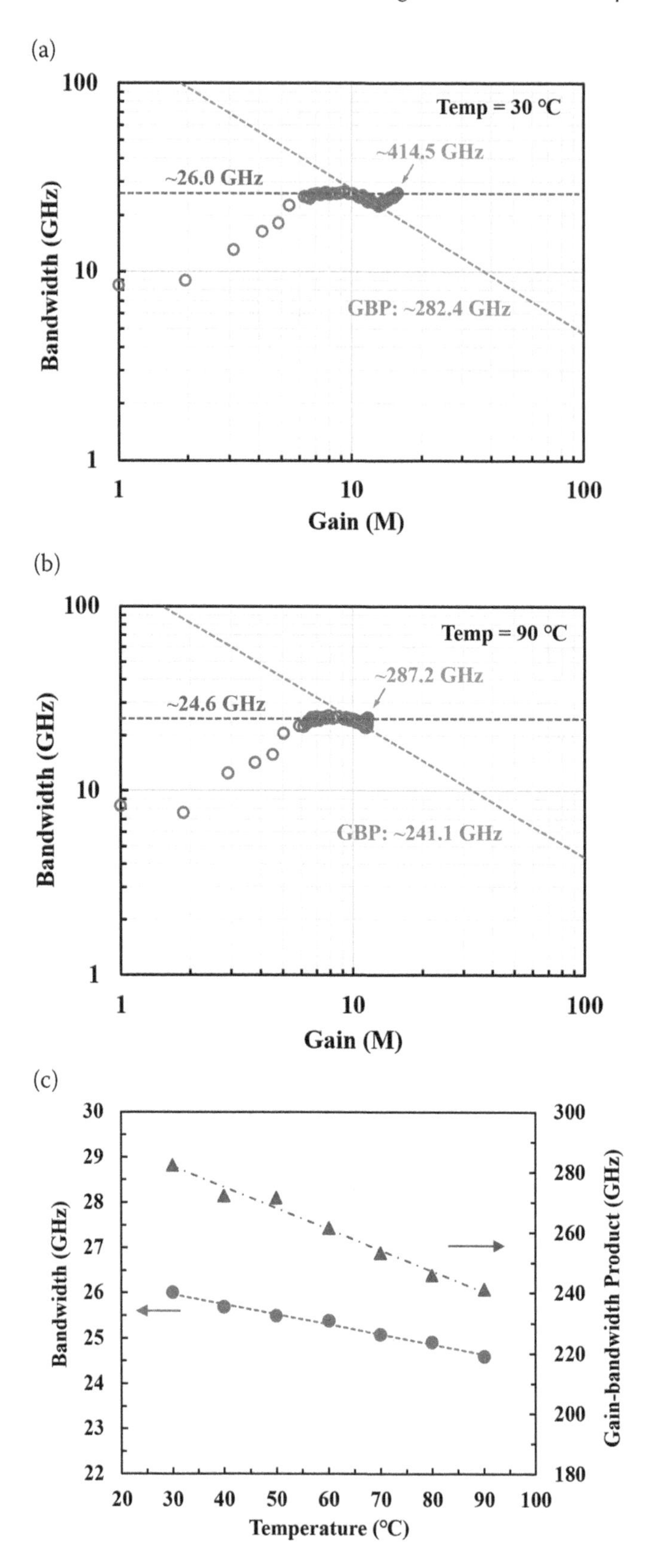

FIGURE 12.9 Bandwidth versus gain at (a) T = 30°C, (b) T = 90°C; (c) bandwidth and GBP versus temperature of the Si-Ge APD.

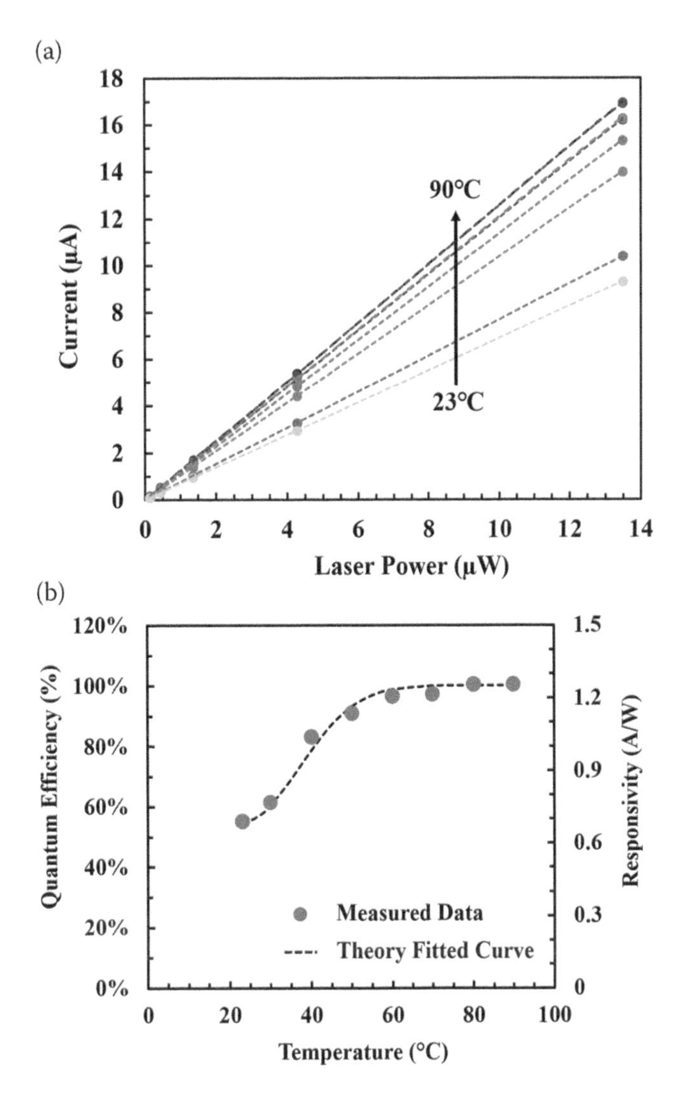

FIGURE 12.10 (a) Photocurrent versus 1,550 nm injected laser power, and (b) internal QE and responsivity of the Si-Ge SACM APDs at different temperatures.

The enhanced QE is benefit to eye diagrams, Figures 12.11(a-b) show the measured eye diagrams at 32 Gb/s NRZ and 64 Gb/s PAM4 modulation (PRBS9) with gain of M ~ 6, 8, and 11.5 at 30°C and 90°C, respectively. As temperature increases, the level height of measured eye diagrams is improved to compensate the drawback from high dark current. Therefore, the Q-factor of the Si-Ge APD is also insensitive to temperature.

In an ongoing effort to develop a product development kit (PDK), a Si-Ge APD behavioral model is necessary so that it can be co-simulated with receiver circuitry based on industry-standard simulation platforms for electronic design such as SPICE and Spectre simulators [35,36]. APDs amplify carriers generated by photon absorption through an internal avalanche gain that depends on the applied bias voltage, however, APD noise current is also amplified by the avalanche gain. Hence, in order to optimize APD-based receiver performance, the best signal-to-noise ratio (SNR) should be achieved by carefully determining the bias voltage and optimizing the noise characteristics of receiver circuits [37]. Thus, APD behavioral models should describe signal as well as noise currents over a wide range of bias voltages.

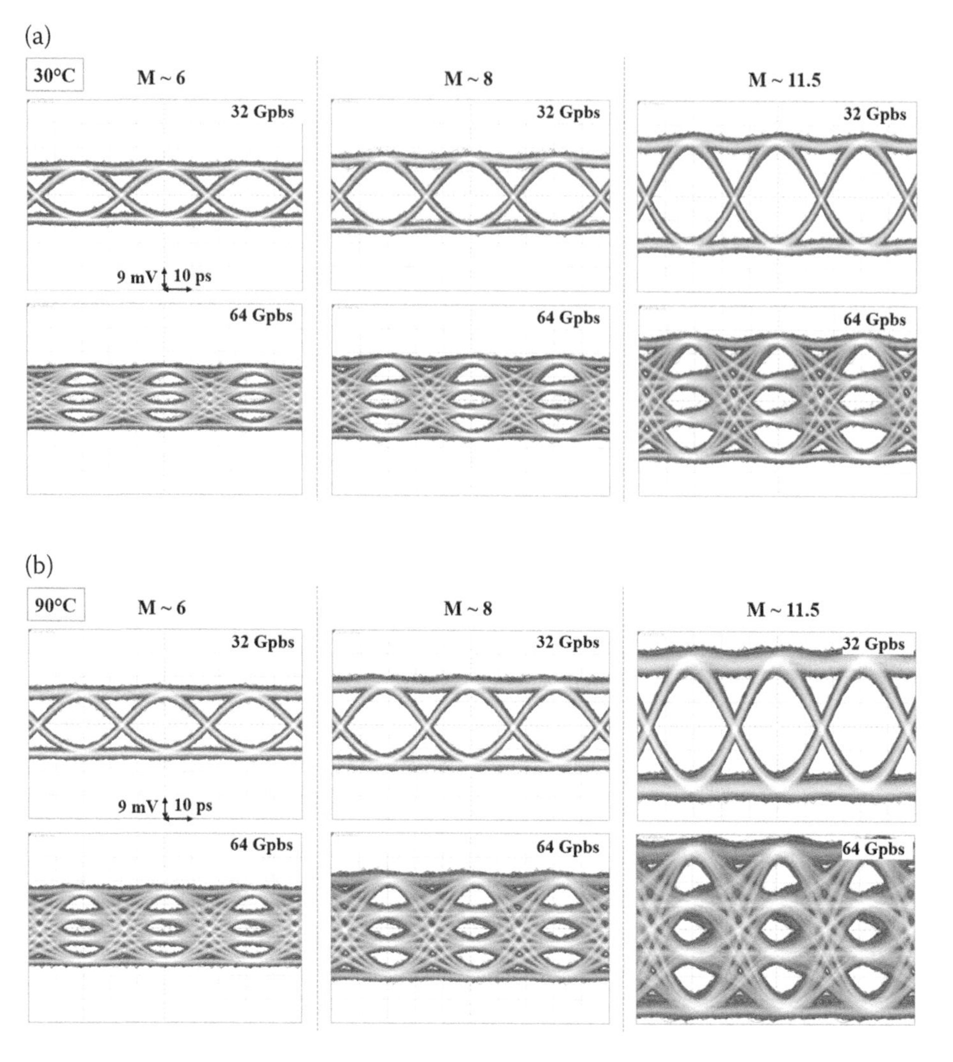

FIGURE 12.11 The 32 Gb/s NRZ and 64 Gb/s PAM4 eye diagrams of Si-Ge waveguide APDs with M ~ 6, 8, and 11.5 at (a) 30°C and (b) 90°C.

Figure 12.12 illustrates a simplified diagram of APD behavioral model which is purely described with Verilog-A language. The APD model has full compatibility with the SPICE-class simulators. The photocurrent (I_{ph}) is determined by an input optical power (P_{opt}), an intrinsic responsivity (R_0), and a multiplication gain (M). The transfer function, H(ω), contains an electro-optic behavior such as avalanche build-up and carrier-transit time as well as parasitic effects, and they have frequency-dependent characteristics. As for the shot noise ($i_{n,\ shot}$), it can be calculated as follows:

$$i_{n,shot}^2 = 2qM^2F\left(R_0P_{opt} + I_{dark}\right)BW_n \tag{12.4}$$

where q is the electron charge, F is the excess noise factor, I_{dark} is the APD dark current, and BW_n is the receiver noise bandwidth. The thermal noise ($i_{n,thermal}$) can be explained as follows:

$$i_{n,thermal}^2 = \left(4kT/R_L\right)BW_n \tag{12.5}$$

where k is the Boltzmann's constant, T is the absolute temperature, and R_L is a load resistance. Also, by incorporating an equivalent input-referred noise of receiver circuits ($I_{n,RX}$) into the shot and thermal noises, the overall SNR at the output of APD can be evaluated.

Figure 12.13 shows the simulated eye diagrams with the APD behavioral model at different avalanche gains and with the different signaling schemes. Figures 12.13(a-c) present both simulation and measurement results with 32 Gb/s NRZ pattern at avalanche gain of about 6, 8, and 11.5, respectively [38]. As shown in the figure, the simulated signal amplitude as well as eye height and width are well-matched with the measured eye diagrams. Besides, Figures 12.13(d-f) show the correlation results with 64 Gb/s PAM4 pattern, and it also has a good agreement. With such an APD behavioral model, silicon photonics and receiver circuit

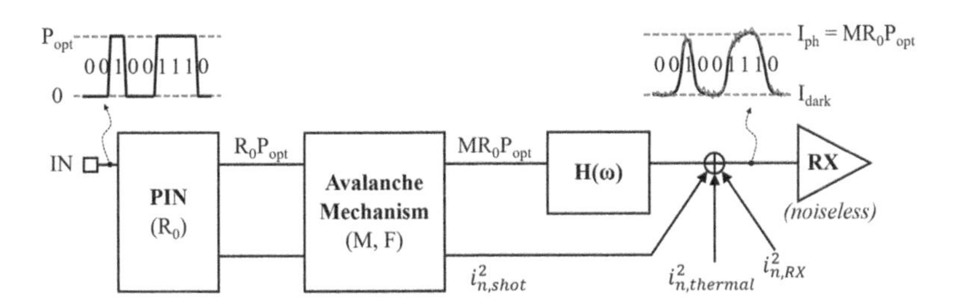

FIGURE 12.12 Simplified block diagram of APD behavioral model.

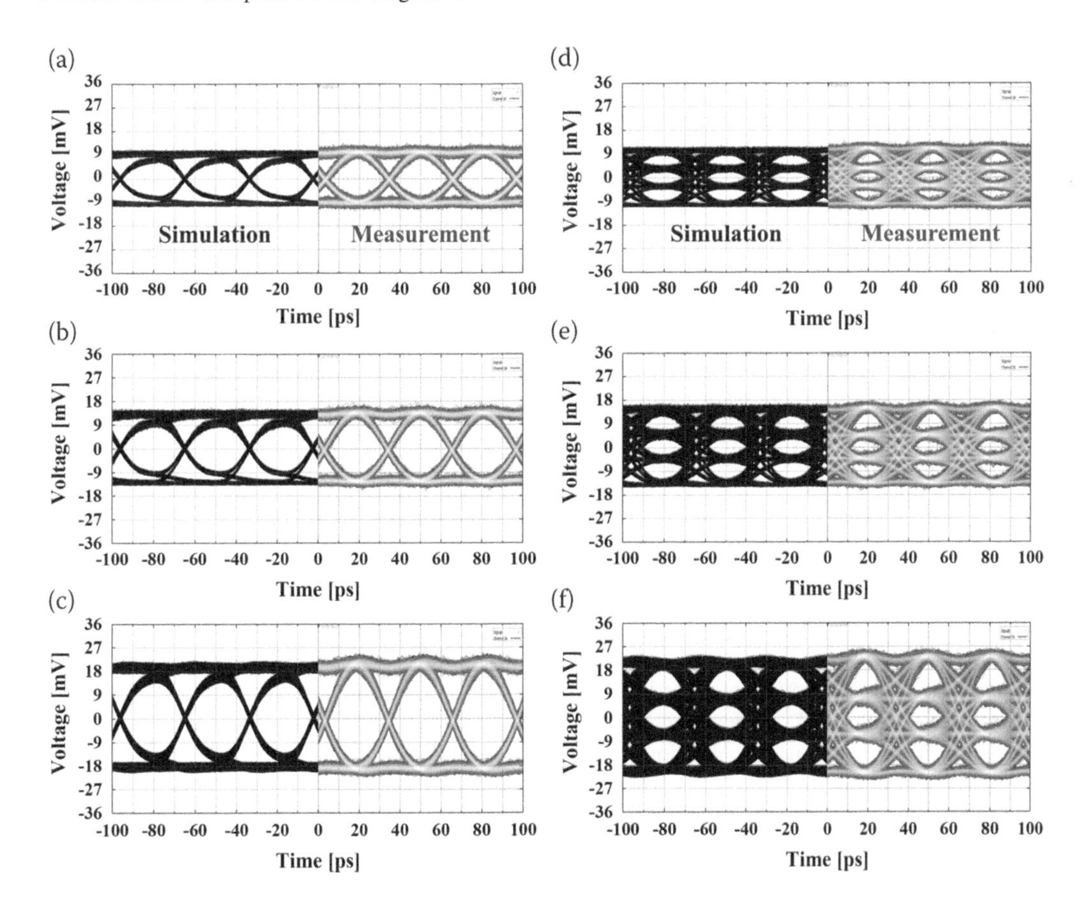

FIGURE 12.13 Simulated and measured eye diagrams with 32 Gb/s NRZ pattern when (a) M ≈ 6, (b) M ≈ 8, and (c) M ≈ 11.5 and 64 Gb/s PAM4 pattern when (d) M ≈ 6, (e) M ≈ 8, and (f) M ≈ 11.5.

designers can optimize not only the SNR of APD-based receivers but also the entire photonic link performance.

12.3.4 High-Gain Heterogeneous Quantum-Dot Avalanche Photodetectors

While Si-Ge APD is our flag-ship receiver choice for excellent efficiency, sensitivity, robustness and CMOS-compatible fabrication process, a recent attempt to study heterogeneous QD photodetectors resulted in a pleasant surprise. These photodiodes were notably made from the same epitaxial layers and fabrication process as QD comb lasers designed for DWDM links, significantly simplifying the processing for a fully integrated transceiver on silicon.

Fabricated devices were characterized on the same temperature-controlled stage for heterogeneous laser testing. Figure 12.14(a) plots the I-V curves of an 11 μm × 60 μm photodiode at different temperatures. A record-low dark current of 10 pA (3.3×10^{-7} A/cm^2) at −1 V at 300 K was demonstrated from a 11 μm × 30 μm photodiode, which is three orders of magnitude lower than the lowest reported dark current (1×10^{-3} A/cm^2 at -1 V) in p-i-n Ge-on-Si detectors [39]. The dark current density scales linearly with device area, signifying that the main contribution to the dark current is from surface leakage current and not from the bulk of the device. Such record-low dark current density can be attributed to the high crystal quality and low dislocation density of the III-V material, as well as sufficient surface passivation of the PD mesa. When reverse bias increases, photocurrent rises up quickly, which is a sign of avalanche gain. The dark current was measured to be 50 μA around -18 V, which is near the breakdown voltage. Figure 12.14(b) shows temperature-dependent avalanche gain. A maximum gain of 45 was measured at 20°C and avalanche gain was still possible up to 60°C. The temperature dependence on the breakdown voltage reveals that impact ionization of free carriers is the primary physical mechanism responsible for the breakdown of the device. As temperature increases, more carriers gather sufficient energy to escape the QDs through thermionic emission and contribute to dark current.

It is also very interesting to discover polarization-dependent gain variation. In Figure 12.15(a), the TE responsivity of 0.06 and 0.34 A/W for an 11 μm × 90 μm waveguide APD were measured at unity gain (-4 V) and -9 V, respectively. The maximum external responsivity of 2.7 A/W was obtained at room temperature. The absorption coefficient was extracted from the responsivity to be about 900 cm^{-1}. A maximum 6 × increase in responsivity and 2 × increase in gain was observed with changes in the input optical polarization from TE to TM. The gain is both wavelength and polarization dependent, suggesting that the carrier injection and multiplication processes may differ with wavelength and polarization [40]. This could be due to differing carrier populations within each energy level of the QDs with a change in the input optical wavelength. At shorter wavelengths, carriers are generated in the excited state within the QDs and require less energy to escape the QDs and impact ionize. At longer wavelengths, carriers are generated at lower energy states within the quantum dots and require more energy in order to escape the QDs and then contribute to gain [41]. We believe that avalanche multiplication occurs in the GaAs spacer layers between the QDs [42,43]. It is also possible that multiplication occurs within the InAs quantum dot material, as suggested in [44]. The change in the absorption process in the QDs with polarization may be due to the generation of light holes in the TM, as opposed to the TE polarization [45]. With more light holes being generated, the absorption and gain both increase because more light holes escape from the QDs and impact ionize. It is also possible that impact ionization also occurs with trapped carriers in the QDs so that a higher number of impact ionization events would occur with more trapped light holes [46].

The output frequency response of the PDs were measured at -16 and -18 V at room temperature with an input wavelength of 1,300 nm of different polarizations in Figure 12.15(b). A maximum 3-dB bandwidth of 15 GHz for an 11 μm x 30 μm PD was observed. With a change from TE to TM

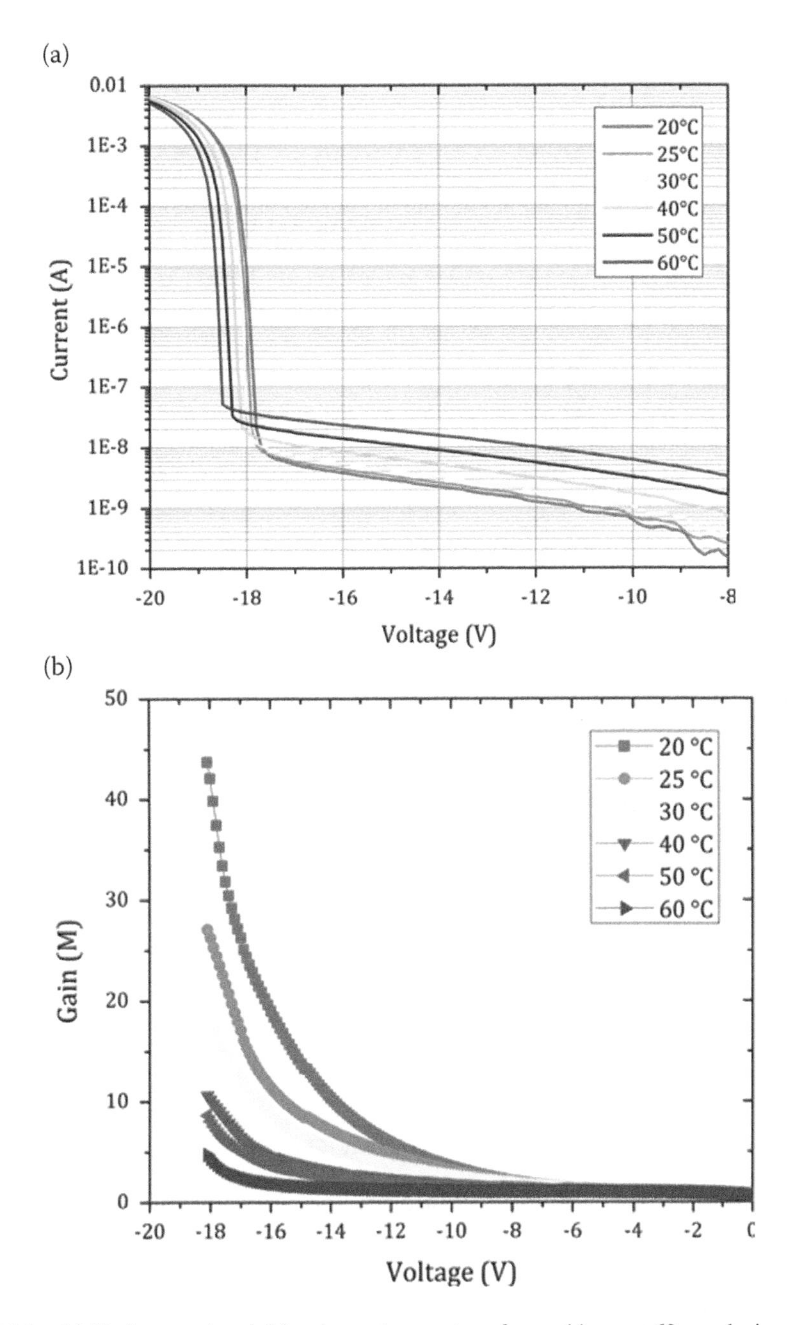

FIGURE 12.14 (a) Dark current and (b) gain vs. temperature for an 11 μm × 60 μm device at 1,310 nm.

polarization while measuring a 12 μm x 50 μm waveguide APD, the bandwidth fell from 2 GHz to 1 GHz, while the gain increased from 150 to 300. The maximum gain-bandwidth product is about 300 GHz at a bias voltage of about -17.9 V for both TE and TM polarizations, which is the record for any QD APD on silicon, and is comparable to Ge-on-Si counterparts [47]. However, the avalanche build-up time is also longer due to higher impact ionizing events pointing to an inherent gain-bandwidth trade-off, which can be extrapolated from the gain and bandwidth at different input optical polarizations.

Large signal characterization and link demo test were also conducted. Light from a previously discussed heterogeneous comb laser was amplified by a Praseodymium-doped fiber

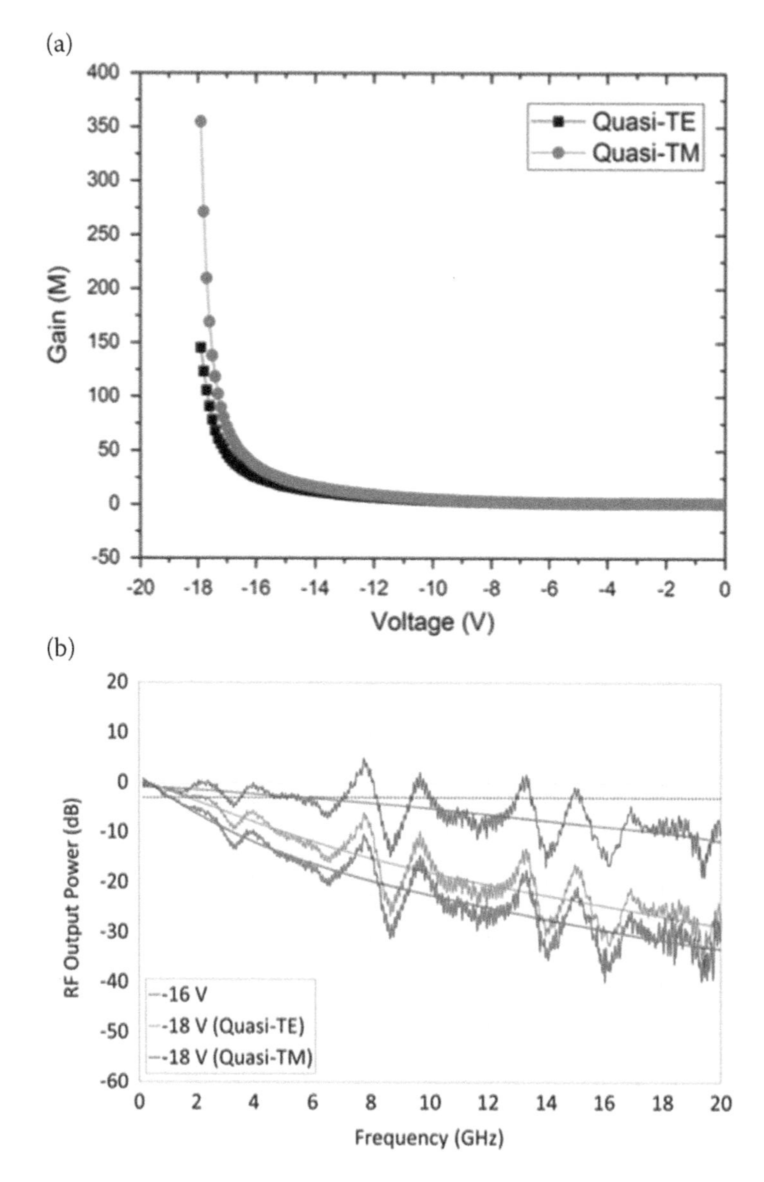

FIGURE 12.15 (a) Gain of quasi-TE mode and quasi-TM mode, and (b) S21 frequency response of a 12 μm × 150 μm device.

amplifier (PDFA) to compensate the passive losses and filtered to select a single comb line. An external modulator encoded NRZ data on the optical carrier before launching into the QD APD. Figures 12.16(a-c) show open electrical eye diagram of a 12 μm × 150 μm APD with a gain of 46.8 at 16, 18, and 20 Gb/s, respectively. A bit error rate (BER) test was conducted using an Anritsu Bit Error Rate Tester at 10 Gb/s. At a gain of 28, the sensitivity was measured to be about -11 dBm at a BER of 1×10^{-12} and -14.6 dBm at a BER of 2.4×10^{-4}, as shown in Figure 12.17. This sensitivity is a few dB higher than that of a typical Ge-on-Si p-i-n PD [48]. The sensitivity can be increased by wire bonding the QD PD to a TIA, as done in a case with a Ge-on-Si APD [47]. Ongoing study on avalanche physics in QD structure will enable more advanced designs to further enhance the performance and make QD APD an attractive receiver alternative besides Si-Ge counterparts.

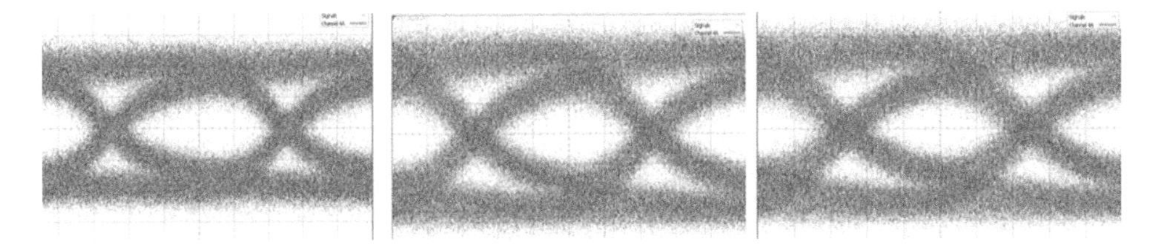

FIGURE 12.16 Measured QD APD electrical eye diagrams at 16, 18, and 20 Gb/s at room temperature.

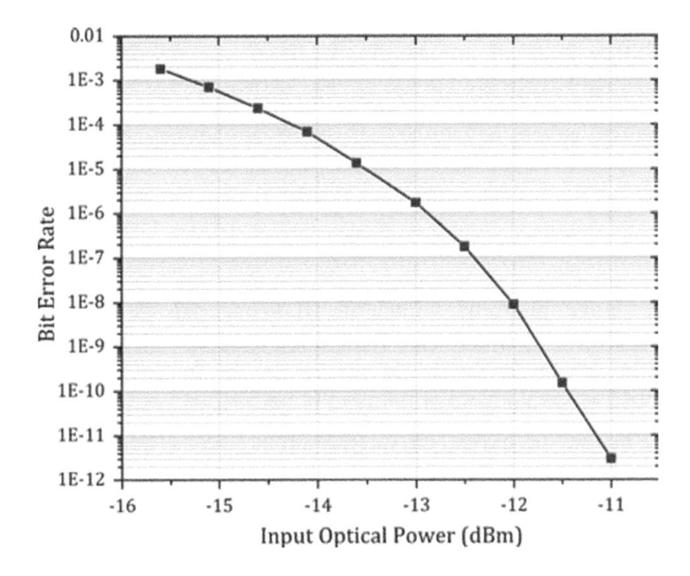

FIGURE 12.17 Bit error rate vs. input optical power of an 11 μm × 90 μm PD at a gain of 28.

12.4 INNOVATIVE INTEGRATION PLATFORM DEVELOPMENT

With all high-performance key building blocks in place, we are pushing the most mass-production ready choice of a pure silicon-based transceiver plus off-chip pure III-V comb source to product development frontline, and a heterogeneous version as the second-generation DWDM solution next. More recently, direct monolithic III-V epitaxy on silicon has regained huge interest as the ultimate silicon-based integration platform choice [49,50]. Compared with the heterogeneous wafer bonding approach, direct monolithic III-V growth on a 300-mm silicon wafer could lead to additional huge cost savings via cheaper substrates and wafer-scale III-V epitaxy. However, when directly growing III-V materials on silicon substrates, the lattice mismatch, the difference in thermal expansion and the different polarities of the materials result in a large number of crystalline defects known as dislocations on the order of 10^8 cm^{-2} [50,51]. This large density of dislocations drastically degrades device performance and lifetime [51].

Combining various dislocation suppressing/filtering methods with a quantum dot active region which is more immune to dislocations than a quantum well is an attractive solution that has recently resulted in tremendous progress. However, it is not ready to integrate with other silicon photonics because of the difficulty in achieving efficient light coupling from the III-V active region to the silicon waveguide due to the inevitable several μm-thick buffer layers and extra optical loss when light propagates in the dense dislocation zones. Therefore, it is desirable to develop a monolithic laser and photonic integration platform on SOI substrates with convenient optical coupling to Si waveguides, a low dislocation density and a scalable, cost-effective manufacturability.

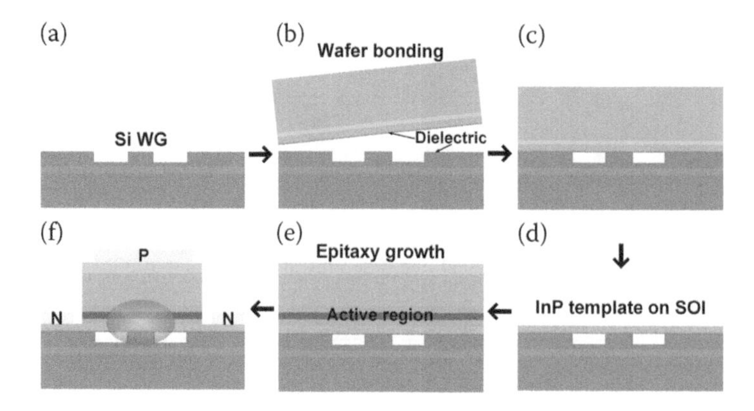

FIGURE 12.18 Schematic process flow to fabricate defect-free heterogeneous platform and silicon light source: (a) silicon waveguide formation, (b) dielectric deposition, (c) III-V-to-silicon wafer bonding, (d) bulk III-V substrate removal, (e) III-V epitaxy growth, and (f) III-V mesa formation and metallization.

By combining the advantages of monolithic growth and wafer bonding approaches, HPE researchers developed a novel platform to integrate III-V materials into SOI substrates, aiming to provide a defect-free, low-cost, wafer-scale integration scheme [52,53]. Figure 12.18 shows the schematic process flow of the defect-free integration scheme. First, silicon waveguides and some bonding-facilitating structures are created in the SOI substrate. Second, a simple 150 nm-thick n-InP layer is transferred onto the SOI wafer by depositing a thin dielectric layer on both surfaces, conducting direct wafer bonding, and then selectively removing the InP substrate. Upon selectively removing the InGaAs etch stop layer to expose the n-InP template layer, the wafer-bonded InP-on-SOI substrate is loaded into a metallic organic chemical vapor deposition (MOCVD) chamber for epitaxy growth at 600°C. Finally, the wafer is processed for laser device fabrication. This bonded template eliminates lattice and polarity mismatches in conventional III-V-on-Si direct epitaxy, so thick intermediate buffer layers are not necessary.

After III-V material regrowth on the bonded template, detailed material characterizations were conducted to investigate the quality of the epitaxy. Figures 12.19(a-b) show the heterogeneous wafer with the transferred n-InP template and the wafer under device fabrication after epitaxy, respectively. The heterogeneous wafer exhibited good yield and robustness after bonding annealing at 300°C, epitaxy growth at 600°C and post-epitaxy device fabrication. Figures 12.19(c-d) are the atomic force microscopy (AFM) images of the epitaxy surface on InP and the bonded substrate, respectively, both showing an identical surface roughness with a root mean square (RMS) value of 0.2 nm. Figures 12.19(e-f) show the cross-sectional transmission electron microscopy (TEM) images of the MQW epitaxy on the bonded InP-on-SOI substrate in the Si waveguide region. No defect was observed in a 10-μm-long and 0.1-μm-thick TEM specimen. MQW layers with good contrast and integrity are clearly exhibited in the high-magnification TEM image in Figure 12.19(f). Since we were unable to find any threading dislocation (TD) from cross-sectional TEM imaging, we then performed plan-view TEM observations on a relatively large area of $30 \times 12\ \mu m^2$. Nevertheless, no TD but some misfit dislocations was observed with the plan-view TEM. Those misfits were observed at the interface between the upper SCH and InP cladding layer and within the InP cladding layer close to the interface by tilting cross-sectional TEM specimens. The existence of the misfits is explained as the result of thermal strain in the InP bonding template and the following epitaxy due to the difference in their thermal expansion coefficients and implies that the TDs are far away across the observed plane-view TEM area. Furthermore, we used electron-channeling contrast imaging (ECCI) to quantify the dislocation density in the plane view. Figure 12.19(g) shows the electron channeling patterns corresponding to the three-beam (400) and (220) imaging conditions that were used. Figure 12.19(h) is a representative image with only one

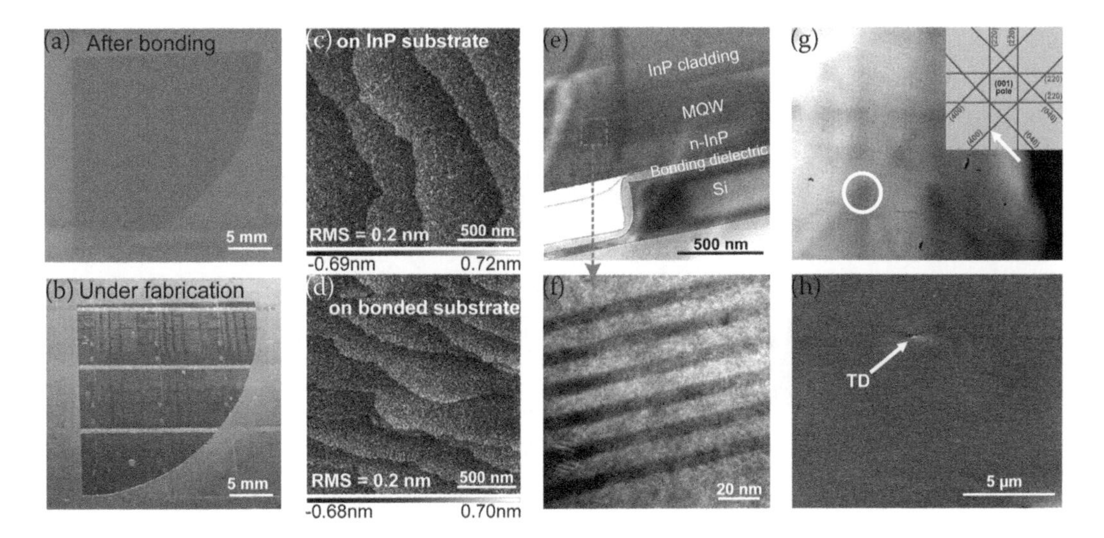

FIGURE 12.19 Pictures of (a) bonded template and (b) post-epitaxy device fabrication; AFM images of the surface of epitaxy on (c) InP and (d) bonded substrates; (e, f) TEM and ECCI (g-h) images of epitaxy on bonded substrate.

TD. A total of 20 TDs were counted in 100 images with a total mapping area of 100 × 14.5 × 14.5 μm^2. This led to a dislocation density of 9.5 × 10^4 cm^2, only one order of magnitude higher than that on native InP substrates and two orders of magnitude lower than state-of-the-art conventional monolithic growth with a thick buffer layer [54]. Additional ECCI investigations observed small areas with more concentrated TDs, which were likely caused by bonding voids or dirt particles. We applied the same ECCI process to an InP witness sample for comparison but could not see any TD due to a very low dislocation density for epitaxy on the native substrate.

High-resolution X-ray diffraction (XRD) ω-2θ (rocking curve) measurements and Photoluminescence (PL) measurements were carried out on the epitaxy samples, the results strongly support the low TDD observation with ECCI and verify that the epitaxy quality is comparable to that on the InP substrate.

Fabry-Perot (FP) lasers were fabricated on the grown epitaxy by treating the regrowth wafer as a conventional heterogeneous wafer and applying the same fabrication procedure. The fabricated FP lasers with diced and polished heterogeneous facets without coating, i.e., facets with a III-V mesa on the SOI waveguides were characterized. The fundamental mode profile is shown in Figure 12.20(d) (inset). Figures 12.20(a-b) show schematic drawing of the device cross-section and SEM of the hybrid facet, respectively. Figures 12.20(c-d) show the respective light-current-voltage (LIV) curves at RT (20°C) and LI curves up to a stage temperature of 40°C, both of which are under the pulsed injection mode (0.5 μs, 0.25% duty cycle). The 1.9-mm-long device starts lasing at 61.8 mA and emits 4.2 mW from a single facet under a 120-mA current injection, corresponding to a decent threshold current density of 813 A/cm^2 and an overall slope efficiency of 0.14 W/A. The observation of lasing at approximately 1,313 nm under the pulsed mode at RT in Figure 12.20(e) matches our MQW design well. Figure 12.20(f) shows CW LI curves up to a stage temperature of 20°C, with a slightly increased threshold due to device joule heating.

To prove convenient integration with other Si photonic circuits, the laser devices with two 50-μm-long III-V-to-silicon taper structures were measured. Silicon waveguide laser facets were formed by the same dicing and polishing procedure without coating. Figures 12.20(g-h) show the LIV at RT and the LI curves up to a stage temperature of 35°C under the same pulsed mode. The threshold current density of a 2.1-mm-long device with a 2.0-mm-long active region

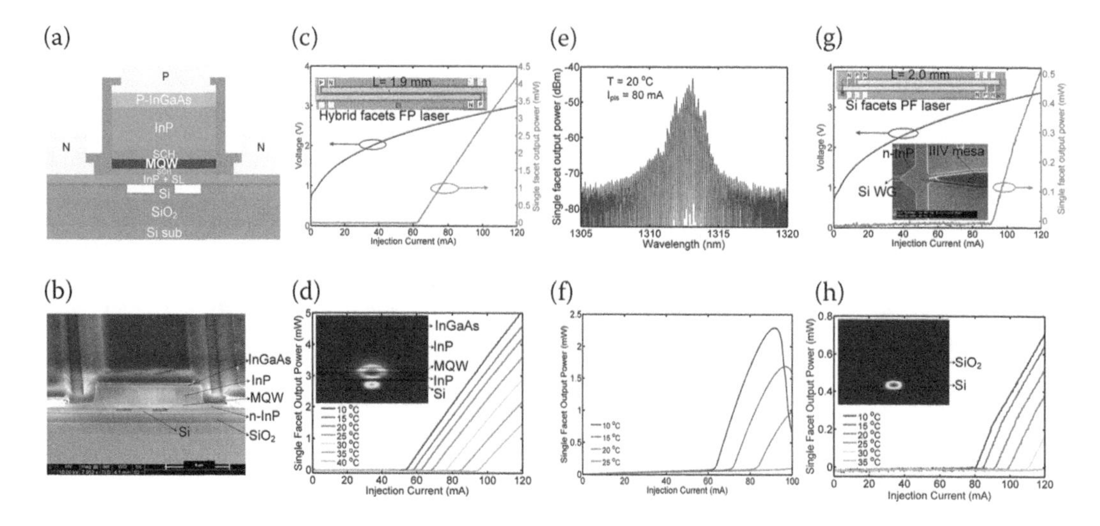

FIGURE 12.20 (a) Device cross-section. (b) SEM of the hybrid facet. Hybrid facet lasers: (c) RT pulsed LIV (inset: microscope image of the device), (d) pulsed LI up to 40°C (inset: mode profile at facets), (e) device spectrum. (f) CW LI up to 25°C. Si facets lasers: (g) RT pulsed LIV (inset: microscope image of the device and SEM of the taper). (h) pulsed LI up to 35°C (inset: mode profile at facets).

was calculated to be 1,125 A/cm^2. Due to the thin bonding template and unnecessary thick buffer layer, the output light from this laser can readily couple from the heterogeneous section to the Si waveguide section via III-V-to-Si tapers. The simulated fundamental output mode profile at the Si facet is shown in the inset of Figure 12.20(g). It is noted that we experienced some fabrication issues that significantly limited the device performance. Despite the growth and fabrication errors, this proof-of-concept demonstration paves a new way for large wafer-scale material and device heterogeneous integration with a number of merits, as discussed in the next section.

The main advantage of conducting epitaxy growth on the bonded template is eliminating two out of three major dislocation root causes: lattice and polarity mismatches between the substrate (e.g., silicon) and the function material (e.g., InP-based III-V) from epitaxial growth. The thermal mismatch between the substrate and template material would still cause dislocations in the regrown materials, but it is measured to be at a significantly low level. According to the aging tests of InAs QD lasers on Si near room temperature [55], a reduction in the TDD from 10^8 cm^{-2} to 10^6 cm^{-2} can extend the laser lifetime from a few months to over 100 years. It is reasonable to expect that lasers from the regrowth on the bonded template with even lower dislocation density would eliminate the defect-induced lifetime concerns for all practical applications. It is noted that the critical thickness of InP on a SiO_2/Si substrate is calculated to be 200~430 nm from a conventional model [56,57]. The thickness from the interface of the bonding dielectric and InP bonding template to the observed misfit locations is approximately 450 nm. It is likely that the observed misfits are formed when the epitaxy thickness (including the InP template) reaches the critical thickness. Further study is expected to confirm this estimation.

Epitaxy of this thickness with such a low dislocation density promises the possibility of growing many standard III-V structures for electronic, photonic, and MEMS applications. Additionally, we believe that this bonding plus epitaxy approach is a generic method for many other heterogeneous material combinations. The substrate could be semiconductors, dielectrics, metals, etc., and the top grown material could be bulk materials, QWs, QDs or other nanostructures. Sequential growth on the same template can be a routine procedure to enable advanced, large wafer-scale, dense photonic integration. A good example in silicon photonics is the integration of light sources,

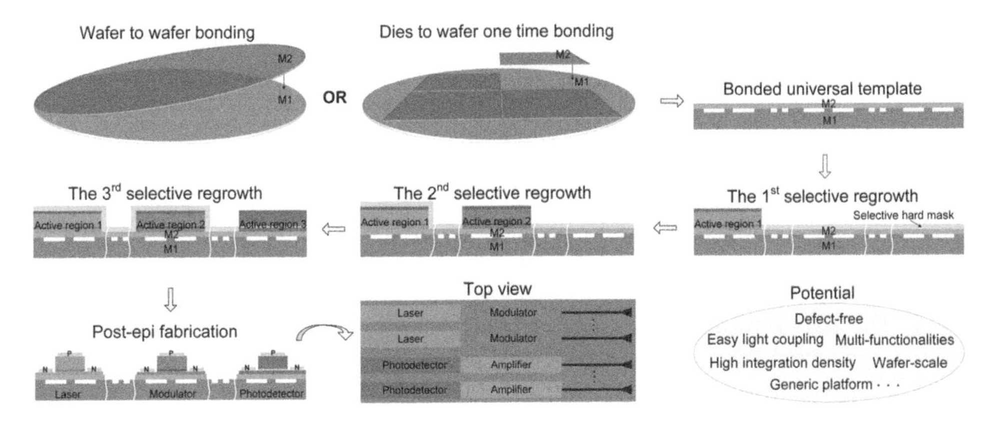

FIGURE 12.21 Schematic of the process of integrating lasers, amplifiers, modulators and photodetectors onto the bonding plus epitaxy integration platform.

amplifiers, modulators and detectors on a single chip with close proximity and low coupling loss by implementing multiple selective regrowth on a single bonding template instead of bonding three or four types of epitaxial structures on each chip [58]. Figure 12.21 schematically shows an example of the process of integrating lasers, amplifiers, modulators and photodetectors onto the bonding plus epitaxy integration platform. The process begins with creating passive waveguide structures on a generic substrate wafer M1, e.g., Si. Then, a one-time bonding of M2, e.g., InP, onto M1, at either the wafer scale or chips-to-wafer scale, is executed to prepare the growth template. Necessary protection and sequential regrowth for the integration of three or four types of active devices are conducted. All regrown materials must be compatible with the template for low-defect growth. Advanced regrowth techniques such as butt-joint regrowth can be applied here to maximize the integration proximity and density and minimize reflection and other undesirable effects associated with abrupt topographic change. Since compound semiconductor substrates may account for significant wafer material costs, particularly for InP substrates, our solution provides the flexibility to reuse M2 substrates, particularly for wafer-scale template transfer. Finally, device processing in sections of different functions in the same material system can share many fabrication steps towards seamless integrated chips.

12.5 ADVANCED WAFER-LEVEL TESTING AND ANALYSIS

Previous sections have addressed our solutions for light sources, phase tuner/modulators, and detectors in either pure silicon or heterogeneous III-V-on-silicon platforms. As described in the beginning, ring resonators are fundamental building blocks of modulators and wavelength (de) multiplexers in our preferred ring-based DWDM architecture. It is challenging to productize ring resonators on SOI substrate. First, tight mode confinement in the silicon photonic waveguide results in high sensitivity to process variation, such that ring resonances of neighboring channels on the same die may shift significantly to overlap or even cross over. Second, reflection from the waveguide sidewall roughness and mode discontinuity at the ring's coupling regions may cause the resonances to have split-peaks instead of a single Lorentzian line shape. Ambiguities arise between missing resonances due to noise and overlapped resonances due to process variation, and between reflection-induced split-peaks of a single ring and two resonances of two neighboring rings. To address these issues, we characterized process variation on 300 mm SOI wafers with high-throughput wafer-level optical testing, and developed an algorithm based on unsupervised machine-learning technique to detect reflection-induced split-peaks and missing resonances that are caused by noise and process variation.

In the following, these techniques are demonstrated with experimental data of a ring resonator Design-of-Experiment (DOE) and ring-based DWDM transceivers that were taped out at STMicroelectronics using the DAPHNE technology on 300 mm SOI wafers with full back-end dielectric stacks. The ring resonators and transceivers are fabricated on SOI wafers with a 300 nm-thick silicon layer on top of a 720 nm-thick buried oxide layer. The top silicon layer is patterned with conventional 193 nm lithography used in 55 nm CMOS node. The ring resonator's ridge waveguide is formed by etching 250 nm into the top silicon layer, while other components including the grating couplers are formed by 140 nm etching. Tapers are inserted between the two etch depths to adiabatically transition the modes with low insertion loss and reflection. The add-drop ring DOE sweeps radius, through gap width, and drop gap width for 255 designs in total. The two gap widths are selected around equal values, where critical coupling is expected to occur. Each ring test loop consists of 4 identical grating couplers that couple light between an array of 4 single-mode fibers and the 4 ports of the add-drop ring resonator, namely input, through, drop and add ports. The fiber array is polished at an 8 degree angle and mounted at 25 μm above the wafer surface. A capacitance sensor is attached to the fiber array to calibrate the spacing between the fibers and the wafers. A tunable laser with wavelength scanning from 1,280 to 1,360 nm at a resolution of 2 pm is coupled to the input port of the ring resonator, and output light at the through, drop, and add ports is collected by three photodetectors respectively.

For each ring resonator, the resonances are analyzed on the three output ports individually. Tested transmission spectra are normalized with respect to the parabolic envelope of the grating couplers' response and converted to linear scale. Continuous-Wavelet Transform (CWT) with Ricker's wavelet is performed on the normalized spectra to detect resonances, which are then partitioned to smaller windows that contain only one resonance in each window for nonlinear fitting to split-peak Lorentzian functions [59]. Reflection in the waveguide's sidewall roughness and mode discontinuity in the coupling region will cause split-peaks in the ring resonances. Due to randomness in the ring waveguide's sidewall roughness and the process, the split-peaks may not occur consistently on all resonances of the same ring within the laser scan range, or at the same wavelength of the same ring design on different dies. Figure 12.22 shows the drop port transmission spectrum of one ring design tested on one die, and it can be seen that two of the three resonances have split-peaks to different extent, while the third resonance does not exhibit split-peaks.

Process information is inferred from wafer-level optical testing data of the ring resonators [60]. All 255 designs of the ring DOE are tested on all testable dies of 8 wafers from 2 lots. Lot-to-lot and wafer-to-wafer variation in resonance wavelengths, even between undoped passive wafers and doped full flow wafers, is 1~2 nm. Majority of the resonance wavelength variability comes from field-to-field variation on the same wafer, which can be more than 10 nm on certain

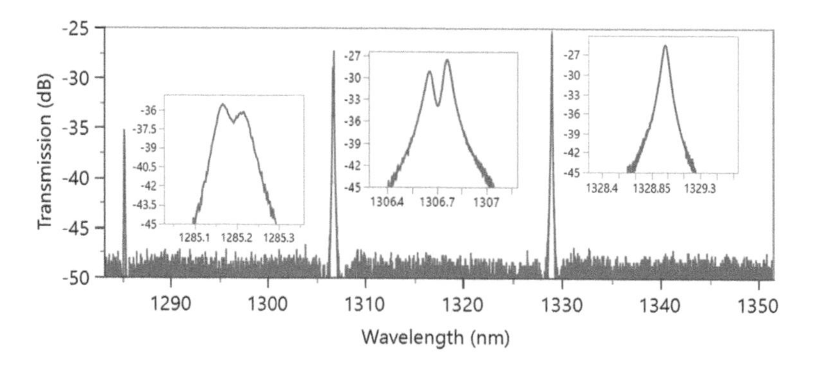

FIGURE 12.22 Drop port resonances with and without split-peaks.

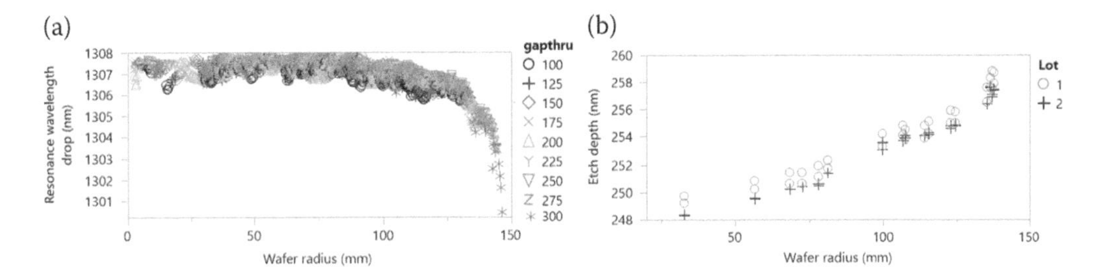

FIGURE 12.23 (a) Resonance wavelengths of all 5 μm radius rings on 1 full flow wafer. (b) Inline metrology data of etch depth.

wafers and designs. Simulations indicate that the ring waveguide's effective index and group index vary with etch depth by -0.572E-3 /nm and +1.432E-3 /nm respectively. Figure 12.23(a) shows the resonance wavelength within the 1,305 nm band for all 5 μm radius rings on one full flow wafer, plotted versus the device's location from the wafer center. It can be seen that the resonance wavelength is driven strongly by the ring's location on the wafer, which implies response to etch depth variation. The resonance wavelength remains largely flat for most of the wafer but increases rapidly near the wafer edge. Inline metrology data of the etch depth on 17 sites across the wafer on 2 wafers from 2 lots is provided by STMicroelectronics. Figure 12.23(b) shows the inline metrology data of etch depth plotted versus the location on the wafer, which agrees reasonably well with optical test results qualitatively and quantitatively. However, due to limited sampling sites on the wafer, the inline metrology did not capture the steep variation in etch depth near the wafer edge, which might cause channel overlapping or crossing over in the ring-based DWDM transceivers.

One transmitter (Tx) and one receiver (Rx) are laid out within each reticle. For each transceiver loop, 24 ring resonators are coupled to the same bus waveguide, which is terminated with two identical grating couplers that couple light between an array of two single-mode fibers and the input/through ports of the loop. The drop ports of all rings are terminated with Ge photodetectors and the add ports are terminated with waveguide terminations. Rings in the Tx loop are critically coupled, while the rings in the Rx loop are slightly under-coupled. The rings' radii increase linearly from 5 μm of the first ring to 5.046 μm of the last ring at a step size of 2 nm, which results in a channel spacing of about 0.35 nm. The ring resonators are doped to form p-i-n junctions for optical modulation by carrier injection. P-type resistive heaters are also formed within the rings to tune the resonances by thermo-optic effect. The transmission spectra of the tested Tx/Rx loops are normalized with respect to the envelope and converted to linear scale, and then CWT is performed on the spectra to detect resonances. Figure 12.24(a) shows part of the Rx spectrum on one die where 29 resonances are detected within one band around 1,310 nm, since the CWT algorithm cannot determine whether the closely spaced twin resonances are split-peaks of a single resonance, or two resonances of two neighboring rings.

One important piece of information that was not utilized in the CWT-based peak detection is the approximately periodic repetition of a ring's resonances. By shifting and overlaying resonances across all bands, the problem of peak detection and labeling is converted to hierarchical clustering that groups the resonance wavelengths into an expected number of clusters, which in this case is the number of rings per Tx/Rx loop [61]. There are three reasons that the resonance detection and labeling can be improved by taking into account all the resonance bands within the laser scan range. First, reflection induced by the ring's waveguide sidewall roughness varies with wavelength and creates different resonance splitting characteristics across different bands of the same ring. In other words, split-peaks of one ring in one band are unlikely to remain split-peaks of the same spacing across all bands, while the spacing between two resonances of two

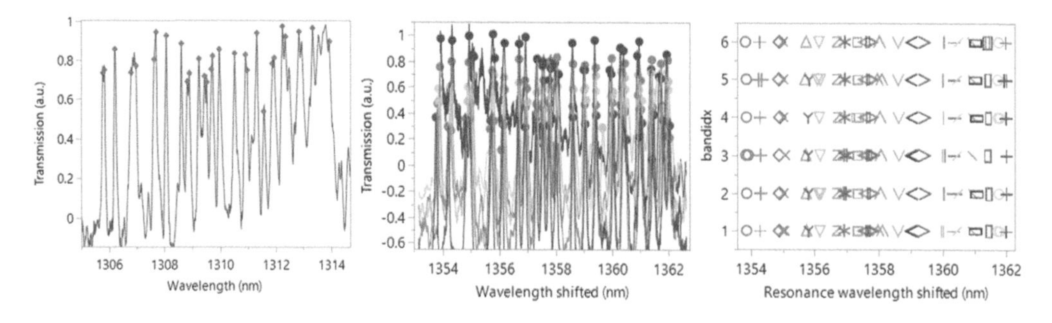

FIGURE 12.24 (a) 29 detected resonance within one band of an Rx loop. (b) All resonances of 6 bands of the same Rx loop. (c) Hierarchical clustering results with the same channels coded by the same color and symbol.

neighboring rings will remain largely constant across all bands barring the ~1% band-to-band shrinking in spacing due to dispersion and change in mode order. By overlaying and comparing resonances of different bands, split-peaks of one resonance can be differentiated from two resonances of two different rings. It is worth noting that such inference is probabilistic instead of deterministic. Second, the CWT peak detection algorithm may occasionally fail to detect one resonance within one band due to noise or spectral distortion, but the probability that multiple resonances of the same ring are missed out across multiple bands diminishes as the number of bands increases. Hence it is possible to label missing resonances by comparing resonances in different bands based on probability. Third, we choose Ward's linkage criterion in the hierarchical clustering, which minimizes the within-cluster variance. More resonance bands and larger number of resonances in each cluster will yield more accurate estimation of variance and more robust clustering.

Figure 12.24(b) shows transmission spectra of the 6 bands of the same Rx loop, where each band is shifted as a whole to minimize the sum of distances of all resonances within the band to their respective nearest neighbors in the previous band. Channel-to-channel spacing increases approximately linearly by ~0.7% between two neighboring bands, which causes a keystone effect. The keystone effect is due to both the dispersion of ring waveguide's effective index and the different resonance orders of different bands. Calculations predict ~1% increase in channel spacing between neighboring bands, which agrees reasonably well with the test results. The keystone effect is corrected for by linearly stretching lower bands and compressing higher bands. Figure 12.3c shows the results of the hierarchical clustering using Ward's linkage criterion and the number of clusters equal to 24, where the resonances are color coded and marked by the cluster number. It can be seen that split-peaks such as channel 1 band 3 (red circle) can be distinguished from two resonances of neighboring rings such as channels 3 and 4 (blue diamond and brown cross); missing resonances such as channel 5 band 4 (cyan triangle) are also identified.

In summary, better understanding of process variation and more accurate resonance detection and labeling enable us to simplify the optical sorting procedures and save testing time.

12.6 NOVEL FIBER ATTACHMENT SOLUTION

After wafer-level characterization to help with design and fabrication optimization and known-good die qualification, the silicon photonic wafer is moved to the packaging stage. Though silicon provides a higher-integration density over conventional III-V substrate for integrated photonics, packaging remains a critical and expensive process for final practical applications. There is a large amount of experience and resources a photonic engineer can leverage from the electronic integrated circuit industry for packaging electronic logic or driver chips with silicon photonic chips.

However, optical fiber attachment is unique to photonics and directly affects link budget, module reliability, production throughput, and total solution cost.

We chose surface coupling through grating coupler (GC) as the standard optical I/O scheme over edge coupling due to convenience in wafer-level testing. The industry standard method to interface GC and single mode (SM) fiber arrays is pigtailing with glass or silicon v-groove ferrules. This impedes the possibility for surface mount assembly of co-packaged silicon photonic transceivers by standard high-volume solder reflow due to the presence of the dangling fiber pigtail. Our approach utilizes a low-profile, small form factor 1 × 8 detachable expanded beam optical connector. The ability to remove and re-attach optical connectors directly to silicon photonic chips and interposers ensures flexibility in component placement on system printed circuit boards (PCB), easy serviceability of damaged fibers and components, and optimization of fiber length to eliminate additional optical jumper cables and connectors.

This solution, shown in Figure 12.25, is composed of a silicon photonic GC array, glass microlens array chip, wedge adapter, optical socket, expanded beam single mode (SM) light turn

(a)

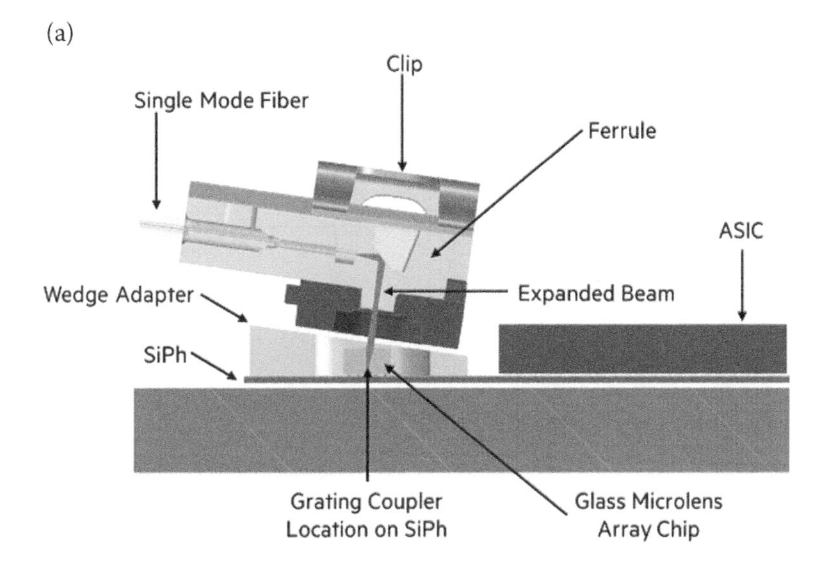

(b)

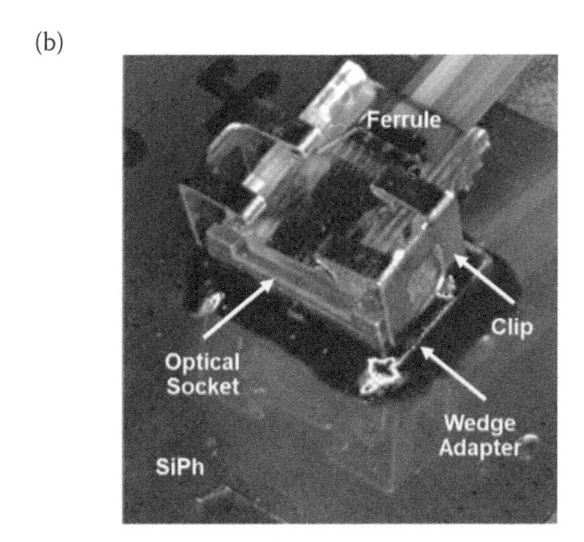

FIGURE 12.25 Optical connector (a) cross section and (b) assembled onto a silicon photonic interposer.

ferrule, and clip [62]. The ferrule redirects the SM fiber mode by 90° and collimates it to ~50 μm diameter in the expanded beam space [63]. The wedge adapter seats the optical socket at 8°, which corresponds to the light beam exit angle from the GC and microlens chip relative to the plane of the silicon photonic GC array. The metal clip latches the ferrule into the socket. The glass lens chip is solder reflow compatible and coefficient of thermal expansion (CTE) matched to silicon, while the optical socket is injection molded with a high-temperature resin capable of withstanding multiple solder reflow cycles.

The assembly process leverages existing surface mount methods. The glass microlens chip is vision aligned with a FineTech Fineplacer Lambda, and affixed to the silicon photonic surface using light cure optical adhesives. After the lens chip is surface mounted, the wedge adapter is coarsely vision aligned and glued to the silicon photonic surface. Next, the ferrule is held stationary in a fixture above the GC array while they are "semi-actively" aligned using a hexapod with 6 degrees of freedom. Light coupled to fiber channel 8 is directed through the ferrule, lens chip, a pair of GCs, an on-chip looped back waveguide, and back to fiber channel 1. This alignment technique is deemed 'semi-active' because the optical power and feedback signal are supplied by an external light source and photodetector rather than on-chip active optical devices. Performing the alignment in the expanded beam space also compensates, to a certain extent, the lateral and angular misalignment between the microlens and GC arrays. At this point, the optical jumper is detached from the optical socket, and a secondary adhesive is dispensed to more rigidly hold the socket to the wedge adapter.

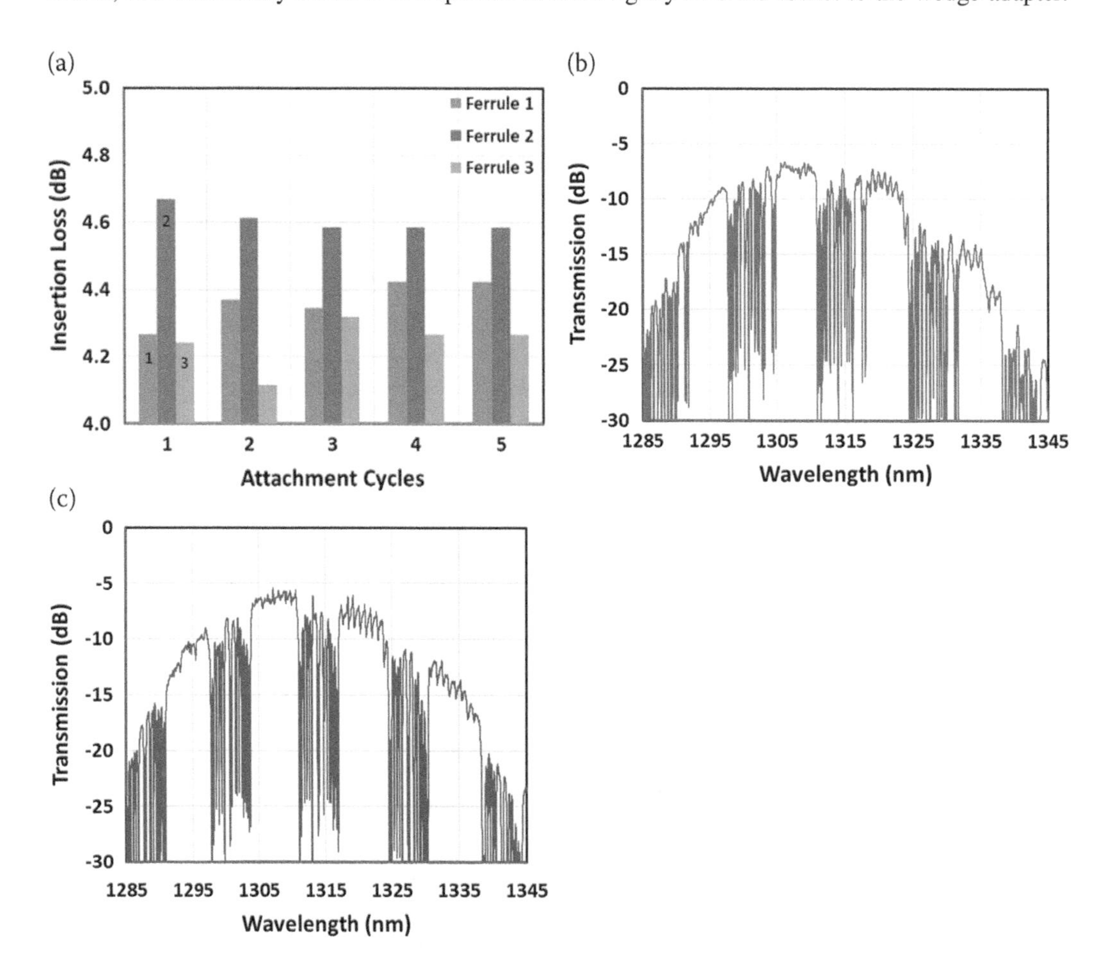

FIGURE 12.26 Fiber to fiber (a) insertion loss repeatability, and (b, c) transmission spectra of two looped-back fiber channels.

To demonstrate insertion loss repeatability, three different ferrules were attached and detached five times. The results are shown in Figure 12.26(a). The fiber-to-fiber insertion loss between the outermost fiber channels (1 and 8) was measured to be <4.7 dB with a maximum variation of 0.2 dB after multiple reattach cycles. The fiber-to-fiber transmission spectra of two looped back fiber channels, with the optical connector attached, are shown in Figures 12.26(b-c), respectively. The transmission curve includes the spectral response of two GCs, a long SOI waveguide and 24 microring resonators, as evidenced by the curvature of the overall envelope, added loss, and multiple resonant dips, respectively.

12.7 SUMMARY

We reviewed our most recent research and product development effort to deploy disruptive DWDM optical transceiver technology for HPC applications. A novel DWDM architecture was conceived to enable simultaneous high energy efficiency, bandwidth scalability and low latency. Silicon-based key building blocks from heterogeneous comb laser source with record-large comb width, ultra-efficient MOS-type phase tuner/modulator, to robust Si-Ge and high-gain heterogeneous avalanche photodetectors were discussed in detail from the experiment and modeling perspective. A promising generic integration platform to combine merits in monolithic and heterogeneous III-V-on-silicon integration was demonstrated. Innovations in wafer-level testing and data analysis and low-loss optical fiber attachment were also discussed as two critical steps towards high-yield volume production. We believe that such a DWDM architecture backed by high-performance components and integration and testing solutions will be very competitive to other traditional CWDM approaches, and particularly shines in upcoming co-packaged optics applications in HPCs and high-end datacenters.

REFERENCES

[1] Ortner, Gerhard, Claudine Nì Allen, C. Dion, Pedro Barrios, Daniel Poitras, Dan Dalacu, Grzegorz Pakulski, et al. "External Cavity InAs/InP Quantum Dot Laser with a Tuning Range of 166 nm." *Applied Physics Letters* 88, no. 12 (2006): 121119. 10.1063/1.2187431.

[2] Arakawa, Yasuhiko, and Hiroyuki Sakaki. "Multidimensional Quantum Well Laser and Temperature Dependence of Its Threshold Current." *Applied Physics Letters* 40, no. 11 (1982): 939–941. 10.1063/1.92959.

[3] Kageyama, Takeo, Kenichi Nishi, Masaomi Yamaguchi, Reio Mochida, Yasunari Maeda, Keizo Takemasa, Yu Tanaka, Tsuyoshi Yamamoto, Mitsuru Sugawara, and Yasuhiko Arakawa. "Extremely High Temperature (220°C) Continuous-Wave Operation of 1300-nm-Range Quantum-Dot Lasers." *2011 Conference on Lasers and Electro-Optics Europe and 12th European Quantum Electronics Conference (CLEO EUROPE/EQEC)*, 2011. 10.1109/cleoe.2011.5943701.

[4] Wojcik, Gregory L., Dongliang Yin, Alexey R. Kovsh, Alexey E. Gubenko, Igor L. Krestnikov, Sergey S. Mikhrin, Daniil A. Livshits, David A. Fattal, Marco Fiorentino, and Raymond G. Beausoleil. "A Single Comb Laser Source for Short Reach WDM Interconnects." *Novel In-Plane Semiconductor Lasers VIII*, 2009. 10.1117/12.816278.

[5] Moore, Stephen A., Liam O'Faolain, Maria Ana Cataluna, Michael B. Flynn, Maria V. Kotlyar, and Thomas F. Krauss. "Reduced Surface Sidewall Recombination and Diffusion in Quantum-Dot Lasers." *IEEE Photonics Technology Letters* 18, no. 17 (2006): 1861–1863. 10.1109/lpt.2006.881206.

[6] Liu, Alan Y., Tin Komljenovic, Michael L. Davenport, Arthur C. Gossard, and John E. Bowers. "Reflection Sensitivity of 1.3 μm Quantum Dot Lasers Epitaxially Grown on Silicon." *Optics Express* 25, no. 9 (2017): 9535. 10.1364/oe.25.009535.

[7] Helms, Jochen, and Klaus Petermann. "A Simple Analytic Expression for the Stable Operation Range of Laser Diodes with Optical Feedback." *IEEE Journal of Quantum Electronics* 26, no. 5 (1990): 833–836. 10.1109/3.55523.

[8] Dong, Bozhang, Heming Huang, Jianan Duan, Geza Kurczveil, Di Liang, Raymond G. Beausoleil, and Frédéric Grillot. "Frequency Comb Dynamics of a 1.3 μm Hybrid-Silicon Quantum Dot

Semiconductor Laser with Optical Injection." *Optics Letters* 44, no. 23 (2019): 5755. 10.1364/ol.44.005755.

[9] Columbo, Lorenzo Luigi, Stefano Barbieri, Carlo Sirtori, and Massimo Brambilla. "Dynamics of a Broad-Band Quantum Cascade Laser: From Chaos to Coherent Dynamics and Mode-Locking." *Optics Express* 26, no. 3 (2018): 2829. 10.1364/oe.26.002829.

[10] Fang, Alexander W., Hyundai Park, Ying-Hao Kuo, Richard Jones, Oded Cohen, Di Liang, Omri Raday, Matthew N. Sysak, Mario J. Paniccia, and John E. Bowers. "Hybrid Silicon Evanescent Devices." *Materials Today* 10, no. 7-8 (2007): 28–35. 10.1016/s1369-7021(07)70177-3.

[11] Kurczveil, Geza, Paolo Pintus, Martijn R., Heck, Jonathan D. Peters, and John E. Bowers. "Characterization of Insertion Loss and Back Reflection in Passive Hybrid Silicon Tapers." *IEEE Photonics Journal* 5, no. 2 (2013): 6600410–6600410. 10.1109/jphot.2013.2246559.

[12] Kurczveil, Géza, Di Liang, Marco Fiorentino, and Raymond G. Beausoleil. "Robust Hybrid Quantum Dot Laser for Integrated Silicon Photonics." *Optics Express* 24, no. 14 (2016): 16167. 10.1364/oe.24.016167.

[13] Kurczveil, Geza, Chong Zhang, Antoine Descos, Di Liang, Marco Fiorentino, and Raymond Beausoleil. "On-Chip Hybrid Silicon Quantum Dot Comb Laser with 14 Error-Free Channels." *2018 IEEE International Semiconductor Laser Conference (ISLC)*, 2018. 10.1109/islc.2018.8516175.

[14] Stern, Brian, Xingchen Ji, Yoshitomo Okawachi, Alexander L. Gaeta, and Michal Lipson. "Battery-Operated Integrated Frequency Comb Generator." *Nature* 562, no. 7727 (2018): 401–405. 10.1038/s41586-018-0598-9.

[15] Dong, Mark, Steven T. Cundiff, and Herbert G. Winful. "Physics of Frequency-Modulated Comb Generation in Quantum-Well Diode Lasers." *Physical Review A* 97, no. 5, 053822 (2018). 10.1103/physreva.97.053822.

[16] Wu, Rui, Chin-Hui Chen, Jean-Marc Fedeli, Maryse Fournier, Kwang-Ting Cheng, and Raymond G. Beausoleil. "Compact Models for Carrier-Injection Silicon Microring Modulators." *Optics Express* 23, no. 12 (2015): 15545. 10.1364/oe.23.015545.

[17] Liang, Di, Geza Kurczveil, Marco Fiorentino, Sudharsanan Srinivasan, John E. Bowers, and Raymond G. Beausoleil. "A Tunable Hybrid III-V-on-Si MOS Microring Resonator with Negligible Tuning Power Consumption." *Optical Fiber Communication Conference*, 2016. 10.1364/ofc.2016.th1k.4.

[18] Liang, Di, and John E. Bowers. "Recent Progress in Lasers on Silicon." *Nature Photonics* 4, no. 8 (2010): 511–517. 10.1038/nphoton.2010.167.

[19] Huang, Xue, Di Liang, Chong Zhang, Geza Kurczveil, Xuema Li, Jiaming Zhang, Marco Fiorentino, and Raymond Beausoleil. "Heterogeneous MOS Microring Resonators." *2017 IEEE Photonics Conference (IPC)*, 2017. 10.1109/ipcon.2017.8116031.

[20] Liang, Di, Geza Kurczveil, Xue Huang, Chong Zhang, Sudharsanan Srinivasan, Zhihong Huang, M. Ashkan Seyedi, et al. "Heterogeneous Silicon Light Sources for Datacom Applications." *Optical Fiber Technology* 44 (2018): 43–52. 10.1016/j.yofte.2017.12.005.

[21] Hiraki, Tatsurou, Takuma Aihara, Koichi Hasebe, Koji Takeda, Takuro Fujii, Takaaki Kakitsuka, Tai Tsuchizawa, Hiroshi Fukuda, and Shinji Matsuo. "Heterogeneously Integrated III–V/Si MOS Capacitor Mach–Zehnder Modulator." *Nature Photonics* 11, no. 8 (2017): 482–485. 10.1038/nphoton.2017.120.

[22] Webster, Mark, Prakash Gothoskar, Vipul Patel, David Piede, Steve Anderson, Ravi S. Tummidi, Dennis A. Adams, et al. "An Efficient MOS-Capacitor Based Silicon Modulator and CMOS Drivers for Optical Transmitters." *11th International Conference on Group IV Photonics (GFP)*, 2014. 10.1109/group4.2014.6961998.

[23] Huang, Zhihong, Cheng Li, Di Liang, Kunzhi Yu, Charles Santori, Marco Fiorentino, Wayne Sorin, Samuel Palermo, and Raymond G. Beausoleil. "25 Gbps Low-Voltage Waveguide Si–Ge Avalanche Photodiode." *Optica* 3, no. 8 (2016): 793. 10.1364/optica.3.000793.

[24] Yuan, Yuan, Zhihong Huang, Binhao Wang, Wayne V. Sorin, Xiaoge Zeng, Di Liang, Marco Fiorentino, Joe C. Campbell, and Raymond G. Beausoleil. "64 Gbps PAM4 Si-Ge Waveguide Avalanche Photodiodes With Excellent Temperature Stability." *Journal of Lightwave Technology* 38, no. 17 (2020): 4857–4866. 10.1109/jlt.2020.2996561.

[25] Jones, Andrew H., Yuan Yuan, Min Ren, Scott J. Maddox, Seth R. Bank, and Joe C. Campbell. "Al_xIn_1-xAs_ySb_1-y Photodiodes with Low Avalanche Breakdown Temperature Dependence." *Optics Express* 25, no. 20 (2017): 24340. 10.1364/oe.25.024340.

[26] Levine, Barry, Robert Sacks, J. Ko, Matthew Jazwiecki, Janis Valdmanis, D. Gunther, and J. H.

Meier. “A New Planar InGaAs–InAlAs Avalanche Photodiode.” *IEEE Photonics Technology Letters* 18, no. 18 (2006): 1898–1900. 10.1109/lpt.2006.881684.

[27] Tan, Lionel Juen Jin, Daniel Swee Guan Ong, Jo Shien Ng, Chee Hing Tan, Stephen K. Jones, Yahong Qian, and John Paul Raj David. “Temperature Dependence of Avalanche Breakdown in InP and InAlAs.” *IEEE Journal of Quantum Electronics* 46, no. 8 (2010): 1153–1157. 10.1109/jqe.2010.2044370.

[28] Ishimura, Eitaro, Eiji Yagyu, Masaharu Nakaji, Susumu Ihara, Kiichi Yoshiara, Toshitaka Aoyagi, Yasunori Tokuda, and Takahide Ishikawa. “Degradation Mode Analysis on Highly Reliable Guardring-Free Planar InAlAs Avalanche Photodiodes.” *Journal of Lightwave Technology* 25, no. 12 (2007): 3686–3693. 10.1109/jlt.2007.909357.

[29] Rouvie, Anne, Daniéle Carpentier, Nadine Lagay, Jean Decobert, Frédéric Pommereau, and Mohand Achouche. “High Gain x Bandwidth Product Over 140-GHz Planar Junction AlInAs Avalanche Photodiodes.” *IEEE Photonics Technology Letters* 20, no. 6 (2008): 455–457. 10.1109/lpt.2008.918229.

[30] Goh Yu Ling, Desiree Ong, Shiyong Zhang, Jo Shien Ng, Chee Hing Tan, and John P. R. David. “InAlAs Avalanche Photodiode with Type-II Absorber for Detection beyond 2 μm.” *Infrared Technology and Applications XXXV* 7298, (2009): 729837. 10.1117/12.819818.

[31] Zhao, Yanli, Junjie Tu, Jingjing Xiang, Ke Wen, Jing Xu, Yang Tian, Qiang Li, et al. “Temperature Dependence Simulation and Characterization for InP/InGaAs Avalanche Photodiodes.” *Frontiers of Optoelectronics* 11, no. 4 (2018): 400–406. 10.1007/s12200-018-0851-8.

[32] Ma, C. L. Forrest, M. Jamal Deen, Larry E. Tarof, and Jeffrey C. H. Yu. “Temperature Dependence of Breakdown Voltages in Separate Absorption, Grading, Charge, and Multiplication InP/InGaAs Avalanche Photodiodes.” *IEEE Transactions on Electron Devices* 42, no. 5 (1995): 810–818. 10.1109/16.381974.

[33] Sidhu, Rubin, Lijuan Zhang, Ning Tan, Ning Duan, Joe C. Campbell, Archie L. Holmes, Chia-Fu Hsu, and Mark Itzler. “2.4 μm Cutoff Wavelength Avalanche Photodiode on InP Substrate.” *Electronics Letters* 42, no. 3 (2006): 181. 10.1049/el:20063415.

[34] Harris Thomas R., Optical properties of Si, Ge, GaAs, GaSb, InAs, and InP at elevated temperatures,” *Air Force Inst of Tech Wright-Patterson AFB OH School of Engineering and Management*, 2010. xx, 64.

[35] Wang, Binhao, Zhihong Huang, Xiaoge Zeng, Wayne V. Sorin, Di Liang, Marco Fiorentino, and Raymond G. Beausoleil. “A Compact Model for Si-Ge Avalanche Photodiodes Over a Wide Range of Multiplication Gain.” *Journal of Lightwave Technology* 37, no. 13 (2019): 3229–3235. 10.1109/jlt.2019.2913179.

[36] Lee, Jeong-Min, Seong-Ho Cho, and Woo-Young Choi. “An Equivalent Circuit Model for a Ge Waveguide Photodetector on Si.” *IEEE Photonics Technology Letters* 28, no. 21 (2016): 2435–2438. 10.1109/lpt.2016.2598369.

[37] Youn, Jin-Sung, Myung-Jae Lee, Kang-Yeob Park, Holger Rücker, and Woo-Young Choi. “SNR Characteristics of 850-Nm OEIC Receiver with a Silicon Avalanche Photodetector.” *Optics Express* 22, no. 1 (2014): 900. 10.1364/oe.22.000900.

[38] Yuan, Yuan, Zhihong Huang, Binhao Wang, Wayne V. Sorin, Xiaoge Zeng, Di Liang, Marco Fiorentino, Joe C. Campbell, and Raymond G. Beausoleil. “64 Gbps PAM4 Si-Ge Waveguide Avalanche Photodiodes With Excellent Temperature Stability.” *Journal of Lightwave Technology* 38, no. 17 (2020): 4857–4866. 10.1109/jlt.2020.2996561.

[39] Colace, Lorenzo, Pasquale Ferrara, Gaetano Assanto, Dom Fulgoni, and Lee Nash. “Low Dark-Current Germanium-on-Silicon Near-Infrared Detectors.” *IEEE Photonics Technology Letters* 19, no. 22 (2007): 1813–1815. 10.1109/lpt.2007.907578.

[40] Pinel, Lucas L. G., Simon J. Dimler, Xinxin Zhou, Salman Abdullah, Shiyong Zhang, Chee Hing Tan, and Jo Shien Ng. “Effects of Carrier Injection Profile on Low Noise Thin Al085Ga015As056Sb044 Avalanche Photodiodes.” *Optics Express* 26, no. 3 (2018): 3568. 10.1364/oe.26.003568.

[41] Yang, Fujun, Karin Hinzer, Claudine Nì Allen, Simon Fafard, Geof C. Aers, Yan Feng, John P. Mccaffrey, and Sylvain Charbonneau. “Quantum Dot p-i-n Structure in An Electric Field.” *Superlattices and Microstructures* 25, no. 1-2 (1999): 419–424. 10.1006/spmi.1998.0669.

[42] Sandall, Ian, Jo Shien Ng, John P. R. David, Chee Hing Tan, Ting Wang, and Huiyun Liu. “1300 nm Wavelength InAs Quantum Dot Photodetector Grown on Silicon.” *Optics Express* 20, no. 10 (2012): 10446. 10.1364/oe.20.010446.

[43] Tossoun, Bassem, Géza Kurczveil, Chong Zhang, Antoine Descos, Zhihong Huang, Andreas Beling, Joe C. Campbell, Di Liang, and Raymond G. Beausoleil. “Indium Arsenide Quantum Dot

Waveguide Photodiodes Heterogeneously Integrated on Silicon." *Optica* 6, no. 10 (2019): 1277. 10.1364/optica.6.001277.

[44] Umezawa, Toshimasa, Koichi Akahane, Atsushi Matsumoto, Atsushi Kanno, Naokatsu Yamamoto, and Tetsuya Kawanishi. "Polarization Dependence of Avalanche Multiplication Factor in 1.5 Mm Quantum Dot Waveguide Photodetector." *Conference on Lasers and Electro-Optics*, 2016. 10.1364/cleo_si.2016.sm4e.8.

[45] Ma, Ying-Jie, Yong-Gang Zhang, Yi Gu, Xing-You Chen, Peng Wang, Bor-Chau Juang, Alan Farrell, et al. "Enhanced Carrier Multiplication in InAs Quantum Dots for Bulk Avalanche Photodetector Applications." *Advanced Optical Materials* 5, no. 9 (2017): 1601023. 10.1002/adom.201601023.

[46] Capasso, Federico, Jeremy Allam, Alfred Y. Cho, Khalid Mohammed, Roger J. Malik, Albert L. Hutchinson, and Deborah L. Sivco. "New Avalanche Multiplication Phenomenon in Quantum Well Superlattices: Evidence of Impact Ionization across the Band-Edge Discontinuity." *Applied Physics Letters* 48, no. 19 (1986): 1294–1296. 10.1063/1.96957.

[47] Huang, Zhihong, Cheng Li, Di Liang, Kunzhi Yu, Charles Santori, Marco Fiorentino, Wayne Sorin, Samuel Palermo, and Raymond G. Beausoleil. "25 Gbps Low-Voltage Waveguide Si–Ge Avalanche Photodiode." *Optica* 3, no. 8 (2016): 793. 10.1364/optica.3.000793.

[48] Yu, Kunzhi, Cheng Li, Hao Li, Alex Titriku, Ayman Shafik, Binhao Wang, Zhongkai Wang, et al. "A 25 Gb/s Hybrid-Integrated Silicon Photonic Source-Synchronous Receiver With Microring Wavelength Stabilization." *IEEE Journal of Solid-State Circuits* 51, no. 9 (2016): 2129–2141. 10.1109/jssc.2016.2582858.

[49] Chen, Siming, Wei Li, Jiang Wu, Qi Jiang, Mingchu Tang, Samuel Shutts, Stella N. Elliott, et al. "Electrically Pumped Continuous-Wave III–V Quantum Dot Lasers on Silicon." *Nature Photonics* 10, no. 5 (2016): 307–311. 10.1038/nphoton.2016.21.

[50] Wan, Yating, Justin Norman, Qiang Li, M. J. Kennedy, Di Liang, Chong Zhang, Duanni Huang, et al. "13 Mm Submilliamp Threshold Quantum Dot Micro-Lasers on Si." *Optica* 4, no. 8 (2017): 940. 10.1364/optica.4.000940.

[51] Bolkhovityanov, Yu B., and Oleg Petrovich Pchelyakov. "GaAs Epitaxy on Si Substrates: Modern Status of Research and Engineering." *Physics-Uspekhi* 51, no. 5 (2008): 437–456. 10.1070/pu2008v051n05abeh006529.

[52] Hu, Yingtao, Di Liang, Chong Zhang, Geza Kurczveil, Xue Huang, Kunal Mukherjee, and Raymond Beausoleil. "Electrically-Pumped 1.31 μm MQW Lasers by Direct Epitaxy on Wafer-Bonded InP-on-SOI Substrate." *2018 IEEE Photonics Conference (IPC)*, 2018. 10.1109/ipcon.2018.8527345.

[53] Hu, Yingtao, Di Liang, Kunal Mukherjee, Youli Li, Chong Zhang, Geza Kurczveil, Xue Huang, and Raymond G. Beausoleil. "III/V-on-Si MQW Lasers by Using a Novel Photonic Integration Method of Regrowth on a Bonding Template." *Light: Science & Applications* 8, no. 1 (2019). 10.1038/s41377-019-0202-6.

[54] Norman, Justin C., Daehwan Jung, Zeyu Zhang, Yating Wan, Songtao Liu, Chen Shang, Robert W. Herrick, Weng W. Chow, Arthur C. Gossard, and John E. Bowers. "A Review of High-Performance Quantum Dot Lasers on Silicon." *IEEE Journal of Quantum Electronics* 55, no. 2 (2019): 1–11. 10.1109/jqe.2019.2901508.

[55] Jung, Daehwan, Robert Herrick, Justin Norman, Katherine Turnlund, Catherine Jan, Kaiyin Feng, Arthur C. Gossard, and John E. Bowers. "Impact of Threading Dislocation Density on the Lifetime of InAs Quantum Dot Lasers on Si." *Applied Physics Letters* 112, no. 15 (2018): 153507. 10.1063/1.5026147.

[56] Fujii, Takuro, Koji Takeda, Shinji Matsuo, Koichi Hasebe, Tomonari Sato, and Takaaki Kakitsuka. "Epitaxial Growth of InP to Bury Directly Bonded Thin Active Layer on SiO2/Si Substrate for Fabricating Distributed Feedback Lasers on Silicon." *IET Optoelectronics* 9, no. 4 (2015): 151–157. 10.1049/iet-opt.2014.0138.

[57] People, R., and J. C. Bean. "Calculation of Critical Layer Thickness versus Lattice Mismatch for GexSi1−x/Si Strained-Layer Heterostructures." *Applied Physics Letters* 47, no. 3 (1985): 322–324. 10.1063/1.96206.

[58] Aihara, Takuma, Tatsurou Hiraki, Koji Takeda, Koichi Hasebe, Takuro Fujii, Tai Tsuchizawa, Takaaki Kakitsuka, and Shinji Matsuo. "Lateral Current Injection Membrane Buried Heterostructure Lasers Integrated on 200-Nm-Thick Si Waveguide." *Optical Fiber Communication Conference*, 2018. .10.1364/ofc.2018.w3f.4

[59] Li, Ang, Thomas Van Vaerenbergh, Peter De Heyn, Peter Bienstman, and Wim Bogaerts. "Backscattering

in Silicon Microring Resonators: a Quantitative Analysis." *Laser & Photonics Reviews* 10, no. 3 (2016): 420–431. 10.1002/lpor.201500207.

[60] Sun, Peng, Raymond G. Beausoleil, Jared Hulme, Thomas Van Vaerenbergh, Jinsoo Rhim, Charles Baudot, Frederic Boeuf, Nathalie Vulliet, Ashkan Seyedi, and Marco Fiorentino. "Statistical Behavioral Models of Silicon Ring Resonators at a Commercial CMOS Foundry." *IEEE Journal of Selected Topics in Quantum Electronics* 26, no. 2 (2020): 1–10. 10.1109/jstqe.2019.2945927.

[61] Sun, Peng, Jared Hulme, Ashkan Seyedi, Marco Fiorentino, and Raymond Beausoleil. "Data-Mining-Assisted Resonance Labeling in Ring-Based DWDM Transceivers." *Optical Fiber Communication Conference (OFC) 2020*, 2020. 10.1364/ofc.2020.m3a.3.

[62] Mathai, Sagi, Paul K. Rosenberg, George Panotopoulos, Dan Kurtz, Darrell Childers, Thomas Van Vaerenbergh, Peng Sun, et al. "Detachable 1x8 Single Mode Optical Interface for DWDM Microring Silicon Photonic Transceivers." *Optical Interconnects XX* 11286, (2020): 112860A. 10.1117/12.2544400.

[63] Hughes, Mike, Darrell Childers, Dan Kurtz, Dirk Schoellner, Shubhrangshu Sengupta, and Ke Wang. "A Single-Mode Expanded Beam Separable Fiber Optic Interconnect for Silicon Photonics." *Optical Fiber Communication Conference (OFC) 2019*, 2019. 10.1364/ofc.2019.tu2a.6.

13 Silicon Photonic Bragg Grating Devices

Mustafa Hammood, Lukas Chrostowski, and Nicolas A. F. Jaeger

CONTENTS

13.1 INTRODUCTION

Silicon photonics is becoming an increasingly attractive photonic integration technology that enables a variety of photonic devices and new applications beyond data-communication. Such applications include quantum computing, LiDAR sensing, and optical computing. With the current surge in such new applications, the designers of silicon photonic integrated circuits are demanding increasingly stringent performance metrics of the individual components and building blocks in their systems, beyond what was thought of as "good enough" to meet tele-com and data-com device standards. One such class of critical components, for any photonic integrated circuit, is the class of wavelength selective devices made using integrated waveguide Bragg gratings.

The Bragg grating is a key component in numerous optical devices, such as lasers, fiber-optic sensors, wavelength filters, and dispersion engineering devices. The Bragg grating is known for its ubiquity in optical fiber devices, in which the grating is written into the fiber core. In addition to their use in optical fibers, Bragg gratings are of considerable interest to the photonic integrated circuits community for their potential to be compacted into small areas and integrated into complex devices and systems.

In this chapter, we present the theory and design fundamentals of integrated waveguide Bragg gratings and we will review recent research developments aimed at achieving complex Bragg grating based devices in silicon photonics. We discuss practical challenges in designing such

DOI: 10.1201/9780429292033-13

Bragg gratings and discuss approaches to model and mitigate them. In addition to integrated waveguide Bragg gratings, we present the contra-directional coupler, a Bragg grating based device that consists of two (or more) coupled waveguides that can be used to design multi-port optical add-drop filters.

13.2 INTEGRATED WAVEGUIDES BRAGG GRATINGS

13.2.1 Bragg Gratings Theory

A Bragg grating is a type of one-dimension photonic crystal, which consists of periodic perturbations in the effective refractive index of a waveguide. The perturbations can be made by varying the geometry or the material of the waveguide structure. Such periodic perturbations of the effective refractive index, shown in Figure 13.1, result in reflections along the length of the perturbed waveguide. Bragg gratings can be engineered to create wavelength-selective reflectors, i.e., ones that only reflect wavelengths that are near to the Bragg wavelength (λ_B).

The Bragg wavelength is given by the Bragg condition [1]:

$$\lambda_B = 2\Lambda_B n_{eff} \tag{13.1}$$

in which Λ is the grating period and n_{eff} is the average effective index. The reflection coefficient of a "uniform" grating (one in which the periodic perturbations are of equal strength everywhere along the length of the grating) with length L can be derived using coupled-mode theory [2]:

$$r = \frac{-j\kappa sinh(\gamma L)}{\gamma cosh(\gamma L) + j\Delta\beta sinh(\gamma L)} \tag{13.2}$$

and

$$\gamma^2 = \kappa^2 - \Delta\beta^2 \tag{13.3}$$

in which $\Delta\beta$ is the propagation constant mismatch from the propagation constant at the Bragg wavelength, which is given by:

$$\Delta\beta = \beta - \beta_0 = \frac{2\pi n_{eff}(\lambda)}{\lambda} - \frac{2\pi n_{eff}(\lambda_B)}{\lambda_B} \approx -\frac{2\pi n_g}{\lambda_B^2}\Delta\lambda \tag{13.4}$$

here, n_g is the group index and κ is the coupling coefficient (or coupling strength). The coupling coefficient is a measure of the amount of reflection per unit length and has the units of 1/m. For a periodic perturbation profile that can be described by a square function, the reflection at each interface can be derived using Fresnel equations and shown to be $\Delta n/2n_{eff}$ (where $\Delta n = n_{eff2} - n_{eff1}$, n_{eff1} and n_{eff2} being the effective indices of the two waveguide sections making up a period of the square function, as shown in Figure 13.1) and each grating period, Λ, contributes two reflecting interfaces. In this case, the coupling coefficient can be written as [2,3]:

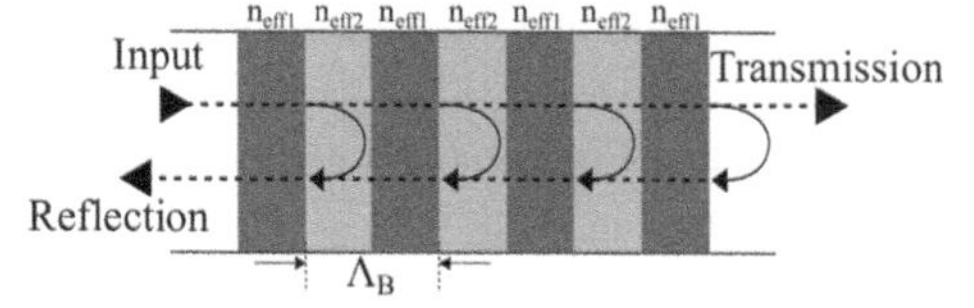

FIGURE 13.1 An illustration of the periodic perturbations in the effective refractive index making up a Bragg grating.

$$\kappa = 2\frac{\Delta n}{2n_{eff}}\frac{1}{\Lambda} = \frac{2\Delta n}{\lambda_B} \tag{13.5}$$

It can be seen that at the Bragg wavelength ($\Delta\beta = 0$), the reflectivity, r, can be described by the Bragg grating's peak power reflectivity ($R_{peak} = |r|^2$):

$$R_{peak} = \tanh^2(\kappa L) \tag{13.6}$$

In addition to the peak reflectivity, the bandwidth of a Bragg grating is also an important metric. We define the bandwidth as the wavelength range between the first nulls on either side of the reflection peak, as highlighted in Figure 13.2, in which a typical response of a Bragg grating is shown. The bandwidth can be determined by [2]:

$$\Delta\lambda = \frac{\lambda_B^2}{\pi n_g}\sqrt{\kappa^2 + \left(\frac{\pi}{L}\right)^2} \tag{13.7}$$

13.2.2 Bragg Gratings Design and Fabrication

In this section, we describe the design process of integrated Bragg gratings, with a focus on silicon photonic integrated Bragg gratings formed in 220 nm thick top-silicon, silicon-on-insulator (SOI) wafers. Such wafers are commonly used in many silicon photonics fabrication platforms. This section will also discuss the fabrication and practical deployment of such devices by addressing the real life challenges associated with designing integrated Bragg gratings.

Fiber-based Bragg gratings are typically made by exposing sections of the fiber core to intense ultraviolet (UV) light in such a way as to periodically modulate the refractive index of the core [4]. In contrast, integrated Bragg gratings are formed by introducing periodic perturbations to the structure of an integrated photonic waveguide. Such perturbations can take many forms, as shown in Figure 13.3(a). Early demonstrations of integrated Bragg gratings on silicon photonic platforms were done by periodically etching the top surface of the of silicon [5] waveguides, however, additional approaches were demonstrated using sidewall corrugations [6,7], cladding modulation [8], and ion implantation [9].

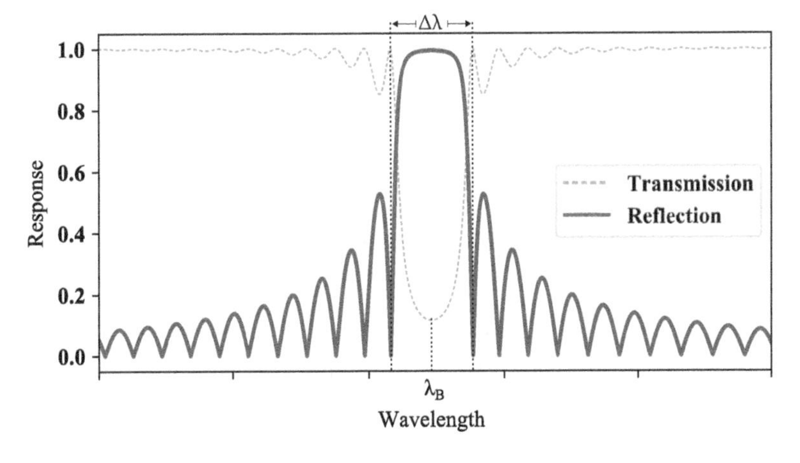

FIGURE 13.2 Typical transmission response and reflection response of a Bragg grating. The Bragg wavelength and bandwidth are highlighted.

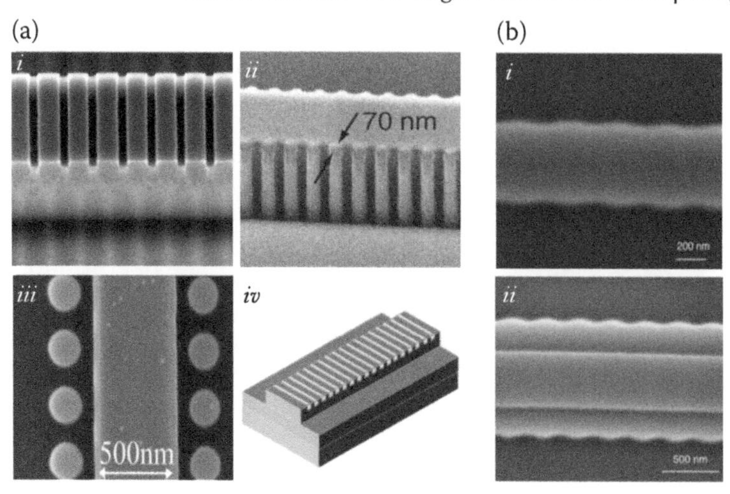

FIGURE 13.3 (a) Micrographs and illustrations of various types of Bragg gratings made on silicon photonic platforms such as (i) top-surface etched Bragg gratings [10]. (ii) Sidewall corrugated Bragg gratings (Reproduced from [6], with the permission of the American Vacuum Society.) (iii) cladding modulated Bragg gratings [8] (iv) periodically doped waveguide Bragg gratings [9]. (b) Micrographs of silicon photonic Bragg gratings made using sidewall corrugations on a (i) strip waveguide and (ii) rib waveguide (Reprinted with permission from [7] Copyright of The Optical Society).

In this chapter, we limit our discussion to integrated Bragg gratings made using sidewall corrugations. Such corrugations can be formed on both strip waveguides and rib (ridge) waveguides, as shown in Figure 13.3(b). The design parameters of a sidewall corrugated Bragg grating including the height of the waveguide (H), the average width of the waveguide (W), The widths of the corrugations (ΔW), and the period of the corrugations (Λ). All of the design parameters affect the response of the device by changing the central wavelength and optical bandwidth of the Bragg grating. Strip waveguide based Bragg gratings are desirable because they allow the device to be patterned using a single-etch process, minimizing the number of lithography masks required for fabrication. In contrast, rib-waveguide-based Bragg gratings require two lithography masks, one to pattern the shallow etched region of the waveguide, and one to pattern the fully etched region of the waveguide. However, due to the larger profile of the mode that can be obtained using rib waveguides, such waveguides allow for precise control of the coupling coefficient and can be made to be more tolerant to fabrication imperfections [3,7].

The measurements and designs we report on in this chapter are typically of sidewall corrugated Bragg gratings made on an SOI platform with a 2 m-thick buried oxide, 220 nm-thick silicon device layer, and with a 2 m-thick cladding oxide layer. The reflected signal can be extracted using an external circulator, however, the extracted reflected signal may be distorted and difficult to analyze since this signal also contains the signals reflected from the optical grating couplers (or edge couplers). As shown in Figure 13.4(a), without an on-chip circulator available, a waveguide Y-branch 3 dB splitter can be placed at the input port of the Bragg waveguide to direct half (3 dB) of the reflected response to an output port. However, this is a not an ideal method to obtain the reflected signal as the input power is also reduced by another 3 dB at the input of the splitter. The reflected signal can also be extracted using external optical vector-network-analyzers or reflectometers, both being relatively expensive equipment. The measured, normalized response of an integrated Bragg grating waveguide is shown in Figure 13.4(b). In this particular device, 300 periodic corrugations were used, the corrugation period was 320 nm, the width of the corrugations was 40 nm, and the average width of the waveguide was 500 nm.

A designer can tailor the spectrum of such a Bragg grating by changing the available design parameters. For example, the reflectively of the grating can be increased by increasing the number of perturbations and the strength of each perturbation, this in turn results in a flat-topped reflected signal. Apodizing the perturbations profile is used to tailor the response of a Bragg grating and can be used to suppress the side-lobes the reflected signal. Apodization refers to a design technique in which the strength of the perturbations are gradually tapered, in accordance with a mathematical function, along the length of the grating. Several functions can be used to describe an apodized grating perturbation profile: Gaussian cosine, raised cosine, and sinc functions are the ones most commonly used [11]. Figure 13.5 highlights the effects apodization

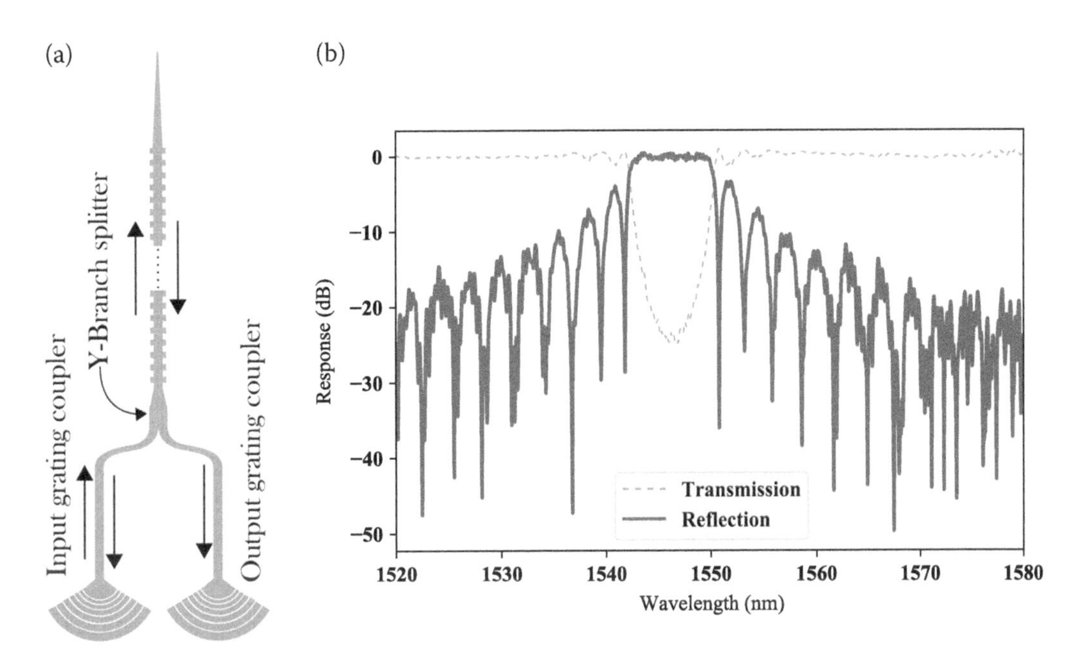

FIGURE 13.4 (a) A schematic of a configuration used to measure both, the transmitted and reflected signal of a Bragg grating using on-chip Y-branch 3 dB splitters. (b) Measured, normalized spectra of the transmission and reflection signals of a typical Bragg grating made on a silicon photonic platform.

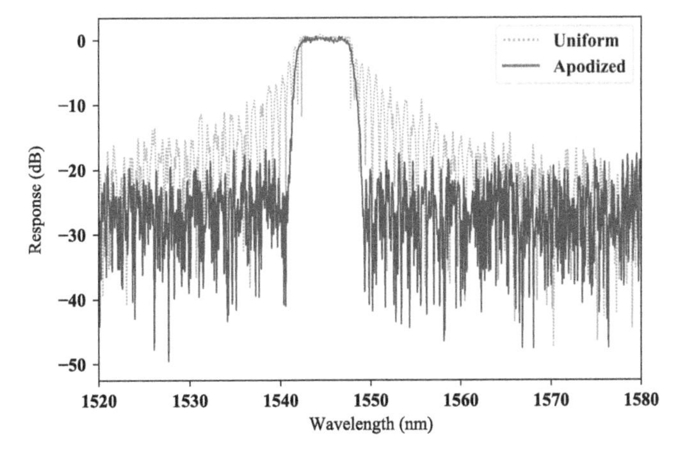

FIGURE 13.5 Measured, normalized spectra of two Bragg grating devices, one in which its corrugations were uniform (cyan) and the other having its corrugations apodized along its length (red).

on integrated Bragg grating devices. The two measured reflected spectra are of two Bragg grating devices made using 1000 corrugation periods, in which one of them had uniform corrugation strengths along the length of the grating and the other had a Gaussian apodized corrugation strength profile [12].

However, due to the change in the waveguide's structure, apodization of the Bragg grating introduces effective index variations along the waveguide grating, resulting in unwanted phase noise [13]. Such apodization phase noise can be characterized and compensated for using advanced apodization techniques [14].

In addition to understanding the effects of a Bragg grating's design parameters, a designer must also account for several factors that affect the response of a Bragg grating when fabricated on an integrated photonics platform. Some of these factors include manufacturing variability, temperature sensitivity, and lithography effects.

13.2.2.1 Manufacturing Variability

Due to the high index contrast between silicon and silicon dioxide, that is typical of silicon photonics platforms, the effective index of a waveguide is strongly governed by the geometry of the waveguide [15]. Therefore, manufacturing variability/non-uniformity can be detrimental to the performance of many devices fabricated on such platforms. Bragg-grating based devices are no exception, manufacturing variability is a challenge that must be well understood and accounted for prior to designing any Bragg grating or Bragg-grating based device. Manufacturing variability can result in changes of the thickness, average width, and/or sidewall angle of the device, as well as introduce sidewall roughness to the waveguides. In-turn, this will change the effective index and, therefore, the central wavelength of the device. In addition to the change of the central wavelength, the effective index non-uniformity along the length of the grating can result in increased phase-noise, which distorts the spectrum and results in high side-lobes at the reflection port [16,17]. The simulated sensitivities of Bragg gratings to manufacturing variability/non-uniformity in terms of changes to the fabricated waveguides' widths $\Delta\lambda/\Delta w$ and heights $\Delta\lambda/\Delta t$ are $\Delta\lambda/\Delta w = 1.2$ nm/nm and $\Delta\lambda/\Delta t = 2.6$ nm/nm for the strip waveguide-based Bragg gratings and $\Delta\lambda/\Delta w = 0.065$ nm/nm and $\Delta\lambda/\Delta t = 2.5$ nm/nm for rib waveguide-based Bragg gratings [3].

In addition to active tuning of the gratings through the use of thermo-optic and electro-optic effects [18], the adverse effects of manufacturing variability can be mitigated using clever design choices. As previously discussed, some waveguide geometries have lower sensitivity to manufacturing variability. Due to their low mode-confinement, sub-wavelength-structures have been studied and used to design devices that are insensitive to dispersion and temperature [19,20]. Sub-wavelength-based Bragg gratings [21] can be used to design Bragg gratings that are less sensitive to manufacturing variability as compared to either strip waveguide or rib waveguide-based Bragg gratings ($\Delta\lambda/\Delta w = 0.433$ nm/nm and $\Delta\lambda/\Delta t = 0.827$ nm/nm). Similarly, due to the TM mode's low confinement, Bragg gratings designed for, and operated using, the TM mode can result in devices that are less sensitive to manufacturing variability as compared to Bragg gratings that are designed for, and operated using, the TE mode [22].

13.2.2.2 Lithography Effects

Due to the small features that are typically used to create sidewall corrugations in integrated waveguides, to make Bragg grating waveguides, lithography effects must be well understood and accounted for prior to fabrication. Lithography effects include smoothing [23] and proximity effects [24], which can both affect the responses of Bragg gratings. The smoothing effects will typically change the shape of square-shaped corrugations to sinusoidal-shaped corrugations with lower corrugation widths and, hence, will result in a grating with a lower coupling strength (κ) and a reduced bandwidth and reflectivity, see Equations 13.2 and 13.7, respectively. The extents of these lithography effects depend on the type of lithography used to pattern a structure. Electron-beam (e-beam) lithography is commonly used to pattern fine device features. However, e-beam

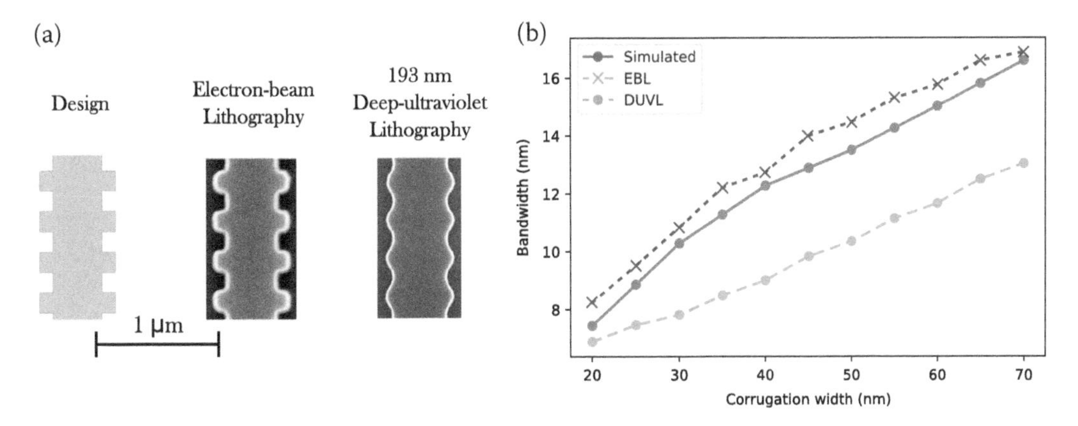

FIGURE 13.6 (a) A comparison of an as-designed Bragg grating to its resultant, fabricated shape as shown by the scanning-electron-microscope micrographs. (b) A comparison between the simulated and measured bandwidths of a set of Bragg gratings. It can be seen that the deep ultra-violet lithography process has narrower measured bandwidths as compared to the simulated ones and ones that were fabricated using an electron-beam lithography process.

lithography is a low-throughput process that is often useful for prototyping devices but not for mass production. In contrast, deep-ultraviolet lithography (DUVL) is another lithography type that can be used in the manufacturing of silicon photonics, and has long been popular in the manufacturing of electronics. DUVL processes typically have high throughputs and are more suitable for mass production. However, DUVL fabrication processes typically result in lower resolution as compared to EBL fabrication processes. Therefore, the effects of DUVL must be well understood and compensated for prior to fabrication.

The extents of the effects of DUVL can be highlighted by comparing the bandwidths of Bragg gratings fabricated using EBL processes to those fabricated using DUVL processes. Two sets of Bragg grating devices, with varying corrugation widths, were fabricated, one using an EBL process and one using a DUVL process, and were measured to extract their bandwidths. Scanning-electron-microscope micrographs of two fabricated devices are shown in Figure 13.6(a). The effects of DUVL smoothing can be seen when comparing the sizes and shapes of the corrugations of the DUVL fabricated device to those of the EBL fabricated device and to those of the design. The extracted bandwidths are illustrated and compared in Figure 13.6(b). When comparing the extracted, as-fabricated bandwidths to the simulated, as-designed bandwidths of the gratings, it can be seen that impact of the DUVL process on the bandwidths is significantly greater than that of the EBL process, this can be attributed to the greater severity of the smoothing effects of the DUVL process.

Work has been done to model and mitigate the adverse effects of DUVL processes by the use of computational lithography techniques [23,25]. A computational lithography model can be built by using scanning-electron-microscope micrographs measurement data from fabricated test patterns to simulate the lithography effects on an input shape. Such a predictive model is process-specific and must be reconfigured for use in each fabrication process. Figure 13.7(a) compares a section of a Bragg grating fabricated using DUVL to that of an ideal design and to that of a predicted outcome as generated by the computational lithography model. It can be seen that the computational lithography model is able to capture the smoothing effects accurately. The responses of the fabricated devices are then measured and compared to simulations of ideal devices and to simulations of devices with geometries predicted by the computational lithography model. The extracted bandwidths of the Bragg gratings are illustrated in Figure 13.7(b). It can be seen that the simulated, ideal bandwidths are much larger than those of the actual, fabricated devices. It can also be seen that the bandwidths of the simulated, predicted devices are comparable to that of the actual,

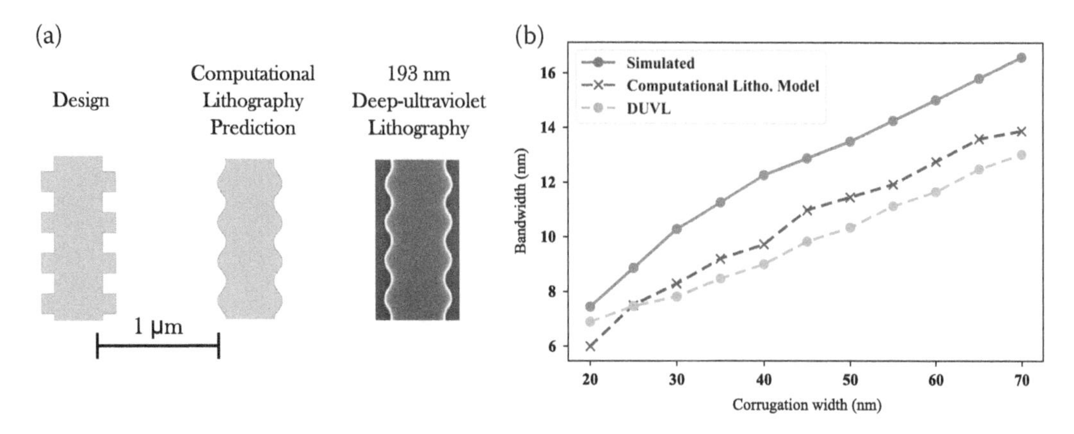

FIGURE 13.7 (a) A comparison of an as-designed Bragg grating to its resultant, fabricated shape as shown by the scanning-electron-microscope micrographs and the predicted shape by the computational lithography model. (b) A comparison between the simulated and measured bandwidths of a set of Bragg gratings. It can be seen that simulated computational lithography model Bragg gratings can accurately predict the bandwidth of the devices fabricated by the deep ultra-violet lithography process.

measured devices. Such lithography predictive techniques can be used to preemptively correct and compensate for the lithography smoothing effects prior to fabrication, enabling the use of Bragg gratings in mass-production, CMOS-compatible, and DUVL fabrication processes.

13.2.3 Applications of Bragg Gratings

Bragg gratings are key to enabling numerous applications in photonic integrated circuits. In this section, we highlight some of these applications including laser cavities, biosensors, and dispersion compensation systems. We encourage the reader to further investigate applications not discussed in this chapter, including polarization rotators [26], polarization-independent transmission filters [27], and optical beam steering [28].

A phase-shifted Bragg grating is a type of a Bragg grating in which an unperturbed waveguide section is introduced between two perturbed sections, as shown in Figure 13.8(a). Such a configuration creates a Fabry-Perot cavity, resulting in resonant modes around the Bragg wavelengths [29]. The number of modes depends on the length of the cavity. Such resonant, phase-shifted Bragg gratings can be designed to have higher quality factors as compared to other silicon photonic resonant devices such as microring resonators [30,31]. Figure 13.8(b) shows a measurement of the transmitted response of a phase-shifted Bragg grating which consisted of a 150 micron long, unperturbed waveguide resonant cavity and two 63.4 micron long Bragg reflectors on each side of the cavity, each creating a mirror that is 99.98% reflective. Such a configuration resulted in a Fabry-Perot cavity with a measured quality factor of 499,700 as shown in Figure 13.8(c).

13.2.3.1 Hybrid-Integrated Lasers

Phase-shifted Bragg gratings are essential devices for many systems and applications that are common to integrated photonics, distributed Bragg reflector (DBR) lasers being an important example. Lasers are key devices for any photonic system and, therefore, much research has been done to integrate them in silicon photonics platforms. Many types of laser have been demonstrated to be integrated on a silicon photonic chip, these include Raman lasers [32,33], III-V hybrid-integrated lasers [34], and Germanium-on-silicon lasers [35]. Currently, hybrid-integrated, silicon and III-V based lasers formed, via bonding techniques [34,36,37], demonstrate the best performance and are currently considered the best candidates for commercial usage in the near future

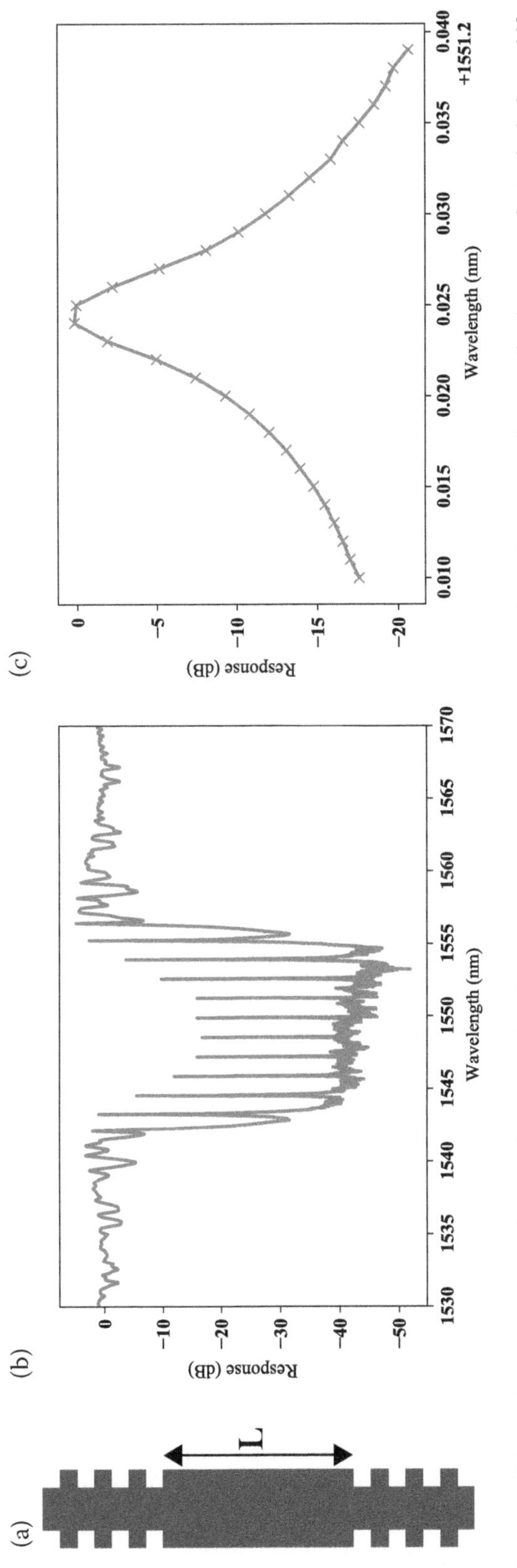

FIGURE 13.8 (a) Schematic drawing of a Fabry-Perot cavity made using phase-shifted Bragg gratings. (b) Measured, normalized spectra of a typical phase-shifted Bragg grating made on a silicon photonic platform. (c) Zoomed in plot around one of the resonances around 1,551.2 nm, showing a high-quality factor of 499,700.

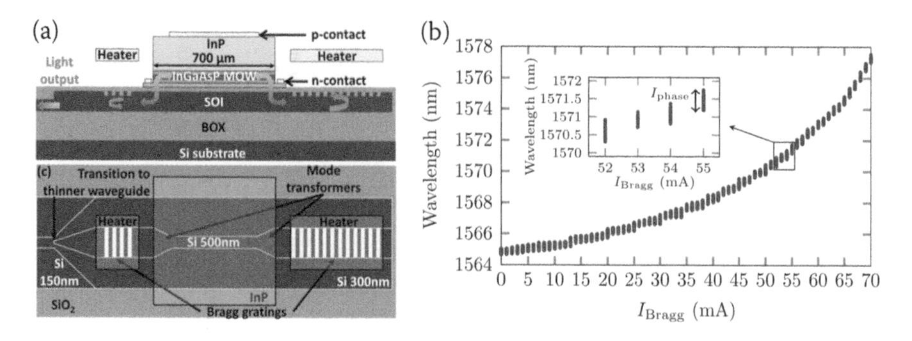

FIGURE 13.9 (a) Top: Cross-sectional view of the hybrid-integrated laser showing the the III-V layers on silicon. Middle top view of the hybrid integrated laser. Bottom: Micrograph image of the complete, fabricated device [37]. (b) Lasing wavelength as a function of the current used to tune the resonant wavelength of the Bragg grating structure [36].

[38]. Such lasers are typically made by bonding a III-V material stack such as InP or GaAs to an SOI wafer. The III-V material is used as the gain section of a laser, with the resonant cavity and mirrors created by a DBR structure made on silicon using phase-shifted Bragg gratings [34]. A DBR laser is shown in Figure 13.9(a) and has been demonstrated to lase at a threshold current of 65 mA. In this particular work [34], one mirror in this laser was a 300 micron long Bragg grating, resulting in a 97% reflective mirror, and the other was a 100 micron long Bragg grating, resulting in a 44% reflective mirror. The DBR hybrid silicon laser can be also tuned in wavelength by tuning the Bragg wavelength using either thermo-optic or electro-optic effects [36] as shown in Figure 13.9(b), in which the Bragg grating sections of the laser were tunable over a range of 12 nm.

13.2.3.2 Biosensors

In addition to being key devices in hybrid-integrated DBR lasers, Bragg gratings have applications in lab-on-chip optical biosensors. Silicon photonic lab-on-chip biosensors are an application of photonic integrated circuit in which the detection and analysis of biological samples is done on a silicon photonic chip [39]. Silicon photonic biosensors have been a focus of research over the past decade for their potential use in environmental monitoring, infectious viruses detection, healthcare, and biomedical research [40]. Such optical sensors often include a photonic system in which the analyte in the biological sample interacts with the evanescent field of the guided wave confined in an integrated photonic waveguide. The effects of such interactions can induce changes in the effective index of the mode. These changes in the effective index are quantified and measured most easily through the use of resonant devices such as microring [41] and microdisk resonators [42]. Bragg gratings can be also used as resonant devices to quantify and measure such effective refractive changes induced by the analyte [43]. As shown in Figure 13.10(a), through the use of a microfluidic channel, the analyte can interact with the evanescent field of the mode guided in the Bragg grating, changing its effective refractive index and, in turn, changing the central wavelength of the grating. Similarly, due to their sharp and narrow resonant peaks, phase-shifted Bragg gratings can also be used to measure the changes in effective refractive index [31]. The high quality factors that can be obtained using phase-shifted Bragg gratings allow for increased detection limits, improved peak tracking efficiencies, and lower sensor costs [40]. One example of the application of such a sensor is shown in Figure 13.10(b), in which the resonant peak of a phase-shifted Bragg grating is plotted as a function of the concentration of the analyte (NaCl in a liquid). The effect of changing the concentration of the analyte is seen as a wavelength red-shift in the resonant peak of the phase-shifted Bragg grating. The sensitivity of such a structure is reported to be 340 nm/RIU (refractive index units) [31]. The sensitivity of phase-shifted Bragg gratings can be further improved by

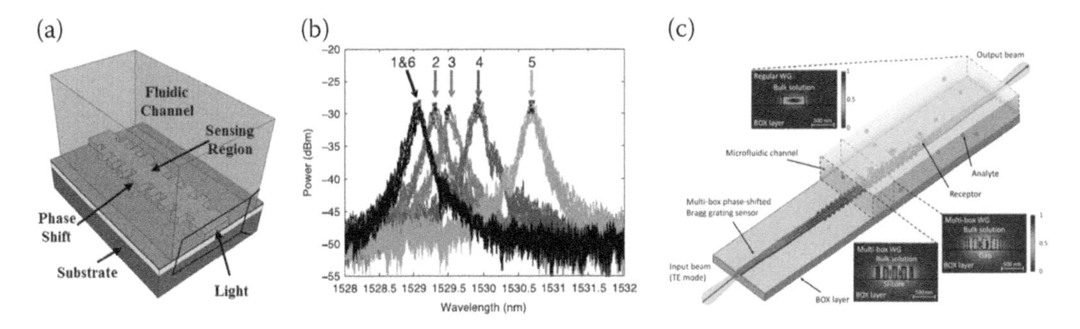

FIGURE 13.10 (a) Illustration of a phase-shifted Bragg grating in which a microfluidic channel over the resonant cavity is used to sense the analyte [40]. (b) Measurement results of the resonant mode in a phase-shifted Bragg gratings using various concentrations of NaCl analyte. Step 1 and 6 correspond to 0 mM, and Step 2 to 5 correspond to 62.5 mM, 125 mM, 250 mM, and 500 mM solutions, respectively [31] (Copyright Wiley-VCH GmbH. Reproduced with permission.). (c) Illustration of a multi-box phase-shifted Bragg grating sensor integrated with a microfluidic channel in which the analyte interacts with the guided mode in the optical waveguide [45].

the use of sub-wavelength structures and meta-materials. Sub-wavelength structure based phase-shifted Bragg gratings provide enhanced sensitivity over regular strip waveguide-based Bragg gratings due to the weak mode confinement and the significant portion of field intensity overlap with the cladding [40,44]. One such unique sub-wavelength structures are multi-box phase-shifted Bragg gratings, shown in Figure 13.10. Compared to traditional sub-wavelength waveguides, multi-box waveguides support an even less confined optical mode and provide more surface area for the analyte to interact with the optical mode propagated in the structure [45]. Such multi-box devices have been reported to have sensitivities as high as 579 nm/RIU.

In addition to the resonant peak tracking approaches discussed above, another approach to detect the analyte in biosensors is using phase-shifted Bragg grating based Mach-Zehnder interferometers. These structures can be used in intensity interrogation schemes [46]. In such an intensity interrogation scheme, the change in the refractive index, due to the presence of the analyte, is detected as a change in the amplitude of the output signal being interrogated [47]. Intensity interrogation schemes allow for the fabrication of low-cost sensors due to their use of relatively inexpensive, broadband optical sources such as light emitting diodes or super-luminescent light emitting diodes. In such a phase-shifted Bragg grating based Mach-Zehnder interferometer biosensor, the sensitivity was measured to be 810 dB/RIU [46].

13.2.3.3 Dispersion Compensation

In a single mode waveguide, dispersion is the phenomenon that refers to the wavelength dependence of the guided mode's phase velocity, i.e., in a dispersive, single mode waveguide, the mode's effective index is a function of wavelength. Furthermore, in a dispersive, single mode waveguide, a pulse's group velocity, or group index, can also be a function of wavelength. The wavelength dependence of the group index causes a temporal broadening of the pulse as it propagates along the waveguide [48]. In many optical communication systems, dispersion is a challenge that may limit the achievable data-rate in a communication link, particularly those that are used for long-reach data-communications. Dispersion can be compensated and corrected for in optical links using dispersion compensation devices. Photonic dispersion compensation devices typically involve the use of dispersion equalization techniques in optical fibers [49,50]. One such technique makes use of chirped fiber Bragg gratings to realize dispersion compensating fibers [51,52]. Chirped Bragg gratings are Bragg gratings in which the periodicity of the refractive index modulation varies, in accordance to a mathematical function, along the length of a

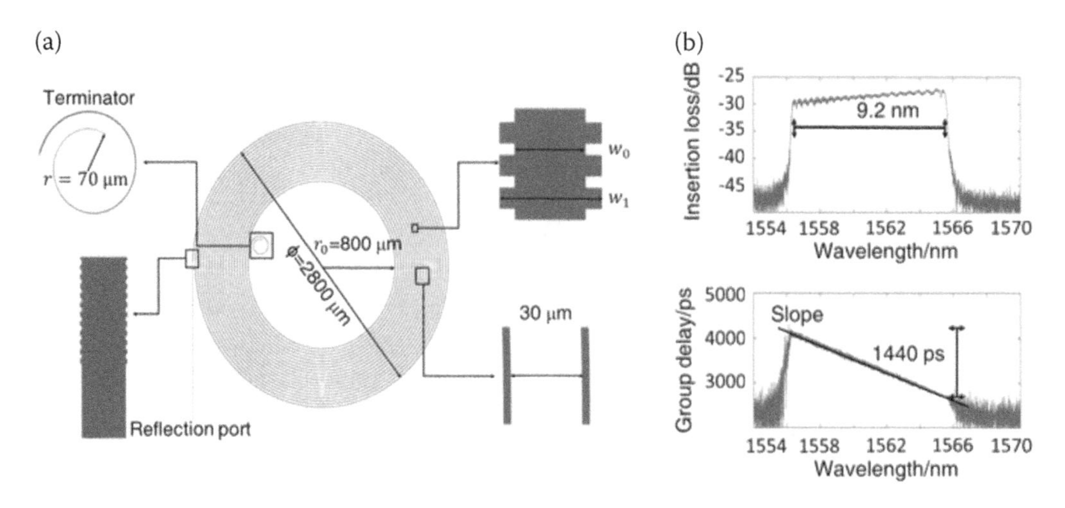

FIGURE 13.11 (a) Illustration of a chirped Bragg grating device designed in the shape of a spiral [56]. (b) The measured reflection spectrum and group delay of a chirped Bragg grating [56].

grating. The variation in the grating's periodicity results in different wavelengths being reflected by different sections of the grating and, therefore, the reflection of each wavelength obtains a specific time delay. Using this approach at the end of an optical link, the wavelength-dependent time delay effectively results in "re-compressing" an optical pulse after it has undergone temporal broadening due to dispersion [48].

Much research has been done to realize dispersion compensation devices on silicon photonics platforms [53]. Microring resonators and Mach-Zehnder interferometers can be used for dispersion compensation [53,54]. However, such approaches often have narrow operating bandwidths and high insertion losses. Dispersion compensating, chirped, fiber Bragg gratings can also be replicated on silicon photonics platforms to design low loss, broadband, on-chip dispersion compensating devices. Similar to their fiber counterparts, chirped Bragg gratings can be made on integrated photonics platforms by gradually changing the periodicity of the Bragg grating structure. Chirped Bragg gratings are ideally very long devices, such that the group delay slope and the bandwidth of the chirped Bragg grating are large [55]. However, long, straight devices are often impractical to implement on silicon photonics platforms. Nevertheless, a long, low-loss, chirped Bragg grating can be made by using a compact, spiral design, as shown in Figure 13.11(a). For example, in [56] a 13.8 cm long, chirped Bragg grating is compacted into a spiral with an outer-diameter of only 2.8 mm. As shown in Figure 13.11(b), this compact, spiral grating was measured to have a group delay slope (dispersion) of 156.5 ps/nm over a wavelength range of 9.2 nm [56]. Another possible method to chirp the Bragg gratings on integrated photonics platforms is to apply a temperature gradient along the length of a uniform Bragg grating [57].

13.3 CONTRA-DIRECTIONAL COUPLERS

Contra-directional couplers (contra-DCs), shown in Figure 13.12(a), are Bragg grating based waveguide couplers in which two, asymmetric, periodically perturbed waveguides are coupled. The concept of such a device was first proposed by C. Elachi and C. Yeh in 1973 [58]. The gratings in such devices are used to contra-directionally couple specific wavelengths between two waveguides that are, otherwise, dissimilar and uncoupled. A contra-DC couples the forward-propagating mode of one waveguide into the backward-propagation mode of the other waveguide. The operating principle of a contra-DC is similar to that of a two-port Bragg grating, however, the filtered wavelengths in such devices couple backwards into a second, output waveguide, instead of the same input waveguide [59]. Having a second, output waveguide eliminates the need to extract the reflected signal from the common, input-output waveguide of a single-waveguide Bragg grating, such as could be done using a circulator. However, an on-chip circulator remains a challenging device to integrate on a silicon photonics platform [60]. Therefore, contra-DCs allow one to selectively add and drop wavelengths, making them a suitable candidate for realizing add-drop filters on a silicon photonics platform.

13.3.1 Contra-Directional Couplers' Theory

Contra-DCs couple the forward propagating mode of one waveguide into the backward propagating mode of the other, at those wavelengths at which the phase-match condition is achieved. The phase-match condition depends on the average effective index of the two waveguides and the period of the perturbations, as illustrated by the phase-match dispersion plot in Figure 13.13(a). For contra-DCs, the phase matching condition is given by [1]:

$$\beta_1 + \beta_2 = \frac{2\pi}{\Lambda_B} \tag{13.8}$$

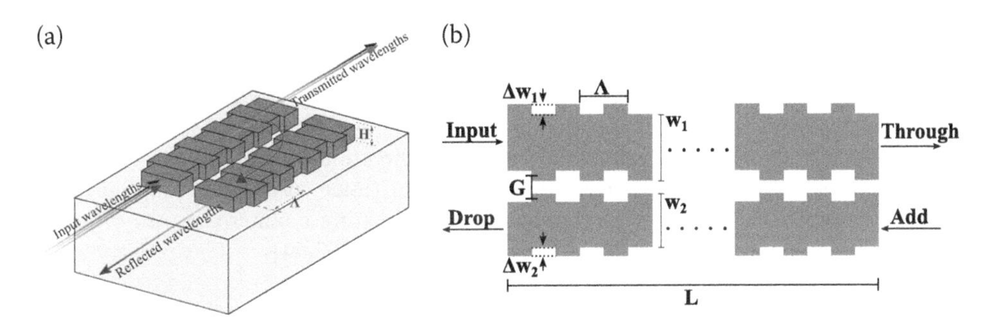

FIGURE 13.12 (a) Illustration of a contra-directional coupler device highlighting the device's operation principle. (b) Top-view illustration of a contra-directional coupler with the design parameters highlighted. Note: both designs illustrated use the "out-of-phase" gratings design to suppress the waveguides' self-reflections as discussed in [61].

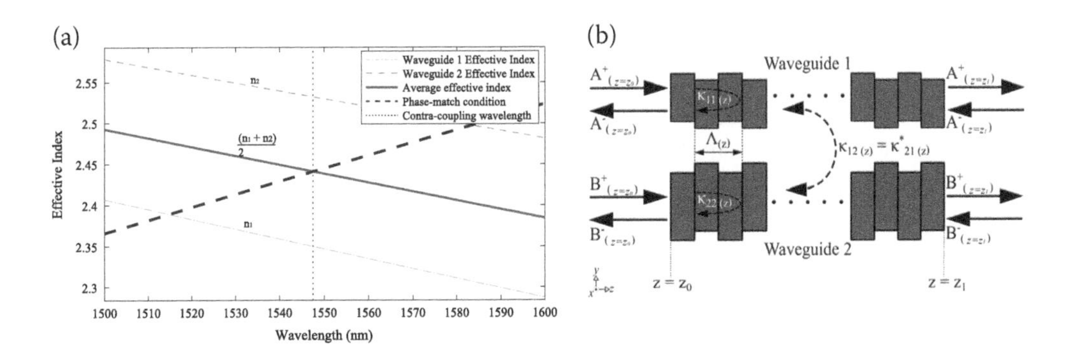

FIGURE 13.13 (a) Dispersion plot highlighting the phase-match condition obtained from a two-waveguide system contra-directional coupler. (b) Top-view schematic of a contra-directional coupler showing the forward and backward propagating electric fields in the two-waveguide system.

in which Λ_B is the period of the perturbations, β_1 and β_2 are the propagation constants of the fundamental mode and the next higher-order mode of the two-waveguide system, respectively, and are given by:

$$\beta_m = \frac{2\pi n_{eff,m}}{\lambda} \tag{13.9}$$

in which $n_{eff,m}$ is the effective index of the mth mode of the two-waveguide system. Using Equation 13.8, the phase-match condition can be derived to be at:

$$\lambda_B = (n_{eff,1} + n_{eff,2})\Lambda_B \tag{13.10}$$

To formulate an expression that describes the fields in a contra-DC, we need to consider the transverse fields of the modes, $E_1(x, y)$ and $E_2(x, y)$, of the unperturbed coupler, which are mainly confined in waveguide 1 and waveguide 2, respectively. As shown in Figure 13.13(b), for each of the two transverse field distributions there can be a forward propagating mode (indicated by a "+" superscript) and a backward propagating mode (indicated by a "–" superscript) and, therefore, the electric field in the contra-DC can be expressed as [62]:

$$\begin{aligned} E(x, y, z) = {} & [A^+(z)e^{-j\hat{\beta}_1 z} + A^-(z)e^{j\hat{\beta}_1 z}]E_1(x, y, z) + \\ & [B^+(z)e^{-j\hat{\beta}_2 z} + B^-(z)e^{j\hat{\beta}_2 z}]E_2(x, y, z) \end{aligned} \tag{13.11}$$

where $\hat{\beta} = \beta - j\alpha$ is the complex propagation constant and $A^+(z)$, $A^-(z)$, $B^+(z)$, and $B^-(z)$ are the field amplitudes as functions of the longitudinal position, z, see Figure 13.13(b).

Due to the periodic dielectric perturbations, contra-directional coupling can occur between the two transverse modes of the two-waveguide system (inter-waveguide coupling) and also between either transverse mode and its oppositely propagating counterpart (self-reflection, or back-reflection, coupling). The coupled-mode equations are: [63,64]

$$\begin{aligned} \frac{dA^+}{dz} &= -j\kappa_{11}A^- e^{j2\Delta\hat{\beta}_1 z} - j\kappa_{12}B^- e^{j(\Delta\hat{\beta}_1+\Delta\hat{\beta}_2)z} \\ \frac{dB^+}{dz} &= -j\kappa_{12}A^- e^{j(\Delta\hat{\beta}_1+\Delta\hat{\beta}_2)z} - j\kappa_{22}B^- e^{j2\Delta\hat{\beta}_2 z} \\ \frac{dA^-}{dz} &= j\kappa_{11}^* A^+ e^{-j2\Delta\hat{\beta}_1 z} + j\kappa_{12}^* B^+ e^{-j(\Delta\hat{\beta}_1+\Delta\hat{\beta}_2)z} \\ \frac{dB^-}{dz} &= j\kappa_{12}^* A^+ e^{-j(\Delta\hat{\beta}_1+\Delta\hat{\beta}_2)z} + j\kappa_{22}^* B^+ e^{-j2\Delta\hat{\beta}_2 z} \end{aligned} \tag{13.12}$$

where $\Delta\hat{\beta}_{1,2} = \hat{\beta}_{1,2} - \frac{\pi}{\Lambda}$ and $\Delta\varepsilon(x, y)$ is the first-order Fourier-expansion coefficient of the dielectric perturbation and κ_{11} and κ_{22} are the back-reflection coupling coefficients (or strengths) for $E_1(x, y)$ and $E_2(x, y)$, respectively. κ_{12} and $\kappa_{21} = \kappa_{12}^*$ are the inter-waveguide, contra-directional coupling coefficients between $E_1(x, y)$ and $E_2(x, y)$. The coupling coefficients are given by:

$$\begin{aligned} \kappa_{11} &= \frac{\omega}{4} \iint E_1^*(x, y) + \Delta\varepsilon(x, y) + E_1(x, y)dxdy \\ \kappa_{22} &= \frac{\omega}{4} \iint E_2^*(x, y) + \Delta\varepsilon(x, y) + E_2(x, y)dxdy \\ \kappa_{12} = \kappa_{21}^* &= \frac{\omega}{4} \iint E_1^*(x, y) + \Delta\varepsilon(x, y) + E_2(x, y)dx\,dy \end{aligned} \tag{13.13}$$

To solve the above coupled-mode equations, one can apply the transfer-matrix method. Given that $E(z)$ is a vector of the amplitudes of the forward propagating and backward propagating modes at any position z along the device, i.e.:

$$E(z) = \begin{bmatrix} A^{+}(z) \\ B^{+}(z) \\ A^{-}(z) \\ B^{-}(z) \end{bmatrix} \tag{13.14}$$

then, the relationship between the fields at z_0 and z_1 (see Figure 13.13(b)) is given by:

$$E(z_1) = C(z_0, z_1)E(z_0) \tag{13.15}$$

where $C(z_0, z_1)$ is the transfer matrix of the contra-DC, for which a solution can be obtained by solving the coupled-mode equations. The transfer matrix can, in general, be described as the product of the n segments (here, in order to allow for design flexibility, a segment can be used to describe a single period or a number of related periods, depending on the profile of the device) making up the contra-DC [64–66]:

$$C(z_0, z_1) = \prod_{i=1}^{n} C(z_{i-1}, z_i = \prod_{i=1}^{n} e^{(z_i - z_{i-1})S_{1,i}} e^{(z_i - z_{i-1})S_{2,i}} \tag{13.16}$$

in which the matrices $S_{1,i}$ and $S_{2,i}$ are given by:

$$S_{1,i} = \begin{bmatrix} j\Delta\hat{\beta}_{1,i} & 0 & 0 & 0 \\ 0 & j\Delta\hat{\beta}_{2,i} & 0 & 0 \\ 0 & 0 & -j\Delta\hat{\beta}_{1,i} & 0 \\ 0 & 0 & 0 & -j\Delta\hat{\beta}_{2,i} \end{bmatrix} \tag{13.17}$$

$$S_{2,i} = \begin{bmatrix} -j\Delta\hat{\beta}_{1,i} & 0 & -j\kappa_{11} e^{j2\Delta\hat{\beta}_{1,i} z_n} & -j\kappa_{12} e^{j(\Delta\hat{\beta}_{1,i}+\Delta\hat{\beta}_{2,i}) z_n} \\ 0 & -j\Delta\hat{\beta}_{2,i} & -j\kappa_{12} e^{j(\Delta\hat{\beta}_{1,i}+\Delta\hat{\beta}_{2,i}) z_n} & -j\kappa_{22} e^{j2\Delta\hat{\beta}_{2,i} z_n} \\ j\kappa_{11}^{*} e^{-j2\Delta\hat{\beta}_{1,i} z_n} & j\kappa_{21} e^{-j(\Delta\hat{\beta}_{1,i}+\Delta\hat{\beta}_{2,i}) z_n} & j\Delta\hat{\beta}_{1,i} & 0 \\ j\kappa_{12} e^{-j(\Delta\hat{\beta}_{1,i}+\Delta\hat{\beta}_{2,i}) z_n} & j\kappa_{22}^{*} e^{-j2\Delta\hat{\beta}_{2,i} z_n} & 0 & j\Delta\hat{\beta}_{2,i} \end{bmatrix} \tag{13.18}$$

in which $z_n = z_i - z_{i-1}$ and $\Delta\hat{\beta}_{1/2,i} = \frac{2\pi n_{eff,1/2}}{\lambda} - \frac{\pi}{\Lambda_i}$ are calculated individually for each segment of the grating and are considered constant within their segments. Solving the matrix exponential $e^{(z_i - z_{i-1})S_{2,i}}$ is a non-trivial process and can be computationally solved using Pad's approximation [67,68].

13.3.2 Applications of Contra-Directional Couplers

In this section, we discuss applications of contra-DCs, namely those used in wavelength-division-multiplexing (WDM) systems. We discuss the design of high-performance WDM optical add-drop multiplexers made using contra-DCs, as well as the design of broadband optical add-drop multiplexers. We encourage the reader to investigate additional applications that we believe are also relevant but not discussed in this chapter such as tunable filters [69], polarization rotators [70], FSR-free microring resonators [71], and FSR-free microring modulators [72].

13.3.2.1 Wavelength-Division-(de)-Multiplexer

A technology commonly used to increase the aggregate data rate of an optical link is WDM. In WDM, multiple wavelengths of light, each carrying a set of data, are combined and simultaneously transmitted through a single optical waveguide. The multiple wavelength channels are combined (added) at the transmitter side of the optical link by the use of an optical multiplexer. Similarly, the wavelengths can be separated (filtered) into individual wavelength channels (the dropped channels) at the receiver side of the optical link by the use of an optical demultiplexer. Therefore, such optical add-drop (de)-multiplexers (OADMs) are important devices for silicon photonics platforms. Typical figures-of-merit used to define OADM performance are bandwidth, insertion loss, group delay (dispersion), extinction ratio, and isolation (out-of-band rejection ratio). Some of these figures-of-merit are more emphasized (or relevant) than others, depending on the application the OADM is being used for. For example, Dense-WDM (DWDM) communication links often require demultiplexers with narrow filter bandwidths, while Coarse-WDM (CWDM) communication links require demultiplexers with wider filter bandwidths [73]. The design of OADMs is a well-studied field in silicon photonics technology; however, increased adoption of silicon photonics technology, for many applications, is driving the need for filters with increasingly stringent specifications. Several types of silicon photonic optical filter have been studied; these include micro-ring resonators, lattice filters, arrayed waveguide gratings, and Echelle gratings [74–76].

Contra-DCs can also be used as a single channel OADM or in multi-channel OADMs. One can design the central wavelength and bandwidth of the grating to match the specifications required for the filter. For example, one can design a flat-top, add-drop filter centered at 1,550 nm by engineering the design parameters such that the Bragg phase-match condition occurs at the desired wavelength. To design such a filter on a 220 nm silicon thick SOI platform, two waveguides, one that is 440 nm wide and the other that is 560 nm wide, with an average gap of 160 nm, can be used. For these waveguides, the period of the filter's corrugations should be 318 nm, such that the phase match condition occurs at the desired wavelength. The corrugations should be 50 nm wide for the wide waveguide and 30 nm wide for the narrow waveguide. The corrugations can also be apodized, as previously discussed, to suppress the side-lobes of the reflected signal. Such a device was fabricated on an SOI platform using a 193 nm DUVL process, as can be seen from the scanning-electron-microscope micrograph in Figure 13.14(a). The measured response of the fabricated device is shown in Figure 13.14(b). The device is measured to have a response centered at 1,549 nm, a 3 dB bandwidth bandwidth of 4 nm (at the reflected, dropped signal), a maximum of 0.22 dB loss at the drop port, and a maximum of 24 dB extinction ratio at the through port. A minor dip in the transmitted signal can be seen at around 1,528 nm. This dip is due to the narrow waveguide's self-reflection Bragg condition, which was used as the input waveguide.

To further improve the response of the device discussed above, one can use a combination of grating apodization techniques and multiple, series-cascaded contra-DCs. Cascading multiple contra-DCs can significantly enhance the signal extinction, around the Bragg wavelength, at the overall through port response of such a device [77]. Similarly, cascading multiple contra-DCs can significantly enhance the overall drop port's signal suppression, outside the Bragg wavelength, enhancing the out-of-band rejection ratio [69,78]. The configuration of a device designed to enhance both the overall through port's extinction ratio and the overall drop port's out-of-band rejection is shown in Figure 13.15(a). In this device, though only 7 contra-DCs are used, it has 4 cascaded contra-DC stages at the through port and another 4 cascaded contra-DC stages at the drop port. All of the contra-DCs used in this device are identical. The measured response of the device is shown in Figure 13.15(b). Due to using cascaded contra-DCs, the through port response

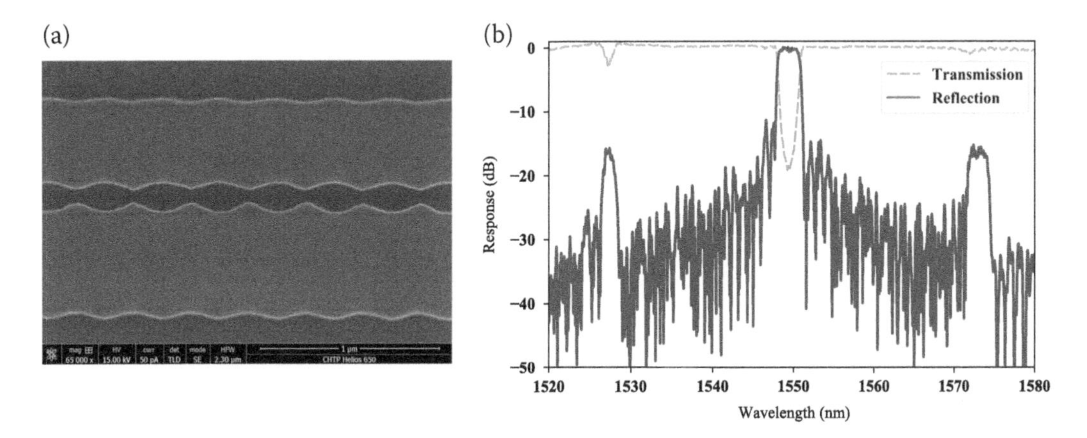

FIGURE 13.14 (a) Scanning-electron-microscope micrograph (top-view) of a section of a contra-directional coupler made on a 193 nm deep ultra-violet lithography fabrication process. (b) Measured, normalized spectra of a typical contra-directional coupler made on a silicon photonic platform.

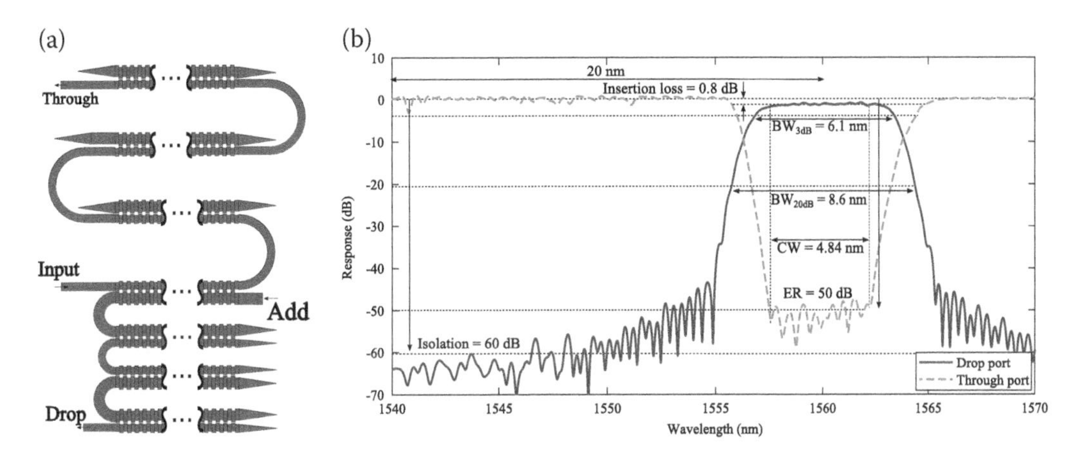

FIGURE 13.15 (a) Schematic of a cascaded contra-directional coupler to increase the extinction ratio and out-of-band rejection ratio. (b) Measured, calibrated spectra of the through port and drop port of a cascaded contra-directional coupler highlighting the devices' figures-of-merit.

is measured to have a high extinction ratio of 50 dB within a clear window (CW) of 4.84 nm and the drop port is measured to have a flat-topped response with a maximum insertion loss of 0.8 dB, a 3 dB bandwidth of 6.1 nm, and an ultra-high, out-of-band rejection ratio in excess of 60 dB, limited by the measurement equipment and the experimental setup.

Such cascaded drop port designs can be further extended to create multi-channel OADMs with high channel isolations, minimizing the detrimental effects of inter-channel crosstalk [79]. For example, a 4-channel OADM can be designed in which each channel filter is made of a set of identical, cascaded contra-DCs, each with its phase-match condition occurring at the desired channel's center wavelength, as shown in the dispersion plot in Figure 13.16(a). In this case, the Bragg phase-match condition for each channel filter is achieved by setting the corrugation period of the set of cascaded contra-DCs constituting the filter. With the exception of the last channel filter, the through port of each channel filter is fed into the input port of the next channel filter, as shown in Figure 13.16(b). In this device, the channels were designed to be 12 nm apart, each with a 3 dB bandwidth of 7 nm.

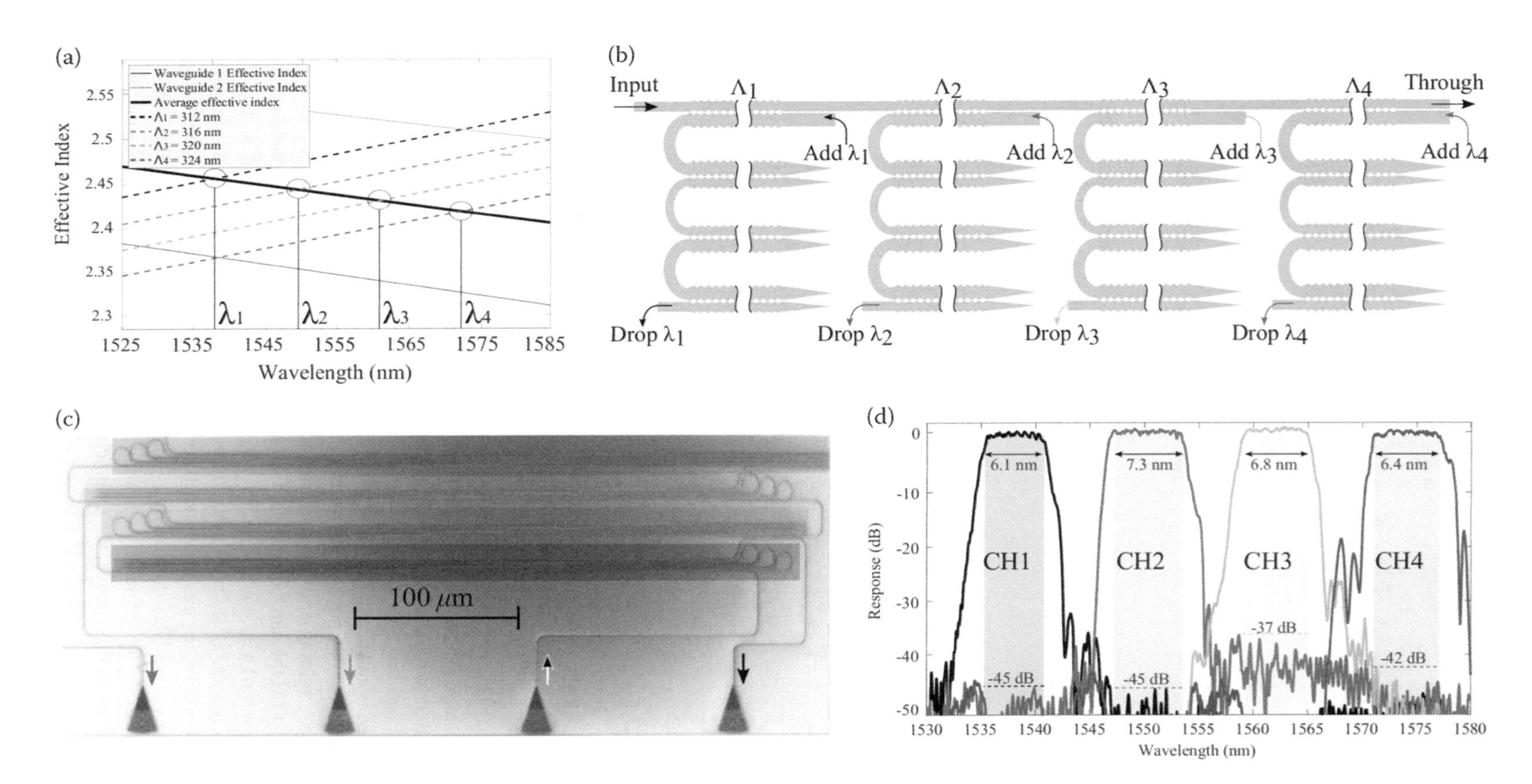

FIGURE 13.16 (a) Dispersion plot highlighting the four phase-match conditions obtained by each channel. (b) Schematic of the four-channel, cascaded contra-directional coupler based optical add-drop (de)-multiplexer. (c) False-color scanning-electron-microscope micrograph of the fabricated device highlighting each set of cascaded contra-directional couplers. (d) Measured, calibrated spectra of four-channel, cascaded contra-directional coupler based optical add-drop (de)-multiplexer when operated as a demultiplexer. [79].

A false color scanning-electron-microscope micrograph of a fabricated device, based on the above design, is shown in Figure 13.16(c) in which each set of contra-DCs is highlighted. The total footprint of the device is 571 micron-by-158 micron. Using the device as a demultiplexer, the channel filters were measured and the response of each is shown in Figure 13.16(d). The device was measured without the use of any post-fabrication tuning or correction. The minimum adjacent channel isolation was measured to be 37 dB and the minimum non-adjacent channel isolation was measured to be 45 dB. The measured, maximum insertion loss was 0.72 dB (seen on channel 4). The measured 3 dB bandwidths and each channel's minimum isolation are highlighted in Figure 13.16(d). Such cascaded contra-DC based, multi-channel OADMs can be designed to meet various communication standards/recommendations, such as those for CWDM and LAN-WDM, while having high performances and compact footprints.

13.3.2.2 Broadband Optical Add-Drop (de)-Multiplexers

As previously discussed, the strong confinement of the guided modes of the waveguides on silicon photonic platforms is often beneficial since it allows for the design of compact photonic integrated circuits. However, such strong confinement of the guided modes may also limit a contra-DC's coupling coefficient (κ) since the coupling coefficient depends on the overlap between the two coupled modes of the two-waveguide system forming the contra-DC, as presented in Equations 13.12. Limiting the coupling coefficients correspondingly limits the bandwidth, as presented for the Bragg grating in Equation 13.7. Therefore, it is challenging to design broadband (or wideband) OADMs with bandwidths in excess of 12 nm using typical contra-DCs made using strip or rib waveguides on 220 nm silicon thick SOI platforms [80]. Such broadband OADMs are necessary to create wavelength band splitters, i.e., ones that can multiplex and demultiplex entire communication bands, which are necessary for the design of ultra-broadband silicon photonic integrated circuits. The modal overlap in a contra-DC can be significantly increased using sub-wavelength grating (SWG) based waveguides, instead of strip or rib waveguides. SWGs are periodic structures in which the periods are less than $\frac{\pi}{\beta}$, β being the propagation constant of the mode [80,81]. The mode propagating in an SWG waveguide can, therefore, propagate as it would in a typical strip or rib waveguide, however, with its effective refractive index being determined by the effective-medium theory [19,82]. The effective refractive index of SWG waveguide can be engineered to support guided modes that are less confined within the waveguide structure, as compared to strip and rib waveguides. Such SWG based waveguides can be used in contra-DCs to design ones that have high coupling coefficients and, in turn, broad bandwidths at their drop ports [80,83,84].

A design of such an SWG based contra-DC on a 220 nm silicon thick SOI platform is shown in Figure 13.17(a). In this design [80], strip waveguide to SWG based waveguide transition tapers, see Region *I* in Figure 13.17(a), are used to gradually transition the modes from the strip waveguide modes to those of the SWG based waveguide modes. Region *II* in Figure 13.17(a) highlights a section of the coupling region in the SWG based contra-DC in which the Bragg condition perturbations are created by periodically offsetting (with the period Λ_{Bragg}) the SWG structures. To design an SWG based contra-DC to have a Bragg phase-match condition occurring at 1,550 nm, the SWG waveguides can be chosen to have widths of 560 nm and 440 nm for the wide waveguide and narrow waveguide, respectively, with an average gap of 200 nm. The period of the perturbation can be set to $\Lambda_{Bragg} = 468$ nm, in which each perturbation can be made by periodically offsetting the SWG structures making up each of the waveguides by 60 nm and 40 nm for the wide waveguide and narrow waveguide, respectively. Similar to previously discussed Bragg reflector and contra-DC designs, the perturbations can be apodized along the length of the grating to suppress the side lobes of the drop port signal.

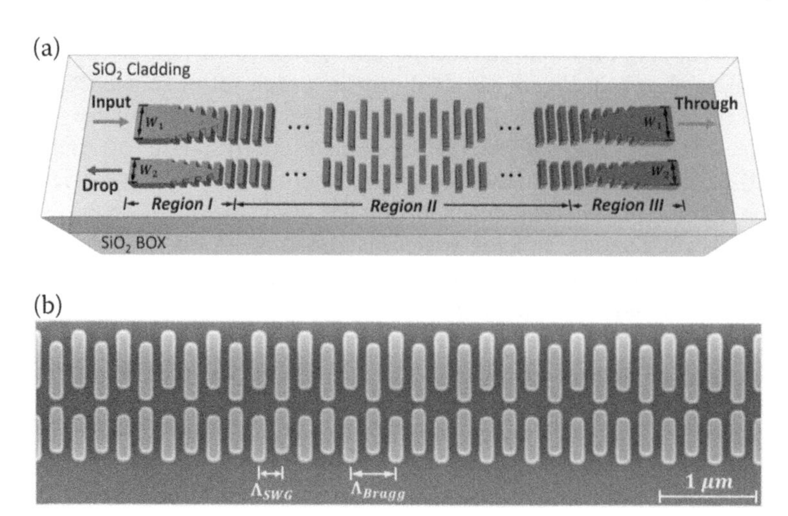

FIGURE 13.17 (a) Illustration of a sub-wavelength grating based contra-directional coupler with the its mode transition and coupling regions highlighted. (b) Scanning-electron-microscope micrographs of a section of the coupling region of the sub-wavelength grating based contra-directional coupler [80].

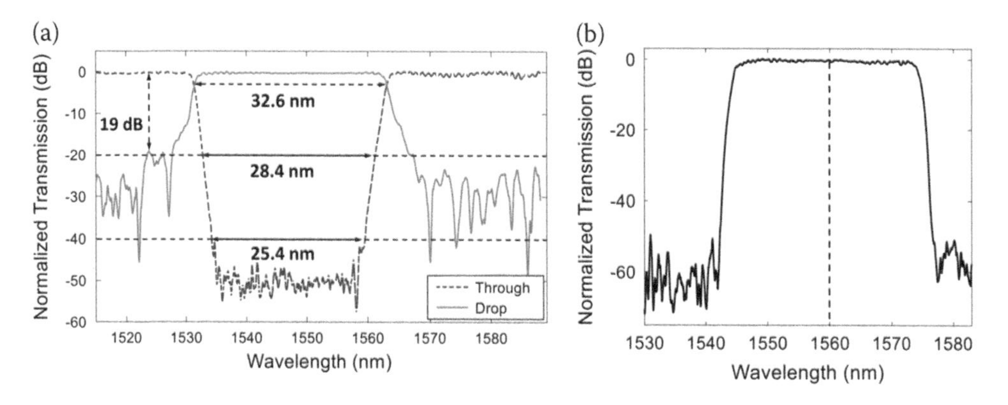

FIGURE 13.18 (a) Measured, calibrated spectra of the through and drop ports of a sub-wavelength grating based contra-directional coupler highlighting the broad bandwidth obtained at the drop port. (b) Measured, calibrated spectrum of the drop port of a cascaded, sub-wavelength grating based contra-directional coupler showing the flat-topped response and high out-of-band rejection ratio [80].

A scanning-electron-microscope micrograph of a section of the fabricated device, based on the above design, is shown in Figure 13.17(b). The device response was measured and is shown in Figure 13.18(a). The measured drop port response is flat-topped with a broad 3 dB bandwidth of 32.6 nm, a side lobe suppression ratio of 19 dB, and an insertion loss of 0.26 dB. Additionally, an extinction ratio in excess of 40 dB is seen at the through port within a clear window of 25 nm. To further suppress the side lobes at the drop port, and increase the out-of-band rejection ratio of such a device, one can apply the cascaded contra-DCs design discussed in the previous section. In the design presented in [80], three identical SWG-based contra-DCs were cascaded, in which the drop port of one contra-DC is input to the next contra-DC in a configuration similar to that shown in Figure 13.15(a). The measured drop port response of the overall, cascaded device is shown in Figure 13.18(b). The drop port response is flat-topped and is measured to have an insertion loss of <0.85 dB, a 3 dB bandwidth of >30 nm, and an ultra-high side lobe suppression ratio in excess of 50 dB. Such high-performing broadband OADMs should find numerous applications in the design of broadband silicon photonic systems.

13.4 CONCLUSION

In this chapter, we have presented the theory of and design techniques for both, Bragg gratings and Bragg grating based devices such as the contra-directional coupler. We have also discussed how such devices can be used in various systems such as lasers, dispersion compensators, bio-sensors, and WDM systems. We have described design and fabrication challenges encountered when deploying such devices in practice, such as manufacturing variability and lithography effects and approaches for understanding and mitigating these challenges. Due to the design versatility and potential of such devices, we highly encourage the reader to further investigate applications of Bragg gratings and Bragg grating based devices in silicon photonics systems.

ACKNOWLEDGMENTS

The authors would like to acknowledge the contributions of the following people and organizations in making the work presented possible. The authors acknowledge colleagues at The University of British Columbia: Dr. Han Yun, Dr. Minglei Ma, Dr. Enxiao Luan, Dr. Hossam Shoman, Dr. Xu Wang, Stephen Lin, Ajay Mistry, and Abdelrahman Afifi. The authors would also like to acknowledge the colleagues at Laval University: Dr. Wei Shi and Jonathan Cauchon for insightful discussions and continued collaboration. The authors would like to acknowledge Dr. Ryan Scott and Dr. Chris Coleman of Keysight Technologies for their contributions, insightful discussions, and facilitating the fabrication of some of the devices presented. We thank CMC Microsystems for providing access to the software tools and brokering some of the fabrication services at the DUV foundry. We thank AppliedNanotools, SiEPICfab, and The University of Washington Microfabrication/Nanotechnology User Facility, a member of the NSF National Nanotechnology Infrastructure Network, for providing E-beam fabrication services and SEM imaging services. We thank Dr. Gethin Owen at The Centre for High-Throughput Phenogenomics for assisting with the SEM imaging. We thank Ansys Lumerical for providing the design and simulation software. This work was funded by the Natural Sciences and Engineering Research Council (NSERC) of Canada and Keysight Technologies.

REFERENCES

[1] Amnon Yariv, Pochi Yeh, and Amnon Yariv. *Photonics: Optical Electronics in Modern Communications*, volume 6. Oxford University Press, 2007.

[2] Jens Buus, Markus-Christian Amann, and Daniel J Blumenthal. *Tunable Laser Diodes and Related Optical Sources*. John Wiley & Sons, 2005.

[3] Lukas Chrostowski and Krzysztof Iniewski. *High-speed Photonics Interconnects*. CRC Press, 2017.

[4] Kenneth O Hill, Y Fujii, Derwyn C Johnson, and BS Kawasaki. Photosensitivity in optical fiber waveguides: Application to reflection filter fabrication. *Applied Physics Letters*, 32(10):647–649, 1978.

[5] Thomas Edward Murphy, Jeffrey Todd Hastings, and Henry I Smith. Fabrication and characterization of narrow-band bragg-reflection filters in silicon-on-insulator ridge waveguides. *Journal of Lightwave Technology*, 19(12):1938, 2001.

[6] Jeffery Todd Hastings, Michael H Lim, James G Goodberlet, and Henry I Smith. Optical waveguides with apodized sidewall gratings via spatial-phase-locked electron-beam lithography. *Journal of Vacuum Science & Technology B: Microelectronics and Nanometer Structures Processing, Measurement, and Phenomena*, 20(6):2753–2757, 2002.

[7] Xu Wang, Wei Shi, Han Yun, Samantha Grist, Nicolas Jaeger, and Lukas Chrostowski. Narrow-band waveguide bragg gratings on SOI wafers with CMOS-compatible fabrication process. *Optics Express*, 20(14):15547–15558, 2012.

[8] Dawn T H Tan, Kazuhiro Ikeda, and Yeshaiahu Fainman. Cladding-modulated bragg gratings in silicon waveguides. *Optics Letters*, 34(9):1357–1359, 2009.

[9] Renzo Loiacono, Graham T Reed, Goran Z Mashanovich, Russell Gwilliam, Simon J Henley, Youfang Hu, Ran Feldesh, and Richard Jones. Laser erasable implanted gratings for integrated silicon photonics. *Optics Express*, 19(11):10728–10734, 2011.

[10] Ivano Giuntoni, Andrzej Gajda, Michael Krause, Ralf Steingrüber, Jürgen Bruns, and Klaus Petermann. Tunable bragg reflectors on silicon-on-insulator rib waveguides. *Optics Express*, 17(21):18518–18524, 2009.

[11] Łukasz Zychowicz, Jacek Klimek, and Piotr Kisała. Methods of producing apodized fiber bragg gratings and examples of their applications. *Informatyka, Automatyka, Pomiary w Gospodarce i Ochronie Środowiska*, 60–63, 2018.

[12] Alexandre D Simard, Nezih Belhadj, Yves Painchaud, and Sophie LaRochelle. Apodized silicon-on-insulator bragg gratings. *IEEE Photonics Technology Letters*, 24(12):1033–1035, 2012.

[13] Dorothea Wiesmann, C David, Roland Germann, D Emi, and Gian-Luca Bona. Apodized surface-corrugated gratings with varying duty cycles. *IEEE Photonics Technology Letters*, 12(6):639–641, 2000.

[14] Rui Cheng, Ya Han, and Lukas Chrostowski. Characterization and compensation of apodization phase noise in silicon integrated bragg gratings. *Optics Express*, 27(7):9516–9535, 2019.

[15] Wim Bogaerts, Martin Fiers, and Pieter Dumon. Design challenges in silicon photonics. *IEEE Journal of Selected Topics in Quantum Electronics*, 20(4):1–8, 2013.

[16] Rui Cheng and Lukas Chrostowski. Spectral design of silicon integrated bragg gratings: A tutorial. *Journal of Lightwave Technology*, 39(3):712–729, 2020.

[17] Alexandre D Simard, Guillaume Beaudin, Vincent Aimez, Yves Painchaud, and Sophie LaRochelle. Characterization and reduction of spectral distortions in silicon-on-insulator integrated bragg gratings. *Optics Express*, 21(20):23145–23159, 2013.

[18] Ashok V Krishnamoorthy, Xuezhe Zheng, Guoliang Li, Jin Yao, Thierry Pinguet, Attila Mekis, Hiren Thacker, Ivan Shubin, Ying Luo, Kannan Raj, et al. Exploiting CMOS manufacturing to reduce tuning requirements for resonant optical devices. *IEEE Photonics Journal*, 3(3):567–579, 2011.

[19] Pavel Cheben, Robert Halir, Jens H Schmid, Harry A Atwater, and David R Smith. Subwavelength integrated photonics. *Nature*, 560(7720):565–572, 2018.

[20] Jens H Schmid, Marc Ibrahim, Pavel Cheben, Jerome Lapointe, Siegfried Janz, Przemek J Bock, Adam Densmore, Boris Lamontagne, Rubin Ma, Winnie N Ye, et al. Temperature-independent silicon subwavelength grating waveguides. *Optics Letters*, 36(11):2110–2112, 2011.

[21] Junjia Wang, Ivan Glesk, and Lawrence R Chen. Subwavelength grating bragg grating filters in silicon-on-insulator. *Electronics Letters*, 51(9):712–714, 2015.

[22] Zhitian Chen, Jonas Flueckiger, Xu Wang, Fan Zhang, Han Yun, Zeqin Lu, Michael Caverley, Yun Wang, Nicolas A F Jaeger, and Lukas Chrostowski. Spiral bragg grating waveguides for TM mode silicon photonics. *Optics Express*, 23(19):25295–25307, 2015.

[23] Xu Wang, Wei Shi, Michael Hochberg, Kostas Adam, Ellen Schelew, Jeff F Young, Nicolas A F Jaeger, and Lukas Chrostowski. Lithography simulation for the fabrication of silicon photonic devices with deep-ultraviolet lithography. In *The 9th International Conference on Group IV Photonics (GFP)*, pages 288–290. IEEE, 2012.

[24] Jonathan St-Yves, Sophie Larochelle, and Wei Shi. O-band silicon photonic bragg-grating multiplexers using uv lithography. In *Optical Fiber Communication Conference*, pages Tu2F-7. Optical Society of America, 2016.

[25] Stephen Lin, Mustafa Hammood, Han Yun, Enxiao Luan, Nicolas A F Jaeger, and Lukas Chrostowski. Computational lithography for silicon photonics design. *IEEE Journal of Selected Topics in Quantum Electronics*, 26(2):1–8, 2019.

[26] Han Yun, Zhitian Chen, Yun Wang, Jonas Fluekiger, Michael Caverley, Lukas Chrostowski, and Nicolas A F Jaeger. Polarization-rotating, bragg-grating filters on silicon-on-insulator strip waveguides using asymmetric periodic corner corrugations. *Optics Letters*, 40(23):5578–5581, 2015.

[27] Han Yun, Lukas Chrostowski, and Nicolas A F Jaeger. Narrow-band, polarization-independent, transmission filter in a silicon-on-insulator strip waveguide. *Optics Letters*, 44(4):847–850, 2019.

[28] Mathias Prost, Yi-Chun Ling, Semih Cakmakyapan, Yu Zhang, Kaiqi Zhang, Junjie Hu, Yichi Zhang, and SJ Ben Yoo. Solid-state MWIR beam steering using optical phased array on germanium-silicon photonic platform. *IEEE Photonics Journal*, 11(6):1–9, 2019.

[29] Carlos Angulo Barrios, Vilson R Almeida, Roberto R Panepucci, Bradley S Schmidt, and Michal Lipson. Compact silicon tunable fabry-perot resonator with low power consumption. *IEEE Photonics Technology Letters*, 16(2):506–508, 2004.

[30] Xiaoyang Cheng, Jianxun Hong, and Shiyoshi Yokoyama. Design and fabrication of bragg-grating-coupled high Q-factor ring resonator using liquid-source CVD-deposited Si3N4 film at 150°c

(conference presentation). In *Integrated Optics: Devices, Materials, and Technologies XXII*, volume 10535, page 1053510. International Society for Optics and Photonics, 2018.

[31] Xu Wang, Jonas Flueckiger, Shon Schmidt, Samantha Grist, Sahba T Fard, James Kirk, Matt Doerfler, Karen C Cheung, Daniel M Ratner, and Lukas Chrostowski. A silicon photonic biosensor using phase-shifted bragg gratings in slot waveguide. *Journal of Biophotonics*, 6(10):821–828, 2013.

[32] Ozdal Boyraz and Bahram Jalali. Demonstration of a silicon raman laser. *Optics Express*, 12(21):5269–5273, 2004.

[33] Haisheng Rong, Richard Jones, Ansheng Liu, Oded Cohen, Dani Hak, Alexander Fang, and Mario Paniccia. A continuous-wave raman silicon laser. *Nature*, 433(7027):725–728, 2005.

[34] Alexander W Fang, Brian R Koch, Richard Jones, Erica Lively, Di Liang, Ying-Hao Kuo, and John E Bowers. A distributed bragg reflector silicon evanescent laser. *IEEE Photonics Technology Letters*, 20(20):1667–1669, 2008.

[35] Rodolfo Ernesto Camacho-Aguilera. *Ge-on-Si laser for silicon photonics*. PhD thesis, Massachusetts Institute of Technology, 2013.

[36] Sören Dhoore, Gunther Roelkens, and Geert Morthier. III-V-on-silicon three-section DNR laser with over 12 nm continuous tuning range. *Optics Letters*, 42(6):1121–1124, 2017.

[37] Thomas Ferrotti, Benjamin Blampey, Christophe Jany, Hélène Duprez, Alain Chantre, Frédéric Boeuf, Christian Seassal, and Badhise Ben Bakir. Co-integrated 1.3 µm hybrid III-V/silicon tunable laser and silicon Mach-Zehnder modulator operating at 25 Gb/s. *Optics Express*, 24(26):30379–30401, 2016.

[38] Zhiping Zhou, Bing Yin, and Jurgen Michel. On-chip light sources for silicon photonics. *Light: Science & Applications*, 4(11):e358, 2015.

[39] Xudong Fan, Ian M White, Siyka I Shopova, Hongying Zhu, Jonathan D Suter, and Yuze Sun. Sensitive optical biosensors for unlabeled targets: A review. *Analytica Chimica Acta*, 620(1-2):8–26, 2008.

[40] Shon Schmidt, Jonas Flueckiger, WenXuan Wu, Samantha M Grist, Sahba Talebi Fard, Valentina Donzella, Pakapreud Khumwan, Emily R Thompson, Qian Wang, Pavel Kulik, et al. Improving the performance of silicon photonic rings, disks, and bragg gratings for use in label-free biosensing. In *Biosensing and Nanomedicine VII*, volume 9166, page 91660M. International Society for Optics and Photonics, 2014.

[41] Matthew S Luchansky, Adam L Washburn, Melinda S McClellan, and Ryan C Bailey. Sensitive on-chip detection of a protein biomarker in human serum and plasma over an extended dynamic range using silicon photonic microring resonators and sub-micron beads. *Lab on a Chip*, 11(12):2042–2044, 2011.

[42] Robert W Boyd and John E Heebner. Sensitive disk resonator photonic biosensor. *Applied Optics*, 40(31):5742–5747, 2001.

[43] Aju S Jugessur, James Dou, J Stewart Aitchison, Richard M De La Rue, and Marco Gnan. A photonic nano-bragg grating device integrated with microfluidic channels for bio-sensing applications. *Microelectronic Engineering*, 86(4-6):1488–1490, 2009.

[44] Hai Yan, Lijun Huang, Xiaochuan Xu, Swapnajit Chakravarty, Naimei Tang, Huiping Tian, and Ray T Chen. Unique surface sensing property and enhanced sensitivity in microring resonator biosensors based on subwavelength grating waveguides. *Optics Express*, 24(26):29724–29733, 2016.

[45] Enxiao Luan, Han Yun, Minglei Ma, Daniel M Ratner, Karen C Cheung, and Lukas Chrostowski. Label-free biosensing with a multi-box sub-wavelength phase-shifted bragg grating waveguide. *Biomedical Optics Express*, 10(9):4825–4838, 2019.

[46] Enxiao Luan, Han Yun, Stephen H Lin, Karen C Cheung, Lukas Chrostowski, and Nicolas A F Jaeger. Phase-shifted bragg grating-based mach-zehnder interferometer sensor using an intensity interrogation scheme. In *Optical Fiber Communication Conference*, pages M1C-5. Optical Society of America, 2020.

[47] Chung-Yen Chao, Wayne Fung, and L Jay Guo. Polymer microring resonators for biochemical sensing applications. *IEEE Journal of Selected Topics in Quantum Electronics*, 12(1):134–142, 2006.

[48] İsa Navruz and Ahmet Altuncu. Design of a chirped fiber bragg grating for use in wideband dispersion compensation. In *New Trends In Computer Networks*, pages 114–123. World Scientific, 2005.

[49] Chinlon Lin, Herwig Kogelnik, and Leonard G Cohen. Optical-pulse equalization of low-dispersion transmission in single-mode fibers in the 1.3–1.7-µm spectral region. *Optics Letters*, 5(11):476–478, 1980.

[50] Craig D Poole, JM Wiesenfeld, Alfred R McCormick, and Katherine T Nelson. Broadband dispersion compensation by using the higher-order spatial mode in a two-mode fiber. *Optics Letters*, 17(14):985–987, 1992.

[51] Sanggeon Lee, Reza Khosravani, Jiangde Peng, Victor Grubsky, Dmitry S Starodubov, Alan E Willner, and Jack Feinberg. Adjustable compensation of polarization mode dispersion using a high-birefringence nonlinearly chirped fiber bragg grating. *IEEE Photonics Technology Letters*, 11(10):1277–1279, 1999.

[52] Francois Ouellette. Dispersion cancellation using linearly chirped bragg grating filters in optical waveguides. *Optics Letters*, 12(10):847–849, 1987.

[53] Richard Jones, Jonathan Doylend, Paniz Ebrahimi, Simon Ayotte, Omri Raday, and Oded Cohen. Silicon photonic tunable optical dispersion compensator. *Optics Express*, 15(24):15836–15841, 2007.

[54] Christi K Madsen, Gadi Lenz, Alan J Bruce, MA Cappuzzo, LT Gomez, and RE Scotti. Integrated all-pass filters for tunable dispersion and dispersion slope compensation. *IEEE Photonics Technology Letters*, 11(12):1623–1625, 1999.

[55] Tao He, Jeffrey Demas, and Siddharth Ramachandran. Ultra-low loss dispersion control with chirped transmissive fiber gratings. *Optics Letters*, 42(13):2531–2534, 2017.

[56] Zhenmin Du, Chao Xiang, Tingzhao Fu, Minghua Chen, Sigang Yang, John E Bowers, and Hongwei Chen. Silicon nitride chirped spiral bragg grating with large group delay. *APL Photonics*, 5(10):101302, 2020.

[57] Charalambos Klitis, Marc Sorel, and Michael J Strain. Active on-chip dispersion control using a tunable silicon bragg grating. *Micromachines*, 10(9):569, 2019.

[58] Charles Elachi and C Yeh. Frequency selective coupler for integrated optics systems. *Optics Communications*, 7(3):201–204, 1973.

[59] Kazuhiro Ikeda, Maziar Nezhad, and Yeshaiahu Fainman. Wavelength selective coupler with vertical gratings on silicon chip. *Applied Physics Letters*, 92(20):201111, 2008.

[60] Yuya Shoji and Tetsuya Mizumoto. Magneto-optical non-reciprocal devices in silicon photonics. *Science and Technology of Advanced Materials*, 15(1):014602, 2014.

[61] Wei Shi, Mark Greenberg, Xu Wang, Yun Wang, Charlie Lin, Nicolas A F Jaeger, and Lukas Chrostowski. Single-band add-drop filters using anti-reflection, contra-directional couplers. In *The 9th International Conference on Group IV Photonics (GFP)*, pages 21–23. IEEE, 2012.

[62] Shi Wei. *Silicon photonic filters for wavelength-division multiplexing and sensing applications.* PhD thesis, University of British Columbia, 2012.

[63] Jin Hong and Weiping Huang. Coupled-waveguide exchange-bragg resonator filters: Coupled-mode analysis with loss and gain. *Journal of Lightwave Technology*, 11(2):226–233, 1993.

[64] Christos Riziotis and Mikhail N Zervas. Design considerations in optical add/drop multiplexers based on grating-assisted null couplers. *Journal of Lightwave Technology*, 19(1):92, 2001.

[65] Jonathan St-Yves. *Contra-directional couplers as optical filters on the silicon on insulator platform.* PhD thesis, Laval University, 2017.

[66] J-P Weber. Spectral characteristics of coupled-waveguide bragg-reflection tunable optical filter. *IEE Proceedings J (Optoelectronics)*, 140(5):275–284, 1993.

[67] Jagdish J Modi. *Parallel Algorithms and Matrix Computation.* University of Michigan, Clarendon Press, 1988.

[68] Cleve Moler and Charles Van Loan. Nineteen dubious ways to compute the exponential of a matrix. *SIAM Review*, 20(4):801–836, 1978.

[69] Jonathan St-Yves, Hadi Bahrami, Philippe Jean, Sophie LaRochelle, and Wei Shi. Widely bandwidth-tunable silicon filter with an unlimited free-spectral range. *Optics Letters*, 40(23): 5471–5474, 2015.

[70] Hideaki Okayama, Yosuke Onawa, Daisuke Shimura, Hiroyuki Takahashi, Hiroki Yaegashi, and Hironori Sasaki. Asymmetric directional coupler type contra-directional polarization rotator bragg grating: design. *Japanese Journal of Applied Physics*, 58(6):068002, 2019.

[71] Ajay Mistry, Mustafa Hammood, Hossam Shoman, Lukas Chrostowski, and Nicolas A F Jaeger. Bandwidth-tunable, fsr-free, microring-based, soi filter with integrated contra-directional couplers. *Optics Letters*, 43(24):6041–6044, 2018.

[72] Ajay Mistry, Mustafa Hammood, Hossam Shoman, Stephen Lin, Lukas Chrostowski, and Nicolas A F Jaeger. Free-spectral-range-free microring-based coupling modulator with integrated contra-directional-couplers. In *Optical Components and Materials XVII*, volume 11276, page 1127607. International Society for Optics and Photonics, 2020.

[73] Quentin Wilmart, Houssein El Dirani, Nicola Tyler, Daivid Fowler, Stéphane Malhouitre, Stéphanie Garcia, Marco Casale, Sébastien Kerdiles, Karim Hassan, Christelle Monat, et al. A versatile silicon-silicon nitride photonics platform for enhanced functionalities and applications. *Applied Sciences*, 9(2):255, 2019.

[74] Wim Bogaerts, Peter De Heyn, Thomas Van Vaerenbergh, Katrien De Vos, Shankar Kumar Selvaraja, Tom Claes, Pieter Dumon, Peter Bienstman, Dries Van Thourhout, and Roel Baets. Silicon microring resonators. *Laser & Photonics Reviews*, 6(1):47–73, 2012.

[75] Folkert Horst, William MJ Green, Bert Jan Offrein, and Yurii A Vlasov. Silicon-on-insulator echelle grating wdm demultiplexers with two stigmatic points. *IEEE Photonics Technology Letters*, 21(23):1743–1745, 2009.

[76] Hiroshi Takahashi, Yoshinori Hibino, and Isao Nishi. Polarization-insensitive arrayed-waveguide grating wavelength multiplexer on silicon. *Optics Letters*, 17(7):499–501, 1992.

[77] Abdelrahman Afifi, Mustafa Hammood, Nicolas A F Jaeger, Sudip Shekhar, Jeff F Young, and Lukas Chrostowski. Contra-directional couplers as pump rejection and recycling filters for on-chip photon-pair sources. In *2019 IEEE 16th International Conference on Group IV Photonics (GFP)*, pages 1–2. IEEE, 2019.

[78] Mustafa Hammood, Ajay Mistry, Minglei Ma, Han Yun, Lukas Chrostowski, and Nicolas A F Jaeger. Compact, silicon-on-insulator, series-cascaded, contradirectional-coupling-based filters with > 50 db adjacent channel isolation. *Optics Letters*, 44(2):439–442, 2019.

[79] Mustafa Hammood, Ajay Mistry, Han Yun, Minglei Ma, Lukas Chrostowski, and Nicolas A F Jaeger. Four-channel, silicon photonic, wavelength multiplexer-demultiplexer with high channel isolations. In *Optical Fiber Communication Conference*, pages M3F-5. Optical Society of America, 2020.

[80] Han Yun, Mustafa Hammood, Stephen Lin, Lukas Chrostowski, and Nicolas A F Jaeger. Broadband flat-top soi add–drop filters using apodized sub-wavelength grating contradirectional couplers. *Optics Letters*, 44(20):4929–4932, 2019.

[81] Der Van and JP Van Der Ziel. Phase-matched harmonic generation in a laminar structure with wave propagation in the plane of the layers. *Applied Physics Letters*, 26:60–61, 1975.

[82] Joseph N Mait and Dennis W Prather. Selected papers on subwavelength diffractive optics, SPIE milestone series, V, 2001.

[83] Dominique Charron, Jonathan St-Yves, Omid Jafari, Sophie LaRochelle, and Wei Shi. Subwavelength-grating contradirectional couplers for large stopband filters. *Optics Letters*, 43(4):895–898, 2018.

[84] Behnam Naghdi and Lawrence R Chen. Silicon photonic contradirectional couplers using sub-wavelength grating waveguides. *Optics Express*, 24(20):23429–23438, 2016.

14 Silicon Photonic Integrated Circuits for OAM Generation and Multiplexing

Yuxuan Chen, Leslie A. Rusch, and Wei Shi

CONTENTS

14.1 INTRODUCTION

Space division multiplexing (SDM) has been intensely investigated in the past decade for next-generation fiber-optic transmission systems [1–3]. Each spatial mode in SDM scheme is an independent data-carrying channel, can be exploited in parallel with multiplexing techniques in the domain of time, frequency and polarization. An SDM system can linearly scale the transmission capacity with the number of spatial channels [2,3]. In fiber-optic communications, there are two common strategies: a waveguide per channel or many modes in a single waveguide. Multi-core fiber (MCF), as its name implies, possess multiple waveguides (cores) in a single fiber, each carrying one mode. Transmission capacity exceeding 1 Pb/s has been demonstrated using a 12-core MCF [4]. Capacity is limited by the upper limit on packing cores in an MCF without introducing inter-channel crosstalk. Few mode fiber (FMF), on the other hand, usually has only one core, but can support multiple fiber modes. Few mode fibers can achieve higher channel density than MCF. The combination of these two techniques forms a few-mode multi-core fiber (FM-MCF). A 10.16-peta-bit/s capacity has been demonstrated in an SDM+WDM transmission experiment using a single FM-MCF, establishing the tremendous potential of such systems [5]. Mode coupling occurs within an FMF, so in practice, multi-input multi-output (MIMO) processing is required to resolve these coupling on the receiver side. Unfortunately, the complexity of MIMO algorithms quickly scales with the number of modes. A lower mode coupling may greatly relieve the burden of digital signal processing, which favors a large mode count.

For modal multiplexing, researchers have considered both linearly polarized (LP) modes and orbital angular momentum (OAM) modes. While LP modes are formed from a combination of eigenmodes with varying effective refractive indices, OAM modes have one effective refractive

DOI: 10.1201/9780429292033-14

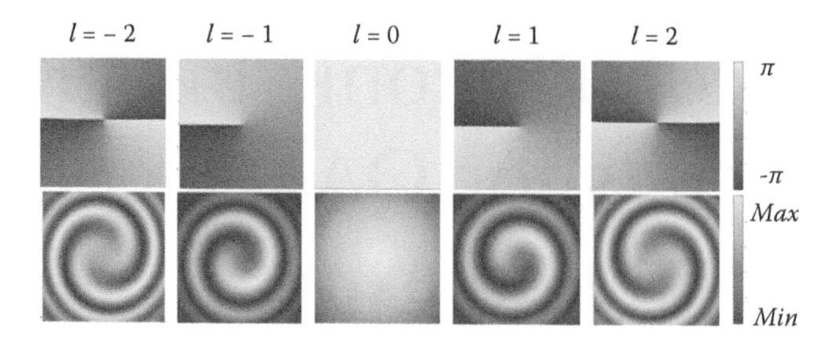

FIGURE 14.1 Phase distributions (first row) and interference pattern with reference Gaussian (second row) of −2 to 2 order OAM.

index in their constituent eigenmodes. This give the OAM the advantage of a lower modal crosstalk [6]. It is the gaps in the propagation constants between OAM modes of different order help reduce crosstalk, making MIMO-free OAM transmission possible over a certain transmission distance [7].

The OAM modes have azimuthal phase dependency of $e^{jl\phi}$, where l is the topological charge of the OAM mode. In Figure 14.1 we illustrate the phase distributions of 5 OAM modes, as well as the interference pattern produced when combined with a Gaussian beam. The topological charges of the OAM modes are above the columns. The sign of l indicates the direction of the rotating phase front of the OAM mode. An OAM mode coming out of the plane with positive topological charge has counter-clockwise π to $-\pi$ phase variation. The absolute value of l indicates the number of π to $-\pi$ rotations. The interference of OAM mode with a co-propagating Gaussian beam produces spiral pattern with the number of spirals equal to the absolute value of l.

In optical fibers, the electrical field of the OAM modes can be expressed by [8]:

$$E(r, \phi, z, t) = E(r)e^{jl\phi}e^{j(\omega t-\beta z)} \tag{14.1}$$

where $E(r)$ is the radial mode profile, ω is the angular frequency and β is the propagation constant. Fiber OAM modes are circularly polarized. There are multiple approaches to generating OAM modes with free-space setups using, for example, spatial light modulators [9], spiral phase plates [10], and q-plates [11]. However, scaling such free-space setups to support a large number of modes is challenging, as it requires precise optical alignment (that is insensitive to environment) at every optical interface, not to mention their large size and high cost.

To generate and multiplex OAM modes in a more compact and robust way, researchers have actively explored various deigns in silicon photonics integrated circuits (PICs) and with impressive results. These designs can be classified into four categories: grating-assisted ring-resonators, phased antenna arrays, in-plane mode convertors, and waveguide surface holographic gratings. In this chapter, we review the principles and recent progress of silicon photonic OAM generators and multiplexers and discuss their future development for OAM-based SDM.

14.2 GRATING-ASSISTED RING-RESONATORS

14.2.1 Principle

A ring resonator is a widely used building block for the silicon on insulator (SOI) platform. It has been used to produce compact optical filters and modulators with very low power consumption [12]. Ring resonators have two typical configurations: all-pass and add-drop. As shown in Figure 14.2, the all-pass ring resonator has one bus waveguide; the two ends of the bus waveguide are used as the input port and through port. The add-drop ring resonator has a second bus for the add and drop ports. Light going to the input port is partially coupled into the ring waveguide;

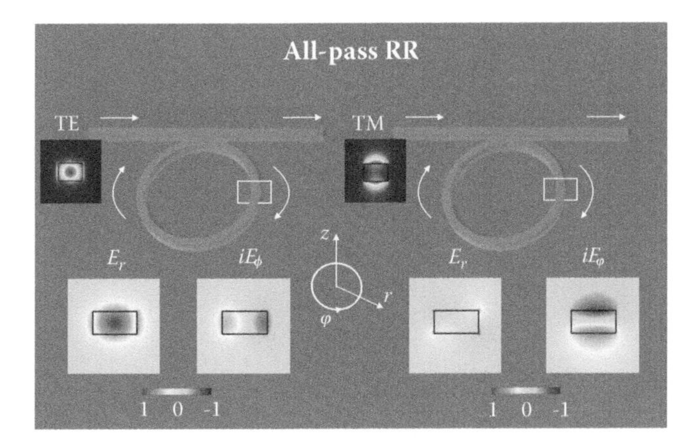

FIGURE 14.2 Electric field distribution in the ring waveguide when input mode is TE (left)/TM (right).

the coupling coefficient is determined by the geometry of the bus waveguide and the ring, as well as by the their spacing. When light propagating in the ring has a round-trip phase equal to an integer multiple of 2π, the ring cavity is in resonance. The mode excited in the resonator is a whispering gallery mode (WGM) and has a propagation constant of $\beta_{\mathrm{WGM}} = \frac{2\pi}{\lambda} n_{eff}$. The PIC configuration and resultant fields from simulation are displayed in Figure 14.2.

By inserting a grating (i.e., periodic dielectric perturbation) into the ring cavity, we can excite a diffracted mode from the WGM when the Bragg condition $\beta_{\mathrm{WGM}} - k_{\mathrm{dif}} = m \cdot K$ is satisfied, where $k_{\mathrm{dif}} = \frac{2\pi}{\lambda} n_{\mathrm{dif}}$ is the wave vector of the diffracted light, n_{dif} is the refractive index of the diffracted medium ($n_{\mathrm{dif}} = 1$, if in air), $K = \frac{2\pi}{\Lambda}$ is the periodicity of the grating and m is the diffraction order. Let R be the radius of the ring cavity, from the Bragg condition equation, we have

$$\frac{2\pi R}{\lambda} n_{eff} - \frac{2\pi R}{\lambda} = m \cdot \frac{2\pi R}{\Lambda} \tag{14.2}$$

An OAM mode is generated with a ring resonator: $\frac{2\pi R}{\lambda} n_{eff}$ is the number of periods of WGM in the ring; we denote it as p. The number of angular periods of the diffracted mode is $\frac{2\pi R}{\lambda}$, in other words the topological charge of the generated OAM, which we denote as l. The number of periodic gratings along the ring is $\frac{2\pi R}{\Lambda}$, which we denote as q. The OAM mode order can then be expressed as

$$l = p - m \cdot q \tag{14.3}$$

The formation of the OAM helix wave front is visualized heuristically in Figure 14.3(b). Gratings on a straight waveguide divert a portion of the incoming wave into free-space with an angle φ (Figure 14.3(b) top-left). Bending the straight waveguide into a ring skews the diffracted light into a vortex (Figure 14.3(b) bottom-left).

The polarization of the diffracted OAM mode is determined by the ring waveguide geometry and grating position. The electric field at the cross section of the ring waveguide (see the white rectangle in Figure 14.2) is simulated using a finite difference eigenmode solver. The bus waveguide has a 500-nm width and a 220-nm height. The ring waveguide has the same cross-sectional geometry as the bus waveguide and 4-μm radius. The gap between the bus waveguide and the ring is 200 nm. The radial and azimuthal components of the first two modes in the ring

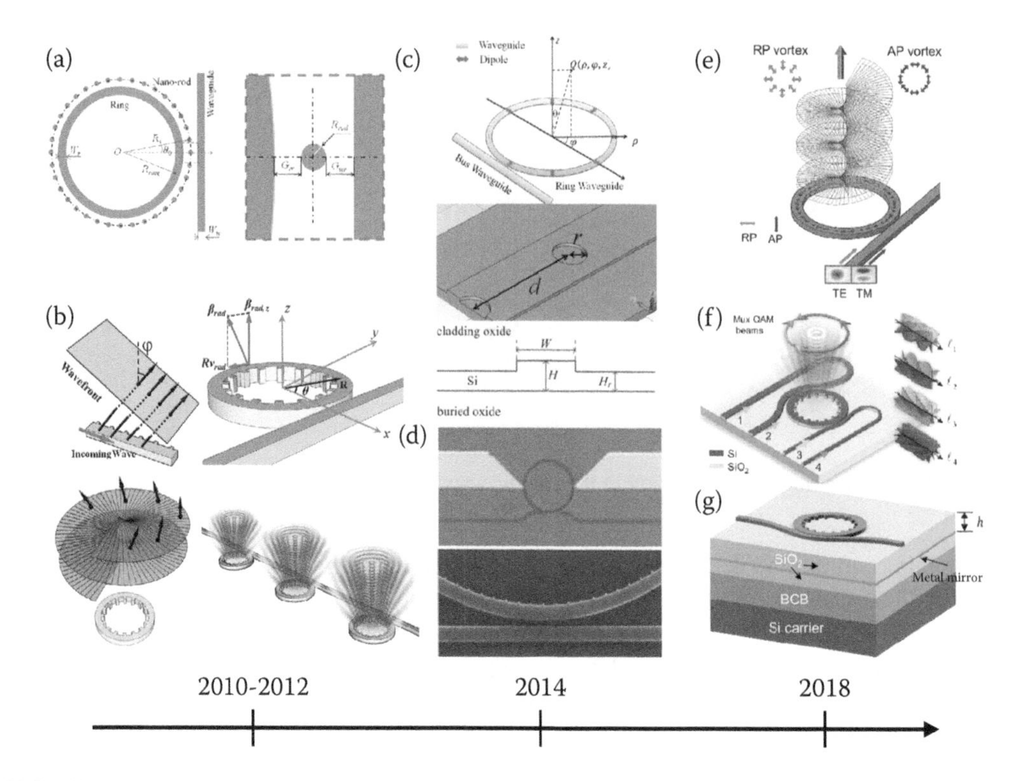

FIGURE 14.3 Grating-assisted ring-resonators: (a) Ring-resonator with nano-rods outside the ring, the nano-rod between the ring and the bus waveguide is shown on the right [13]; (b) scattered wave front transition from straight waveguide to ring-resonator, and an array of three identical emitters [14]; (c) ridge waveguide ring-resonator with angular grating on-top [15]; (d) angular grating embedded ring-resonator with metal heater [16]; (e) switchable radially/azimuthally polarized generator [17]; (f) ring-resonator based multiplexer that supports 4 OAM modes [18]; (g) high-efficiency generator with backside metal mirror [19].

waveguide are presented in Figure 14.2 (color bar shown on the bottom, normalized separately). The first mode is excited by the TE mode in the bus waveguide; the second mode is excited by the TM mode. When gratings are formed onto the ring waveguide, the radial component E_r diffracts a radially polarized (RP) beam while the azimuthal component E_φ generates an azimuthally polarized (AP) beam. As for the grating position, adding gratings on the central part of the top surface can effectively circumvent the extraction of E_φ/E_r component and generate predominant RP/AP OAM modes [17]. When using a grating-assisted ring-resonator device in an OAM fiber transmission link, additional optical operations are needed to first isolate one circular polarization and then combine various OAM modes before their coupling into an OAM fiber.

14.2.2 Demonstrations

In [13], the grating is introduced through nano-rods outside the ring resonator. Light in the straight waveguide first couples to the ring resonator (Figure 14.3(a)). Then, due to the existence of the nano-rods outside the ring, the evanescent field of the mode is scattered at each rod. The nano-rods are placed outside this ring resonator, as the evanescent field lies more outside the ring than inside the ring. The radius of the nano-rods R_{rod} = 120nm. Fabricating such nano-rods without overlapping with the ring is challenging.

An experimental demonstration of an inner-wall grating is detailed in [14]. A side view of the structure is shown on the top right of Figure 14.3(b). On the bottom, three identical grating-assisted ring-resonators are placed along the straight waveguide, each having the ability to generated OAM order from −4 to +4 at the corresponding resonance wavelength ranging from 1,480 nm to 1,580 nm. The emission efficiency varies from mode to mode, with a peak value of 13% at OAM −3 order.

Gratings are applied on the top surface of the ring to generate RP OAM beam [15]. Ridge waveguide is used in the ring to reduce the propagation loss [20]. The grating elements are designed to have the same height as the slab after etching, for the convenience of fabrication. By optimizing the radiation efficiency of a single grating element, the highest total efficiency is simulated to be 49%.

The device in [16] adds thermal optic elements to tune the n_{eff} of the WGM, thus enabling tuning of the OAM order generated. By adding thermo-optical control to the ring cavity (Figure 14.3(d)), the device is able to change the refractive index of the ring waveguide, which in turn changes the WGM propagation constant p. The topological charge of the emitted OAM, equal to the difference between the WGM order and the angular grating number inside the ring, can thus be switched by tuning the voltage. Measurement result show that the emission efficiency of the device falls into the range of 6.7 to 14.9%, when the resonance wavelength varies between 1,525 nm to 1,580 nm.

In [17], a design that supports both RP and AP OAM generation is proposed and experimentally demonstrated using a multi-mode ring waveguide. By placing the gratings on top of the strip silicon nitride waveguide, TE input light from the bus waveguide produces RP OAM, while TM input light produces AP OAM. The effective refractive index for TE and TM modes are different, so the RP and AP OAM are of different wavelengths. A total of 17 emitting peaks are observed in each polarization over the 1,530 nm to 1,580 nm wavelength range. The measured emission efficiency for TE mode varies from 5.5% to 10.1%, the efficiency for TM is in the range of 5.2% to 11.9%.

To generate multiple OAM simultaneously at one wavelength, researchers in [18] use an add-drop configuration to generate 4 OAM states. The bus waveguides have different widths. In Figure 14.3(f) input#1 and input#2 belong to the wider waveguide and excite TE_0 mode in the ring resonator, input#3 and input#4 from the narrower waveguide excite TE_1 mode in the ring. Again, as the refractive index of TE_0 and TE_1 are different, the geometry of the resonator needs to coincide resonant wavelengths from these two modes. With proper design, OAM of $\pm$ 6 are derived with inputs from input#1 and #2, while input#3 and #4 give $\mp$ 9 order OAM, all at wavelength 1,550 nm. The emission efficiency for TE_1 is around 20%, a few percentage points greater than TE_0. The paper also conducted a chip-to-chip transmission experiment. The crosstalk ranged from −13 dB to −17 dB at coincident wavelengths.

Adding a metal layer underneath the buried oxide layer during fabrication has been proven to increase the emission efficiency of surface grating couplers [21]. The emission efficiency of the grating-assisted ring-resonator is increased to 37% with this technique [19]. Using flip-chip bonding and substrate removal, a 100 nm thick aluminum exists under the buried oxide layer, as in Figure 14.3(g), to reflect the originally down-propagating light back upward for emission. The comparison between devices with/without a mirror confirms the significant efficiency improvement when using a mirror. Adding a backside mirror also has a positive effect on suppressing side lobes during emission.

14.3 PHASED ANTENNA ARRAYS

14.3.1 PRINCIPLE

A general phased antenna array OAM generator has three building blocks: input fiber-to-chip coupler, amplitude and phase assigning structure (star coupler in most of the cases) and antenna array emitter. Conventionally, star couplers are used in pairs connected through an array of waveguides of calculated lengths to multiplex/de-multiplex wavelengths in WDM systems. For OAM generation, the star coupler is used to achieve desired phase distribution over the output waveguides.

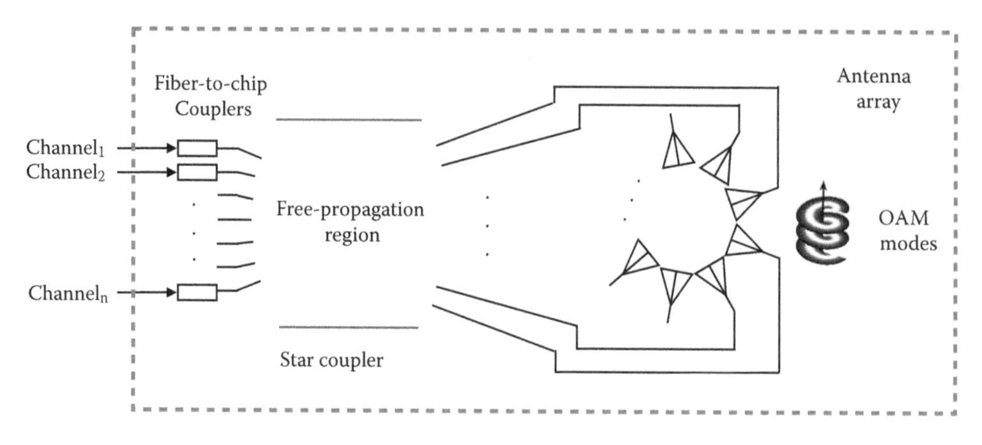

FIGURE 14.4 Schematic of a general phased antenna array that supports n OAM modes.

The star coupler has two typical configurations: confocal and Rowland circle. As shown in Figure 14.4, the input and output waveguides of the star coupler are placed on two circular arcs with tangential orientations. For both the confocal and Rowland circle configuration, the center of the output arc always lands on the very middle of the input arc. In confocal star coupler, the radius of the input arc equals to the radius of the output arc. With Rowland circle geometry, the radius of the input arc is half the radius of the output arc. Figure 14.4 includes a confocal star coupler that has n inputs and m outputs. Light from each input diffracts in the free-propagation region and is split into output waveguides. As the center of the output arc lands on the middle of the input arc, light entering through the central horizontal input travels identical length optical paths before coupling into the output waveguides; light in the output waveguides are thus in phase. Light entering through other inputs travel non-identical optical paths, creating phase differences in output waveguides.

The antenna array connected to the output waveguides converts the created phase difference into helical phase front and emits OAM modes. The polarization state of the emitted OAM modes is decided by the antenna design: 1D antenna generates RP or AP OAM modes, 2D antenna can generate circularly polarized OAM modes. As each of the star coupler inputs is an independent channel and the antenna array is shared among all these channels, designs in this category can support a relatively large number of OAM modes. In addition, the bandwidth is usually wide as no wavelength-sensitive component is involved (optical I/O not considered), the main difference between these designs is at the emitter.

The on-chip transformation of light in a general phased antenna array circuit can be described by the following matrix equation:

$$E_{\text{out}} = M_{AA} M_{SC} M_C E_{\text{in}} \tag{14.4}$$

where E_{in} is the amplitude vector of the input modes and E_{out} is the amplitude vector of free-space OAM modes. The matrices for input couplers, star coupler and antenna array are denoted as M_C, M_{SC} and M_{AA}, respectively. Consider the case where there are n channels at the input and m antennas in the antenna array.

The coupling efficiency of the fiber-to-chip couplers is given by η_{in}. The star coupler has a number of input ports equal to the number of input channels n and output ports equals to the number of antennas m. The phase difference between adjacent output ports should be a multiple of $2\pi/m$. The accumulated phase over the emitter circumference can thus be multiples of 2π. The amplitudes at the output waveguides follow a Gaussian distribution, therefore, the center output port always has more power than other output ports. As a uniform power distribution is favored for

OAM generation, on-chip variable optical attenuator is sometimes used for power equalization. We assume the uniform amplitude of $\frac{1}{\sqrt{m}+\triangle m}$ in the following discussion (Δm is the power attenuation due to splitting and equalization). The element from row p, column q in M_{SC}

$$M_{\mathrm{SC}_{p,q}} = \frac{e^{i\varphi_{p,q}}}{\sqrt{m} + \triangle m} \tag{14.5}$$

The phase component $\varphi_{p,q}$ represents the contribution from q^{th} input port to p^{th} output port. For an intended l^{th} order OAM generation from input q,

$$\varphi_{p,q} = \frac{2\pi l_q (p-1)}{m} \tag{14.6}$$

The coupling efficiency of the antenna array is another constant η_{out}. The on-chip transfer matrix for a general phased antenna array can be expressed as

$$M_{\mathrm{on-chip}} = M_{\mathrm{AA}} M_{\mathrm{SC}} M_{\mathrm{C}} = \begin{bmatrix} t_{1,1} & \cdots & t_{1,2n} \\ \vdots & t_{p,q} & \vdots \\ t_{m,1} & \cdots & t_{m,2n} \end{bmatrix} \tag{14.7}$$

where $t_{p,q} = \eta_{in}\eta_{out}\left(\frac{1}{\sqrt{m}+\triangle m}\right)e^{i\frac{2\pi l_q(p-1)}{m}}$, the polarization of the generated OAM is decided by the polarization of the waveguide mode: TE for azimuthal and TM for radial.

14.3.2 Generating Circularly Polarized OAM Modes

A unique advantage of the phased antenna array lies in the flexibility of the emitter design, as the antenna array accepts already assigned phase and amplitude for OAM generation. Therefore, the antenna array can be designed to generate a circularly polarized beam, eliminating the need for further polarization manipulation.

The schematic of such a structure is illustrated in Figure 14.5[c]. The two dimensional grating coupler (2D-GC) multiplexes x-pol and y-pol, as the two arms of the 2D-GC are spatially orthogonal. Forming circularly polarized light with a 2D-GC, however, requires 90° phase difference between its two arms, in addition to their equal intensity. A well-designed directional coupler could achieve 50/50 splitting ratio while naturally providing a 90° phase difference. The combination of directional coupler and 2D-GC creates and multiplex both left and right circular polarizations. Accordingly, two star couplers are needed, one for left circular polarization ($\mathrm{SC_L}$ in Figure 14.5(c)) and the other for right circular polarization ($\mathrm{SC_R}$ in Figure 14.5(c)). The phase component $\varphi_{p,q}$ in $\mathrm{SC_L}$

$$\varphi_{p,q} = \frac{2\pi (l_q + 1)(p-1)}{m} \tag{14.8}$$

And in $\mathrm{SC_R}$

$$\varphi_{p,q} = \frac{2\pi (l_q - 1)(p-1)}{m} \tag{14.9}$$

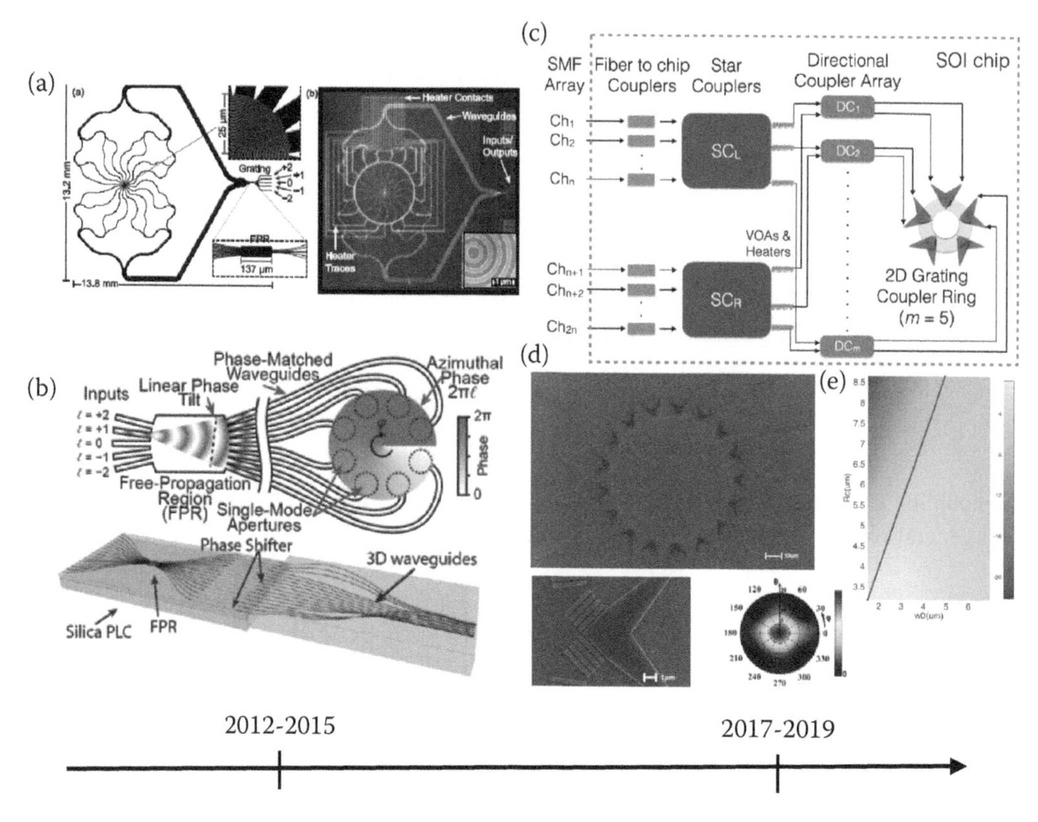

FIGURE 14.5 Phased antenna arrays: (a) Star couplers with circular grating as emitter [22]; (b) star couplers with 3D waveguides block as emitter [23]; (c) star couplers with 2-dimensional gratings as emitter [24]; (d) SEM image of the fabricated 2D-GCs and the far field distribution of one 2D-GC [25]; (e) heat map for optimizing antenna design using transfer matrix method.

An ideal directional coupler offers 50/50 power splitting ratio and ±90° phase difference, positive or negative depending on the input port with a transfer matrix given by

$$M_{\mathrm{DC}} = \frac{\sqrt{2}}{2}\begin{bmatrix} 1 & e^{i\frac{\pi}{2}} \\ e^{i\frac{\pi}{2}} & 1 \end{bmatrix} \tag{14.10}$$

The matrix for 2D-GC implies the local polarization and is position sensitive, as the orientation of the 2D-GC varies with its position in the circumference. For the pth 2D-GC,

$$M_{\mathrm{AA},p} = \eta_{\mathrm{out}}\left[M_{\mathrm{arm1},p} M_{\mathrm{arm2},p}\right] \tag{14.11}$$

Supposing that the orientation of the first arm is θ_p, given by

$$\theta_p = \frac{2\pi(p-1)}{m} \tag{14.12}$$

The expressions for the two arms are

$$M_{\text{arm1}.p} = \begin{pmatrix} -\sin\theta_p \\ \cos\theta_p \end{pmatrix};\; M_{\text{arm2},p} = \begin{pmatrix} -\cos\theta_p \\ -\sin\theta_p \end{pmatrix} \tag{14.13}$$

Multiplying all these matrices in sequence gives the total transfer matrix of the circuit

$$M_{\text{on-chip}} = M_{\text{AA}} M_{\text{DC}} M_{\text{SC}} M_{\text{C}} = \begin{bmatrix} t_{1,1} & \cdots & t_{1,2n} \\ \vdots & t_{p,q} & \vdots \\ t_{m,1} & \cdots & t_{m,2n} \end{bmatrix} \tag{14.14}$$

When the input channel is intended for LCP generation, $q \le n$,

$$t_{p,q} = -\frac{1}{2}\eta_{\text{in}}\eta_{\text{out}} i \begin{pmatrix} 1 \\ i \end{pmatrix} e^{-i\theta_p} \left(\frac{1}{\sqrt{m} + \Delta m} \right) e^{i\varphi_{p,q}} \tag{14.15}$$

Otherwise, $n \le q \le 2n$,

$$t_{p,q} = -\frac{1}{2}\eta_{\text{in}}\eta_{\text{out}} \begin{pmatrix} 1 \\ -i \end{pmatrix} e^{i\theta_p} \left(\frac{1}{\sqrt{m} + \Delta m} \right) e^{i\varphi_{p,q}} \tag{14.16}$$

The $(1\ i)^T$ and $(1\ -i)^T$ component dictate the circularly polarized nature of the generated beam, while the $e^{il\varphi}$, which indicate the generation of OAM, is wrapped in $e^{i\varphi_{p,q}}$.

14.3.3 Free-Space Propagation

The simulation of the emission and propagation of the light at each 2D-GC assumes a Gaussian beam approximation, where the emitted beam from a 2D-GC is written as a Gaussian beam that propagates perpendicularly to the chip surface:

$$E(r, z) = E_0 \hat{c} \frac{w_0}{w(z)} e^{\frac{-r^2}{w(z)^2}} e^{-i\left(kz + k\frac{r^2}{2R(z)} - \psi(z)\right)} \tag{14.17}$$

Here, the origin of the coordinate system is the center of the 2D-GC; r is the radial distance and z is the propagation distance; E_0 is the in-plane field we just calculated; $\hat{c}$ is the polarization of the generated beam, either LCP or RCP; w_0 is the Gaussian beam waist and is directly related to the size of the grating region of 2D-GC; $w(z)$, given by $w(z) = w_0\sqrt{1 + (\frac{z}{z_R})^2}$, reflects the change of the beam waist along z axis, $z_R = \frac{\pi w_0{}^2}{\lambda}$ is the Rayleigh range; k is the wave number; $R(z) = z[1 + (\frac{z_R}{z})^2]$ is the radius of the wave front curvature; $\psi(z) = \arctan(\frac{z}{z_R})$ is the Gouy phase.

The emitted beams from all 2D-GCs will interfere with each other. As the Gaussian waist w_0 of each emitted beam is decided by the size of the grating region, the design parameters of 2D-GC could impact the interfered beam property. Moreover, the arrangements of these 2D-GCs, or, the radius of the circumference where the centers of these 2D-GCs are on, influence how the emitted beams interfere. Thus, the radius of the emitter circumference is another parameter that worth to be optimized.

Aiming at direct coupling with OAM fiber [26], all the supported OAM fiber modes are used as references to optimize on-chip circuit design. The figure of merit is defined as the worst-case coupling loss among all fiber OAM modes. The heat map based on the sweeping of the size of the generated beam waist w_0 (x-axis Figure 14.5(e)) and the radius of the emitter circumference R_c (y-axis Figure 14.5(e)) is plotted. The performance under different 2D-GC number m are evaluated and compared in the paper. The minimal value of m needed to support all the OAM fiber modes in [26] is five. As m increases, the purities of the modes generated will increase. The size of the emitter grows with m, and there is a limit on proximity of the 2D-GCs. If the 2D-GCs were placed too closely, the Gaussian approximation would no longer hold. The red line in the heat map represents this physical limitation on circuit design. The most bottom-left (w_0, R_c) combination above the red line in (Figure 14.5(f)) gives the lowest worst-case coupling loss IL_{max} and thus is the optimal design parameter. Optimal combinations under other m values can be calculated following the same logic. Reference [24] sets up a good example of how to balance trade-offs in SDM targeted silicon photonic circuit designs, the model can be adapted to benefit circuits with other types of emitters which will be discussed in the following section.

14.3.4 Demonstrations

A design that uses a circular grating emitter is proposed [22]. As the scanning electron microscope (SEM) image shown in Figure 14.5(a), concentric circular gratings, acting as the optical output, vertically emit light fed by its connecting waveguides. Light fed by these connecting waveguides follows a phase pattern such that the phase difference between any pair of adjacent waveguides is identical. The accumulated phase difference around the circumference should be integer multiples of 2π. The integer, l, equals the topological charge of the generated OAM. This phase pattern is acquired at the end of the free-propagation-region (FPR) of the star coupler. The connecting waveguides between the star coupler and circular grating are of same length to maintain the phase difference from star coupler. The star coupler in this design has five inputs. From top to bottom, they correspond to the phase pattern needed for OAM of $l = +2$ to $+ 2$. The generated beam is azimuthally polarized. Heaters (Figure 14.5(a), right side) are placed on top of the connecting waveguides to compensate for phase errors through thermal-optics effect. The device is characterized by reflecting the generated beam back to the emitter, using the same device as the de-multiplexer, and minimizes the crosstalk through thermal tuning. After the characterization, a chip-to-chip transmission is done, utilizing all five OAM channels. The loss from the input fiber of the mux chip to the output fiber of the de-mux chip is estimated to be 55 dB.

A follow-up publication to [22] expands the star coupler to support up to 15 OAM states [23]. The outputs of the phase-matched waveguide array are now connected to 3D waveguides fabricated through laser writing. An indication of the 2D-3D hybrid device is shown on the bottom of Figure 14.5(b). The other ends of the 3D waveguides are evenly placed on a circumference of 204 um diameter with 40um spacing [27]. The hybrid PIC is designed to support both TE and TM waveguide modes. Chip-to-chip transmission between two copies of this hybrid PICs achieved a 9.6-b/s/Hz spectral efficiency (14 wavelength channels × 20 Gb/s × 2 polarizations × 3 OAM states), almost five times of that from its previous publication [22]. The characterized worst-case crosstalk, considering all supported modes, is −8 dB for TE polarization and −6 dB for TM polarization.

The OAM generator demonstrated in [28] is continuously tunable and can generate both integer and non-integer OAM states. There is no star coupler in this device; instead, there is a bus waveguide that couples to all emitting antenna cells, assigning a fraction of power and appropriate phase needed for OAM generation. There are 30 antenna cells in the generator. Most of the components in these antenna cells share the same design parameters, except for the directional couplers. The directional coupler in each antenna cell provides a unique splitting ratio, to ensure that the powers at all the antennas are the same. The phase difference between adjacent antenna cells can be attributed to phase differences due to bus waveguide routing $e^{j\varphi}$, and the phase differences introduced by tuning $e^{-j\Delta\varphi}$. The routing of the bus waveguide provides uniform $e^{j\varphi}$, and the parallel-connected integrated heaters among all antenna cells ensure identical $e^{-j\Delta\varphi}$. In this case, the phase difference between any adjacent antenna cells $e^{j(\varphi-\Delta\varphi)}$ can be simultaneously tuned. As a result, non-integer states can be observed as the beam evolves from one integer state to the next. The generated beam is azimuthally polarized. The optical emitter features 51% up-emitting efficiency [29].

Reference [25] is the experimental demonstration of the circularly polarized OAM generator. An SEM image of antenna number $m = 17$ is shown, with a more detailed 2D-GC on the bottom, together with its far field distribution that implies a low divergence angle (Figure 14.5(d)). The measured on-chip loss is around –25 dB, mode-dependent loss is below 2 dB among all 14 OAM states. Worst-case crosstalk characterized by de-multiplexing 0th order OAM is 6 dB.

14.4 IN-PLANE MODE CONVERTOR

In-plane OAM convertors excite higher order waveguide modes or hybrid modes that are combined with a specific phase difference to form an OAM mode at the waveguide output. This mode conversion approach enables compact, inline OAM generation. However, the highest achievable OAM order is limited by the waveguide geometry. Given the symmetric mode OAM profiles, a thicker waveguide is required to generate higher order beyond ±1, which is generally incompatible with standard sub-micrometer SOI wafer processes.

Inspired by Allen's research [30] where two high-order Hermite-Gaussian modes are combined to generate Laguerre-Gaussian modes with a helical phase front, [31] proposed the generation of +1/–1 order OAM by combining E_{21}^{x} and E_{12}^{x} with $\pm\frac{\pi}{2}$ phase difference. Illustrated in Figure 14.6(a), the TE (E_{11}^{x}) and TM (E_{11}^{y}) waveguide modes are first coupled to a pair of hybrid modes. These hybrid modes are then converted to the targeted E_{21}^{x} and E_{12}^{x} modes through adiabatic converting waveguides. Phase shifter is added on the upper branch (input1) to adjust the relative phase between E_{21}^{x} and E_{12}^{x} modes. At the end of the last directional coupler, as E_{21}^{x} and E_{12}^{x} are combined with proper relative phase, OAM of ± 1 order is generated. The simulated loss of this design is 0.7 dB.

Another in-plane OAM generation is proposed through trenched waveguide [33]. Adding trench in rectangular silicon waveguide breaks the waveguide symmetry and splits mode degeneracy. With an optimized width and height of the trench, two orthogonal eigen modes with different propagation constants are equally excited (Figure 14.6(c)). As the propagation constants of two eigen-modes are different, they started beating within the trench waveguide. Adjusting the length of the trench waveguide till $\pm\frac{\pi}{2}$ phase difference is satisfied forms OAM of ± 1 order. The design methodology is compatible with both x-pol and y-pol in waveguide. Facet reflection of around 3 dB is simulated to be the main contribution to the total loss of this device.

14.5 WAVEGUIDE SURFACE HOLOGRAPHIC GRATINGS

Waveguide surface holographic grating (WGSHG) based generators etch the waveguide surface into holographic patterns, where light comes into the waveguide is diffracted by the holographic patterns to synthesize OAM modes. The grating is patterned through shining the time-reversal of the intended OAM mode onto the straight waveguide and interfere with the waveguide mode. The intended OAM mode is produced when exciting the conjugate waveguide mode in the opposite direction [32]. Designs in this category can have above 100 nm bandwidth and straightforward circuit design. The wavelength dependency of the emitting angle (14.4 degree per 100 nm) in some of the demonstrations may raise attention in real applications.

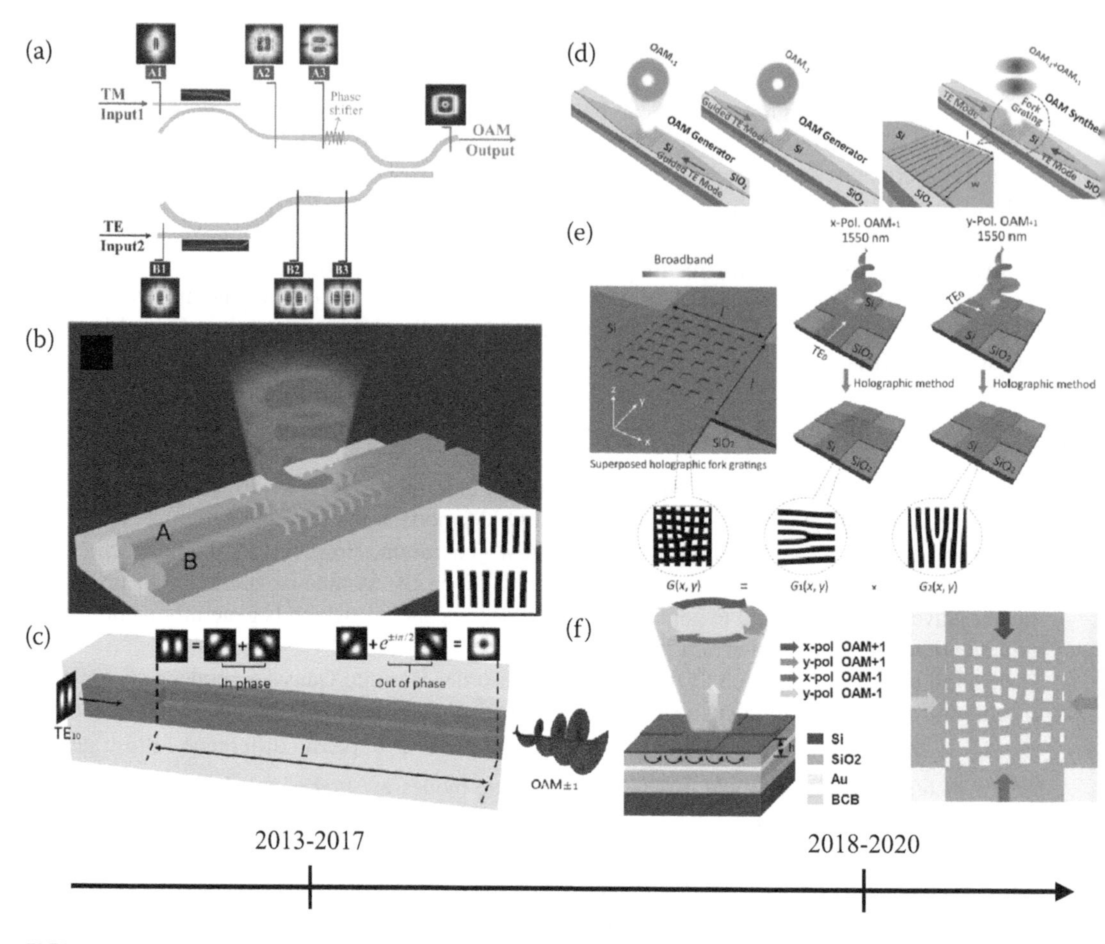

FIGURE 14.6 In-plane mode convertor and waveguide surface holographic gratings: (a) In-plane OAM generator with directional coupler [31]; (b) OAM generator with waveguide surface holographic grating [32]; (c) in-plane OAM generator with trenched waveguide [33]; (d) generating two OAM modes with holographic grating [34]; (e) superimposed holographic grating that supports 4 OAM modes [35]; (f) superimposed holographic grating with backside mirror [36].

In [32], WGSHG is proposed on the Si3N4 waveguide platform to convert waveguide modes into free-space vortex beams. Waveguide A with WGSHG diffracts and generates vortex beam. The generated vortex beam can be modulated by another waveguide with WGSHG (waveguide B, 0.5 um gap). Lights diffracted by waveguide A and waveguide B interfere, so controlling the phase difference $\Delta\theta$ between the two waveguide modes effectively modulates the generated vortex. The bandwidth of the device is around 175 nm.

WGSHG based structure in [34] experimentally demonstrated an efficiency above −11.8 dB covering 1,500 nm to 1,600 nm (Figure 14.3(d)). Despite the fact that the device is broadband in terms of emission efficiency, the emission angle dependence on wavelength is non-negligible. The emission angle varies 14.4° over the 100 nm wavelength span. The offset in emission angle may raise problems in real applications where long free-space traveling distance are involved.

A design that superposes holographic fork gratings to multiplex *x*-pol and *y*-pol is experimentally demonstrated in 2019 [35]. The design generates OAM of ±1 order in both x and y polarization as illustrated in Figure 14.6(e). The measured efficiency for y-pol is −8.2 dB at 1,550 nm. The crosstalk level is characterized by using this design as a multiplexer then de-multiplexes the generated beam with a SLM. The accumulated crosstalk, considering all four channels, is around −9 dB.

TABLE 14.1
Studied silicon photonic circuits for OAM generation and multiplexing

Reference	Principle	OAM order/channel count per λ	Bandwidth	Polarization	Circuit complexity
[14]	RR: all-pass, sidewall grating	+4 ~ −4 / 1	Narrow	Azimuthal	Low
[17]	RR: all-pass, switchable pol	+3, −3 / 2	Narrow	Radial/ Azimuthal	Low
[18]	RR: add-drop, multi-mode ring	±1, ±6, ±9, ±17 / 4	Narrow	Radial	Low
[22]	SC: circular grating	+2 ~ −2 / 5	N/A	Azimuthal	High
[23]	SC: 3D waveguide	+7 ~ −7 / 30	9 nm	Azimuthal/ Radial	High
[25]	SC: 2D-GC	+3 ~ −3 / 14	17 nm	LCP & RCP	High
[33]	In plane: trenched waveguide	+1, −1 / 2	60 nm	Linear	Low
[34]	WGSHG	+1, −1 / 2	110 nm	Linear	Low
[35]	Superimposed WGSHG	+1, −1 / 4	N/A	Linear	Low

To increase the emission efficiency of the superposed holographic grating device, a backside mirror is proposed [36]. By adding backside mirror to the structure proposed in [35], around 5 dB efficiency improvement have been measured for all four OAM modes (Figure 14.6(f)).

14.6 CHALLENGES AND OUTLOOK

Looking back on these publications from the last decade, we have seen impressive attempts in generating OAM modes with compact silicon photonic circuits and continuously improved performance. Using the devices based on grating-assisted ring-resonators, in-plane mode convertors and waveguide surface holographic gratings, OAM modes of high purity can be generated within a tiny footprint, whereas the parallel channel count in each of these designs is relatively low (no greater than 4). Grating-assisted ring-resonators cannot effectively multiplex wavelengths, while the other two categories have trouble scaling OAM order. To increase channel counts, design in [17] multiplexed azimuthal and radial polarization; design in [18] doubled the bus waveguide and could simultaneously generate 4 OAM modes; design in [28] superimposed the holographic pattern to multiplex x and y polarization. The phased antenna arrays, on the other hand, have an advantage in scaling, as expanding the star coupler design immediately equips the phase pattern for higher order OAM generation at the emitters. The choice of the emitter is not yet settled in this category. We have seen circular gratings, 3D waveguides, and 2D-GCs that can multiplex both left-circular and right-circular polarization states. However, they use complex photonic circuits involving tens of metal heaters to compensate for phase errors, which raises practical problems like temperature stabilization.

The designs with backside mirrors appear in both grating-assisted ring resonators and WGSHG devices to increase efficiency, which exemplifies how advanced fabrication technique could help improve performance. Just like the backside mirror technique, exciting possibilities lie in multi-layer platforms [37]. Adapting ring-resonator designs in multiple layers could potentially times the channel count without jeopardizing the beam quality. Apart from using multiple layers, the emission angle dependence on wavelength from waveguide surface holographic gratings can engineered to focus light from parallel waveguides to the same point. Inverse design could be a possible approach to create complex holographic pattern that generates OAM of multiple orders.

Efforts can be made to build more applicable devices. The analysis in [24] is a thorough showcase of how to effectively optimize photonic circuit design towards better coupling with target OAM fiber. The optimization procedure can easily be adapted for more specific yet complex applications. For example, the fiber transmission link could be changed into free-space transmission link with turbulence. As direct coupling between OAM fiber and silicon chip has not been achieved yet, the interaction between circuit design and SDM fiber design is definitely favored.

In this chapter, we focus on the generation and multiplexing of the OAM modes. We have reviewed various types of OAM generators and multiplexers on the silicon platform with their performance summarized in Table 14.1. As design with lower loss, design with higher mode counts and design with multiplexed polarizations arise, we have a good reason to expect new devices with even better performance in the near future thanks to the continuous dedication from the researchers in the active field. OAM generators based on silicon photonic circuits can be integrated with many other photonic components on a single chip for large-scale integrated photonic systems. Its contribution to the realization of an ultra-high-capacity transmission system is worth expecting.

REFERENCES

[1] Shi, Wei, Ye Tian, and Antoine Gervais. "Scaling capacity of fiber-optic transmission systems via silicon photonics." *Nanophotonics* 9.16 (2020): 4629–4633.

[2] Winzer, Peter J., David T. Neilson, and Andrew R. Chraplyvy. "Fiberoptic transmission and networking: the previous 20 and the next 20 years." *Optics Express* 26.18 (2018): 24190–24239.

[3] Winzer, Peter J., and David T. Neilson. "From scaling disparities to integrated parallelism: A decathlon for a decade." *Journal of Lightwave Technology* 35.5 (2017): 1099–1115.

[4] Takara, Hidehiko, et al. "1.01-Pb/s (12 SDM/222 WDM/456 Gb/s) crosstalk-managed transmission with 91.4-b/s/Hz aggregate spectral efficiency." *European Conference and Exhibition on Optical Communication.* Optical Society of America, 2012.

[5] Soma, Daiki, et al. "10.16-Peta-B/s dense SDM/WDM transmission over 6-mode 19-core fiber across the C+ L band." *Journal of Lightwave Technology* 36.6 (2018): 1362–1368.

[6] Rusch, Leslie A., et al. "Carrying data on the orbital angular momentum of light." *IEEE Communications Magazine* 56.2 (2018): 219–224.

[7] Ingerslev, Kasper, et al. "12 mode, WDM, MIMO-free orbital angular momentum transmission." *Optics Express* 26.16 (2018): 20225–20232.

[8] Brunet, Charles, et al. "Design, fabrication and validation of an OAM fiber supporting 36 states." *Optics Express* 22.21 (2014): 26117–26127.

[9] Y. Ohtake et al., "Universal generation of higher-order multiringed Laguerre–Gaussian beams by using a spatial light modulator," *Optics Letters*, vol. 32, no. 11, pp. 1411–1413, 2007.

[10] M. Massari, G. Ruffato, M. Gintoli, F. Ricci, and F. Romanato, "Fabrication and characterization of high-quality spiral phase plates for optical applications," *Applied Optics*, vol. 54, no. 13, pp. 4077–4083, 2015.

[11] Karimi, E., Piccirillo, B., Nagali, E., Marrucci, L. and Santamato, E., 2009. "Efficient generation and sorting of orbital angular momentum eigenmodes of light by thermally tuned q-plates." *Applied Physics Letters*, 94(23), p. 231124.

[12] Dubé-Demers, Raphaël, Sophie LaRochelle, and Wei Shi. "Low-power DAC-less PAM-4 transmitter using a cascaded microring modulator." *Optics Letters* 41.22 (2016): 5369–5372.

[13] Yu, Y. F., et al. "Pure angular momentum generator using a ring resonator." *Optics Express* 18.21 (2010): 21651–21662.

[14] Cai, Xinlun, et al. "Integrated compact optical vortex beam emitters." *Science* 338.6105 (2012): 363–366.

[15] Li, Rui, et al. "Radially polarized orbital angular momentum beam emitter based on shallow-ridge silicon microring cavity." *IEEE Photonics Journal* 6.3 (2014): 1–10.

[16] Strain, Michael J., et al. "Fast electrical switching of orbital angular momentum modes using ultra-compact integrated vortex emitters." *Nature Communications* 5.1 (2014): 1–7.

[17] Shao, Zengkai, et al. "On-chip switchable radially and azimuthally polarized vortex beam generation." *Optics Letters* 43.6 (2018): 1263–1266.

[18] Li, Shimao, et al. "Orbital angular momentum vector modes (de) multiplexer based on multimode micro-ring." *Optics Express* 26.23 (2018): 29895–29905.
[19] Li, Shimao, et al. "Compact high-efficiency vortex beam emitter based on a silicon photonics micro-ring." *Optics Letters* 43.6 (2018): 1319–1322.
[20] Dong, Po, et al. "Low loss shallow-ridge silicon waveguides." *Optics Express* 18.14 (2010): 14474–14479.
[21] Luo, Yannong, et al. "Low-loss two-dimensional silicon photonic grating coupler with a backside metal mirror." *Optics Letters* 43.3 (2018): 474–477.
[22] Su, Tiehui, et al. "Demonstration of free space coherent optical communication using integrated silicon photonic orbital angular momentum devices." *Optics Express* 20.9 (2012): 9396–9402.
[23] Guan, Binbin, et al. "Polarization diversified integrated circuits for orbital angular momentum multiplexing." *2015 IEEE Photonics Conference (IPC)*. IEEE, 2015.
[24] Chen, Yuxuan, Leslie A. Rusch, and Wei Shi. "Integrated circularly polarized OAM generator and multiplexer for fiber transmission." *IEEE Journal of Quantum Electronics* 54.2 (2017): 1–9.
[25] Chen, Yuxuan, et al. "WDM-Compatible Polarization-Diverse OAM Generator and Multiplexer in Silicon Photonics." *IEEE Journal of Selected Topics in Quantum Electronics* 26.2 (2019): 1–7.
[26] Brunet, Charles, et al. "Design of a family of ring-core fibers for OAM transmission studies." Optics Express 23.8 (2015): 10553–10563.
[27] Guan, Binbin, et al. "Free-space coherent optical communication with orbital angular, momentum multiplexing/demultiplexing using a hybrid 3D photonic integrated circuit." Optics Express 22.1 (2014): 145–156.
[28] Sun, Jie, et al. "Chip-scale continuously tunable optical orbital angular momentum generator." *arXiv preprint arXiv:1408.3315* (2014).
[29] Sun, Jie, et al. "Large-scale nanophotonic phased array." *Nature* 493.7431 (2013): 195–199.
[30] Allen, Les, et al. "Orbital angular momentum of light and the transformation of Laguerre-Gaussian laser modes." *Physical Review A* 45.11 (1992): 8185.
[31] Zhang, Dengke, et al. "Generating in-plane optical orbital angular momentum beams with silicon waveguides." *IEEE Photonics Journal* 5.2 (2013): 2201206-2201206.
[32] Liu, Aiping, et al. "On-chip generation and control of the vortex beam." *Applied Physics Letters* 108.18 (2016): 181103.
[33] Zheng, Shuang, and Jian Wang. "On-chip orbital angular momentum modes generator and (de) multiplexer based on trench silicon waveguides." *Optics Express* 25.15 (2017): 18492–18501.
[34] Zhou, Nan, et al. "Generating and synthesizing ultrabroadband twisted light using a compact silicon chip." *Optics Letters* 43.13 (2018): 3140–3143.
[35] Zhou, Nan, et al. "Ultra-compact broadband polarization diversity orbital angular momentum generator with $3.6 \times 3.6\ \mu m^2$ footprint." *Science Advances* 5.5 (2019): eaau9593.
[36] Tan, Heyun, et al. "High-efficiency broadband vortex beam generator with a backside metal mirror." *CLEO: Science and Innovations*. Optical Society of America, 2020.
[37] Sodagar, Majid, et al. "High-efficiency and wideband interlayer grating couplers in multilayer Si/SiO 2/SiN platform for 3D integration of optical functionalities." *Optics Express* 22.14 (2014): 16767–16777.

15 Novel Materials for Active Silicon Photonics

Chi Xiong

CONTENTS

DOI: 10.1201/9780429292033-15

15.1 MOTIVATION AND OVERVIEW

15.1.1 ACTIVE SILICON PHOTONICS

Metal interconnects are expected to be the bottleneck for moving large volume of data between modern computing systems. Optical data transmission, ranging from rack-to-rack down to chip-to-chip and intra-chip interconnections, allows for much higher data rates, much lower power dissipation and would at the same time eliminate problems resulting from electromagnetic interference (Fang et al. 2006).

Traditional optical interconnects, implemented using parallel multimode fiber coupled to vertical cavity surface emitting laser arrays, face modal dispersion-induced limitations in satisfying longer reach (>100 m) requirements within massively parallel data center and high-performance computing systems (Young et al. 2010) (Miller 2000). On the other hand, silicon photonics links offer a scalable solution using wavelength division multiplexing, which can be implemented using compact single-mode silicon photonic modulators (Reed et al. 2010) (Xu et al. 2005) and on-chip Ge photodetectors (Michel, Liu, and Kimerling 2010).

The implementation of optical interconnects relies on micro-sized optical devices that are integrated with microelectronics on silicon chips. Silicon is transparent to light from 1.1 μm to far-infrared region and is suitable for guiding light for telecommunication wavelength (1,550 nm). The current state-of-the-art single-mode silicon rib waveguides exhibit propagation losses less than 0.1 dB/cm (Tran et al. 2018). Another advantage of silicon is its large refractive index ($n = 3.5$ for 1,550 nm). The large refractive index contrast with the cladding materials provides excellent confinement of photons and allows for ultra-compact waveguide with dimensions less than one micrometer to be realized. In addition, the high refractive index contrast enables high-performance photonic bandgap structures such as photonic crystal waveguides and high-quality photonic crystal cavities.

In order for silicon to perform active functionalities and process optical signals other than passively guiding and confining light, optical nonlinear effects in this material need to be explored. However, developing active silicon photonic elements meet immediate material constraints. First, silicon is an indirect bandgap semiconductor, making silicon poor at emitting and amplifying light. Second, silicon has a centrosymmetric crystal structure which prevents the existence of second-order optical nonlinearity ($\chi^{(2)}$) and hence important applications such as electro-optic modulation and parametric amplification. Third, silicon has a small bandgap (1.1 eV) and is highly susceptible to two photon absorption and the subsequent free carrier absorption. The nonlinear losses in silicon limit presents a challenge for its application in nonlinear optics (Lin, Painter et al. 1999).

Among the various optical nonlinear effects, second-order nonlinearity ($\chi^{(2)}$) is particularly important for wavelength conversion and also the basis of the electro-optic effect, which is commonly employed for high-speed modulation. To implement optical emitters, significant progress has been recently on quantum dot lasers (A. Y. Liu et al. 2015). To achieve modulation in silicon devices, the plasma dispersion effect is most commonly exploited to manipulate the refractive index of silicon waveguides through mechanisms such as carrier accumulation, carrier injection and depletion (A. Liu et al. 2004) (Green et al. 2007) (Xu et al. 2005) (Dong et al. 2009). Silicon Mach-Zehnder interferometer (MZI) modulators operating in depletion mode with a modulation data rate exceeding 50 Gb/s have been demonstrated (Streshinsky et al. 2013). Compact silicon modulators based on ring resonators have also been developed to enhance the modulation effects by the optical resonance quality factor (Q) (Xu et al. 2005) (Zheng et al. 2014) (Pantouvaki et al. 2015).

Due to the extended lifetime of the free carriers, a careful balance between the modulation speed, efficiency, doping configuration and carrier induced absorption has to be maintained in silicon modulators. The Pockels effect, on the other hand, is a wideband, intrinsic property of non-centrosymmetric crystals. Modulation based on the Pockels effect does not rely on free carriers and has been widely employed in commercial high speed electro-optic modulators, such as devices made from $LiNbO_3$ (Wooten et al. 2000) and electro-optic organics (Hochberg et al. 2006) (Alloatti et al. 2011). However, common electro-optic materials with large Pockels coefficient are usually not compatible with CMOS semiconductor technologies. Within this context, the search for a suitable electro-optic material with potential of large-scale integration on silicon substrates remains an open issue.

The current industry standard optical modulators employ bulk crystals of lithium niobate ($LiNbO_3$). The modulation replies on the linear electro-optic (Pockels) effect intrinsic to the electro-optic crystals. Because the Pockels effect is intrinsically wideband, integration of electro-optic crystals on silicon substrates is attractive for high data rate applications. Furthermore, DC leakage power is entirely eliminated due to the insulating nature of wide bandgap crystals. Additionally, compared with silicon modulators, crystal-based modulators can have much lower device complexity by eliminating layers of masks for different doping and vias and hence reduce the cost of fabrication.

In this chapter, we consider two crystalline materials that are integrated on silicon substrates: barium titanate and aluminum nitride. Barium titanate is a ferroelectric oxide, having one of the largest reported electro-optic coefficients (Zgonik et al. 1994). AlN, on the other side, is non-ferroelectric, and yet has a non-centrosymmetric lattice and spontaneous polarization which gives rise to significant $\chi^{(2)}$ effects. Both materials have a large electronic bandgap and hence complement silicon for many of the drawbacks we have discussed above.

15.1.2 Overview of the Chapter

In the first section, we describe the electro-optic properties of $BaTiO_3$ and how the crystalline anisotropy and micro-domains affects the choice of the substrate orientation. The design methodology of the partially etched horizontal slot waveguides structure adopted for BTO modulators is given. We describe the experimental details of epitaxy, device fabrication, and how we optimize the process for minimizing the optical insertion loss. We present both DC and high frequency characterization of the electro-optic modulation for both Mach-Zehnder interferometer (MZI) and micro-ring type modulators. Further measures to reduce the optical insertion loss are proposed.

In the second section, we present integrated photonic circuits in GaN and AlN. Compared with bonded GaN, sputtered AlN provides better interface roughness and is more viable for wafer scale production. In this chapter, we focus on the characterization of the linear optical properties of GaN and AlN waveguides. We discuss the engineering of the photonic band gap structure in AlN. We present our measurement of one-dimensional (1-D) nanobeam cavity and two-dimensional (2-D) slow light photonic crystal waveguide structure based on AlN.

At last, we focus on the nonlinear optical aspects of the GaN and AlN photonic circuits. We give an in-depth study of the phase matching condition for the most efficient second harmonic generation (SHG). We present the measurement of the optical harmonics generated out of a GaN ring resibatir. We discuss the design of an integrated electro-optic optical modulators based on AlN. The high data rates operation and limitation of the modulation are discussed.

15.2 BARIUM TITANATE ELECTRO-OPTIC MODULATORS

Thin-film electro-optic (EO) modulators have attracted great interest as a way to achieve compact modulators that can be assembled in combination with integrated photonic circuits.

In this context, waveguide modulators are essential components for high-speed optical communication systems and ultrafast information processing applications. Using EO modulators rather than modulating the driving laser diode directly offers the advantage of reduced chirp and wavelength variation. Thin film modulators rely on the EO effect to induce refractive index changes that alter the properties of propagating light fields. In order to realize low loss and compact designs, it is thus essential to utilize materials with strong EO coefficients. Ferroelectric materials are thus of great interest, because they have a wide transparency window, covering the visible region and often part of the near ultraviolet. In addition, they show good thermal and mechanical stability. Most importantly, however, their electro-optic coefficients are typically one order of magnitude larger than those of non-ferroelectric materials. High-bandwidth modulators fabricated in $LiNbO_3$ with data rates in excess of 40 Gb/s are commercially available (Wooten et al. 2000). However, even higher EO coefficients have been reported for devices based on BaTiO3 (BTO) (Zgonik et al. 1994). High-speed EO modulators have been realized in an integrated fashion using silicon nitride strip waveguides deposited on top of a BTO thin film.

A major objective in the development of EO modulators is minimization of the half-wave voltage-length product ($V_\pi L$). Low V_π decreases power consumption, a particularly valuable attribute in specialized applications such as chip-scale communications. Lower drive voltages also ease amplifier and RF circuit design requirements. Small device lengths allow compact EO modulators to be fabricated for hybrid optical integration. So far, one of the most direct approaches to lower the half-wave voltage–length product at a given operating wavelength is to maximize the EO coefficient of the EO material. However, engineering of the material properties is not always an option when developing components that can potentially be integrated with standard integrated circuits (IC) fabrication procedures. It is therefore desirable to build EO modulators using a material system that is CMOS compatible to a large degree.

15.2.1 Electro-Optic Properties of $BaTiO_3$

Barium titanate exhibits the ferroelectric properties below its Curie point, when the crystal starts to transform from the cubic structure to the tetragonal structure and a permanent dipole results from the broken symmetry of the positive and negative ion charges. For unstrained barium titanate, the Curie point is about 120°C; for epitaxial thin film this transition temperature is expected to be higher due to the intrinsic strain effect between the barium titanate and the underlying substrate. The effective electro-optic modulation efficiency depends on the relation between the light propagation direction and the electric field direction due to the anisotropy of the electro-optic modulation and the multi-domain structure of the thin film material.

15.2.1.1 Electro-Optic Coefficients

The electro-optic coefficient of single crystalline barium titanate can be summarized in the following matrix:

$$\begin{bmatrix} \Delta(1/n^2)_1 \\ \Delta(1/n^2)_2 \\ \Delta(1/n^2)_3 \\ \Delta(1/n^2)_4 \\ \Delta(1/n^2)_5 \\ \Delta(1/n^2)_6 \end{bmatrix} = \begin{bmatrix} 0 & 0 & r_{13} \\ 0 & 0 & r_{13} \\ 0 & 0 & r_{13} \\ 0 & r_{42} & 0 \\ r_{51} & 0 & 0 \\ 0 & 0 & 0 \end{bmatrix} \cdot \begin{bmatrix} E_x \\ E_y \\ E_z \end{bmatrix} \tag{15.1}$$

Typical numbers for bulk BTO: r_{33} is 30 pm/V and $r_{51} = r_{42} = 820$ pm/V. To achieve a low

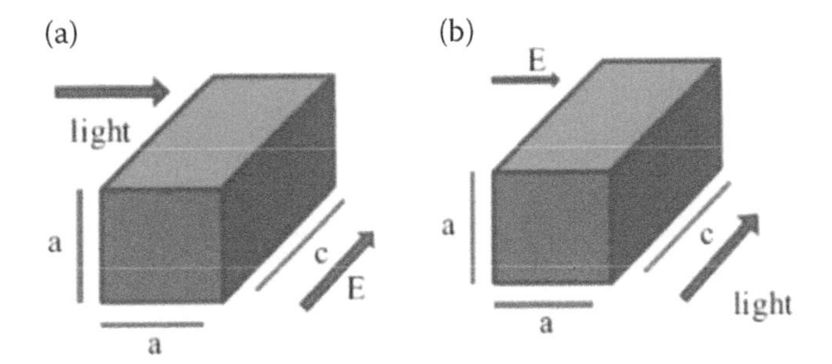

FIGURE 15.1 Schematic drawing of configurations between the polarization of the light and the applied RF field. (a) The RF field is parallel to c-axis; (b) the RF field is perpendicular to c-axis.

electro-optic switching voltage, we need to optimize the use of the largest electro-optic coefficient of the BTO. As such, we consider the following two scenarios (Figure 15.1).

When the direction of the applied RF field is parallel to the c-axis of the crystal and the light is polarized perpendicular to the c-axis. One can derive the refractive index change which arrives at the index modification to light traveling at ordinary and extraordinary index are as follows respectively:

$$\begin{aligned} \Delta n_0 &= -\frac{1}{2} n_o^3 r_{13} E_3 \\ \Delta n_e &= -\frac{1}{2} n_e^3 r_{33} E_3 \end{aligned} \tag{15.2}$$

In the above situation, only the small effect of the r_{33} is utilized. Now consider the condition when the direction of electric field is applied perpendicular to the c-axis and the light is polarized along the *c*-axis. Under this condition, the index ellipsoid will have a rotated principal axes at an angle theta about the x axis with respect to the principal axes of the crystal when a field E is applied along the *a* axis. The new principal refractive indices are given by

$$\begin{aligned} n'_x &= n_0 - \frac{1}{2} n_0^3 r_{51} E_x \tan\theta \\ n'_y &= n_0 \\ n'_z &= n_e + \frac{1}{2} n_e^3 r_{51} E_x \tan\theta \end{aligned} \tag{15.3}$$

Therefore, the change in the refractive indices is of second order in E_x.

$$\Delta n_z = \frac{n_0^2 n_e^5 r_{51}^2}{n_e^2 - n_o^2} - E_x^2 \tag{15.4}$$

The phase change:

$$\Delta\phi = \Delta n_z - \frac{2\pi}{\lambda} - L \tag{15.5}$$

The half-wave voltage at which the phase change is π can thus be determined as

$$V_x = \frac{d}{n_o r_{51} n_e^2} \sqrt{\frac{\lambda (n_e^2 - n_o^2)}{2 \cdot L \cdot n_e}} \tag{15.6}$$

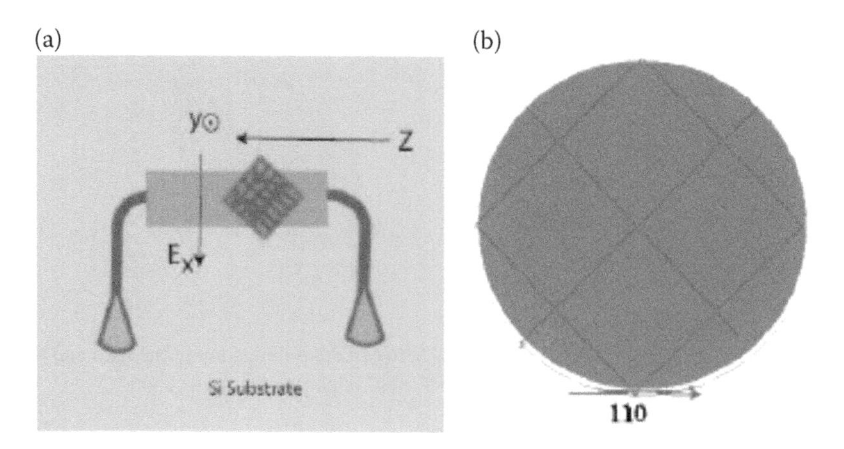

FIGURE 15.2 Choice of the orientation of the SOI substrate. (a) Four in-plane domain variants; (b) the dicing of the 6-in SOI wafer for the BTO epitaxy.

For wavelength $\lambda = 633$ nm, $n_e = 2.365$ and $n_o = 2.437$ and assuming the electrode spacing is 1 μm, $r_{51} = 820$ pm/V and the transverse length of the device is 1 mm, we can calculate the theoretical half-wave voltage $V_\pi = 0.61$ V.

15.2.1.2 Multi-Domain Structure

Ferroelectric samples typically exhibit separate regions so-called "domains" which differ in their spontaneous polarization direction. During the molecular beam epitaxy (MBE) growth, because the discrepancy in the lattice constant of silicon (5.43 Å) and barium titanate (a = 3.99 Å, c = 4.04 Å), the unit cell of the as-grown barium titanate has to rotate about 45° relative to the silicon unit cell. Due to the different thermal expansion of the silicon and barium titanate, the resulting structure is cooled down from the growth temperature of 600°C and the barium titanate unit cells are arranged over the silicon substrate so that the longest dimensions thereof are oriented parallel to the plane of the silicon substrate. Accordingly, the four possible variants of the as-grown ferroelectric barium titanate domains are illustrated in Figure 15.2.

While the analyses in the previous sections were based entirely on a single-domain assumption, the effective electro-optic coefficient has to be considered by taking account of the canceling effects of the oppositely aligned ferroelectric domain. Changing or reversal of the ferroelectric domain direction, known as the poling process, is possible with sufficiently large electric field. Consider the case of aligning the electric field parallel to the c-axis of the crystal, all the domains with spontaneous polarization perpendicular to the electric field will not be pole but cancel each other and hence the effective electro-optic effect is reduced. To optimize the electro-optic modulation efficiency, the direction of the applied electric field should be rotated by about 45° with respect to the c-axis of the epitaxial barium titanate crystal in order that after the poling process each of the four domain structure will be so aligned that each has a contribution to the total modulation (Figure 15.2).

15.2.2 The Waveguide Design: Partially Etched Horizontal Slot Waveguides

The suggested modulator relies on the electro-optical change of the effective refractive index in a ridge waveguide with integrated horizontal slot (Figure 15.3). The horizontal slot structure was demonstrated (Barrios et al. 2007). This type of slot waveguide is ideally suited for the realization of planer electro-optical modulators, because it leads to high electric field concentrations in the

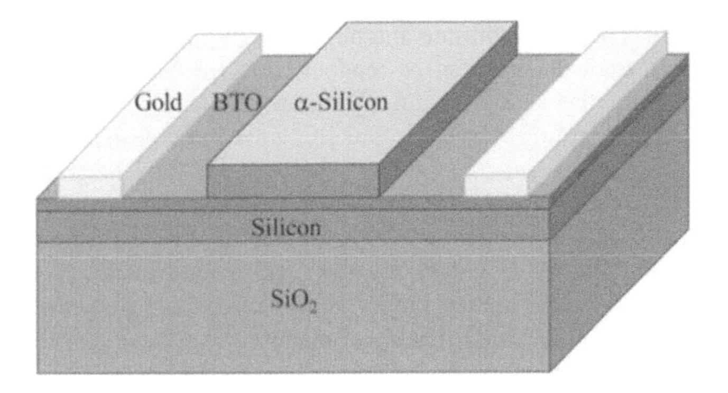

FIGURE 15.3 The layout of the suggested modulator. The modulator is fabricated on standard SOI wafers. Additional layers of BTO and α-silicon are deposited on top of the silicon layer. A ridge waveguide with integrated horizontal slot is realized in the α-silicon layer.

sandwiched oxide layer. Therefore, good field overlap between the modulating electric field and the propagating electromagnetic fields in the waveguide can be achieved.

In recent years slot waveguides have been receiving increasing attention as a new class of waveguides for a multitude of applications. In the context of optical waveguiding, a slot waveguide is composed of one or several narrow low refractive index regions enclosed between areas of high refractive index. The field discontinuity at interfaces between the high and low refractive index regions leads to strong field enhancement in the low index region, particularly close to the interfaces. Initial slot waveguides were realized by fabricating nanophotonic waveguides with small separation, which creates vertical slots. Because of the field concentration near the waveguides sidewalls, these vertical slot waveguides often suffer from enhanced scattering loss and thus large propagation loss. The best reported loss for the quasi-TE mode in a vertical slot waveguide with a single slot of 50 nm or less is greater than 11.6 ± 3.6 dB/cm (Sun et al. 2007). As a result, optical resonators involving vertical slots feature relatively low-quality factors, which constraints their application in certain photonic applications.

Scattering losses can be significantly reduced by employing slot waveguides with a horizontal slot layer instead of a lithographically defined vertical slot. A horizontal slot structure including a horizontal low index slot layer can be fabricated by layered deposition or thermal oxidation. The corresponding slot waveguide devices have virtually no fabrication constraints on slot thickness and can have very low scattering loss due to small surface or interface roughness for the fundamental slot mode. In addition, multiple slot configurations can be used in a horizontal slot waveguide concept to provide enhanced optical confinement in the low index slot region. In this section, we describe a nanophotonic architecture based on horizontal slot waveguides. Employing PECVD multi-layer deposition onto silicon-on-insulator substrates provides a flexible and cost-effective way to tailor the optical properties of the waveguiding structure for specific applications. We deposit an amorphous silicon top layer onto a lower-refractive index slot layer. Optical quality thin films of a-Si:H are deposited by plasma enhanced chemical vapor deposition (PECVD) at low temperatures, typically around 400°C. At these temperatures a significant atomic percentage of hydrogen is incorporated in the material to fill dangling Si bonds and reduce optical loss by absorption. The guiding slot layer can be chosen freely. Here we employ silicon nitride and silicon dioxide slot layers as exemplary materials. Differing from previous slot-designs, we pattern waveguiding geometry solely in the top silicon layer. By etching only the silicon top layer, slotted ridge waveguides are formed and scattering loss in the waveguides is minimized.

The horizontal slot waveguide provides a convenient way to realize efficient grating couplers. Grating couplers are emerging as a promising way to couple light into integrated photonic circuits. They allow for coupling light into and out of on-chip devices through first order Bragg reflection. As a result, optical chips can be readout from the top, which improves alignment tolerance. It is generally the case that efficient grating couplers require a shallow grating structure to achieve high coupling efficiency. Our slot waveguide design lends itself naturally to shallow grating design, because only the top layer is etched and thus the grating is inscribed only into the top layer. In addition to grating couplers we realize on-chip photonic circuitry. We fabricate Mach-Zehnder interferometers with high extinction ratio and ring/racetrack resonators. We achieve propagation loss of less than 2 dB/cm and loaded quality factors in our optical cavities exceed 10^5. These are the highest Q factors ever realized in slot waveguides to date. Our platform holds promise for embedding functional materials that have low refractive index on a silicon photonics platform.

15.2.2.1 Design of the Horizontal Slot Waveguides

Our photonic platform is realized on commercial Silicon-On-Insulator (SOI) wafers, Soitec smart-cut with a buried oxide (BOX) layer of 3 μm thickness. The top silicon layer is thinned down to 110 nm thickness by thermal oxidization and subsequent wet-etch. The dielectric slot layer is deposited by PECVD in silane/nitrous oxide (for silicon dioxide slot) and silane/ammonia (for silicon nitride slot) chemistry with a targeted thickness of 80 nm. The final waveguiding layer composed of amorphous silicon (α-Si) is deposited by PECVD in a 150 sccm flow of silane precursor at 400°C at 2.4 torr. The vertical layer geometry used throughout the paper is shown schematically in Figure 15.4(a). After the deposition process nanophotonic structures are defined by electron-beam lithography and subsequent reactive ion etching (RIE) using chlorine gas chemistry.

We consider nanophotonic waveguides of varying width, ranging from 200 nm to 900 nm. As shown in the mode profiles in Figure 15.4(b), narrow waveguides support a fundamental mode predominantly guided in the bottom silicon layer. Increasing the waveguide width leads to improved confinement of the total mode to the slot region below the α-Si waveguide.

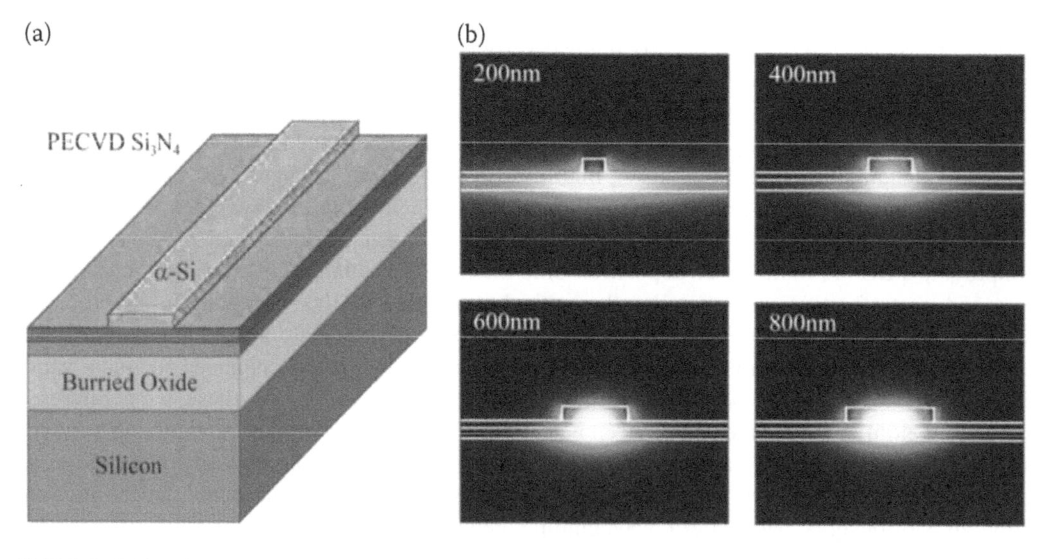

FIGURE 15.4 The vertical layer structure used to define the partially buried horizontal slot waveguides. PECVD silicon nitride (green) and amorphous silicon (blue) layers are deposited onto SOI substrates. (a) The vertical layer structure used to define the partially buried horizontal slot waveguides. (b) Calculated mode profiles for horizontal slot waveguides.

15.2.2.2 On-Chip Grating Couplers

In order to provide convenient access to the fabricated photonic circuits we employ focusing grating couplers. Due to first-order Bragg reflection on the grating structure, light can be coupled into and out of the photonic circuits into the vertical direction. Thus, optical access to the chip is achieved by vertical alignment, which is much easier than conventional optical alignment through butt-coupling with lensed fibers or inverse tapers. Grating couplers have the advantage of not requiring polished facets for coupling, which enables wafer scale testing of the integrated circuits. In addition, grating couplers are very compact and have a large optical bandwidth.

Our photonic platform is designed for high-yield and easy fabrication, thus only a single electron-beam lithography and RIE step is desired. In traditional SOI nanophotonic circuits this approach will result in deep etched grating couplers. Such grating couplers do not allow for good coupling efficiency. The reported fiber coupling efficiency obtained with standard uniform grating structures amounts to roughly 20%. In order to improve the coupling efficiency, one common approach is to employ shallow etch in the grating area. Shallow etching improves the coupling efficiency significantly. Our horizontal slot design lends itself naturally to a shallow grating etching, because only the α-Si layer is removed during the one-step etching.

We design grating couplers employing the finite-difference time-domain method. Because the grating structure is translational invariant the devices are investigated in a two-dimensional cross-section through the grating structure. The couplers are designed for operation at 1,550 nm by adjusting the grating period. As shown in Figure 15.5(a) the coupling loss can be as low as 3.4 dB from the simulation results. From the simulation results we find and optimized coupler period of 710 nm to obtain a central coupling wavelength of 1,550 nm. In particular, the coupler bandwidth measured at the 3 dB points is greater than 55 nm.

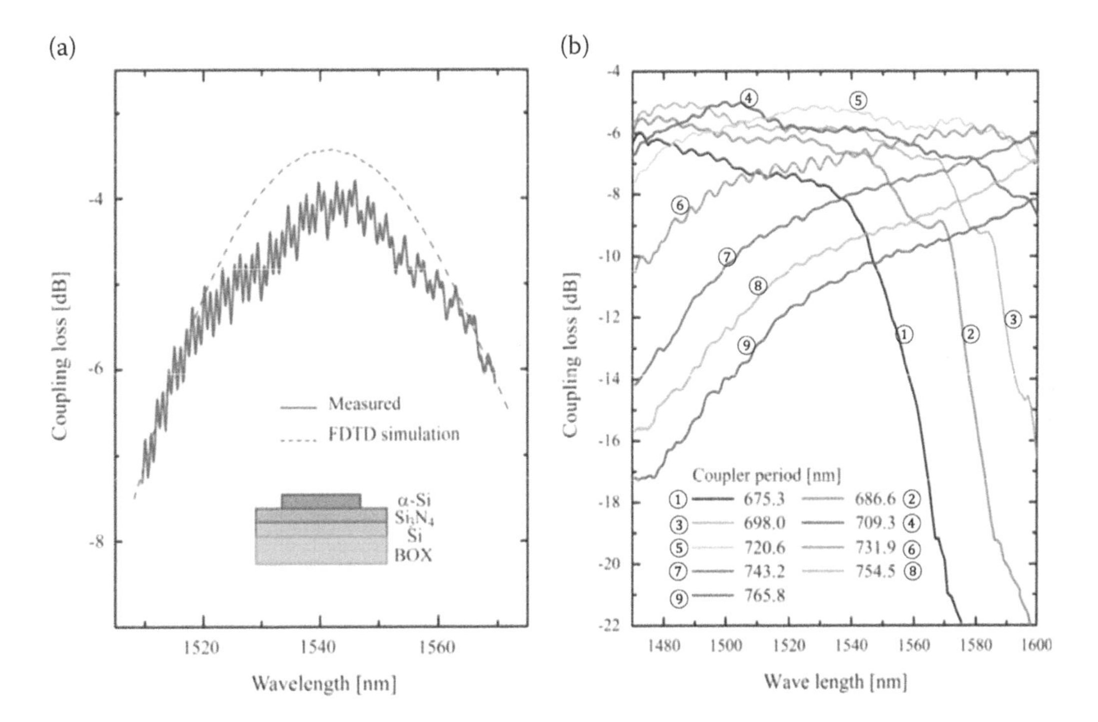

FIGURE 15.5 (a) Simulated and measured transmission profile of a SiNx slot grating coupler with a central coupling wavelength of 1,540 nm for a period of 710 nm. The nitride slot layer is 80 nm thick. The coupler features a minimum coupling loss of ~4 dB and a 3 dB bandwidth exceeding 110 nm. Notable is the wide flat-top regime in the central coupling window. (b) Simulated transmission spectra as a function of the grating coupler period.

In order to verify the theoretical predictions, we fabricate grating couplers from slot substrates with a α-Si layer of 110 nm thickness and a silicon nitride slot layer of 80 nm thickness. As shown in Figure 15.5(a) the measured transmission profile follows the simulated transmission curve well. The measured coupling loss is 4 dB, corresponding to 40% coupling efficiency. Discrepancy to the theoretical result is expected due to fabrication imperfections and the propagation loss of the waveguides within in the device.

In addition to the couplers built on a silicon nitride slot layer we also fabricate grating couplers with a silicon dioxide PECVD slot layer. Oxide based grating couplers feature a higher coupling bandwidth as shown in Figure 15.5(b). We display the measured coupler profiles for different grating periods, shifting the central coupling wavelength to longer wavelengths with increasing coupler period. Striking is the wide coupling bandwidth, which spans beyond the tuning range of our laser. Compared to the nitride grating coupler the coupling loss is slightly higher (5 dB) at the central coupling point. From the measured result we estimate a 3 dB bandwidth of more than 110 nm for the couplers fitting into the tuning range of the laser. The increased optical bandwidth observed for the silicon oxide slot couplers is due to the reduced effective index of the coupling structure. This leads to improved mode matching properties of the coupling region over a wider wavelength range. In addition, the coupler features a flat passband over a wide wavelength range, which makes the design particularly useful for broadband spectroscopic applications. The coupling bandwidth is much improved compared to traditional SOI couplers, which feature a reduced 3 dB bandwidth of roughly 30 nm.

15.2.2.3 Mach-Zehnder Interferometers

The grating couplers described above are employed to provide optical input and output ports to on-chip photonic circuitry. We first investigate Mach-Zehnder interferometers (MZIs), which are commonly used for sensitive metrological applications and nowadays also for the measurement of extended photonic functionalities.

In an MZI, an optical waveguide is split into two separate arms, which are joined together after a chosen propagation distance. A given path difference between the arms leads to a characteristic interference pattern at the output of the interferometer due to constructive and destructive interference. The free spectral range (FSR) of the interferometer is determined by the path difference and the group index of the waveguide. When the optical loss in both arms is equal and the splitting ratio is 50:50, the total loss through the interferometer is small and the extinction ratio between the peaks and the valleys of the fringes is high. In this case the interferometer is considered to be balanced.

We fabricate MZIs from silicon nitride horizontal slot waveguides with a path difference of 100 μm. The waveguide width is fixed at 900 nm, while the thickness of the nitride slot layer is kept at 80 nm. Given the group index of the waveguide of 3.35 obtained from finite-element simulations, the FSR of the MZI amounts to 7.2 nm. Therefore, many fringes will fit into the bandwidth of the grating couplers. In Figure 15.6(a) we show an optical micrograph of a fabricated sample. The image illustrates the layout of the photonic circuit, where the MZI is enclosed between the two focusing grating couplers. The measured response of the device is shown in Figure 15.6(b), featuring nice interference fringes of an MZI. The response is enveloped by the profile of the grating couplers. From the measurement we obtain an extinction ratio of more than 20 dB, which illustrates that the interferometer is well balanced. The transmission loss at the peak of the interference fringes compared to a calibration sample without the interferometer amounts to 1 dB, which further proofs that the two arms of the MZI are balanced and the splitting ratio at the input and output is indeed close to 50:50.

15.2.2.4 Optical Resonators

In addition to MZI devices we also fabricate resonant optical cavities on chip. Because we are investigating waveguide-based geometries we focus here on ring and racetrack resonators. We

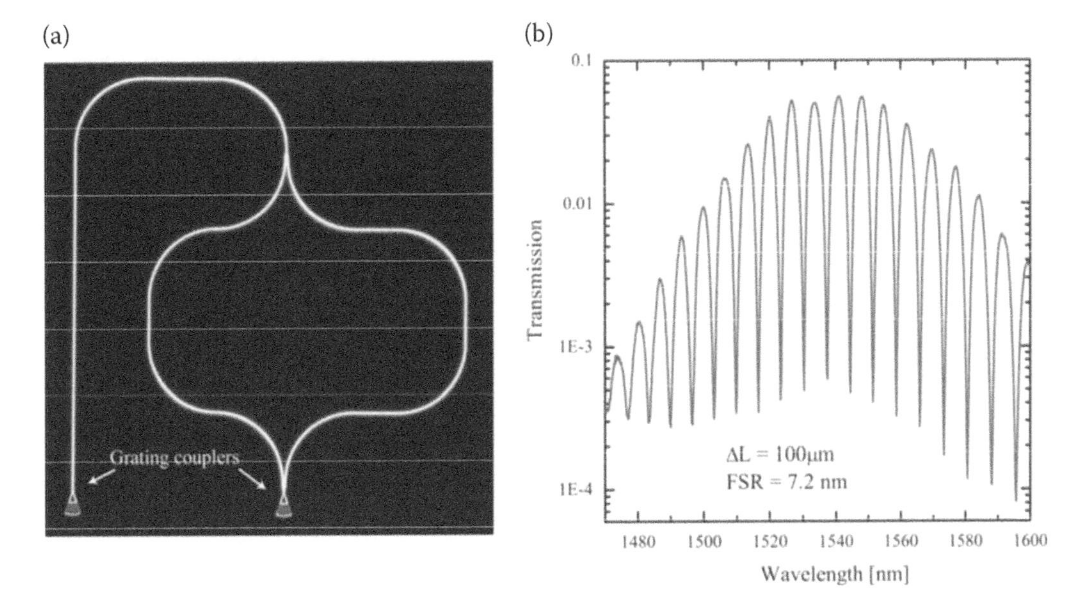

FIGURE 15.6 (a) Optical image of a fabricated photonic circuit with input/output grating couplers and an integrated Mach-Zehnder interferometer with a path difference of 100 m. The waveguide width is 900 nm. The horizontal slot is made of 80 nm PECVD SiNx. (b) The measured response of the fabricated sample shown in (a). The response shows the typical interference fringes of an MZI, enveloped by the profile of the grating coupler. The free-spectral range is 7.2 nm, which implies a waveguide group index of 3.35. The extinction ratio is greater than 20 dB, which illustrates that the arms of the MZI are well balanced.

fabricate samples with different coupling gaps in order to achieve critical coupling as shown in Figure 15.7. The samples are investigated optically through transmission measurements. First, we analyze ring resonators with a radius of 100 μm and a waveguide width of 900 nm, fabricated with a silicon nitride slot layer. An optical micrograph of a fabricated device is shown in the inset of Figure 15.7(a). Maximum extinction ratio in the transmitted optical signal is achieved under critical coupling conditions, which are reached when the optical power coupled into the ring matches the power dissipated inside the ring during one round trip. Under critical coupling conditions the measured loaded optical Q corresponds to half the intrinsic optical quality factor. Under near-critical coupling condition with a coupling gap of 300 nm we find an extinction ratio of ~20 dB, as shown in the transmission spectrum in Figure 15.7(a). From fitting the resonance dips with a Lorentzian curve, we extract loaded optical Q factors of 79,000, as shown in the zoom-in graph in Figure 15.7(b). When measuring devices with larger coupling gap of 600 nm the devices are operated in the under-coupled regime and therefore the extinction ratio is smaller. For the best devices we find a cavity Q of ~100,000. For comparison we also investigate ring resonators with a silicon dioxide slot layer. The transmission profile of a near critical coupled device is shown in Figure 15.7(c). The flat-top response reveals the broad bandwidth of the grating couplers. In this device good extinction of 20 dB is found, comparable to the silicon nitride slot device.

In order to establish the low propagation loss, we consider a further set of ring resonators that are separated from the input waveguide by a larger gap of 500 nm. When measuring these weakly coupled devices, we find high optical Q of 125,000 as shown in Figure 15.7(d). This measured Q is the highest Q-factor obtained in a slot resonator to the best of our knowledge.

The measurements of the resonators allow us to extract the properties of the resonator in more detail, using the method introduced in references. We define the minimum power transmission

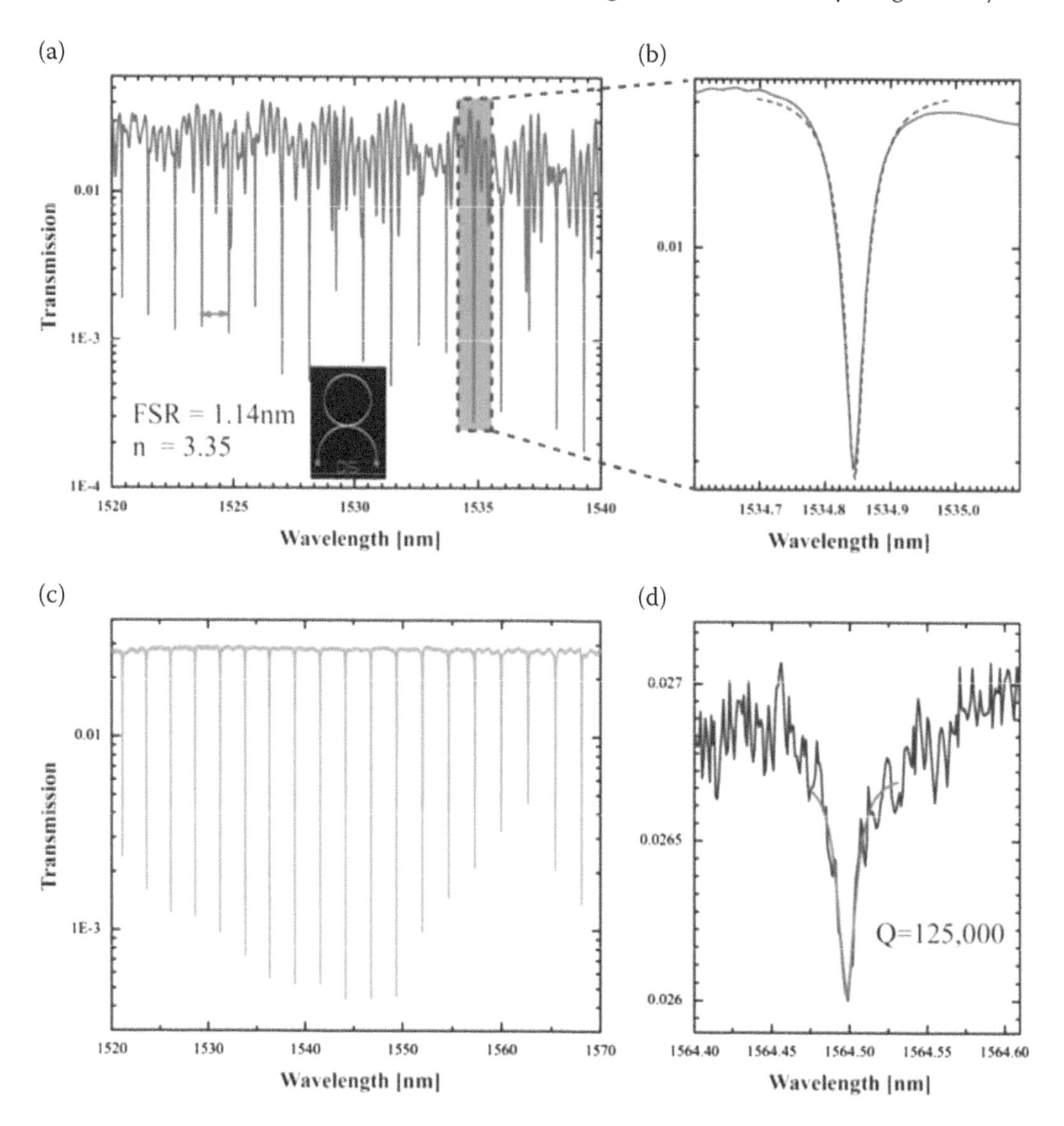

FIGURE 15.7 (a) The transmission profile of a ring resonator fabricated from a silicon nitride horizontal slot waveguide. The fabricated device is shown in the optical micrograph in the inset. (b) Zoom into one of the ring resonances for the near critical coupling coupling case. A Q of 79,000 is found from the fit with a Lorentzian dip. (c) The transmission response of a near critical coupled ring resonator with a silicon dioxide slot layer, showing an extinction ratio of 20 dB. The broadband flat-top response is the envelope of the slot grating couplers. (d) A high-Q resonance measured in a separate, weakly coupled ring resonator with gap of 500 nm. The Lorentzian fit reveals a best optical Q of 125,000.

in the through-port of the ring as γ and the -3 dB bandwidth as $\delta\lambda$. The response period of the resonator is the FSR introduced in the previous sections.

Given the extinction ratio for the device in Figure 15.7(b) of 12.43 dB ($\gamma = 0.0571$), $\delta\lambda = 19.5$ pm and FSR = 1.12 nm, we find a total Q_t of 79,000 at 1,534.84 nm. The extracted power loss coefficient is then 0.1616, which implies a propagation loss of 1.83 dB/cm. The corresponding intrinsic Q_i is 3.3×10^5 and the coupling coefficient is determined to be $\gamma = 0.204$.

For a typical resonance in Figure 15.7(c) for the oxide ring resonator we find an extinction ratio of 17.99 dB ($\gamma = 0.0159$), $\delta\lambda = 50.5$ pm and FSR = 2.63 nm, we find a total Qt of 30,000 at 1,554.68 nm. The extracted power loss coefficient is then $\kappa_p = 0.1229$, which implies a propagation

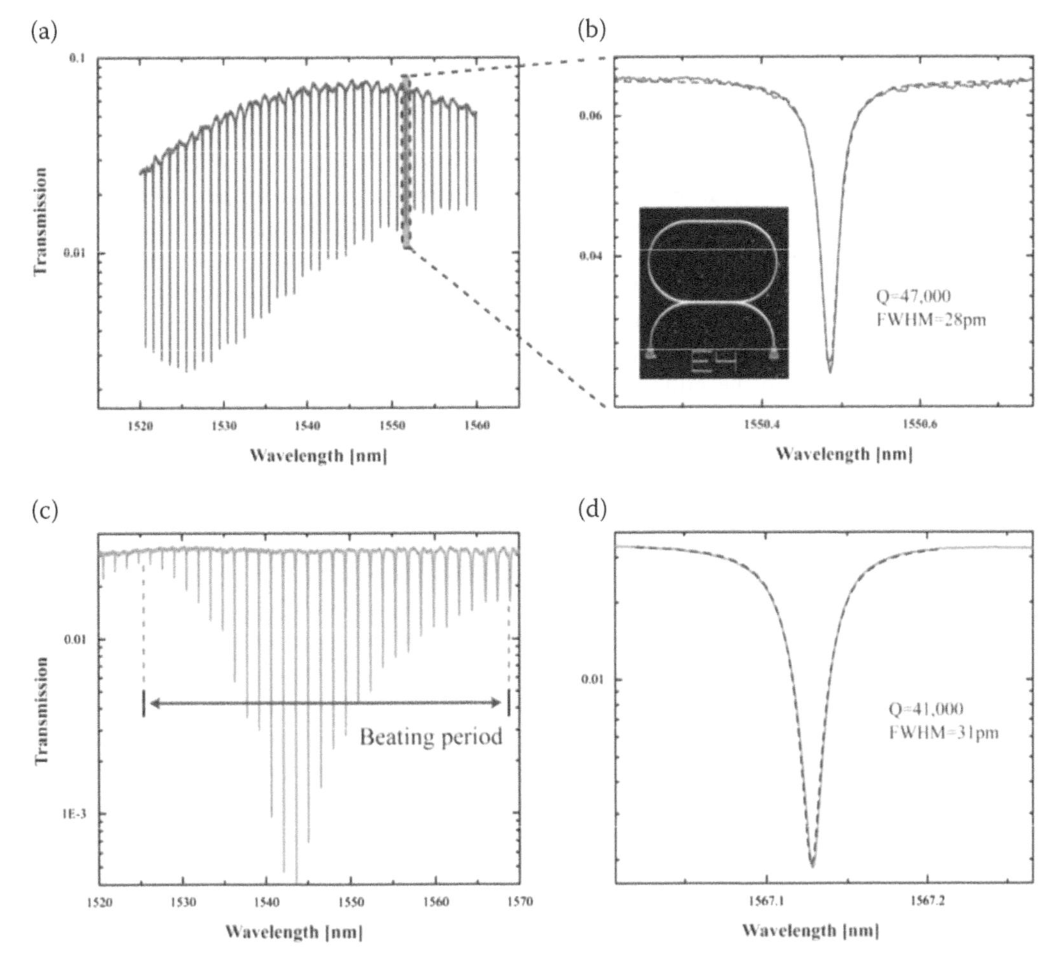

FIGURE 15.8 Shown is the measured transmission response of a racetrack resonator with a silicon nitride slot layer. The waveguide width is 900 nm. The transmission profile is enveloped by the coupler response, whereas the extinction ratio is enveloped by the beating pattern of the input directional coupler. (a) Shown is the measured transmission response of a racetrack resonator with a silicon nitride slot layer. (b) The fitted response shows an optical Q of 47,000, slightly reduced from the Q measured in ring resonators. (c) The transmission profile for a racetrack resonator fabricated with a silicon dioxide slot layer. The waveguide width is 900 nm. The beating pattern reveals a beating period of roughly 43 nm. (d) The fitted response of one of the resonances, showing an optical Q of 41,000, comparable to the Q found in the nitride slot resonators.

loss of 2.1 dB/cm, given the ring radius of 50 μm. The corresponding intrinsic Q_i is 2.44 × 105 and the coupling coefficient is determined to be $\kappa = 0.23$.

In addition to ring resonators we investigate racetrack resonators with a straight coupling section of 100 μm length, as shown in Figure 15.8. Racetrack resonators can be considered extended ring resonators, into which two straight waveguide sections have been inserted. The resonators are coupled to an input waveguide, which is separated from the resonator by a coupling gap *g*. The straight coupling section presents a directional coupler. Therefore, the coupling ratio into the waveguide depends on the optical wavelength. When the length of the straight section is long enough to be close to the coupling length of the directional coupler, the critical coupling condition for the racetrack resonator can always be matched for a given wavelength. This is observed in the transmission spectrum in Figure 15.8(a). Shown is the measured transmission for silicon nitride slot device. A zoom into the resonances reveals an optical Q of 47,000, as shown in Figure 15.8(b). Compared to the ring resonator the Q is slightly reduced which is partly due to the modal mismatch between the straight and bent waveguides inside a racetrack resonator.

Similar behavior is observed for racetrack resonators fabricated with a silicon dioxide slot layer, as shown in Figure 15.8(c, d). In Figure 15.8(c), the beating pattern due to the straight coupling section is clearly visible, with a beating period of roughly 43 nm. Taking into account the measured group index of the waveguide of 3.6, the beating length is close to the numerically expected beating period of 46 nm. By fitting the response in Figure 15.8(d) to a Lorentzian we find a maximum Q of 41,000, which is comparable to the Q found for the nitride slot resonator. Under optimal coupling conditions the extinction ratio for resonances around 1,545 nm is close to 20 dB, which also corresponds to the critical coupling extinction found in the ring resonators in the previous section.

15.2.3 Theoretical Considerations of BTO Modulators

In Figure 15.3, we show the layout of the proposed modulator. The devices are fabricated on standard SOI wafers. A thin layer of BTO is deposited by molecular beam epitaxy onto the silicon layer. Subsequently a layer of α-silicon is deposited onto the BTO layer. The silicon layer is patterned by ebeam lithography to generate the waveguiding structure. By etching the waveguides into the top silicon layer, a ridge waveguide with integrated horizontal slot is realized. Slot waveguides have become increasingly popular because the light can be highly confined inside the slot. In this case much of the propagating light will be concentrated in the BTO layer and thus interact strongly with the driving modulation field. To excite the modulating field, gold electrodes are fabricated alongside the waveguide. Because the waveguide mode is confined between the ridge silicon layer and the bottom silicon layer only a small portion of the field will reach the gold electrodes. Therefore, absorption loss due to optical absorption by the electrode material will be small.

In order to design an efficient EO modulator we investigate the optimal geometry for the slotted ridge waveguide. We employ the finite-element mode solver COMSOL Multiphysics to determine the mode properties of the waveguide. Our SOI wafers have measured silicon thickness of 110 nm.

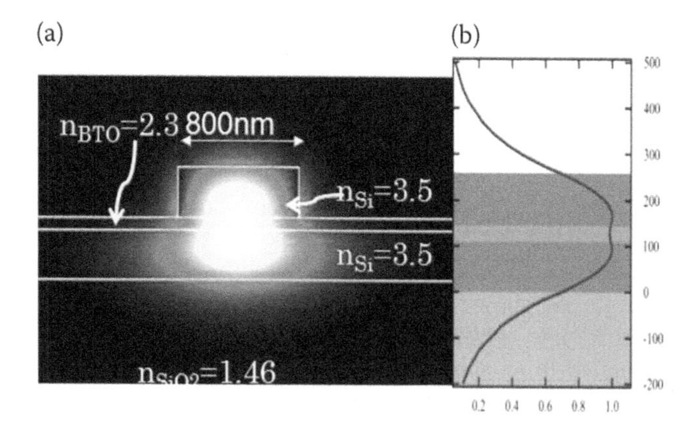

FIGURE 15.9 (a) A plot of the Ex component of the optical mode confined in the Si- BTO/α-Si structure. Here the thickness of both silicon layers is set to 220 nm and the BTO layer is assumed to be 70 nm thick. The waveguide is 800 nm wide. A strong field concentration in the BTO layer is observed. (b) A cross-section view of the mode profile in the slotted ridge waveguide.

To get high-yield deposition of BTO we assume oxide thicknesses of up to 100 nm. A typical mode profile is shown in Figure 15.9(a).

The simulation demonstrates that the optical mode is well confined in the Si- BTO-αSi sandwich structure as shown in Figure 15.9(b). In particular, strong fields are observed in the BTO layer. The confinement to the gap, however, is not as strong as for Si-SiO_2-Si slot waveguides, because of the lower refractive index contrast between BTO and silicon. It is worth pointing out that only a small portion of the optical mode is found outside the ridge structure. This is of importance, because the gold electrodes will absorb optical fields inside the gold layer. Due to the good confinement to the ridge structure it is therefore possible to pattern the gold electrodes in the vicinity of the waveguide. In the current simulation the electrodes are separated from the waveguide by a distance of 1 μm.

15.2.3.1 Transmission Loss

We first investigate the necessary width of the ridge structure that is required to yield low loss propagation. We employ the finite-difference time-domain method using the commercially available package OmniSim (Photon Design). The transmission loss is estimated by calculating the transmission through a 100 μm long ridge waveguide. The transmission is sampled in intervals of 20 μm. Fitting the calculated transmission at each interval at a wavelength of 1.55 μm yields the estimated propagation loss for a given waveguide width. The results are presented in Figure 15.10(a).

From the simulation results it is apparent, that low loss propagation is possible when the waveguide width is properly chosen. The structure will support a single waveguide mode below 900 nm width. When the width is increased from narrow waveguides the propagation loss reaches a minimum of 0.33 dB/cm. For narrower waveguides the propagation loss is larger, because the

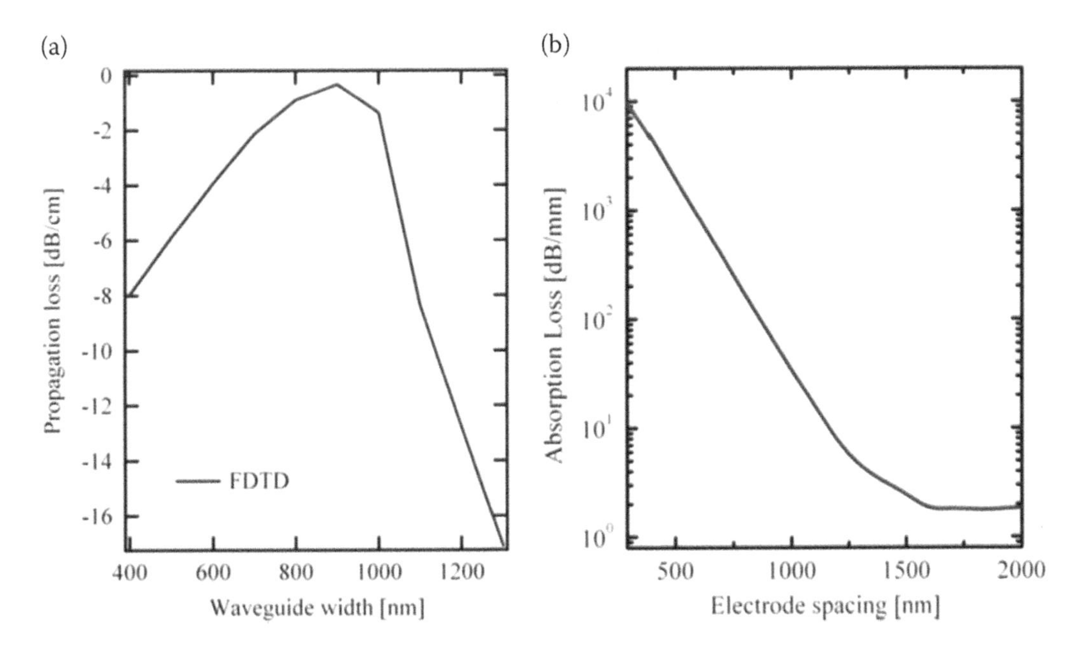

FIGURE 15.10 (a) The propagation loss of a slotted ridge waveguide in dependence of the waveguide width. The thickness of the silicon layers is kept at 110 nm and the thickness of the BTO layer is 70 nm. Minimum propagation loss is obtained for a waveguide width of 900 nm. Wider waveguides support multi-mode propagation with higher loss. Narrow waveguides show increased loss because the waveguide is operated above the cutoff wavelength. (b) The loss due to optical absorption in dependence of the electrode spacing.

waveguide is operated in the cutoff regime. When the waveguide width is increased further, multi-mode propagation is observed. Thus, higher propagation losses due to inter-mode beating and competition are present. In this regime we expect loss values of minus several dB/cm. Therefore, the target width of the waveguide will be 900 nm.

15.2.3.2 Absorption Loss From the Gold Electrodes

The second loss mechanism present in the modulator stems from optical absorption of the propagation fields by the metallic electrodes. This loss mechanism is more pronounced when the electrodes are placed in close proximity to the waveguide. Small separation between electrodes and waveguide is desirable because of the potentially lower modulation voltage. Therefore, we are interested in finding the smallest feasible separation that does not jeopardize the propagation loss significantly.

We simulate the loss due to optical absorption with FDTD by applying a dispersive material model to the gold electrodes. Gold is modeled as a Drude model in the near infrared wavelength regime. Results are shown in Figure 15.10(b). We find that low loss propagation is possible when the electrodes are separated by more than 500 nm from the waveguide. In this case, the propagation loss will not be affected to a significant degree.

15.2.3.3 Bend Loss

A third loss mechanism is of importance for the final layout of the EO modulator. In order to achieve a compact design, the waveguides have to be arranged in a meander or spiral layout. The layout will be more compact if smaller bend radii can be achieved. We therefore calculate the loss resulting from waves propagating along a 180-degree bend.

The loss is obtained for waveguides with a width within the range of desirable propagation loss and the loss is analyzed in dependence of bend radius. We consider waveguides with widths from 700 nm to 1,100 nm. The propagation loss around a 180-degree bend is shown in Figure 15.11(a).

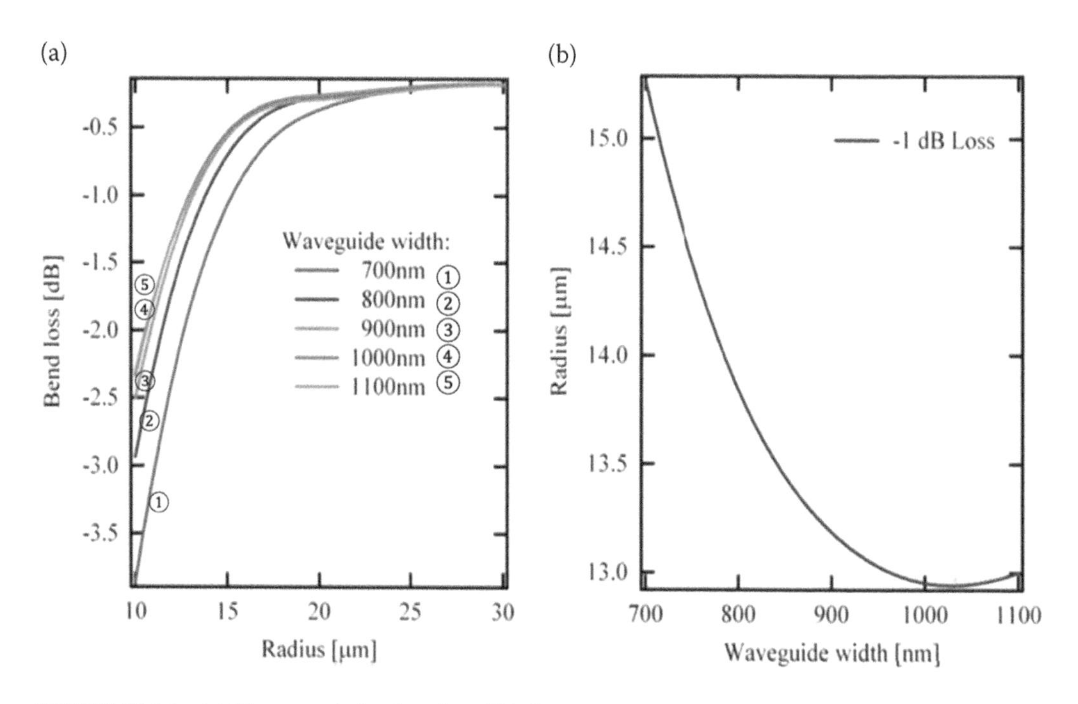

FIGURE 15.11 (a) The transmission loss for a 180-degree turn. The BTO thickness is assumed to be 70 nm. The bend loss decreases with increasing bend radius. We note that for wider waveguides smaller bend radii are feasible. (b) The bend radius required to obtain a transmission loss of 1dB in dependence of waveguide width. The minimum radius is found for a waveguide of 1,000 nm width.

It is apparent that the slotted ridge waveguide supports low loss propagation around a bend only for radii in excess of 20 μm. Furthermore, wider waveguides will yield lower loss at smaller bend radii. This is demonstrated further in Figure 15.11(b). Here we show the required bend radius to obtain a loss of -1 dB. For the target waveguide width of 900 nm we therefore require a bend radius of larger than 13 μm in order to avoid propagation loss in excess of 1 dB.

As a result, we are able to achieve a compact modulator design by introducing proper design rules for the layout of the waveguide. A bend radius of roughly 15 μm allows us to layout the complete modulator in a compact fashion.

15.2.4 Electro-Optical Properties

Having identified a suitable optical structure, we now consider the electro-optical properties of the modulator. Relying on the nonlinear optical properties of the BTO layers we intend to find a structure that achieves a maximum change of the effective refractive index of the propagating mode. Of importance hereby is to achieve good overlap between the driving electric field and the propagating electromagnetic fields. In the following we investigate the influence of three parameters on the EO properties: the width of the ridge, the thickness of the BTO layer and the thickness of the top silicon layer. These three parameters are easily accessible by the fabrication procedure and thus the results can be confirmed experimentally. The parameters sweeps are carried out using a two-stage FEM simulation. In a first step, we calculate the static electric field distribution resulting from a voltage applied to the electrodes. Subsequently we calculate the effective refractive index of the fundamental mode in the waveguide taking into account the modified refractive index of the BTO layer due to the Pockels effect. Because the Pockels effect introduced additional anisotropy to the BTO layer we simulate the mode profile using the anisotropic mode solver available in the COMSOL multiphysics package. Here we assume that the BTO layer is characterized by the following electro-optic coefficients: r_{13} = 8 pm/V, r_{33} = 28 pm/V, r_{51} = 820 pm/V and the refractive index is 2.4. For silicon we assume a refractive index of 3.477 and for the substrate a refractive index of 1.46. We apply a varying voltage of ±V to the left and right electrodes. The resulting electric potential is shown in Figure 15.12(a). When the voltage is increased, the effective refractive index of the fundamental mode decreases linearly. This is illustrated in Figure 15.12(b) for a sample structure with a waveguide width of 900 nm, a BTO layer thickness of 70 nm and an α-silicon layer thickness of 110 nm. The gold electrodes are assumed to be 100 nm thick and 2 μm wide. They are separated from the waveguide by a distance of 1 μm.

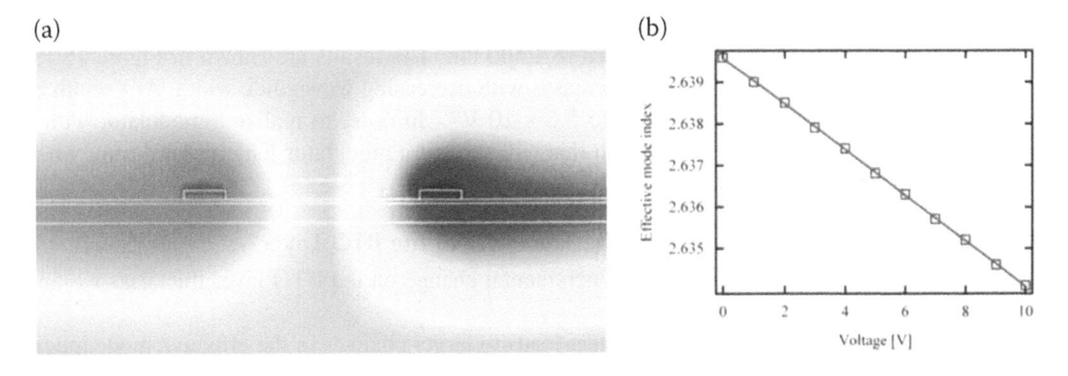

FIGURE 15.12 (a) The electrostatic potential resulting from the application of a modulation voltage of ± 1 V to the left and right gold electrodes. The spacing between the waveguide and the edge of the electrodes is 1 μm. (b) The dependence of the effective refractive index of the fundamental mode on the driving voltage. A clear linear relationship is observed.

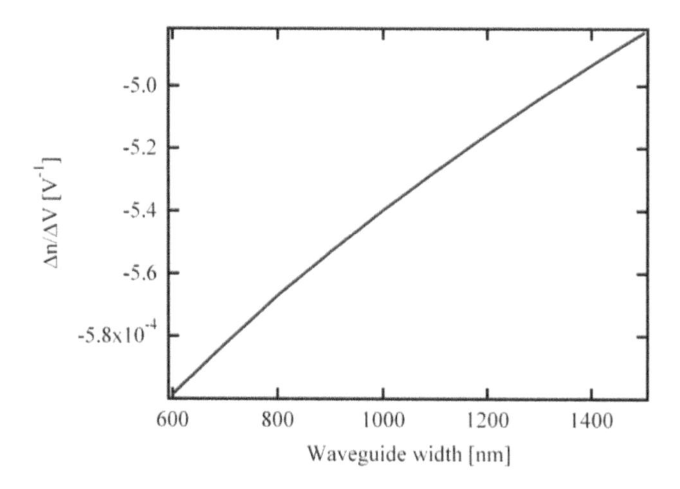

FIGURE 15.13 The calculated incremental change in the effective refractive index of the fundamental mode in dependence of waveguide width. The BTO layer is assumed to be 70 nm thick and the silicon layers are 110 nm thick.

This is due to the fact that the Pockels effect depends linearly on the driving electric field. Therefore, we can extract the incremental change of the effective mode index by fitting the results with a line. In order to build a short modulator, we thus try to maximize the incremental change.

Furthermore, we also note that the sign of the change is reversed when the sign of the driving voltage is reversed. This feature can be exploited when the modulator is realized in a Mach-Zehnder interferometer (MZI) configuration. By modulating both arms of the MZI with reversed voltage signs, the modulator length can be reduced by a factor of two, because each arm of the MZI will contribute to the total phase change.

In the following we perform a study of the dependence of the achievable refractive index changes when the parameters of the slotted waveguide are changed.

15.2.4.1 Dependence of the Δn on the Waveguide Width

Building on the linear behavior of the effective mode change with voltage we first investigate the change of the effective mode index with waveguide width. We are mainly interested in ridge structures that lead to low propagation and bend loss, as found in the previous section. In order to keep optical loss from the electrodes small, we assume a waveguide-electrode separation of 1 μm.

We scan waveguide widths from 600 nm to 1,500 nm. The results are shown in Figure 15.13. We find that the effective index change decreases with decreasing waveguide width. At a width of 900 nm the incremental change amounts to $5.6 \times 10\ V^{-1}$. In order to realize a modulator with a phase change of π at 1,550 nm wavelength this will require a modulator length of 1.4 mm, when a switching voltage of 1 V is applied to the electrodes.

15.2.4.2 Dependence of the Δn on the Thickness of the BTO Layer

When looking into the dependence of the incremental change on the BTO layer thickness we find the relationship shown in Figure 15.14.

Clearly increasing the BTO layer thickness leads to larger changes in the effective mode index. However, thicker BTO layers are more difficult to fabricate. In addition, the change does not vary too much when the waveguide width is modified. This is illustrated in Figure 15.14 for three waveguide widths from 800 nm to 1,000 nm. A slight increase is observed when the waveguide

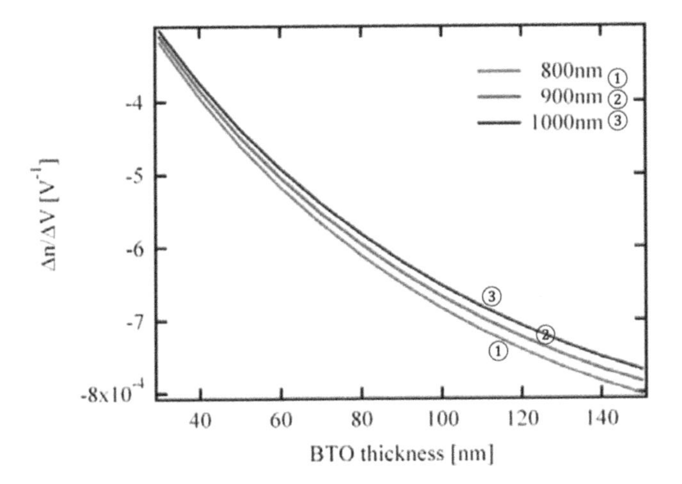

FIGURE 15.14 The calculated incremental change in the effective refractive index of the fundamental mode in dependence of thickness of the BTO layer. The silicon layers are 110 nm thick. The change is calculated for three different waveguide widths ranging from 800 nm.

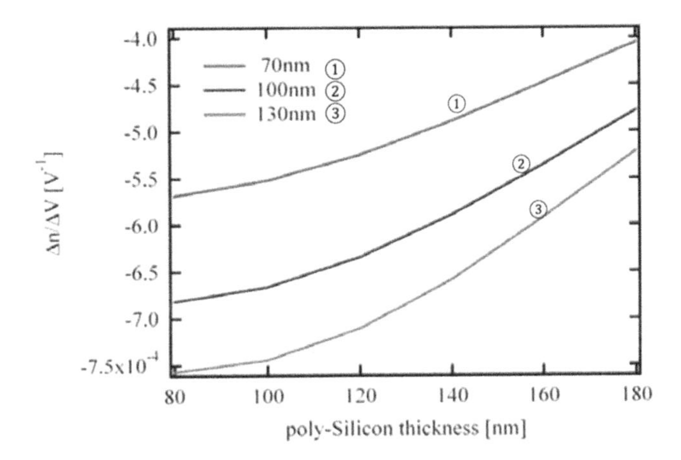

FIGURE 15.15 The calculated incremental change in the effective refractive index of the fundamental mode in dependence of thickness of the poly-silicon. The BTO layer thickness is varied from 70 nm to 130 nm.

width is increased. As a result, we are able to maintain the ideal waveguide width 900 nm (for low propagation loss) without sacrificing incremental change in the refractive index.

15.2.4.3 Dependence of the Δn on the Thickness of the Poly-Silicon Layer

Last, we investigate the dependence on the thickness of the top α-silicon layer. This layer is grown on top of the BTO layer by PECVD. As a result, thinner silicon layers are desirable in order to avoid damage to the underlying wafer structure. The results are shown in Figure 15.15.

From the simulation it is apparent that higher incremental changes are obtained when the thickness of the α-silicon layer is reduced. This trend is unchanged when the thickness of the BTO layer is increased.

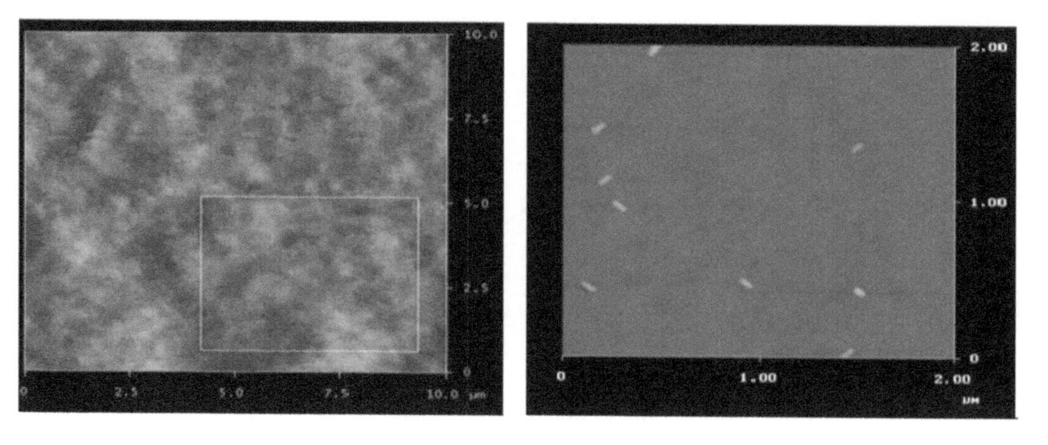

FIGURE 15.16 AFM images of STO(4 nm)/SOI surface (left) and BTO(80 nm)/STO/SOI surface.

15.2.5 Material Growth, Characterization and Device Fabrication

15.2.5.1 Epitaxial Growth

The growth of barium titanate starts with a prototypical system of epitaxial thin films of strontium titanate ($SrTiO_3$; STO) on silicon. Because the waveguide structure adopted is based on SOI substrates, we start the sample preparation from SOI pieces of 25 mm by 25 mm. The substrate undergoes a pre-epitaxy cleaning procedure similar to that used for RCA clean. The substrates first undergo heated bath in Piranha solution to strip off the organic residuals. Then the substrates are dipped in hydrofluoric (HF) acid to strip off the thin oxide layer formed during the Piranha bath. Finally, a fresh native oxide is grown by transferring the substrate promptly for exposure under an ultraviolet (UV) lamp. It is suggested that the thin native oxide layer helps to keep the pristine silicon surface from contamination and is crucial for producing high quality epitaxial thin films.

In ultrahigh vacuum, the SOI substrates are heated up to 450°C to drive off the thin native silicon dioxide formed during the UV exposure. Two monolayer of strontium (Sr) is deposited onto the silicon (100) surface of the SOI substrate. Then the substrate is annealed at 650°C for 15 min to desorb the strontium oxide. This Sr desorption helps to reduce the forming of surface islands. The substrate is subsequently cooled down to 550°C for preparation of the $SrTiO_3$ epitaxy. The Sr flux is tuned on until the ½ ml Sr reconstruction is seen. The substrates then further cool down to room temperature and 1 ml of Sr is added. Oxygen flow is tuned on at this point to oxidize the SrO_2 to $SrTiO_3$. The temperature is raised to 350°C for 15 min for the STO to fully crystallize. After the preparation of the initial layer of STO, further 8 ml of STO is epitaxially grown.

The 10 ml of STO (roughly 4 nm) templates provides a lattice matched perovskite surface for the BTO to deposit. As discussed in the above sections, thicker BTO films will allow for larger electro-optic modulation. In our experiments describe below, we use a BTO film thickness of 80 nm since this thickness ensures a significant EO overlap and a high film quality.

In Figure 15.16, *ex-situ* atomic form microscopic images are taken for the surface of STO grown on the SOI substrates and the final surface morphology of the 80 nm BTO. Figure 15.16 shows that the STO surface has RMS roughness as low as 0.4 nm. On the surface of the 80 nm BTO film, there are islands with dimension on the orders of 100 nm × 50 nm × 50 nm (L × W × H) scattered on the surfaces. The origins of the islands are still under investigation, but it is speculated that the islands are related to the initial silicon dioxide drive-off process and the forming of different crystalline phases in the films. The footprint of the islands is small compared with the

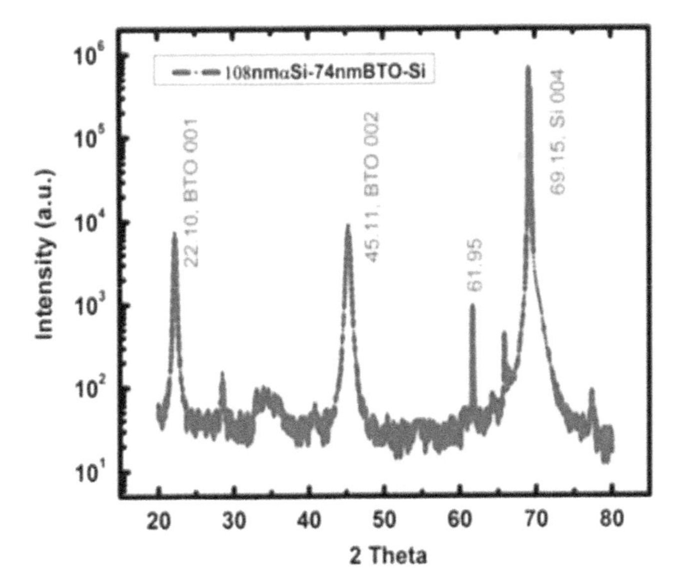

FIGURE 15.17 XRD scanning of BTO thin film stack.

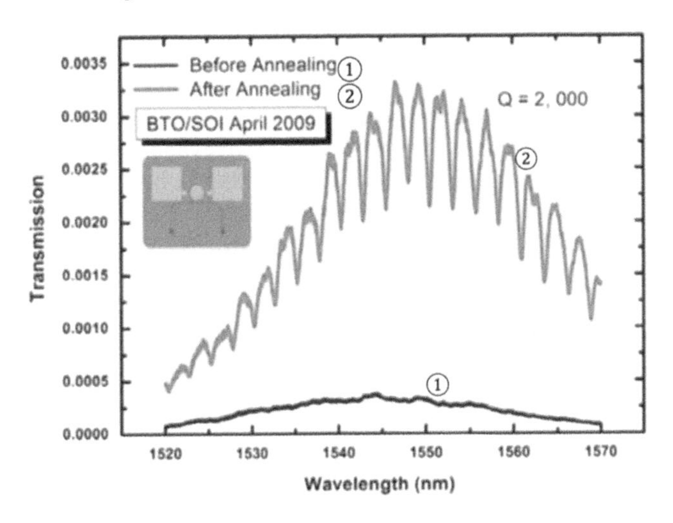

FIGURE 15.18 Effects on oxygen annealing on the optical loss of the BTO waveguide.

optical wavelength of concern (1,550 nm), and hence the scattering from those islands are likely to be minor.

Figure 15.17 shows a typical X-ray diffraction (XRD) θ-2θ scanning measurement of the BTO films. The results show that the films have its c-axis in-plane. Because of the mismatch in thermal expansion coefficient of the silicon and BTO, the long axis (c-axis) of the BTO will tend to align in-plane to relax the strain. As we have discussed in the previous sections, the c-axis in-plane orientation is in fact critical for the optimization of the EO modulation.

15.2.5.2 Oxygen Annealing

Grown in ultrahigh vacuum, the epitaxial perovskites normally have a great concentration of oxygen vacancies inside the films. The vacancies can lead to the change of the transport characteristics of the films and also leads to excessive increase of optical absorption. To fully oxide the film out of the chamber, ex-situ annealing at elevated temperature around 700°C

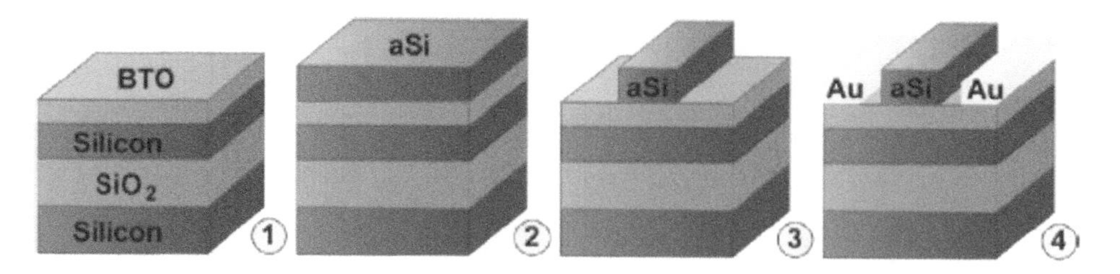

FIGURE 15.19 Process flow for the BTO waveguide modulator

for 30 min can significantly reduce the amount of the oxygen vacancies in the films. In Figure 15.18, we illustrate the effect of the oxygen anneal on the optical absorption of the BTO thin films. The measurement was carried out on a same micro-ring resonator based on BTO horizontal slot waveguide before and after that device is annealed in a tube furnace with oxygen flow at 700°C for 30 min. Before the annealing, no optical resonances can be observed. After the annealing the total optical transmission improves by approximately ten-fold and the optical resonances can be measured to be around 2,000.

15.2.5.3 Device Fabrication

After the BTO is grown onto the SOI substrates, the next step is to deposit the amorphous silicon (α-Si) layer on top of the BTO. We choose an α-Si thickness of 110 nm to match that of the SOI device layer. The α-Si layer is deposited by PECVD using silane (SiH_4) as process gas. In order to prevent the hydrogen produced in the PECVD step from causing the BTO to reduce and form additional oxygen vacancies, we evaporate an additional 13 nm thick of alumina (Al_2O_3) as a capping layer over the BTO. It has been reported that the alumina can serve as a hydrogen diffusion barrier (HDB) which is effective in preventing ferroelectrics from been reduced in ferroelectric memory processing. The addition of the Al_2O_3 (refractive index n = 1.7) does not interfere with the normal operation of the original optical waveguide design.

After the deposition of the α-Si on top of the alumina, we spin coated the substrates with a negative tone ebeam resists HSQ of roughly 100 nm. The optical patterned is transferred to the αSi layer in a chlorine ICP etcher. The titanium/gold (5 nm/100 nm) electrodes are patterned

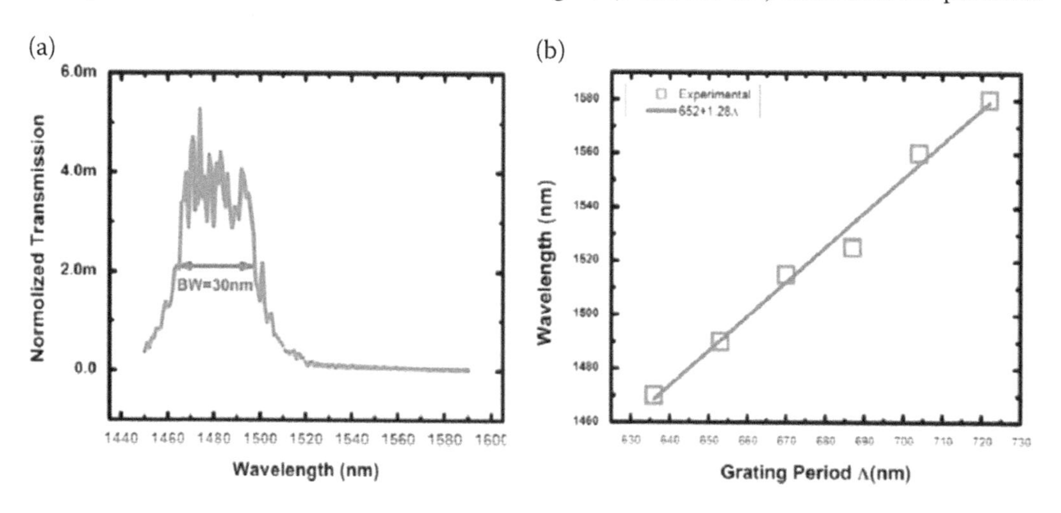

FIGURE 15.20 (a) A typical transmission spectrum a pair of grating couplers made in BTO horizontal slot waveguide. (b) The linear relationship between the central passing wavelength versus the grating period of the couplers.

by the side of the waveguides using a lift-off procedure. The full process flow is illustrated in Figure 15.19.

15.2.6 Experimental Results

15.2.6.1 Grating Couplers

The measurement results (Figure 15.20) from the fabricated calibration coupler pairs show consistent linear relationships between the coupler central wavelength and the grating period and a relatively large bandwidth (3 dB, 30 nm). The typical measured loss per grating coupler is 10 dB.

15.2.6.2 Micro-Ring Resonators

The measured resonances in the one of the micro-ring devices with a coupling gap of 100 nm as illustrated in Figure 15.21 indicates a loaded quality factor Q about 7,000. Using the relation $\alpha = 10\log_{10}e\ 2\pi\ n_g / Q_{int}\lambda$ (where Q_{int} is the intrinsic quality factor, λ is the wavelength and n_g is the group index), we determine a propagation loss α = 44 dB/cm.

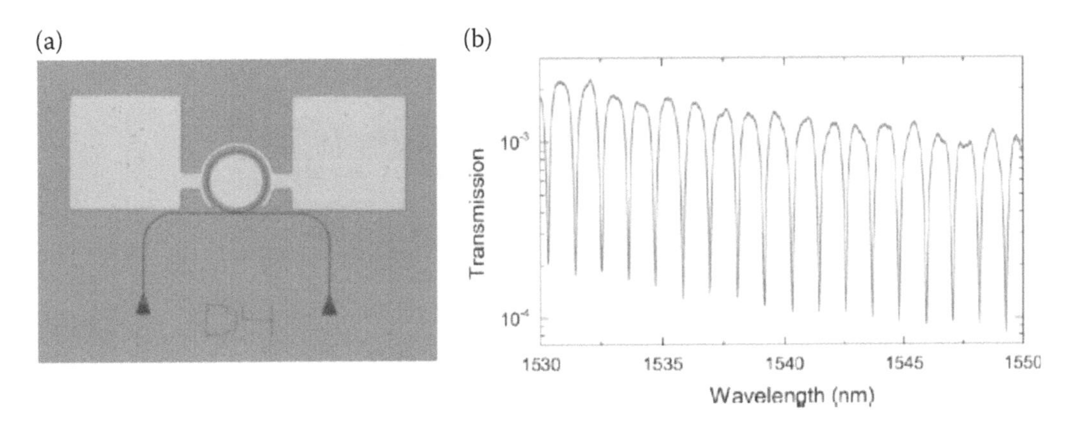

FIGURE 15.21 (a) Micrograph of a BTO microring modulator. (b) Optical transmission spectra of the microring modulator.

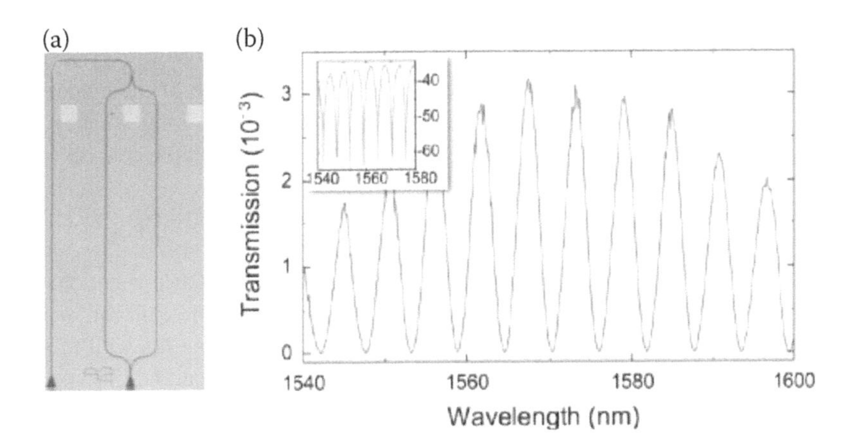

FIGURE 15.22 (a) An optical micrograph of a BTO MZI modulator. (b) Maximum optical transmission as a function of the interferometer length.

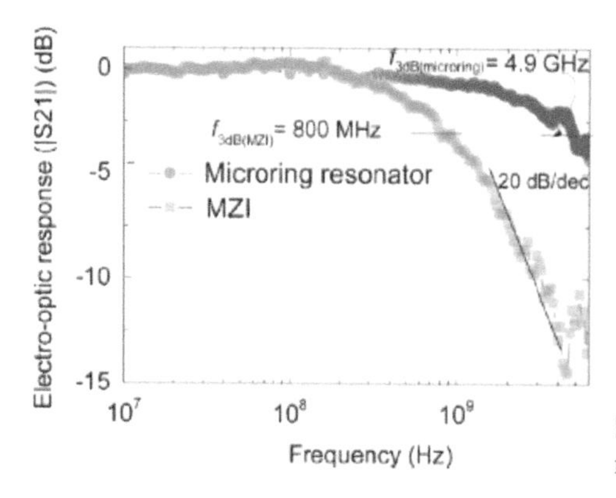

FIGURE 15.23 Frequency response of a BTO MZI modulator (red) and ring modulator (blue).

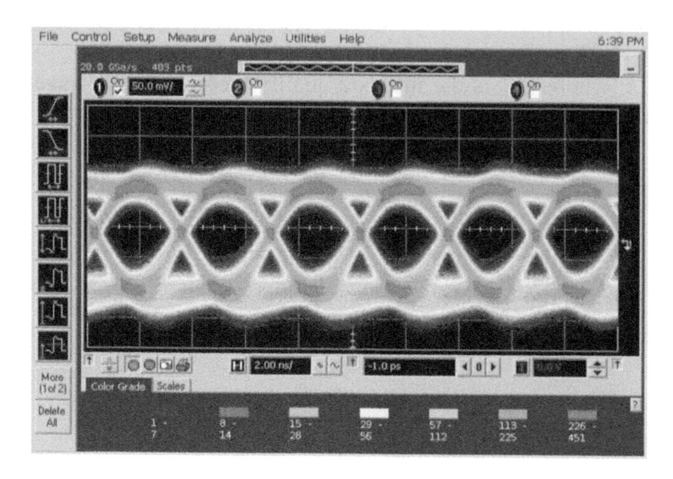

FIGURE 15.24 Eye diagram measurement of the BTO MZI modulator for a 300 Mb/s non-return-zero PRBS 231-1 sequence.

15.2.6.3 Mach-Zehnder Interferometers

In this section, we explore another configuration for the electro-optic modulators. Mach-Zehnder interferometers (MZIs) have two balanced arms, are generally more immune to environmental fluctuations, and can enable wide band operations (Figure 15.22).

15.2.6.4 Frequency Response of the BTO Electro-Optic Modulators

Using a network analyzer, we can characterize the frequency response of the electro-optic modulation. The 3-dB bandwidth of the MZI modulator discussed above is at a frequency of 800 MHz (Figure 15.23). For the frequency range being considered, the modulator can be modeled as a lumped-node device.

Using an LCR meter, we measure the electrode capacitance per unit length of electrode to be C = 10 pF/mm, resulting in an RC time constant limited 3 dB bandwidth of $f_{RC(MZI)}$ = 1.3 GHz for the MZI devices and $f_{RC(ring)}$ = 6.4 GHz for the ring devices. Both estimations are slightly greater than our measured bandwidths, suggesting factors such as RF loss in the metal also contribute to the roll-off. As shown in Figure 15.23, the |S21| roll-off slope fits well to a -20 dB per decade trend predicted by a single-pole RC circuit model. We can see that the ring modulator has a greater bandwidth compared to the MZI modulator. This increased bandwidth is because the ring modulator has a shorter device length and hence smaller capacitance.

Applying a time domain pseudorandom bit sequence (PRBS) to the modulator electrode and sending the output light into an InGaAs photoreceiver with a bandwidth of 125 MHz, we are able to characterize the performance of the modulators in time domain. The eye diagram is shown in Figure 15.24, which is taken at a data rate of 300 Mb/s.

15.2.7 Summary

One of the challenges in developing a new functional material for integrated optics is that the new material's optical properties must first be optimized. In the case of the BTO modulators, the optical absorption in the BTO films is now considered to be the limiting factor preventing us from measuring lower propagation loss than 44 dB/cm. To further reduce the oxygen vacancies induced losses, more aggressive oxidizing recipes need to be experimented. One approach is to anneal the films in a stronger oxidizing gas such as nitrogen dioxide (NO_2). In addition, high-density oxygen plasma annealing in an inductively coupled plasma generator can also be a solution given that positive results have been achieved in obtaining fully oxygenate BTO grown on germanium substrates. Some recent progress has been achieved in reducing the BTO waveguide loss to as low as 6 dB/cm (Eltes et al. 2016). We anticipate that the optical absorption of BTO can still be further reduced. In this way, the switching voltage can be reduced with less than one volt and BTO electro-optic modulators can find use in applications demanding low power and high-speed operations in a compact device footprint.

15.3 INTEGRATED PHOTONIC CIRCUITS IN GALLIUM NITRIDE AND ALUMINUM NITRIDE

Photonic integrated circuits offer excellent prospects for high speed complex optical applications on a chip scale. During the past decade, silicon-on-insulator (SOI) has emerged as one of the most promising integrated optics platforms due to the high refractive index contrast it provides, which allows for building ultra-compact devices (Vlasov, Green, and Xia 2008) (Levy et al. 2010). However, silicon has a narrow indirect bandgap (1.1 eV) and centrosymmetric crystal structure, which not only limits its operation to wavelengths above 1,100 nm but also precludes important active functionalities, such as light emission, optical nonlinearity and the linear electro-optic (Pockels) effect. $\chi^{(2)}$ nonlinearity is particularly important for wavelength conversion, while the electro-optic effect is commonly employed for high-speed modulation (Guarino et al. 2007) (O'Brien 2007).

In this section, we will present our approaches to implement nonlinear crystal especially III-nitride materials on silicon for nonlinear optical applications as well as to achieve high-efficiency electro-optic modulation with very low power consumption.

TABLE 15.1
Comparison of the electronic and optical properties of Si, GaN, and AlN at 300 K

	Si	GaN	AlN
Crystal structure	Diamond cubic	Wurtzite or zinc-blende	
Refractive index	3.47	2.43	2.11
Bandgap (eV)	1.12	3.4	6.2
Thermo-optic coefficient	1.8×10^{-4}	6.0×10^{-5}	2.3×10^{-5}
Thermal conductivity (W/m.K)	130	130	285

[a]All data refer to T=300 K.

15.3.1 Group III-Nitride Semiconductors as Optical Materials

Group III-nitride semiconductors such as gallium nitride and aluminum nitride are wide bandgap semiconductors with important optoelectronic applications. III-nitrides have outstanding electrical, thermal and optoelectronic properties which enable a broad range of technological applications including high speed and high-power electronics (Mishra, Parikh, and Wu 2002), blue/UV light emitting and laser diodes (Ponce and Bour 1997). GaN is particularly notable for its application in LEDs and AlN is an important material for making piezoelectric resonators.

In Table 15.1, the electronic and optical properties of Si, GaN and AlN are compared. Both nitrides have a large refractive index (n = 2.43 for GaN; n = 2.11 for AlN) at the telecom wavelengths (1,550 nm), which can enable compact-sized photonic waveguides. GaN and AlN can be found in either wurtzite or zinc blende structures, both of which are non-centrosymmetric and can hence give rise to significant second order optical nonlinearity. It has been reported that the GaN and AlN have second-order optical susceptibility on the same order of that of lithium niobate (Sato et al. 2009) (Chowdhury et al. 2003) (Long et al. 1995) (Miragliotta et al. 1993). Compared with silicon, III-nitride semiconductors provide an additional benefit by having a large electronic bandgap (3.4 eV for GaN; 6.2 eV for AlN), and thus it not only provides suppression of two-photon absorption but also allows for wide-band operation from ultra-violet to infrared (IR) wavelengths. Additionally, AlN has a superior thermal conductivity ($\kappa_{AlN} = 285\ W\ m^{-1}\ K^{-1}$) and a small thermo-optic coefficient[21] ($dn_{GaN}/dT = 6.0 \times 10^{-5}\ K^{-1}$, $n_{AlN}/dT = 2.32 \times 10^{-5}\ K^{-1}$). Hence, AlN photonic devices are expected to be more tolerant to temperature fluctuations as well.

15.3.2 Harmonic Generation in GaN Photonic Circuits

15.3.2.1 *GaN Photonic Circuits*

Waveguiding in GaN requires cladding layers of lower refractive index to confine light to the GaN layer. Learning from the success of the silicon-on-insulator (SOI) waveguides, we identify SiO_2 as an ideal material to serve as the cladding of GaN waveguides. However, growth of GaN on a CMOS substrate is extremely challenging due to great mismatch in lattice constants.

Here, we tailored a bonding process that allows us to realize photonic structures in GaN thin films atop silicon dioxide on silicon substrates (GaNOI) (Xiong et al. 2011). Photonic waveguides are realized in the GaN top layer by electron beam lithography and subsequent dry etching in inductively coupled chlorine plasma. In Figure 15.25(a) we show a cross-sectional scanning electron microscope (SEM) image of a cleaved fabricated device.

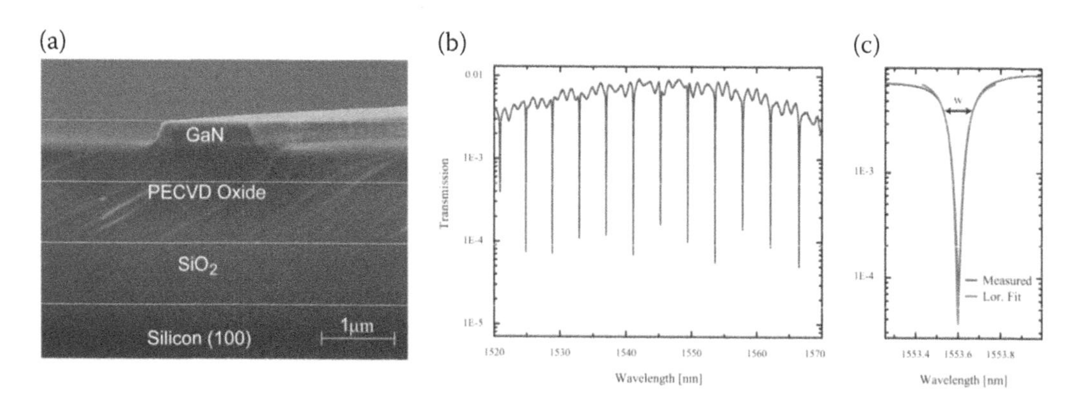

FIGURE 15.25 (a) Cross sectional scanning electron micrographs of a GaN on insulator waveguide. (b) Optical transmission of an GaN microring resonator with radius of 40 μm. (c) Zoomed-in optical resonance near 1,553.6 nm showing an optical Q-factor of 10,000.

We design and fabricate GaN microring resonators with radii of 40 μm, coupled to the input waveguide of 860 nm width, which is separated by a gap of 150 nm from the ring resonator. Such microrings are near critical coupled at wavelengths around 1,550 nm, showing extinction ratio of ~20 dB and typical measured quality factors around 10,000. The typical optical transmission is illustrated in Figure 15.25(b). Figure 15.25(c) shows a zoomed-in of the optical resonance near 1,550 nm.

The optical response of the photonic circuits is obtained by launching light from a tunable IR laser source which is amplified with an erbium doped fiber amplifier (EDFA) and measuring the output light using an InGaAs photodetector. Using the multi-mode matching technique, we are able to achieve the phase matching between the fundamental light and the second harmonic light by tuning the waveguide widths.

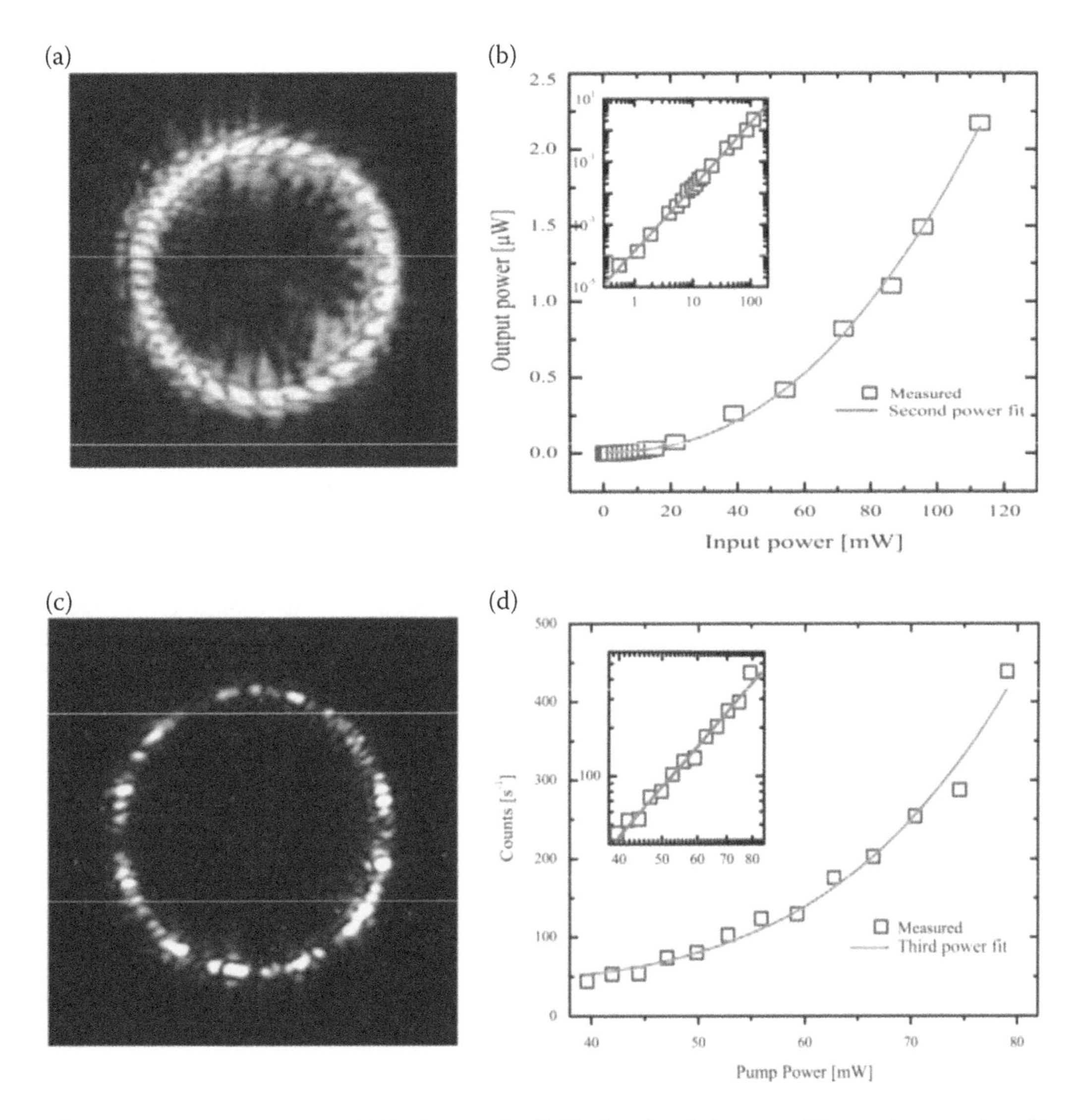

FIGURE 15.26 (a) The captured CCD image of a GaN microring during second harmonic generation. (b) The power of the second harmonic power as a function of the pump power. (c) The captured CCD image of a GaN microring during third harmonic generation. (d) The power of the third harmonic power as a function of the pump power.

15.3.2.2 Second and Third Harmonic Generation from GaN Photonic Circuits

For the properly designed phase-matched GaN waveguides, second harmonic generation can be observed when the waveguides are pumped by a high intensity 1,550 nm light. The wavelength conversion is enhanced in an optical cavity. When the input wavelength is tuned to one of the optical resonances, both the pump power and generated second harmonic inside the optical microring can be greatly enhanced. As shown in Figure 15.26(a), the generated second-harmonic light is strong enough to be observable with a CCD camera. By focusing on the ring with a microscope objective we collect the light scattered out of the microring. By analyzing the collected light with a spectrometer, we verify that the wavelength of the emission corresponds to half the pump laser wavelength.

By monitoring the power dependence of the generated SH around 775 nm we confirm the second order nature of the nonlinear process. In Figure 15.26(b) we plot the power of the SH in dependence of pump power. The expected quadratic dependence of the output power is clearly observed in the data, where the red line is a best fit to a quadratic dependence. By plotting the power dependence on a log-log scale we obtain a best fit to the slope of 2.03 ± 0.02, which is very close to the expected second order scaling law.

Using a bandpass filter around 520 nm, we also observe green light emission from the GaN microring pumped at 1,550 nm, as shown in Figure 15.26(c). The collected green light is sent to a visible spectrometer, and we verify that the light generated has a wavelength of exact one third of the pump wavelength, indicating the occurrence of third harmonic generation. By plotting the power dependence of the third harmonic generation on the fundamental power, we obtain a relationship very close to third order power law as shown in Figure 15.26(d). Because our GaN waveguides are designed to maximize the second harmonic generation efficiency, the observed third harmonic generation is more likely the result of sum-frequency mixing between the second harmonic light and the pump light.

15.3.3 AlN Photonic Circuits and Electro-Optic Modulators

15.3.3.1 *AlN Photonic Circuits*

Compared with GaN, AlN offers more desirable properties such as its larger bandgap, larger thermal conductivity and lower thermal-optic coefficient. Because of its applications in piezoelectric thin film resonators (Piazza, Philip, and Pisano 2006; Mareschal et al. 2010), mature RF magnetron sputtering technology has been developed for full-wafer scale growth of textured AlN

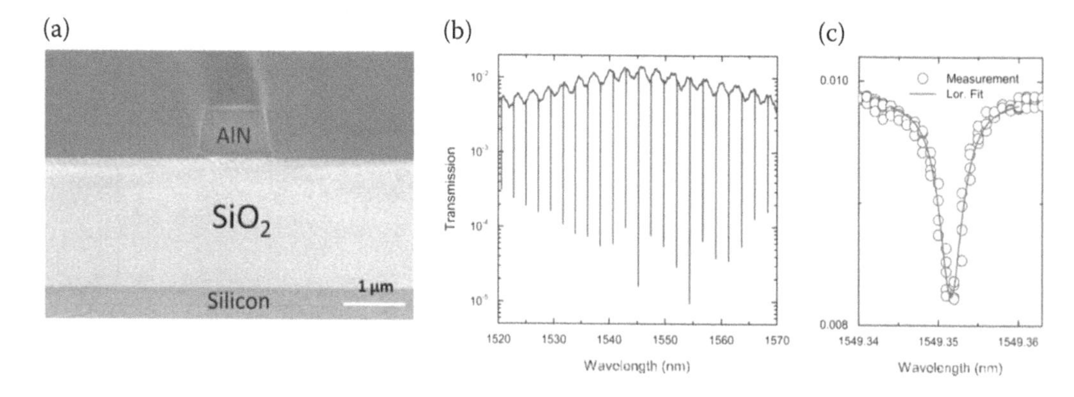

FIGURE 15.27 (a) Cross sectional scanning electron micrographs of AlN on insulator waveguide. (b) Optical transmission of a critically coupled AlN microring resonator with a radius of 40 µm. (c) Zoomed-in optical resonance near 1549.35 nm showing an optical Q-factor of 100,000. The red line is the Lorentzian fitting.

thin films. AlN can be sputter-deposited on a variety of substrates, including silicon, metal and amorphous SiO_2, which ideally lends to the construction of an AlN-on-insulator photonic substrate. The sputtering process can be carried out at ambient temperature, which is CMOS compatible. The AlN sputtering process can avoid many drawbacks of the GaN bonding process by eliminating rough dry-etched surface and issues with surface roughness.

Figure 15.27(a) shows a cross-sectional scanning electron micrograph of the AlN-on-insulator waveguide. We fabricate AlN microring resonators with radii of 40 μm. Critically coupled AlN microrings show extinction ratio of ~20 dB and measured quality factors around 100,000. The typical optical transmission is illustrated in Figure 15.27(b). Figure 15.27(c) shows a zoomed-in of the optical resonance near 1,550 nm. The optical Q-factor of the AlN microring can be used to calculate the propagation loss of the AlN waveguides to be around 0.6 dB/cm, which compares favorably with state-of-the-art SOI waveguides (Nezhad et al. 2011).

15.3.3.2 *AlN Microring Electro-Optic Modulators*

Because of the second order optical nonlinearity in AlN, intrinsic linear electro-optic effect can be found in AlN waveguides. Because Pockels effect is a field effect in nature, it has very high bandwidth and ultralow power consumption compared with the plasma dispersion effects currently employed in silicon optical modulators (Reed et al. 2010). Here in order to utilize the largest electro-optic component of the AlN, we design our AlN waveguide and modulating electrode in such an arrangement so that an effective out-of-plane RF field exists in AlN waveguide.

To introduce an effective out-of-plane electric field (E_Z) overlapping with the guided optical mode, we deposit a layer of SiO2 using plasma-enhanced chemical vapor deposition (PECVD) as a cladding layer on top of the waveguides. A set of ground-signal-ground (GSG) contact pads are placed atop the PECVD oxide. Figure 15.28(a) illustrate the numerically calculated fundamental TE optical mode profile and the electric field distribution (in contour lines). From the finite element simulations, we find with the top electrode configuration a significant Ez component across the AlN waveguide cross section. Since the Ez component of the electric field contributes most effectively to the modulation, it is desirable to fabricate the center electrode close to the AlN waveguides so that a larger electro-optic overlap integral is achieved. Here we define Eavg as the effective out-of-plane RF field inside the AlN waveguide, d as the distance between the center electrode and the AlN optical waveguide, g as the spacing between the edge of the waveguide and

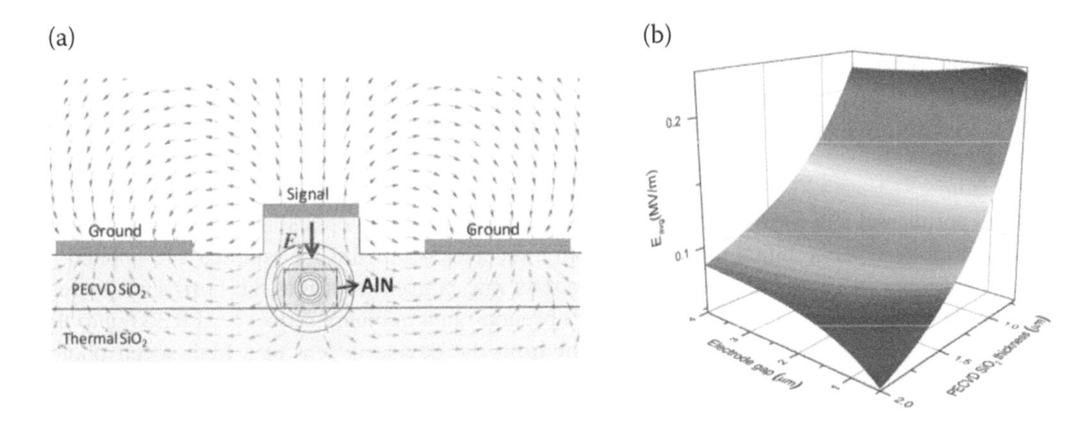

FIGURE 15.28 (a) The simulated RF electric field distribution (arrows) and the optical field (contour plots) distribution. An effective out-of-plane electric field (Ez) is generated. (b) Simulated average effective electric field inside the AlN waveguide as a function of the PECVD SiO_2 thickness and the electrode gap. The simulation was carried out at a fixed deposited SiO_2 thickness of 0.8 μm and +1 V bias voltage on the signal electrode.

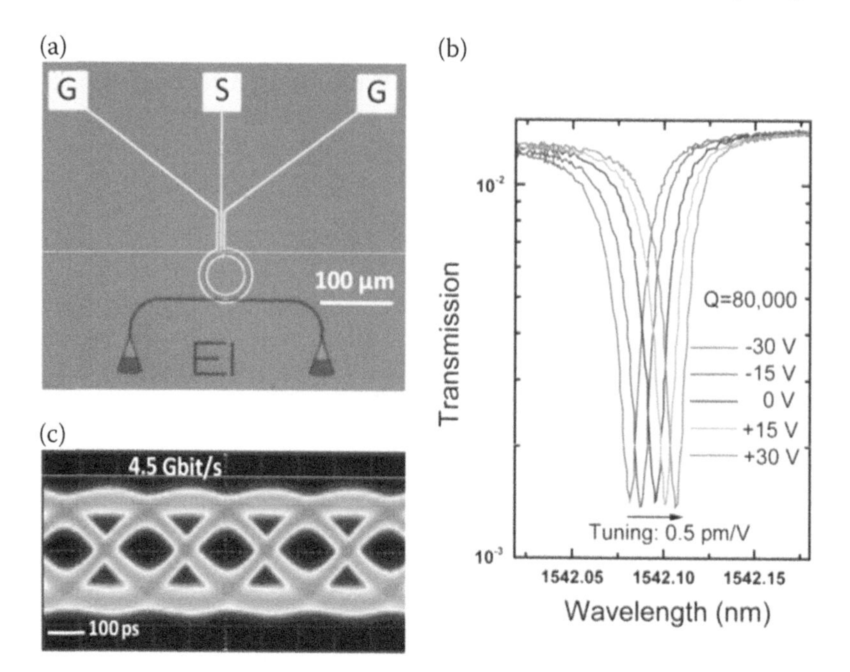

FIGURE 15.29 (a) Optical micrograph of an AlN microring modulator. (b) Tuning of the microring resonance at different DC voltage. (c) 4.5 Gbit/s eye diagram measured on the AlN microring modulator.

the electrode. Figure 15.28(b) shows the simulated average electric field as a function of the electrode gap g and the PECVD SiO_2 thickness *d*. The simulation was carried out at a fixed deposited oxide thickness of 0.8 μm, which ensures negligible absorption of the light by the center electrode. As shown in Figure 15.28(b), the largest average electric field is found when both of the oxide thickness and the electrode gap is the smallest, although the influence of the latter parameter is relatively small.

After the SiO_2 deposition, the GSG metallic Ti/Au electrodes (10 nm/200 nm) were deposited on top of the SiO_2 cladding layer and patterned by a PMMA ebeam lithography recipe and subsequent lift-off procedure. Figure 15.29(a) shows an optical micrograph of a fabricated AlN microring modulator circuit with optical input/output ports in the bottom half and RF electrodes in the top half. When a DC bias voltage is applied on the electrodes, the resonance conditions for the microring resonator will change in response to the modified refractive index. As shown in Figure 15.29(b), we measure a near critically coupled microring TE-mode resonance around 1,542.10 nm with extinction ratio of 10 dB. The resonance shifts 15 pm when the bias voltage is varied from -30 V to +30 V.

An experimental tool for the evaluation of the performance of an optical transmission system is eye diagram, in which a digital data signal from a receiver is repetitively sampled and applied to the vertical input of an oscilloscope, while the data rate is used to trigger the horizontal sweep. Here, we apply 2^{23-1} none-return-to-zero (NRZ) pseudo-random binary sequence (PRBS) onto the AlN microring modulator, and the output light from the device is sent to a high-speed photoreceiver. With a modulating peak-to-peak voltage Vpp of 4 V, we clearly observe open eye at 4.5 Gb/s with an extinction ratio of 3 dB, as shown in the inset of Figure 15.29(c).

15.3.3.3 *Slow Light in AlN 2D Photonic Crystals*

Two-dimensional (2-D) photonic crystals(Benisty et al. 1999), which show arbitrary propagation, can be used in various applications such as optical filters, switches and sensors. Semiconductors such as silicon, gallium arsenide (GaAs) (Combrié et al. 2008) and gallium phosphide (GaP)

(Rivoire, Faraon, and Vuckovic 2008) have been used to realize 2-D photonic crystals operated at telecommunications wavelengths. Recently, photonic crystals based on wide bandgap (Calusine, Politi, and Awschalom 2014) and optically nonlinear materials have attracted interest because they provide not only the suppression of the two-photon absorption at high input power, but also active functionalities such as electro-optic modulation and frequency conversion capabilities. In photonic crystal waveguides, the waveguide photonic dispersion can produce large slow down factors greatly increasing the effective nonlinear interaction length over a broad wavelength range. It has been shown that the circulating intensity in the waveguide is enhanced by the group index n_g, providing a factor of n^2 enhancement in conversion efficiency compared to a conventional waveguide with similar size.

Here we investigate building 2-D photonic crystals on AlN. We calculated a photonic band diagram of the AlN 2-D photonic crystal slab structure using three- dimensional (3-D) photonic bandgap (PBG) simulations. A basic 2-D photonic crystal slab structure consists of a triangular pattern of air holes defined by the lattice constant *a*, as shown in Figure 15.30(a). The radius of the air holes was set to 0.3*a*. The slab thickness was set to 330 nm and the refractive index of AlN was assumed to be 2.1. The calculated band diagram for this structure is shown in Figure 15.30(b). The solid black line is the so- called light line; modes above this line are leaky to air. It is apparent that a photonic bandgap exists in the frequency region from 0.427 to 0.478 (*c/a*). The PBG corresponds to the NIR range of 1,477–1,627 nm, when a = 700 nm. We

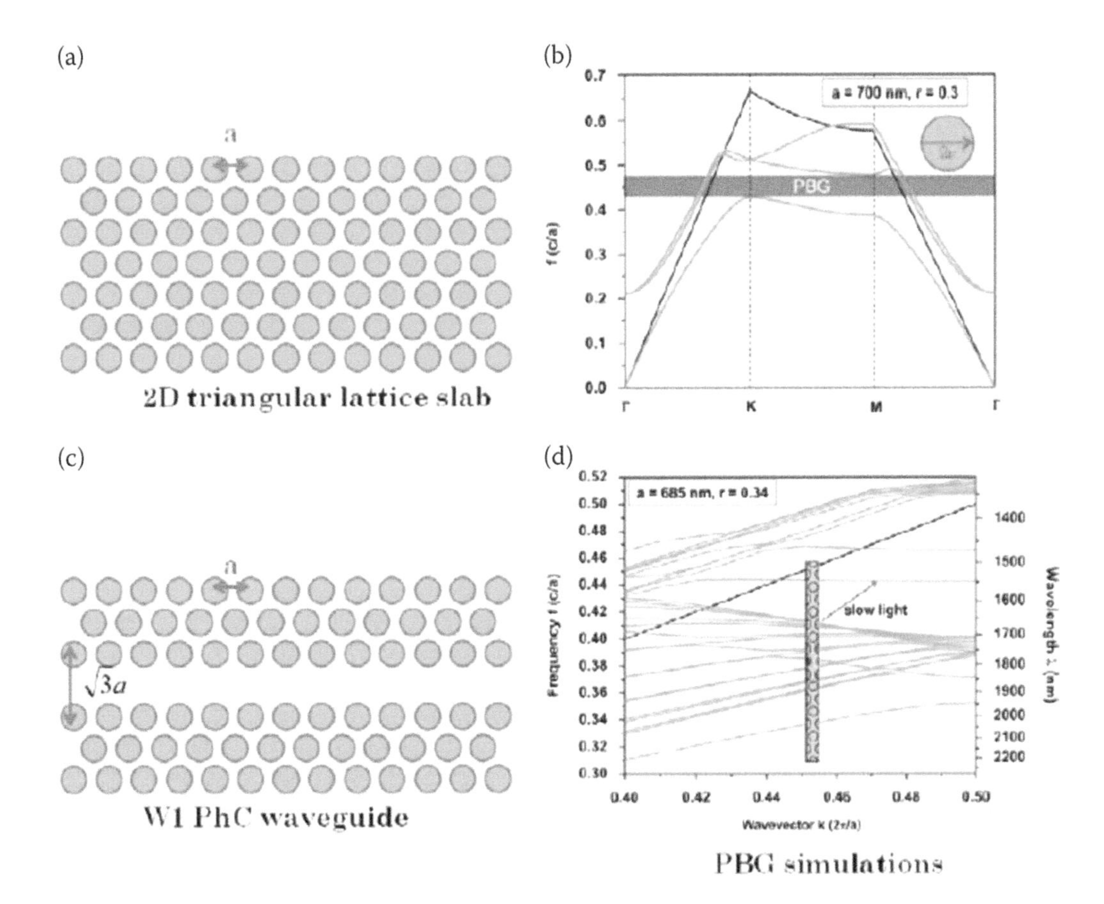

FIGURE 15.30 (a) and (c) schematic drawings of an AlN 2-D photonic crystal structure and its line-defect waveguide. (b) Calculated band diagrams for the structures depicted in (a) and (c). (b) (d) is the calculated band diagrams for the structures depicted in (a) and (c) respectively.

next investigate a line-defect waveguide as shown in Figure 15.30(c). The calculated dispersion relation of this waveguide is shown in Figure 15.30(d). As indicated by line referenced by the arrow, the waveguide formed by a row of missing air holes (width $W = \sqrt{3}a = W1$) has a bandwidth of 2.5 nm in wavelength, which is very narrow compared with that of high refractive index material, such as silicon. By making the radius of the holes smaller, the average refractive index increases. The band will shrink to lower frequency and the bandwidth for the slow light mode will increase. For example, for $r/a = 0.22$, $a = 625$ nm, the PBG simulation predicts a slow mode bandwidth of 25.9 nm centered around 1553 nm.

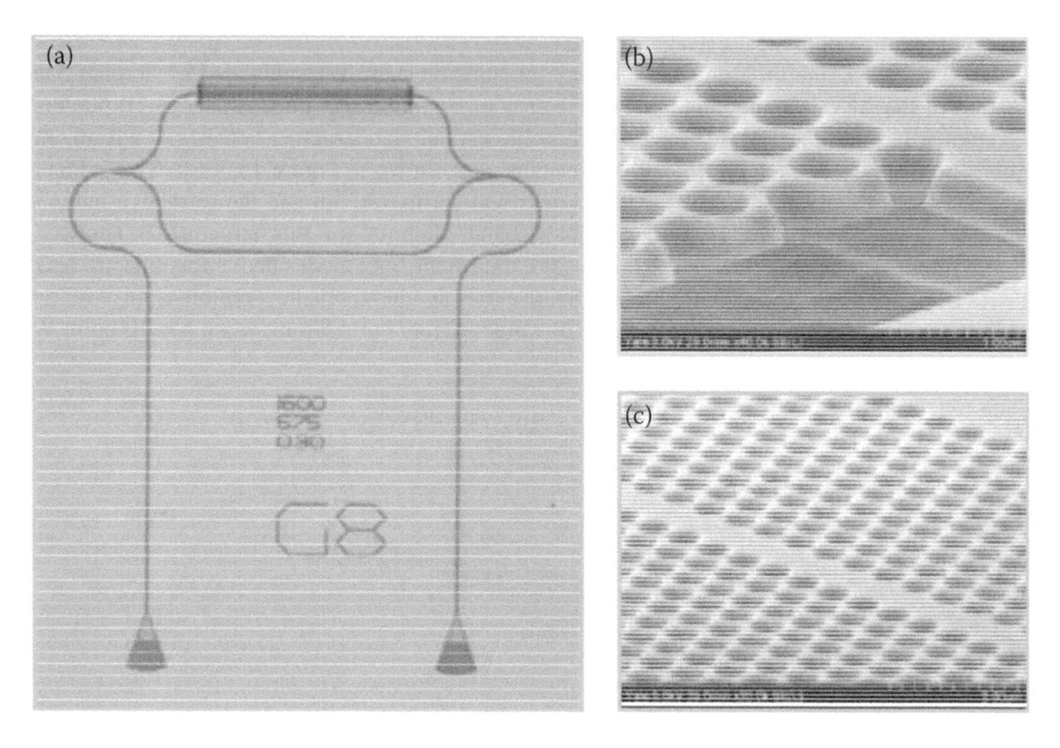

FIGURE 15.31 (a) An optical micrograph of a MZI incorporating 2-D photonic crystal as one arm. (b) and (c) are zoom in electron scanning micrographs of the crystal lattice.

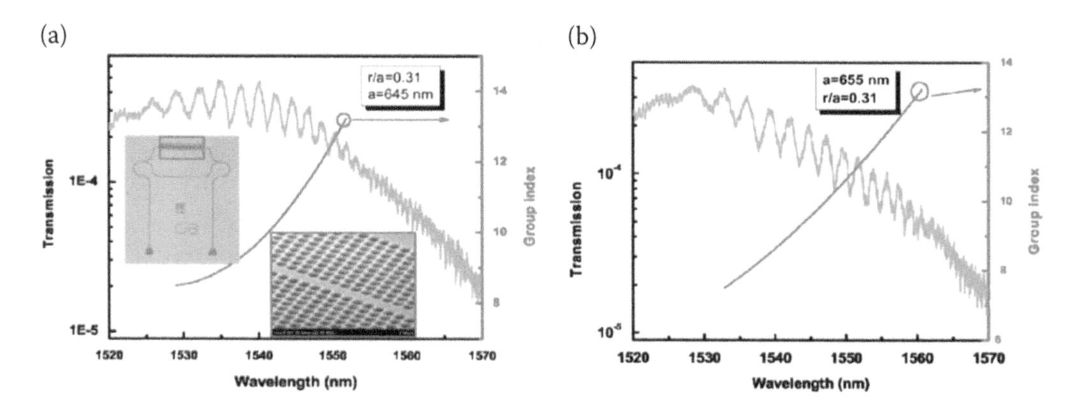

FIGURE 15.32 (a) Optical transmission of a slow light MZI showing a group index increase of seven-fold near the band cutoff near 1,550 nm. The r/a is 0.31. The lattice constant is 645 nm. (b) Optical transmission of a slow light MZI showing a group index increase of seven-fold near the band cutoff near 1,560 nm. The r/a is 0.31. The lattice constant is 655 nm.

By embedding the 2-D photonic crystal into one of the two arms of an MZI, we can extract the group index of the light propagating in the photonic crystal waveguides according to the procedure given in the literature. Figure 15.31(a) shows an optical micrograph of a fabricated MZI with one arm comprising of the photonic crystal waveguide and the other arm the reference straight waveguide. The device was fabricated using the similar two step etching approach adopted for the 1-D photonic crystal cavity to clearly define the areas of release. Figure 15.32 shows typical optical transmission spectra of two AlN slow light interferometers. For c/a=0.31 and a=645 nm, the group index increases by a fold of seven near the slow mode cutoff (Figure 15.32(a)). As predicted by simulations, the slow light band shifts to longer wavelengths when the r/a is fixed constant, while the lattice constant a is decreased (Figure 15.32(b)).

15.3.3.4 *AlN One-Dimensional Nanobeam Photonic Crystal Cavity*

Recently considerable research effort has been devoted to developing high-quality Q factor photonic crystal (PhC) cavities that have dimensions comparable to the wavelength of light (Akahane et al. 2003) (Tanabe et al. 2007). By shrinking the modal volume to near the fundamental limit of V = (λ/2n)3, these cavities have enabled various applications ranging from ultrasmall lasers(Painter et al. 1999) (Park et al. 2004), strong light-matter coupling (Yoshle et al. 2004) (Englund et al. 2007; Srinivasan and Painter 2007) to optical switching (Tanabe et al. 2005). Besides designs in two-dimensional photonic crystal cavities, there has been much interest in cavities realized in suspended nanobeams patterned with a one-dimensional lattice of holes (Sauvan et al. 2005) (Md Zain et al. 2008). These 1D PhC cavities offer exceptional quality factor to modal volume ratios (Q/V) and relative ease of design and fabrication. While such cavity designs have been investigated in stand-alone configurations in which the cavity is read out via free-space optical setups or fiber tapers, their true potential lies in the integratability with nanophotonic circuitry. While integrated photonic circuits have mostly been developed for applications in the telecoms window around 1,550 nm, there are increasing interests in realizing nanophotonic circuits for applications in the near infrared and visible wavelength regime. Therefore, optical materials with wide transparency windows are investigated, such as SiN, GaP, $LiNbO_3$, and III-nitride semiconductors.

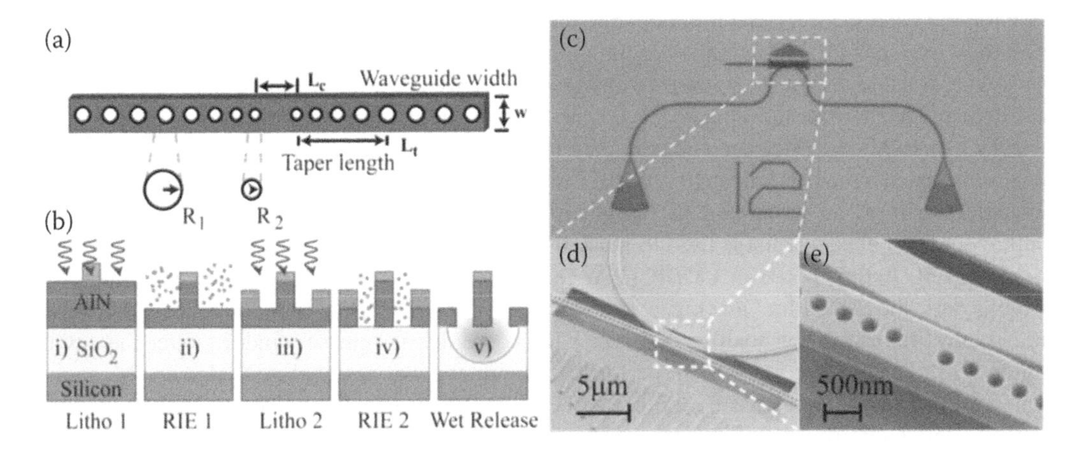

FIGURE 15.33 (a) The design of the one-dimensional AlN photonic crystal cavity. The hole radius is tapered parabolically from R1 to R2. (b) The process flow used to fabricate the PhC devices. (c) An optical micrograph of a fabricated photonic circuit with grating coupler input ports, nanophotonic waveguide and released cavity region. A SEM picture of the released waveguide region, showing the PhC nanobeam as well as the input waveguide. (d) A SEM picture of the released waveguide region, showing the PhC nanobeam as well as the input waveguide. (e) A magnified view of the cavity section of the waveguides.

AlN shows exceptional mechanical and thermal properties and has found wide use in microelectromechanical resonators due to its high piezoelectric transduction efficiency (Cleland, Pophristic, and Ferguson 2001; Lakin 2005). We design and experimentally demonstrate high Q 1D photonic crystal cavities in free-standing AlN nanobeams by employing tapered Bragg mirrors. The cavity is integrated into an on-chip photonic circuit and coupled to a feeding waveguide for convenient optical access. We measure optical quality factors up to 146,000 and achieve high extinction ratio in critically coupled devices. Our AlN platform provides a viable route towards wideband optical applications in an integrated framework.

We investigate the cavity design illustrated schematically in Figure 15.33(a). In our layout the cavity region is enclosed between tapered PhC Bragg mirrors, in which the primary lattice constant of the PhC is tapered down parabolically towards a secondary, smaller lattice constant in the cavity region (Deotare et al. 2009). The cavity is patterned into a free- standing AlN (refractive index of 2.1) nanobeam with a height of 330 nm. The relevant optimization parameters, as shown in Figure 15.33(a), are the length of the taper (L_t), the hole radii at the cavity center (R_2) and in the mirror region (R_1), as well as the waveguide width (w). The whole radii are linked to the lattice constant l_i by keeping a fixed filling ratio, so that $R_i=f^*l_i$. By scanning this multi-dimensional parameter space using a finite- difference time-domain (FDTD) method, we arrive at an optimized design space where the cavity parameters can be investigated experimentally. In the optimized design we use a primary lattice constant of l_1=560 nm which is tapered down to the cavity region parabolically over a length of l_4 lattice periods to a value of l_2=430 nm. The ratio of the hole radius to the PhC lattice constant is kept at f=0.29, while the width of the nanobeam is set to 975 nm. By varying the cavity length in the optimized PhC design intrinsic optical quality factors up to 1.1×10^6 are found numerically. Changing the cavity length, however, also affects the cavity resonance condition, thus allowing for tuning of the resonance wavelength. When the feeding waveguide is included in the simulations, the resulting optical quality factors are reduced depending on the external coupling strength, determined by the gap between the waveguide and the cavity beam. Under critical coupling conditions, the cavity linewidth increases by a factor of roughly 2 and thus the optical Q amounts to half of the intrinsic value.

To confirm the predicted optical properties of the designed PhC structures, nanophotonic devices are fabricated from AlN-on-insulator substrates with a 330 nm thick AlN layer on top of a 2.6 μm thick oxide layer. The film thickness of both the waveguiding layer and the buried oxide layer was chosen such that maximal coupling efficiency of the grating couplers into the feeding waveguide is achieved. In the optimized design constructive interference between incoming light and light reflected from the underlying silicon carrier wafer leads to a maximum coupling efficiency of roughly 30% for each coupler. We use a two-step electron-beam lithography process with subsequent reactive ion etching (RIE) to define the cavity and the supporting on- chip photonic circuitry, as illustrated in Figure 15.33(c). A first iteration of e-beam lithography employing HSQ ebeam resist is used to mask the designed optical circuit against the subsequent etch step (i)). A timed RIE in Cl2/BCl3/Ar inductively coupled plasma is applied to etch the patterns into the AlN film, leaving a residual AlN slab of 70 nm (ii)). In a second e-beam lithography step using the positive resist ZEP520A, we define release windows over the cavity region (iii)). Then a second RIE is performed to remove the residual AlN slab in the cavity region (iv)). During the etch step the cavity beam is still protected by the residual HSQ resist which is left over from the first etch. After removal of the ZEP resist, the 70 nm AlN slab remaining everywhere except for the cavity region provides a natural mask layer against subsequent wet etching. A timed wet etch step in buffered oxide etchant is performed in order to remove the underlying buried oxide and release the nanobeam cavity (v)). The wet etching leads to a sufficient undercut of the cavity beam such that the influence of the substrate is minimized during optical measurements. An optical micrograph of a fabricated sample is shown in Figure 15.33(c), with nanophotonic waveguides as well as alignment markers which indicate the cavity region (vertical bars in the top half of the image). The grating couplers are used to launch light into the feeding waveguide, which is routed to the cavity region as shown in the SEM picture in Figure 15.1(d). A zoom into the cavity region, see Figure 15.33(e), illustrates that the devices can be fabricated with low sidewall roughness and almost vertical angles. The width of the nanophotonic waveguides is 1,000 nm in order to allow for single-mode waveguiding. The width of the cavity beam is 975 nm, which yielded the highest Q values in the numerical simulations.

The optical properties of the fabricated device are investigated by measuring the transmission through the feeding waveguide. Light from a tunable laser source (New Focus 6428) is coupled into the chip using an optical fiber array. The chip is mounted on a motorized stage in order to allow for efficient alignment of the fibers to the grating couplers. After transmission through the chip, the light is collected with a low-noise photodetector (New Focus 2011). In order to optimize

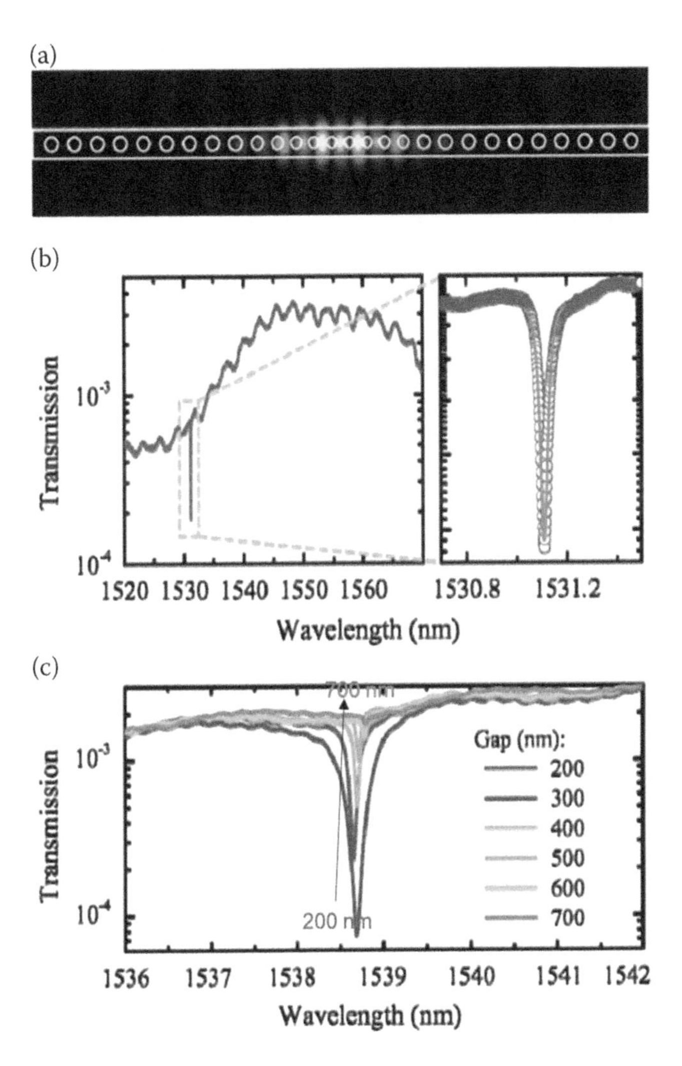

FIGURE 15.34 (a) The simulated optical mode profile for the fundamental cavity mode. Shown is the x-component of the electric field. (b) The measured transmission spectrum of a fabricated device, showing the cavity resonance at 1,530 nm. The Lorentzian fit to the data (red line) in the zoom-in picture on the right reveals an optical Q of 85,000. (c) The measured optical linewidth of the cavity in dependence of the coupling gap to the feeding waveguide.

the coupling to the cavity we fabricate a variety of structures with varying gap between the feeding waveguide and the cavity beam. In the final design, the feeding waveguide is an arc with a radius of 10 um in order to avoid significant bend loss. Because the optical mode profile of the nanobeam is highly concentrated in the cavity region as shown in Figure 15.34(a), the feeding waveguide is designed to only couple to the nanobeam near the point of highest intensity. By measuring the transmission profile of the feeding waveguide the cavity properties can be extracted. A typical result for a device with a coupling gap of 350 nm and a cavity length of 600 nm is shown in Figure 15.34(b). The envelope of the grating coupler's transmission profile exhibits small fringes due to back-reflection from the output grating coupler. A clear dip at 1530.9 nm in the transmission spectrum indicates the cavity resonance wavelength. Fitting the resonance with a Lorentzian function as shown in the detailed dataset of Figure 15.34(b) reveals a loaded cavity Q of 85,000.

In order to evaluate the influence of the feeding waveguide we fabricate several devices with coupling gaps ranging from 200 to 700 nm. The measurement results are shown in Figure 15.34(c). For the smallest coupling gap, we find an overloaded cavity with an expected high extinction ratio greater than 15 dB. By increasing the separation between input waveguide and nanobeam the coupling strength reduces exponentially, which goes in hand with a significant reduction of the cavity linewidth. For the best measured device, we find an optical quality factor of 146,000 at a designed coupling gap of 600 nm. In this case the cavity is operating in the under coupled

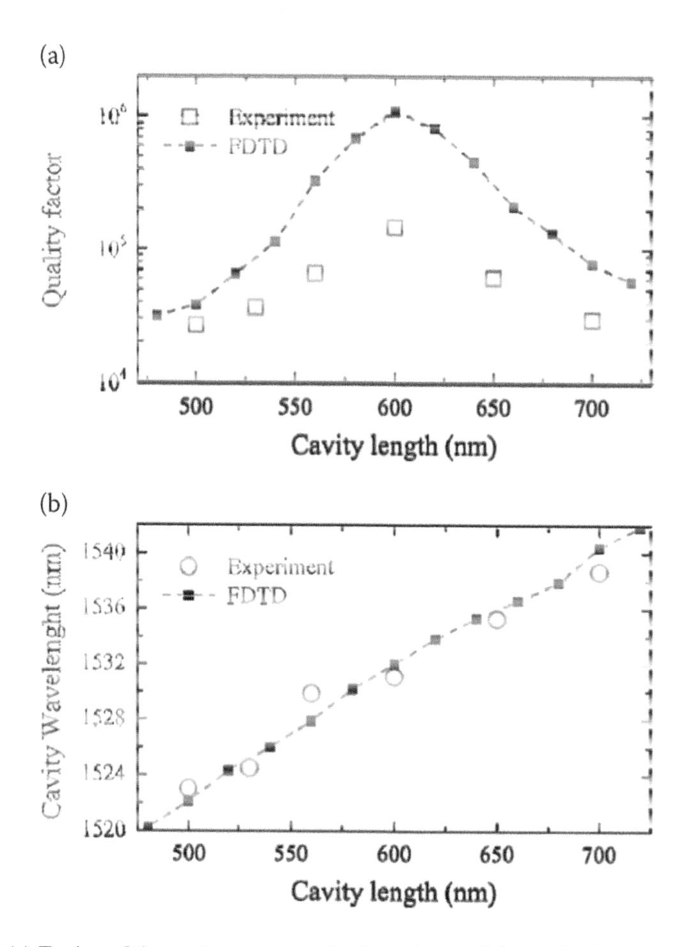

FIGURE 15.35 (a) Tuning of the cavity resonance in dependence of the cavity length. The best optical Q of 146,000 is found for a cavity length of 600 nm. (b) The resonance wavelength increases with cavity length. A length change from 500 to 700 nm shifts the resonance wavelength by 15 nm.

regime. The reduction in coupling strength is accompanied by a cavity red shift as shown in Figure 15.34(c). The red shift is expected due to the decrease of the effective refractive index of the cavity mode when the input beam is moved further away from the cavity region.

To investigate the spectral dependence of the cavity mode on the geometry of the cavity we measure a series of devices with varying cavity length, around the designed optimum gap of 600 nm, as shown in Figure 15.35(a). The experimentally determined Q is indicated by the blue markers, while the numerical results are shown by the dashed lines. The best optical Q is reached for a cavity length of 600 nm, where we measure optical Q of 146,000. When the cavity length is reduced or increased, the optical Q drops significantly. The trend is also confirmed by numerical simulations of the intrinsic cavity design, as shown by the dashed blue line in Figure 15.35(a). When the feeding waveguide is added to the simulation, we expect a critically coupled cavity Q of 340,000 from the FDTD simulations for the optimal cavity length of 600 nm. In the critical coupled case, the intrinsic Q is reduced because the input waveguide provides an additional loss channel to the cavity. The lower quality factor found experimentally is furthermore attributed to the presence of the substrate in the evanescent cavity field and the surface roughness due to fabrication imperfections. On the other hand Figure 15.35(b) illustrates how the resonance wavelength increases with increasing cavity length. For the smallest length of 500 nm we find an optical resonance at 1523 nm while the resonance wavelength shifts to 1,538.7 nm at a cavity length of 700 nm, i.e., by more than 15 nm. The experimentally found results are in good agreement with the numerical predictions, which are shown in Figure 15.35(b) by the dashed red line.

15.4 CHAPTER SUMMARY

In this chapter, we have investigated the electro-optic properties of integrated barium titanate modulators. The barium titanate modulators show promise for realizing low power, low switching voltage modulators on chip as a large effective electro-optic coefficient (100 pm/V) and small voltage device length product (2 V·cm) is demonstrated.

In this chapter, we have explored GaN and AlN thin film as a new material system for building nonlinear optical devices that assist on-chip wavelength conversion and electro-optic modulation. We study the bonded GaN thin films for generating visible second harmonics on chip. The fully CMOS-compatible AlN modulator is a promising candidate for electro-optic signal processing on a silicon photonics platform. AlN-based photonic circuitry could facilitate a plethora of $\chi^{(2)}$ enabled functionalities such as second harmonic generation, parametric downconversion and optical parametric oscillator to be realized on a monolithic In addition, we have realized demonstrated high-Q photonic crystal cavities in free-standing AlN nanobeams with a measured Q factor of 146,000. We also demonstrate a robust 2-D photonic band gap and slow light mode for AlN 2-D photonic crystals despite AlN's relatively small index (n = 2.1 for NIR).

The potential of AlN photonic technologies lies in the fact that wafer-scale deposition of high-quality films can be conducted in a mature procedure compatible with CMOS manufacturing. The nature of the AlN sputtering process also makes it possible to design three dimensional and multi-layer structures which can enable more flexible and efficient integration.

Due to its very wide transparency window, spanning from the mid-IR all the way to visible/UV wavelengths (220 nm~13.6 μm), AlN is a promising candidate for integrated optical applications. Since AlN thin films can be sputter-deposited onto suitable substrates, wafer-scale fabrication is feasible. As shown here, high quality photonic components can be fabricated from AlN with standard nanofabrication routines and are thus compatible with traditional CMOS manufacturing. The potential for seamless integration on silicon substrates using our AlN-on-insulator approach provides a viable route towards a multitude of wideband optical applications. Low propagation loss makes it possible to generate large-area photonic circuits for applications in nonlinear and traditional photonic devices.

REFERENCES

Akahane, Yoshihiro, Takashi Asano, Bong Shik Song, and Susumu Noda. 2003. "High-Q Photonic Nanocavity in a Two-Dimensional Photonic Crystal." *Nature*, *425*(6961), 944–947. https://doi.org/10.1038/nature02063

Alloatti, L., D. Korn, R. Palmer, D. Hillerkuss, J. Li, A. Barklund, R. Dinu, et al. 2011. "427 Gbit/s Electro-Optic Modulator in Silicon Technology." *Optics Express*, *19*(12), 11841. https://doi.org/10.1364/oe.19.011841

Barrios, C. A., B. Sánchez, K. B. Gylfason, A. Griol, H. Sohlström, M. Holgado, and R. Casquel. 2007. "Demonstration of Slot-Waveguide Structures on Silicon Nitride/Silicon Oxide Platform." *Optics Express*, *15*(11), 6846. https://doi.org/10.1364/oe.15.006846

Benisty, H., C. Weisbuch, D. Labilloy, M. Rattier, C. J.M. Smith, T. F. Krauss, Richard M. De La Rue, et al. 1999. "Optical and Confinement Properties of Two-Dimensional Photonic Crystals." *Journal of Lightwave Technology*, *17*(11), 2063–2077. https://doi.org/10.1109/50.802996

Calusine, Greg, Alberto Politi, and David D. Awschalom. 2014. "Silicon Carbide Photonic Crystal Cavities with Integrated Color Centers." *Applied Physics Letters*, *105*(1). https://doi.org/10.1063/1.4890083

Chowdhury, Aref, Hock M. Ng, Manish Bhardwaj, and Nils G. Weimann. 2003. "Second-Harmonic Generation in Periodically Poled GaN." *Applied Physics Letters*, *83*(6), 1077–1079. https://doi.org/10.1063/1.1599044

Cleland, A. N., M. Pophristic, and I. Ferguson. 2001. "Single-Crystal Aluminum Nitride Nanomechanical Resonators." *Applied Physics Letters*, *79*(13), 2070–2072. https://doi.org/10.1063/1.1396633

Combrié, Sylvain, Alfredo De Rossi, Quynh Vy Tran, and Henri Benisty. 2008. "GaAs Photonic Crystal Cavity with Ultrahigh Q: Microwatt Nonlinearity at 155 Mm." *Optics Letters*, *33*(16), 1908. https://doi.org/10.1364/ol.33.001908

Deotare, Parag B., Murray W. McCutcheon, Ian W. Frank, Mughees Khan, and Marko Lončar. 2009. "High Quality Factor Photonic Crystal Nanobeam Cavities." *Applied Physics Letters*, *94*(12). https://doi.org/10.1063/1.3107263

Dong, Po, Shirong Liao, Dazeng Feng, Hong Liang, Dawei Zheng, Roshanak Shafiiha, Cheng-Chih Kung, et al. 2009. "Low V_pp, Ultralow-Energy, Compact, High-Speed Silicon Electro-Optic Modulator." *Optics Express*, *17*(25), https://doi.org/10.1364/oe.17.022484

Eltes, Felix, Daniele Caimi, Florian Fallegger, Marilyne Sousa, Eamon O'Connor, Marta D. Rossell, Bert Offrein, Jean Fompeyrine, and Stefan Abel. 2016. "Low-Loss BaTiO3-Si Waveguides for Nonlinear Integrated Photonics." *ACS Photonics*, *3*(9), 1698–1703. https://doi.org/10.1021/acsphotonics.6b00350

Englund, Dirk, Andrei Faraon, Ilya Fushman, Nick Stoltz, Pierre Petroff, and Jelena Vučković. 2007. "Controlling Cavity Reflectivity with a Single Quantum Dot." *Nature*, *450*(7171), 857–861. https://doi.org/10.1038/nature06234

Fang, Alexander W., Hyundai Park, Oded Cohen, Richard Jones, Mario J. Paniccia, and John E. Bowers. 2006. "Electrically Pumped Hybrid AlGaInAs-Silicon Evanescent Laser." *Optics Express*, *14*(20), 9203. https://doi.org/10.1364/oe.14.009203

Green, William M., Michael J. Rooks, Lidija Sekaric, and Yurii A. Vlasov. 2007. "Ultra-Compact, Low RF Power, 10 Gb/s Silicon Mach-Zehnder Modulator." *Optics Express*, *15*(25), 17106. https://doi.org/10.1364/oe.15.017106

Guarino, Andrea, Gorazd Poberaj, Daniele Rezzonico, Riccardo Degl'Innocenti, and Peter Günter. 2007. "Electro-Optically Tunable Microring Resonators in Lithium Niobate." *Nature Photonics*, *1*(7), 407–410. https://doi.org/10.1038/nphoton.2007.93

Hochberg, Michael, Tom Baehr-Jones, Guangxi Wang, Michael Shearn, Katherine Harvard, Jingdong Luo, Baoquan Chen, et al. 2006. "Terahertz All-Optical Modulation in a Silicon-Polymer Hybrid System." *Nature Materials*, *5*(9), 703–709. https://doi.org/10.1038/nmat1719

Lakin, Kenneth M. 2005. "Thin Film Resonator Technology." *IEEE Transactions on Ultrasonics, Ferroelectrics, and Frequency Control*, *52*(5), 707–716. https://doi.org/10.1109/TUFFC.2005.1503959

Levy, Jacob S., Alexander Gondarenko, Mark A. Foster, Amy C. Turner-Foster, Alexander L. Gaeta, and Michal Lipson. 2010. "CMOS-Compatible Multiple-Wavelength Oscillator for on-Chip Optical Interconnects." *Nature Photonics*, *4*(1), 37–40. https://doi.org/10.1038/nphoton.2009.259

Lin, Q., Oskar J. Painter, and Govind P. Agrawal. 2007. "Nonlinear Optical Phenomena in Silicon Waveguides: Modeling and Applications." *Optics Express*, *15*(25), 16604. https://doi.org/10.1364/oe.15.016604

Liu, Alan Y., Sudharsanan Srinivasan, Justin Norman, Arthur C. Gossard, and John E. Bowers. 2015. "Quantum Dot Lasers for Silicon Photonics." *Photonics Research 3*(5), B1–B9.

Liu, Ansheng, Richard Jones, Ling Liao, Dean Samara-Rubio, Doron Rubin, Oded Cohen, Remus Nicolaescu, and Mario Paniccia. 2004. "A High-Speed Silicon Optical Modulator Based on a Metal-Oxide-Semiconductor Capacitor." *Nature*, *427*(6975), 615–618. https://doi.org/10.1038/nature02310

Long, X. C., R. A. Myers, S. R.J. Brueck, R. Ramer, K. Zheng, and S. D. Hersee. 1995. "GaN Linear Electro-Optic Effect." *Applied Physics Letters*, *67*, 1349. https://doi.org/10.1063/1.115547

Mareschal, Olivier, Sébastien Loiseau, Aurélien Fougerat, Laurie Valbin, Gaëlle Lissorgues, Sebastien Saez, Christophe Dolabdjian, Rachid Bouregba, and Gilles Poullain. 2010. "Piezoelectric Aluminum Nitride Resonator for Oscillator." In *IEEE Transactions on Ultrasonics, Ferroelectrics, and Frequency Control*, *57*(3), 513–517. https://doi.org/10.1109/TUFFC.2010.1441

Md Zain, Ahmad R., Nigel P. Johnson, Marc Sorel, and Richard M. De La Rue. 2008. "Ultra High Quality Factor One Dimensional Photonic Crystal/Photonic Wire Micro-Cavities in Silicon-on-Insulator (SOI)." *Optics Express*, *16*(16), 12084. https://doi.org/10.1364/oe.16.012084

Michel, Jurgen, Jifeng Liu, and Lionel C. Kimerling. 2010. "High-Performance Ge-on-Si Photodetectors." *Nature Photonics*, *4*(8), 527–534. https://doi.org/10.1038/nphoton.2010.157

Miller, David A.B. 2000. "Rationale and Challenges for Optical Interconnects to Electronic Chips." *Proceedings of the IEEE*, *88*(6), 728–749. https://doi.org/10.1109/5.867687

Miragliotta, J., D. K. Wickenden, T. J. Kistenmacher, and W. A. Bryden. 1993. "Linear- and Nonlinear-Optical Properties of GaN Thin Films." *Journal of the Optical Society of America B*, *10*(8), 1447. https://doi.org/10.1364/josab.10.001447

Mishra, Umesh K., Primit Parikh, and Yi Feng Wu. 2002. "AlGaN/GaN HEMTs - An Overview of Device Operation and Applications." *Proceedings of the IEEE*, *90*(6), 1022–1031. https://doi.org/10.1109/JPROC.2002.1021567

Nezhad, Maziar P., Olesya Bondarenko, Mercedeh Khajavikhan, Aleksandar Simic, and Yeshaiahu Fainman. 2011. "Etch-Free Low Loss Silicon Waveguides Using Hydrogen Silsesquioxane Oxidation Masks." *Optics Express*, *19*(20), 18827. https://doi.org/10.1364/oe.19.018827

O'Brien, Jeremy L. 2007. "Optical Quantum Computing." *Science*, *318*(5856), 1567–1570. https://doi.org/10.1126/science.1142892

Painter, O., R. K. Lee, A. Scherer, A. Yariv, J. D. O'Brien, P. D. Dapkus, and I. Kim. 1999. "Two-Dimensional Photonic Band-Gap Defect Mode Laser." *Science*, *284*(5421), 1819–1821. https://doi.org/10.1126/science.284.5421.1819

Pantouvaki, M., P. Verheyen, J. De Coster, G. Lepage, P. Absil, and J. Van Campenhout. 2015. "56 Gb/s Ring Modulator on a 300 mm Silicon Photonics Platform." In *European Conference on Optical Communication, ECOC, 2015-November*. https://doi.org/10.1109/ECOC.2015.7341888

Park, Hong Gyu, Se Heon Kim, Soon Hong Kwon, Young Gu Ju, Jin Kyu Yang, Jong Hwa Baek, Sung Bock Kim, and Yong Hee Lee. 2004. "Electrically Driven Single-Cell Photonic Crystal Laser." *Science*, *305*(5689), 1444–1447. https://doi.org/10.1126/science.1100968

Piazza, Gianluca, Philip J. Stephanou, and Albert P. Pisano. 2006. "Piezoelectric Aluminum Nitride Vibrating Contour-Mode MEMS Resonators." *Journal of Microelectromechanical Systems*, *15*(6), 1406–1418. https://doi.org/10.1109/JMEMS.2006.886012

Ponce, F. A., and D. P. Bour. 1997. "Nitride-Based Semiconductors for Blue and Green Light-Emitting Devices." *Nature*, *386*(6623), 351–359. https://doi.org/10.1038/386351a0

Reed, G. T., G. Mashanovich, F. Y. Gardes, and D. J. Thomson. 2010. "Silicon Optical Modulators." *Nature Photonics*, *4*(8), 518–526. https://doi.org/10.1038/nphoton.2010.179

Rivoire, Kelley, Andrei Faraon, and Jelena Vuckovic. 2008. "Gallium Phosphide Photonic Crystal Nanocavities in the Visible." *Applied Physics Letters*, *93*(6), https://doi.org/10.1063/1.2971200

Sato, Hiroaki, Ichiro Shoji, Jun Suda, and Takashi Kondo. 2009. "Accurate Measurements of Second-Order Nonlinear-Optical Coefficients of Silicon Carbide." In *Materials Science Forum*, *615–617*, 315–318. https://doi.org/10.4028/www.scientific.net/MSF.615-617.315

Sauvan, C., G. Lecamp, P. Lalanne, and J. P. Hugonin. 2005. "Modal-Reflectivity Enhancement by Geometry

Tuning in Photonic Crystal Microcavities." *Optics Express, 13*(1), 245. https://doi.org/10.1364/opex.13.000245

Srinivasan, Kartik, and Oskar Painter. 2007. "Linear and Nonlinear Optical Spectroscopy of a Strongly Coupled Microdisk-Quantum Dot System." *Nature, 450*(7171), 862–865. https://doi.org/10.1038/nature06274

Streshinsky, Matthew, Ran Ding, Yang Liu, Ari Novack, Yisu Yang, Yangjin Ma, Xiaoguang Tu, et al. 2013. "Low Power 50 Gb/s Silicon Traveling Wave Mach-Zehnder Modulator near 1300 Nm." *Optics Express, 21*(25), 30350. https://doi.org/10.1364/oe.21.030350

Sun, Rong, Po Dong, Ning-ning Feng, Ching-yin Hong, Jurgen Michel, Michal Lipson, and Lionel Kimerling. 2007. "Horizontal Single and Multiple Slot Waveguides: Optical Transmission at λ = 1550 Nm." *Optics Express, 15*(26), 17967. https://doi.org/10.1364/oe.15.017967

Tanabe, Takasumi, Masaya Notomi, Eiichi Kuramochi, Akihiko Shinya, and Hideaki Taniyama. 2007. "Trapping and Delaying Photons for One Nanosecond in an Ultrasmall High-Q Photonic-Crystal Nanocavity." *Nature Photonics, 1*(1), 49–52. https://doi.org/10.1038/nphoton.2006.51

Tanabe, Takasumi, Masaya Notomi, Satoshi Mitsugi, Akihiko Shinya, and Eiichi Kuramochi. 2005. "All-Optical Switches on a Silicon Chip Realized Using Photonic Crystal Nanocavities." *Applied Physics Letters, 87*(15), 1–3. https://doi.org/10.1063/1.2089185

Tran, Minh, Duanni Huang, Tin Komljenovic, Jonathan Peters, Aditya Malik, and John Bowers. 2018. "Ultra-Low-Loss Silicon Waveguides for Heterogeneously Integrated Silicon/III-V Photonics." *Applied Sciences 8*(7), 1139. https://doi.org/10.3390/app8071139

Vlasov, Yurii, William M.J. Green, and Fengnian Xia. 2008. "High-Throughput Silicon Nanophotonic Wavelength-Insensitive Switch for on-Chip Optical Networks." *Nature Photonics, 2*(4), 242–246. https://doi.org/10.1038/nphoton.2008.31

Wooten, Ed L., Karl M. Kissa, Alfredo Yi-Yan, Edmond J. Murphy, Donald A. Lafaw, Peter F. Hallemeier, David Maack, et al. 2000. "Review of Lithium Niobate Modulators for Fiber-Optic Communications Systems." *IEEE Journal on Selected Topics in Quantum Electronics, 6*(1), 69–82. https://doi.org/10.1109/2944.826874

Xiong, Chi, Wolfram Pernice, Kevin K Ryu, Carsten Schuck, King Y Fong, Tomas Palacios, and Hong X Tang. 2011. "Integrated GaN Photonic Circuits on Silicon (100) for Second Harmonic Generation." *Optics Express 19*(11), 10462–10470. https://doi.org/10.1364/oe.19.010462

Xu, Qianfan, Bradley Schmidt, Sameer Pradhan, and Michal Lipson. 2005. "Micrometre-Scale Silicon Electro-Optic Modulator." *Nature, 435*(7040), 325–327. https://doi.org/10.1038/nature03569

Yoshle, T., A. Scherer, J. Hendrickson, G. Khitrova, H. M. Gibbs, G. Rupper, C. Ell, O. B. Shchekin, and D. G. Deppe. 2004. "Vacuum Rabi Splitting with a Single Quantum Dot in a Photonic Crystal Nanocavity." *Nature, 432*(7014), 200–203. https://doi.org/10.1038/nature03119

Young, Ian A., Edris Mohammed, Jason T.S. Liao, Alexandra M. Kern, Samuel Palermo, Bruce A. Block, Miriam R. Reshotko, and Peter L.D. Chang. 2010. "Optical I/O Technology for Tera-Scale Computing." In *IEEE Journal of Solid-State Circuits, 45*(1), 235–248. https://doi.org/10.1109/JSSC.2009.2034444

Zgonik, M., P. Bernasconi, M. Duelli, R. Schlesser, P. Günter, M. H. Garrett, D. Rytz, Y. Zhu, and X. Wu. 1994. "Dielectric, Elastic, Piezoelectric, Electro-Optic, and Elasto-Optic Tensors of BaTiO3 Crystals." *Physical Review B, 50*(9), 5941–5949. https://doi.org/10.1103/PhysRevB.50.5941

Zheng, Xuezhe, Eric Chang, Philip Amberg, Ivan Shubin, Jon Lexau, Frankie Liu, Hiren Thacker, et al. 2014. "A High-Speed, Tunable Silicon Photonic Ring Modulator Integrated with Ultra-Efficient Active Wavelength Control." *Optics Express, 22*(10), 12628. https://doi.org/10.1364/oe.22.012628

Section V

Emerging Computing Technologies and Applications

16 Neuromorphic Silicon Photonics

S. Bilodeau, T. Ferreira de Lima, C. Huang, B. J. Shastri, and P. R. Prucnal

CONTENTS

16.1 INTRODUCTION

Neural networks have enjoyed renewed popularity over the last decade under the appellation of "deep learning" [1,2]. The idea of mimicking the brain to process information, however, can be traced back half a century prior to Rosenblatt's perceptron [3], and the first experimental models of biological neurons to Hodgkin and Huxley a few years prior [4]. The artificial neurons that make up neural networks take many forms, some more closely related to this biological inspiration. Yet all neural networks take the form of simple nodes that (a) perform a linear operation on multiple other neurons' outputs, (b) integrate the resulting signals, and (c) perform a nonlinear transformation on the summed, weighted inputs. Various interconnection topologies—feedforward, feedback (recurrent), close-neighbor translationally-invariant (convolutional), etc.—endow the network with different computational properties.

Such an asynchronous, parallel framework is at odds with the digital von Neumann architecture that electronic microprocessors often employ for their emulation. This mismatch was recognized early on, leading to pioneering work by VLSI engineers starting in the 1980s to map the physics of transistors to neuronal models for gains in computational density, energy efficiency, and speed [5]. However, Moore's Law and Dennard scaling kept such "neuromorphic" architecture outside of the limelight in favor of general-purpose digital processors. Today, this scaling nears its end, and researchers turn to ever more specialized hardware such as graphical processing units [6], tensor processing units [7], and specially-configured field-programmable gate arrays [8] to run demanding neural network models. This is renewing interest in neuromorphic application-specific integrated circuits (ASICs), the extrapolated conclusion of this trend.

Since the requirements of neuromorphic hardware differ from von Neumann digital computing, it is not obvious that silicon microelectronics must provide the best substrate for neuromorphic ASICs [9,10]. The reliance of neural networks on simple networked nodes suggests that a platform suited for communications, such as photonics, might have an advantage. This was recognized in the 80s [11], yet the lack of integrability limited investigations at the time. The commercial silicon

DOI: 10.1201/9780429292033-16

photonic platforms that have arisen over the last few years, however, now offer high index contrast, low-loss waveguides integrated with high bandwidth optoelectronics for signal modulation and detection [12]. Furthermore, the reuse of materials and processes from microelectronics allows the platform to enjoy its economies of scale. This, combined with the intrinsic appeal of photonics to emulate neural models, is one of the reasons that the newly termed field of *neuromorphic photonics* has attracted considerable attention [10,13–17].

In this chapter, we summarize past work and outline future directions of neuromorphic photonics, with a focus on silicon photonics implementations. We take a hierarchical approach. First, suitable neuronal models and the instantiation of their components is reviewed. Networking techniques to achieve neural networks proper is then discussed. Next, we review applications of neuromorphic photonics. We finish with a short outlook.

16.2 SILICON PHOTONIC NEURONS

A major impetus of the resurgence of neuromorphic photonics in the 2010s was the recognition that the dynamics of some active photonic components are mathematically equivalent to leaky integrate-and-fire spiking neuron models [18]. Silicon's indirect bandgap precludes efficient light sources and amplifiers that would easily allow such a quantum-level spiking neuron model to be implemented. This can be overcome by, for example, depositing optically active films like phase-change materials [19] or by combining emission from emissive centers with single-photon detectors at cryogenic temperatures [20].

There are, however, alternate and easier to program neuron models that lend themselves almost perfectly to the room-temperature high-bandwidth optoelectronics of silicon photonic. Such continuous artificial neurons encode information in an analog property of the light instead of spike timing:

$$\mathrm{ds(t)/dt} = \mathbf{W}\cdot \mathrm{f}(\mathbf{s}(\mathrm{t})) - \mathrm{s(t)}/\tau + \mathrm{w_{in}u(t)}. \qquad (16.1)$$

Here, a neuron's internal state s drives others through a continuous (non-spiking) nonlinear transfer function. This model has two important components: (1) matrix-vector multiplication $\mathbf{W} \cdot \mathbf{x}$ (equivalently multiply-accumulate (MAC) operations) between a neuron's inputs x and its weights w, the non-biological equivalent of a synapse, and (2) a nonlinear transformation of the input state to the broadcast output $f(s)$. τ captures the time constant of the nonlinear unit, and u is the external drive. Stripped of temporal components, this reduces to the non-dynamical artificial neuron model ubiquitous in deep learning for the neuron's output $f(s) = f(\mathbf{W} \cdot f(\mathbf{s}) + w_{in}u)$. This artificial neural network model, while simple, has been immensely successful in applications and is almost exclusively used. Implemented in silicon photonics, it offers a way to do complex neural computation with nanosecond latencies, opening up a wealth of new application domains.

16.2.1 Multiply-Accumulate Operation

Two broad philosophies have been explored for multiply-accumulate operations in silicon photonics: coherent and incoherent. In the coherent framework pictured in Figure 16.1a-b, beamsplitters and phase shifters control the interference of light of a well-defined wavelength, mode, and polarization. When meshed appropriately, any unitary transformation can be performed on a path-encoded coherent input beam, directly implementing matrix-vector multiplication at the speed of light in the waveguide. This approach was originally considered for linear photonic quantum information processing, and was demonstrated with a mesh of Mach-Zehnder modulators [22]. Since the coherent approach is isomorphic to a vector-matrix multiplication, summing occurs naturally.

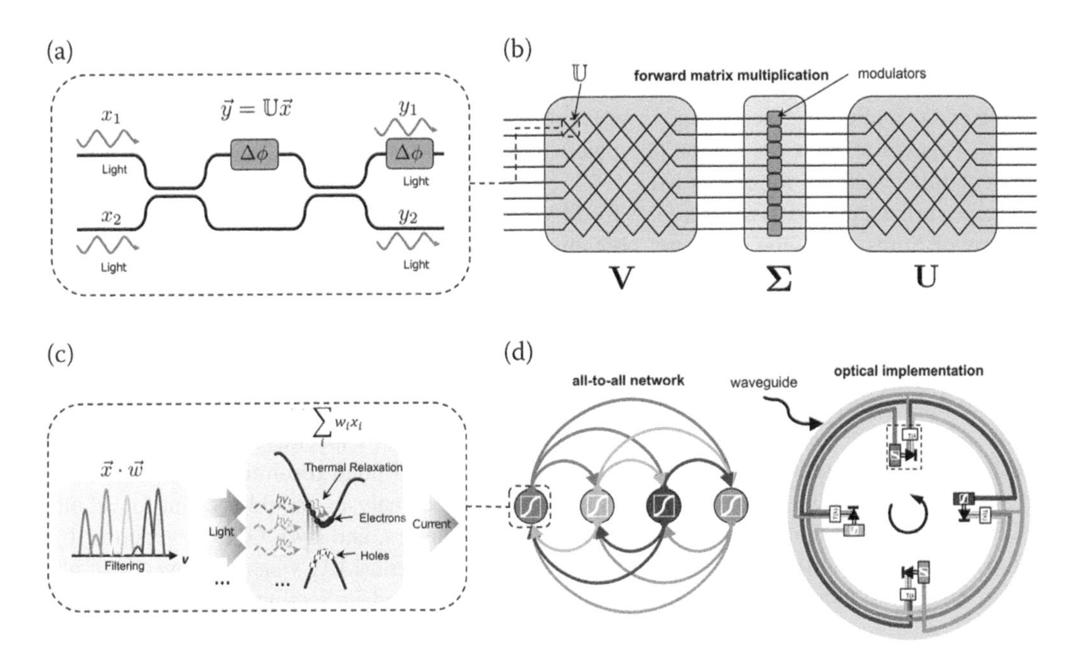

FIGURE 16.1 Coherent (top) and incoherent (bottom) multiply-accumulate operations. (a) Coherent approaches can apply unitary rotation to incoming lightwaves. This unit can perform a tunable 2 × 2 unitary rotation denoted by U. (b) Example of scaling the system to perform a matrix operation in a feedforward topology, using a U unit at each crossing together with singular value decomposition. (c) Incoherent approaches can directly perform dot products on optically multiplexed signals. However, they rely on detectors and O/E conversion for summation. (d) The ability to multiplex allows for network flexibility, which can enable larger-scale networks with minimal waveguide usage. Figure and caption adapted from [21], with permission

The incoherent approach, in contrast, is presented in Figure 16.1c-d. There, values are represented through the relative intensities of light in a collection of wavelengths, modes and/or polarization. Linear operations are performed through selective filtering and/or attenuation. In this approach, a single photodetection step implements summing by yielding a photocurrent proportional to the sum of the optical powers across the incoming modes. Termed "broadcast-and-weight", this is explored in [23] and first demonstrated in silicon by controlling the transmission of microring resonators in [24]. Resonators can also only be used for filtering and followed by electro-absorption modulators to achieve the same effect [25].

To be able to adjust a neuron's weights, the beamsplitters, phase shifters, filters, and attenuators mentioned above must be tunable. This constitutes the main source of complexity in a photonic neural network, since e.g., a fully connected network of N neurons will require N^2 weights to be controlled. Silicon exhibits a strong thermo-optic effect, and local metal or doped heaters are often used for "slow" index changes to implement this reconfigurability. Phase-change materials have also been considered for non-volatile control of transmission [26,27]. A technique worth mentioning and used for incoherent networks is resonator photoconductive control. First applied to silicon photonic neuron weights in [28], this technique leverages the measurable photoabsorption-induced change of a doped ring resonator's resistance to "lock" the filter transmission to the desired point. Using photoresistance as a proxy for optical power further has the advantage of not requiring access to the optical signal for calibration, enabling large-scale actuation [29]. Control of this transmission is what ultimately limits the effective fixed-point bit resolution that can be achieved for the multiply-accumulate operation, an important metric for comparison with digital alternatives. Currently, record 7.2 bits of accuracy and precision were demonstrated [29], close to the 8-bit reduced precision popular in digital deep learning, and above some electronic neuromorphic architectures (for instance, IBM TrueNorth's 4 + 1 (sign) bits [30]). In practice, controlling and reconfiguring these weights will require a full-scale neuromorphic photonic processing system. This will include the silicon photonic chip itself, copackaged with requisite laser sources, microcontroller, and RF interfaces [31].

The performance of passive photonics for MAC operations was compared to electronics in [21]. In terms of limits of analog compute, photonics shares a similar implementation strategy as resistive (or memristive) crossbar arrays, with one part of the MAC held fixed (weight) and the other fast changing (input). This leads to similar fundamental limits. For state-of-the-art crossbar and photonic components, on-chip aJ/MAC efficiencies and 100s of PMACs/s/mm compute densities are possible. A photonic core scales better in terms of energy per MAC and compute density for (1) >100 μm core sizes, since crossbars see reduced bandwidth with length unlike waveguides, (2) for >500 channels due to the $O(N)$ scaling of photodetector capacitance compared to $O(N^2)$ crossbar capacitance scaling, and (3) for low (<4 bits) of fixed-point resolution, due to photonics having extra (shot) noise increasing with power. Future such analyses should account for all other power consumption (control electronics, memory access, lasers, and E/O, O/E conversions if required). In any case, the end-to-end latency of passive photonic MACs can be lower than electronics, leading to unique application areas that will be explored in Section 16.3.

16.2.2 Nonlinear Transformation

The core of the neuron is its internal dynamics, leading to the nonlinear transformation it performs on its weighted, summed inputs. For path-encoded coherent beams, in theory any optical nonlinearity could be used to perform an all-optical nonlinear transformation. Materials deposited on waveguides such as phase-change [32] have demonstrated such functionality. They, however, lack reconfigurability once fabricated. This can be remedied with tunable silicon photonic devices at the cost of footprint [33,34]. The all-optical nonlinear approaches above require high optical powers, however, and so local opto-electronic conversions with a tap and detector that self-modulates phase were demonstrated [35]. The incoherent approach, on the other hand, already relies on photodetection for summing. The photocurrent can be used to actuate a wavelength (or mode, or polarization)-selective amplitude modulator, whose transfer function implements an effective optical-optical nonlinearity. In [36], this is achieved with a microring modulator driven by a photodetector output and is reproduced in Figure 16.2. While the full microring transfer function is

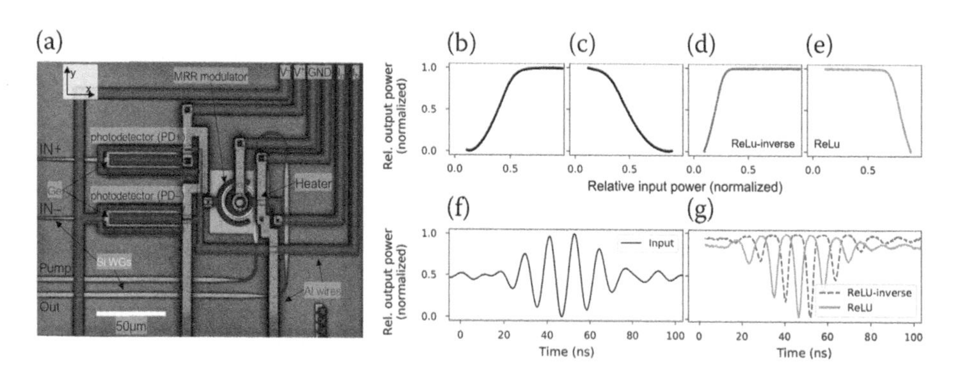

FIGURE 16.2 Example of a silicon photonic continuous nonlinear unit and its experimentally measured transfer function. (a) False-color confocal micrograph of a fabricated neuron comprising a balanced photodetector pair and electro-optic modulator. (b-e) A variety of relevant O-E-O transfer functions seen from the PD-modulator pair, taken at different bias conditions: (b,c) sigmoids; (d,e) rectified linear units ("ReLU"). (f,g) Time resolved pictures of transfer functions: (f) the input is a 40-ns burst of a 100-MHz carrier; (g) both ReLUs. Figure and caption adapted from [36], with permission.

characterized by a Lorentzian, proper restriction of the current swings allows emulation of popular deep learning activation functions such as rectified-linear (ReLu), sigmoid, and quadratic (not shown). Similar to the weighting case, electro-absorption modulators were also proposed for this purpose [37].

An important property of the nonlinear unit is its cascadability. Physical cascadability requires that the output of a neuron be compatible with its input i.e., that the nodes described be networked. The incoherent units of Figure 16.2 do this by matching the resonator modulator's wavelengths to the resonant filters used for the MAC operations. Gain cascadability quantifies if the output of a neuron suffices to drive all of its fan-out. The specific amount of gain required within a single nonlinear unit ultimately depends on signal levels and network characteristics such as fan-in, fan-out, and level of attenuation by the weights. Finally, for cascadability to be present from the point of view of noise, we require a signal-to-noise ratio >1 after the signal has propagated through the network. This means the neuronal nonlinearity must counteract the amplitude and phase noise due to e.g., imperfect control elements that are present in both coherent and incoherent approaches.

A quantified measure of cascadability can be obtained through an autapse, or self-connection, experiment. Such experiments have been performed in both laser spiking [38] and silicon modulator neurons [36]. For the forward-biased pn-junction modulator neuron of Figure 16.2, for instance, a minimum optical pump power of $2V_\pi/\pi R_{pd}R_b$ is required to have a gain larger than unity, and this is seen in experiments. The balance of signal degradation and noise trimming from the Lorentzian transmission in a pn-junction system with realistic component values was studied theoretically [39]. The results yielded over 50 dB of calculated signal-to-noise after an arbitrarily large network, hinting that such a system offers an amount of cascadability from the point of view of noise.

16.3 SILICON PHOTONIC NEURAL NETWORKS AND APPLICATIONS

Given a cascadable nonlinear unit (or layer) that can take in and perform weighted summation of input signals, neural networks proper can be considered. Coherent layers, by preserving their path-encoding scheme, are straightforwardly cascaded to form a deep feedforward network, although the mesh routing must be maintained [22]. For incoherent neurons where every node emits on its own channel (wavelength, mode, polarization), all of which can exist in the same single physical waveguide, the "broadcast-and-weight" protocol was introduced in [23]. The O/E/O conversion can be leveraged to isolate different waveguides on the same chip, enabling spectrum reuse. Which such "broadcast loop(s)" a neuron outputs into determines the network topology, with outputs back to a neuron's inputs allowing recurrence. The specific network instantiation given this physical topology, which neuron connects to which and how strongly, is then dictated by the synaptic weight described in Section 16.2.1. For microring-based neurons as previous-discussed, the microring resonator filters can be assembled into what are called weight banks as displayed in Figure 16.3a [40]. Temporal multiplexing can be used to go beyond limited amount of hardware. Fast implementation of convolutional neural networks has been proposed this way by [41,42].

An approach related to recurrent neural networks operating at the network level, called reservoir computing, is briefly mentioned here since it was also demonstrated in silicon photonics [43]. The idea is to create a network with (semi) random connections such that it exhibits nontrivial dynamics, send time-series data through the system, and train e.g., the output layer of this "reservoir" [44,45]. This is attractive since it is easier than training a full recurrent neural network. It is a popular approach for photonics in general [46–48]. While sharing with deep learning the training of network parameters conditional on data, the requirement on the "neurons" as described in the preceding section is relaxed in this case, since the only requirement of the network is that it "lifts" the time-series input to a higher-dimensional space, effectively performing feature extraction. For instance, in silicon photonics, interference in passive structures and nonlinearity from photodetection are found to be enough to successfully perform computing tasks [43,49].

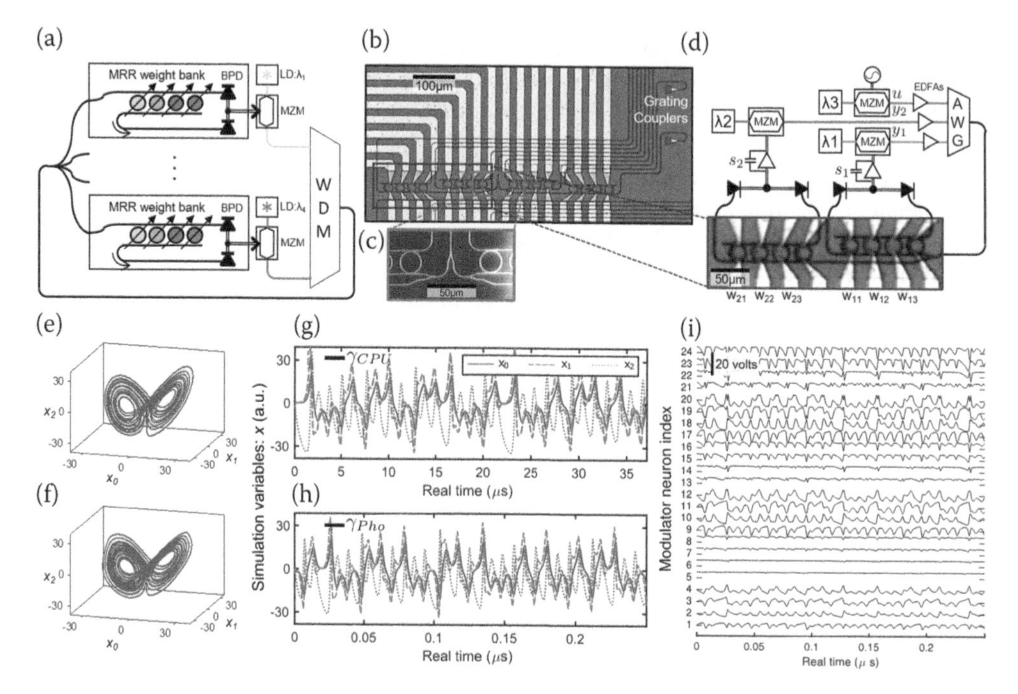

FIGURE 16.3 Photonic neural network benchmarking against a CPU. (a) Concept of a broadcast-and-weight network with modulators used as neurons. MRR: microring resonator, BPD: balanced photodiode, LD: laser diode, MZM: Mach-Zehnder modulator, WDM: wavelength-division multiplexer. (b) Micrograph of 4-node recurrent broadcast-and-weight network with 16 tunable microring (MRR) weights and fiber-to-chip grating couplers. (c) Scanning electron micrograph of 1:4 splitter. (d) Experimental setup with two off-chip MZM neurons and one external input. Signals are wavelength-multiplexed in an arrayed waveguide grating (AWG) and coupled into a 2×3 subnetwork with MRR weights, w_{11}, w_{12}, etc. Neuron state is represented by voltages s_1 and s_2 across low-pass filtered transimpedance amplifiers, which receive inputs from the balanced photodetectors of each MRR weight bank. (e,f) Phase diagrams of the Lorenz attractor simulated by a conventional CPU (e) and a photonic CTRNN (b). (g,h) Time traces of simulation variables for a conventional CPU (g) and a photonic CTRNN (h). The horizontal axes are labeled in physical real time, and cover equal intervals of virtual simulation time, as benchmarked by γ_{CPU} and γ_{Pho}. The ratio of real-time values of γs indicates a 294-fold acceleration. (i) Time traces of modulator voltages s_i (minor y-axis) for each modulator neuron i (major y-axis) in the photonic CTRNN. The simulation variables, x, in (h) are linear decodings of physical variables, s, in (i). Figure and caption adapted from [24], with permission.

16.3.1 Application I: Neural ODE Solver

With an interconnection strategy, networks can be created and actual processing tasks considered. For instance, using the neural compiler Nengo [50], incoherent silicon photonic modulator neurons in a broadcast-and-weight configuration were configured for time series prediction [24]. The neural network emulates three coupled ODEs (Lorenz attractor), with parameters set to produce a chaotic output. The physical device and experimental setup, reproduced in Figure 16.3a-d, is used to extract experimental behavior for a single node. A 24-node network is then emulated, and its performance is favorably benchmarked against a CPU solving the same problem, with a predicted speedup of 294x.

16.3.2 Application II: Nonlinear Programming and Model-Predictive Control

If a neural network can model time dynamics quickly, model-predictive control is an interesting next step. Model-predictive control is a nonlinear scheme that, in opposition to linear control such

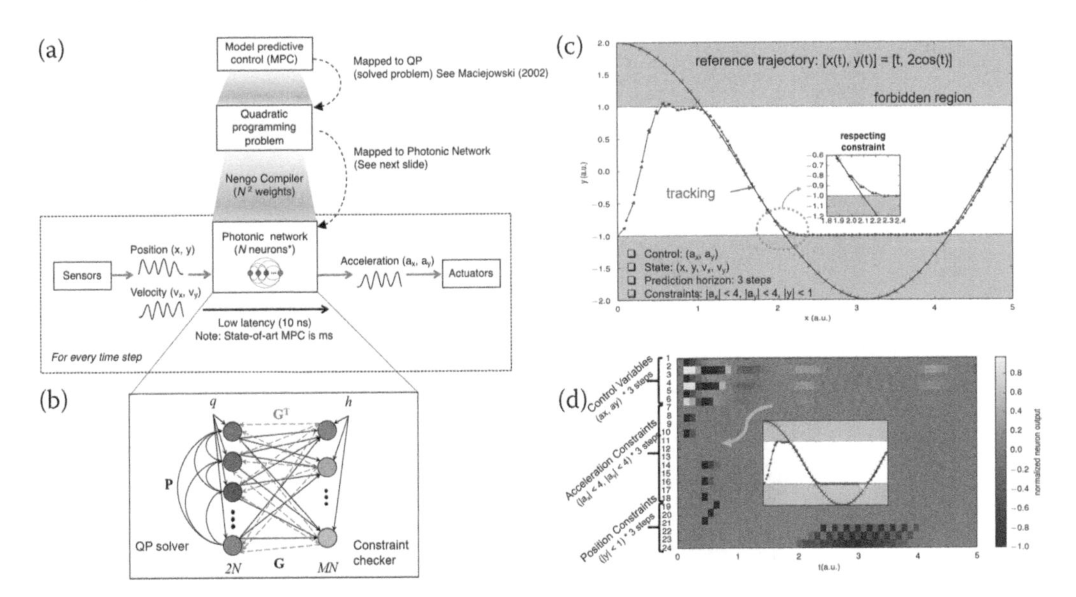

FIGURE 16.4 (a) Schematic figure of the procedure to implement the MPC algorithm on a neuromorphic photonic processor. Firstly, map the MPC problem to QP. Then, construct a QP solver with continuous-time recurrent neural networks (CT-RNN). Finally, build a neuromorphic photonic processor to implement the CT-RNN. (b) Schematic figure of construction of a QP solver with CT-RNN. In this example, $N = 3$, which is the prediction horizon, $M = 6$, which is the number of inequalities, and 2 is the vector dimension. (c) The trajectory of the moving target is shown in the black curve, and the blue dots and blue arrows are the simulated results of the position and velocity of the tracker at each time step respectively. The inset shows that the controller predicts a constraint violation and starts turning the tracker to avoid violating the acceleration's constraint. (d) The "constraint checker" neurons fire around $t = 0.5$ and between $t = 2$ and $t = 4$, inhibiting the output of the "QP solver neurons" such that the outcome of the system does not violate the acceleration and position constraints, respectively. Figure and caption adapted from [31], with permission.

as PID where only the immediate past is responded to, models future trajectories and proactively corrects its course towards an optimal solution that avoids constraints. Electronic processors implementing these algorithms for e.g., chemical plant process these are capped at kHz speeds, whereas photonic neural networks can potentially reach hundreds of MHz, enabling new control regimes [31]. Figure 16.4 shows this process. The task must first be mapped to a quadratic programming problem, which can then be compiled by a neural compiler into a neural network. The network contains a layer dedicated to solving the quadratic problem, and another to enforce the constraints. A photonic neural network can converge to a solution on the order of its neuron timescale, about 10 ns for the neuron of Figure 16.2. Figure 16.4c-d shows a set of 24 neurons solving the model-predictive problem, while predicting in advance if the constraints will be violated and reacting accordingly.

16.3.3 Application III: Intelligent Signal Processing

An interesting use case is when data is already in the optical (and analog) domain. Fiber nonlinearity compensation is such a situation. Figure 16.5 displays microring modulator silicon photonic neurons described in Section 16.2.2 used to learn the transmission characteristics of a 10,080 km trans-pacific transmission link. A two-layer feedforward network of such units is trained to compensate the nonlinear transmission impairment and achieves a Q-factor improvement of 0.51 dB. The results with the network trained with experimental data are only 0.06 dB from a numerical simulation with the same parameters. Since the neuron operates at comparable

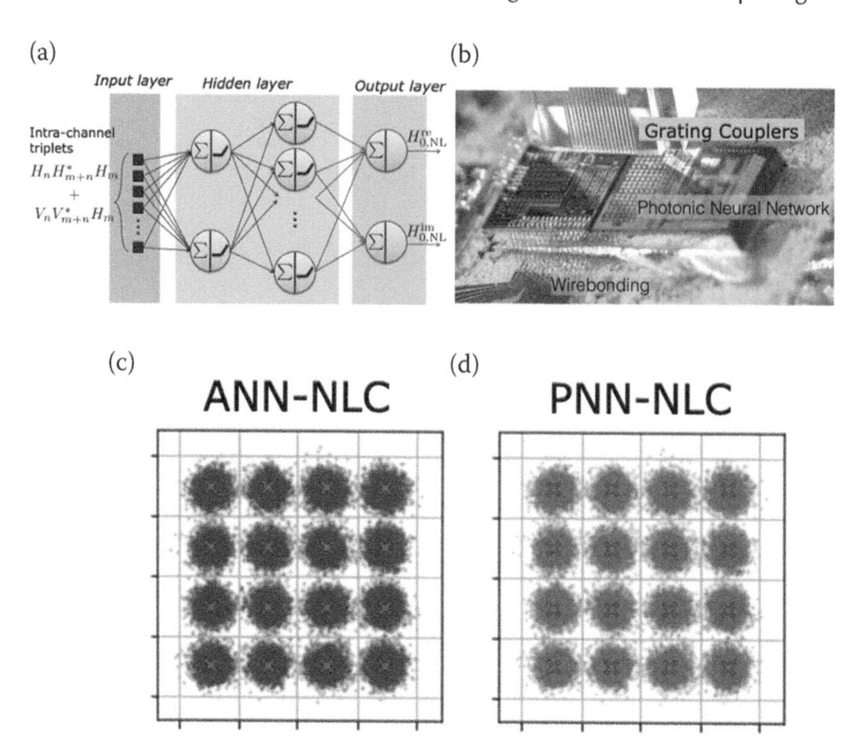

FIGURE 16.5 Photonic silicon photonic neural network for fiber nonlinearity compensation. (a) Schematic of ANN-NLC structure. (b) Image of the PNN chip under test and experimental setup for optical coupling and wirebonding. Constellations of X-polarization of a 32 Gbaud PM-16QAM, with the ANN-NLC gain of 0.57 dB in Q-factor (c) and with the PNN-NLC gain of 0.51 dB in Q-factor (d). Figure and caption adapted from [53], with permission.

bandwidths to the data rata, this suggests that real-time processing is within reach to improve optical communications.

Another intelligent signal processing application enabled by silicon neuromorphic photonic technology is in wideband RF signal processing. The linear part of silicon photonic neurons, for instance the weight banks displayed in Figure 16.3a-d, can be used on their own to perform e.g., principal component analysis [51] and independent component analysis [52]. These algorithms can perform dimensionality reduction on GHz modulated data in the optical domain to perform e.g., blind source separation. Doing this in an analog, wideband, passive photonic domain obviates the need for digital signal processing. It further requires only a single analog-digital conversion at the output instead of one per narrowband channel considered per input at the front end.

16.4 CONCLUSION AND FUTURE DIRECTIONS

Neuromorphic photonics has experienced rapid growth over the last few years. Still, there are many outstanding questions. The introduction of new electro-optic materials to silicon photonics would improve performance and functionality of silicon photonic neural networks. Tighter co-integration of light sources and amplifiers is a general aim of silicon photonics, and would also increase opportunities in neuromorphic engineering as well as ease of deployment [54]. The advent of zero-change CMOS silicon photonics platforms [55] is exciting to break the gain-bandwidth tradeoff in current silicon modulator neurons and improve density of control electronics. In the short term, this may be accomplished with separate dedicated CMOS chips wirebonded or flip-chip bonded to the silicon photonic chips. The applications reviewed were focused on inference tasks, but on-chip learning is also a very promising area. Spike timing–dependent plasticity was demonstrated with phase-change materials [26], and in the coherent approach, time-reversal symmetry to obtain gradients from intensity measurements was suggested to perform in-situ backpropagation [56]. In any case, ongoing investigations in neuromorphic photonics enabled by silicon photonics promise to bring machine intelligence to unexplored regimes, a salutary direction for a society increasingly dependent on neural network processors.

REFERENCES

[1] Y. LeCun, Y. Bengio, and G. Hinton, 'Deep learning', *Nature*, vol. 521, no. 7553, pp. 436–444, May 2015, doi: 10.1038/nature14539.

[2] J. Schmidhuber, 'Deep learning in neural networks: An overview', *Neural Netw.*, vol. 61, pp. 85–117, Jan. 2015, doi: 10.1016/j.neunet.2014.09.003.

[3] F. Rosenblatt, 'The perceptron: A perceiving and recognizing automaton', Jan. 1957. Accessed: Oct. 16, 2020. [Online]. Available: https://blogs.umass.edu/brain-wars/files/2016/03/rosenblatt-1957.pdf.

[4] A. L. Hodgkin and A. F. Huxley, 'A quantitative description of membrane current and its application to conduction and excitation in nerve', *J. Physiol.*, vol. 117, no. 4, pp. 500–544, 1952, doi: 10.1113/jphysiol.1952.sp004764.

[5] C. Mead and M. Ismail, Eds., *Analog VLSI Implementation of Neural Systems*. Springer US, 1989.

[6] V. K. Pallipuram, M. Bhuiyan, and M. C. Smith, 'A comparative study of GPU programming models and architectures using neural networks', *J. Supercomput.*, vol. 61, no. 3, pp. 673–718, Sep. 2012, doi: 10.1007/s11227-011-0631-3.

[7] N. P. Jouppi *et al.*, 'In-datacenter performance analysis of a tensor processing unit', *ArXiv170404760 Cs*, Apr. 2017, Accessed: Oct. 24, 2020. [Online]. Available: http://arxiv.org/abs/1704.04760.

[8] J. Duarte *et al.*, 'Fast inference of deep neural networks in FPGAs for particle physics', *J. Instrum.*, vol. 13, no. 07, p. P07027, Jul. 2018, doi: 10.1088/1748-0221/13/07/P07027.

[9] D. Marković, A. Mizrahi, D. Querlioz, and J. Grollier, 'Physics for neuromorphic computing', *Nat. Rev. Phys.*, vol. 2, no. 9, pp. 499–510, Sep. 2020, doi: 10.1038/s42254-020-0208-2.

[10] K. Berggren *et al.*, 'Roadmap on emerging hardware and technology for machine learning', *Nanotechnology*, vol. 32, no. 1, p. 012002, Jan. 2021, doi: 10.1088/1361-6528/aba70f.

[11] D. Psaltis and N. Farhat, 'Optical information processing based on an associative-memory model of neural nets with thresholding and feedback', *Opt. Lett.*, vol. 10, no. 2, pp. 98–100, Feb. 1985, doi: 10.1364/OL.10.000098.

[12] L. Chrostowski and M. Hochberg, *Silicon Photonics Design: From Devices to Systems*. Cambridge: Cambridge University Press, 2015.

[13] P. R. Prucnal and B. J. Shastri, *Neuromorphic Photonics*. CRC Press, 2017.

[14] L. D. Marinis, M. Cococcioni, P. Castoldi, and N. Andriolli, 'Photonic neural networks: A survey', *IEEE Access*, vol. 7, pp. 175827–175841, 2019, doi: 10.1109/ACCESS.2019.2957245.

[15] T. F. de Lima *et al.*, 'Primer on silicon neuromorphic photonic processors: Architecture and compiler', *Nanophotonics*, vol. 9, no. 13, pp. 4055–4073, Aug. 2020, doi: 10.1515/nanoph-2020-0172.

[16] X. Sui, Q. Wu, J. Liu, Q. Chen, and G. Gu, 'A review of optical neural networks', *IEEE Access*, vol. 8, pp. 70773–70783, 2020, doi: 10.1109/ACCESS.2020.2987333.

[17] B. J. Shastri *et al.*, 'Photonics for artificial intelligence and neuromorphic computing', *ArXiv201100111 Phys.*, Oct. 2020, Accessed: Nov. 05, 2020. [Online]. Available: http://arxiv.org/abs/2011.00111.

[18] D. Rosenbluth, K. Kravtsov, M. P. Fok, and P. R. Prucnal, 'A high performance photonic pulse processing device', *Opt. Express*, vol. 17, no. 25, pp. 22767–22772, Dec. 2009, doi: 10.1364/OE.17.022767.

[19] J. Feldmann, N. Youngblood, C. D. Wright, H. Bhaskaran, and W. H. P. Pernice, 'All-optical spiking neurosynaptic networks with self-learning capabilities', *Nature*, vol. 569, no. 7755, pp. , 208–214, May 2019, doi: 10.1038/s41586-019-1157-8.

[20] J. M. Shainline, S. M. Buckley, R. P. Mirin, and S. W. Nam, 'Superconducting optoelectronic circuits for neuromorphic computing', *Phys. Rev. Appl.*, vol. 7, no. 3, p. 034013, Mar. 2017, doi: 10.1103/PhysRevApplied.7.034013.

[21] M. A. Nahmias, T. F. de Lima, A. N. Tait, H. Peng, B. J. Shastri, and P. R. Prucnal, 'Photonic multiply-accumulate operations for neural networks', *IEEE J. Sel. Top. Quantum Electron.*, vol. 26, no. 1, pp. 1–18, Jan. 2020, doi: 10.1109/JSTQE.2019.2941485.
[22] Y. Shen *et al.*, 'Deep learning with coherent nanophotonic circuits', *Nat. Photonics*, vol. 11, no. 7, pp. 441–446, Jul. 2017, doi: 10.1038/nphoton.2017.93.
[23] A. N. Tait, M. A. Nahmias, B. J. Shastri, and P. R. Prucnal, 'Broadcast and weight: An integrated network for scalable photonic spike processing', *J. Light. Technol.*, vol. 32, no. 21, pp. 3427–3439, Nov. 2014.
[24] A. N. Tait *et al.*, 'Neuromorphic photonic networks using silicon photonic weight banks', *Sci. Rep.*, vol. 7, no. 1, pp. 1–10, Aug. 2017, doi: 10.1038/s41598-017-07754-z.
[25] M. Miscuglio and V. J. Sorger, 'Photonic tensor cores for machine learning', *Appl. Phys. Rev.*, vol. 7, no. 3, p. 031404, Jul. 2020, doi: 10.1063/5.0001942.
[26] Z. Cheng, C. Ríos, W. H. P. Pernice, C. D. Wright, and H. Bhaskaran, 'On-chip photonic synapse', *Sci. Adv.*, vol. 3, no. 9, p. e1700160, Sep. 2017, doi: 10.1126/sciadv.1700160.
[27] C. Ríos *et al.*, 'In-memory computing on a photonic platform', *Sci. Adv.*, vol. 5, no. 2, p. eaau5759, Feb. 2019, doi: 10.1126/sciadv.aau5759.
[28] A. N. Tait *et al.*, 'Feedback control for microring weight banks', *Opt. Express*, vol. 26, no. 20, pp. 26422–26443, Oct. 2018, doi: 10.1364/OE.26.026422.
[29] C. Huang *et al.*, 'Demonstration of scalable microring weight bank control for large-scale photonic integrated circuits', *APL Photonics*, vol. 5, no. 4, p. 040803, Apr. 2020, doi: 10.1063/1.5144121.
[30] F. Akopyan *et al.*, 'TrueNorth: Design and tool flow of a 65 mW 1 million neuron programmable neurosynaptic chip', *IEEE Trans. Comput.-Aided Des. Integr. Circuits Syst.*, vol. 34, no. 10, pp. 1537–1557, Oct. 2015, doi: 10.1109/TCAD.2015.2474396.
[31] T. F. de Lima *et al.*, 'Machine learning with neuromorphic photonics', *J. Light. Technol.*, vol. 37, no. 5, pp. 1515–1534, Mar. 2019, doi: 10.1109/JLT.2019.2903474.
[32] W. Zhang, R. Mazzarello, M. Wuttig, and E. Ma, 'Designing crystallization in phase-change materials for universal memory and neuro-inspired computing', *Nat. Rev. Mater.*, vol. 4, no. 3, pp. 150–168, Mar. 2019, doi: 10.1038/s41578-018-0076-x.
[33] C. Huang, A. Jha, T. F. de Lima, A. N. Tait, B. J. Shastri, and P. R. Prucnal, 'On-chip programmable nonlinear optical signal processor and its applications', *IEEE J. Sel. Top. Quantum Electron.*, vol. 27, no. 2, pp. 1–11, Mar. 2021, doi: 10.1109/JSTQE.2020.2998073.
[34] A. Jha, C. Huang, and P. R. Prucnal, 'Reconfigurable all-optical nonlinear activation functions for neuromorphic photonics', *Opt. Lett.*, vol. 45, no. 17, pp. 4819–4822, Sep. 2020, doi: 10.1364/OL.398234.
[35] M. M. P. Fard *et al.*, 'Experimental realization of arbitrary activation functions for optical neural networks', *Opt. Express*, vol. 28, no. 8, pp. 12138–12148, Apr. 2020, doi: 10.1364/OE.391473.
[36] A. N. Tait *et al.*, 'Silicon photonic modulator neuron', *Phys. Rev. Appl.*, vol. 11, no. 6, p. 064043, Jun. 2019, doi: 10.1103/PhysRevApplied.11.064043.
[37] R. Amin *et al.*, 'ITO-based electro-absorption modulator for photonic neural activation function', *APL Mater.*, vol. 7, no. 8, p. 081112, Aug. 2019, doi: 10.1063/1.5109039.
[38] H.-T. Peng *et al.*, 'Autaptic circuits of integrated laser neurons', in *Conference on Lasers and Electro-Optics (2019), paper SM3N.3*, May 2019, p. SM3N.3, doi: 10.1364/CLEO_SI.2019.SM3N.3.
[39] T. F. de Lima *et al.*, 'Noise analysis of photonic modulator neurons', *IEEE J. Sel. Top. Quantum Electron.*, vol. 26, no. 1, pp. 1–9, Jan. 2020, doi: 10.1109/JSTQE.2019.2931252.
[40] A. N. Tait *et al.*, 'Microring weight banks', *IEEE J. Sel. Top. Quantum Electron.*, vol. 22, no. 6, pp. 312–325, Nov. 2016, doi: 10.1109/JSTQE.2016.2573583.
[41] A. Mehrabian, Y. Al-Kabani, V. J. Sorger, and T. El-Ghazawi, 'PCNNA: A photonic convolutional neural network accelerator', in *2018 31st IEEE International System-on-Chip Conference (SOCC)*, Sep. 2018, pp. 169–173, doi: 10.1109/SOCC.2018.8618542.
[42] V. Bangari *et al.*, 'Digital electronics and analog photonics for convolutional neural networks (DEAP-CNNs)', *ArXiv190701525 Phys.*, Apr. 2019, Accessed: Oct. 30, 2020. [Online]. Available: http://arxiv.org/abs/1907.01525.
[43] K. Vandoorne *et al.*, 'Experimental demonstration of reservoir computing on a silicon photonics chip', *Nat. Commun.*, vol. 5, no. 1, pp. 1–6, Mar. 2014, doi: 10.1038/ncomms4541.
[44] W. Maass, T. Natschläger, and H. Markram, 'Real-time computing without stable states: A new framework for neural computation based on perturbations', *Neural Comput.*, vol. 14, no. 11, pp. 2531–2560, Nov. 2002, doi: 10.1162/089976602760407955.

[45] H. Jaeger and H. Haas, 'Harnessing nonlinearity: Predicting chaotic systems and saving energy in wireless communication', *Science*, vol. 304, no. 5667, pp. 78–80, Apr. 2004, doi: 10.1126/science.1091277.

[46] D. Brunner, M. C. Soriano, C. R. Mirasso, and I. Fischer, 'Parallel photonic information processing at gigabyte per second data rates using transient states', *Nat. Commun.*, vol. 4, no. 1, pp. 1–7, Jan. 2013, doi: 10.1038/ncomms2368.

[47] L. Appeltant *et al.*, 'Information processing using a single dynamical node as complex system', *Nat. Commun.*, vol. 2, no. 1, pp. 1–6, Sep. 2011, doi: 10.1038/ncomms1476.

[48] L. Larger *et al.*, 'Photonic information processing beyond Turing: an optoelectronic implementation of reservoir computing', *Opt. Express*, vol. 20, no. 3, pp. 3241–3249, Jan. 2012, doi: 10.1364/OE.20.003241.

[49] A. Katumba *et al.*, 'Neuromorphic computing based on silicon photonics and reservoir computing', *IEEE J. Sel. Top. Quantum Electron.*, vol. 24, no. 6, pp. 1–10, Nov. 2018, doi: 10.1109/JSTQE.2018.2821843.

[50] T. Bekolay *et al.*, 'Nengo: A Python tool for building large-scale functional brain models', *Front. Neuroinformatics*, vol. 7, Jan. 2014, doi: 10.3389/fninf.2013.00048.

[51] A. N. Tait *et al.*, 'Demonstration of multivariate photonics: Blind dimensionality reduction with integrated photonics', *J. Light. Technol.*, vol. 37, no. 24, pp. 5996–6006, Dec. 2019, doi: 10.1109/JLT.2019.2945017.

[52] P. Y. Ma *et al.*, 'Photonic independent component analysis using an on-chip microring weight bank', *Opt. Express*, vol. 28, no. 2, pp. 1827–1844, Jan. 2020, doi: 10.1364/OE.383603.

[53] C. Huang *et al.*, 'Demonstration of photonic neural network for fiber nonlinearity compensation in long-haul transmission systems', in *Optical Fiber Communication Conference Postdeadline Papers 2020 (2020), paper Th4C.6*, Mar. 2020, p. Th4C.6, doi: 10.1364/OFC.2020.Th4C.6.

[54] D. Liang and J. E. Bowers, 'Recent progress in lasers on silicon', *Nat. Photonics*, vol. 4, no. 8, pp. 511–517, Aug. 2010, doi: 10.1038/nphoton.2010.167.

[55] V. Stojanović *et al.*, 'Monolithic silicon-photonic platforms in state-of-the-art CMOS SOI processes [Invited]', *Opt. Express*, vol. 26, no. 10, pp. 13106–13121, May 2018, doi: 10.1364/OE.26.013106.

[56] T. W. Hughes, M. Minkov, Y. Shi, and S. Fan, 'Training of photonic neural networks through in situ backpropagation and gradient measurement', *Optica*, vol. 5, no. 7, pp. 864–871, Jul. 2018, doi: 10.1364/OPTICA.5.000864.

17 Logic Computing and Neural Network on Photonic Integrated Circuit

Zheng Zhao, Zhoufeng Ying, Chenghao Feng, Ray T. Chen, and David Z. Pan

CONTENTS

17.1 INTRODUCTION OF OPTICAL COMPUTING

Optical computing uses photons generated by lasers or diodes for computation as opposed to the more traditional electron-based computation. Further assisted with optical interconnects, optical computing is able to achieve a computational speed enhancement of at least two orders of magnitude over the state-of-the-art and three orders of magnitude in power efficiency [1,2]. Even though optics is generally less compact than electronics, as a result of the physical principles, the promise of computational efficiency has found its niche in data centers and cloud computing services that have less physical space limitation. Compared with computing, optical interconnects have been more intensively investigated, which manifests the advantages over metal interconnects especially in intra- and inter-chip communications. To catch up with the advancement with optical interconnects, previous works on optical computing have demonstrated upon two computing paradigms: digital and analog computing. In digital computing paradigm, signals are interpreted in binary values, 0 and 1; and the computation observes the rule of Boolean algebra, where optical switches are the core of this paradigm. In analog optical computing, on the other hand, interprets light signals as continuous values in the real or complex domain and performs on the basis of linear optics. For the first category, intensive study has been made on basic bitwise operation gates, algebraic functions such as 1-bit half and full adders and switchers both optical logic units [3–7]. In order to implement general and larger-scale logic functions and pave the way for design-space exploration, automated design methods are proposed generic logic synthesis algorithms [8–12]. For the second category, research efforts have been made on matrix multiplication [13–16], and optical neural networks (ONNs), evolving from the former [17,18]. The ONN paradigm

DOI: 10.1201/9780429292033-17

distinguishes itself by directly exploiting linear optics to perform neuromorphic operations. Especially, the core and performance-critical computation of neural networks, matrix multiplication, is a computationally expensive operation for electronics, but for optics, this computation has been studied and successfully demonstrated on chip [16,17], by which matrix multiplication can be performed with near-zero energy. In this chapter, we introduce the principles of the two computing paradigms as well as recent breakthrough towards automated design for better scalability and robustness.

17.2 OPTICAL LOGIC SYNTHESIS

Optical logic synthesis aims to provide automated, quality design of large-scale photonic integrated circuits (PICs) of arbitrary logic functions and pave the way for design-space exploration. As a pioneer attempt, a synthesis scheme based on virtual gates (VGs) and optical splitters were proposed in [8]. Every literal of a Boolean function is implemented by a VG. While the concept of VG is worthwhile for functional cascading, the proposed method usually generates a large number of optical components, including VGs and splitters, which inevitably leads to great optical power loss.

Compared to VG-based synthesis, binary decision diagram (BDD)-based synthesis [9,10,12] shows both area and optical power efficiency in the generated design. BDD [19] is a widely used data structure for logic synthesis and verification. As shown the BDD in Figure 17.1(a), each round node (labeled by variable *a*, *b*, or *c*) has a function of a 1 × 2 crossbar switch, controlled by the corresponding variable. The squared 1-terminal node corresponds to logic one: if there is a directed path from the top function node *f* to it, then the function is evaluated to be logic 1 and vice versa. A BDD can be directly mapped to optical implementation using optical switches [9,10]. As can be seen in the implementation in Figure 17.1(c), it simply replaces each BDD node by an optical crossbar switch (S), each controlled by a primary input and the output label 0/1 denotes the port switched by controlling variable being 0/1. The switch can be implemented with either a Mach-Zehnder interferometer (MZI) or microresonator. The light (λ) from a laser (or from the output of the previous optical network) is sourced from the function node. A photodetector (PD) (or optical amplifier to the next computation stage) is located at the 1-terminal. If there is light detected at the PD, the output of the optical network is logical 1, otherwise logical 0. Waveguides and optical combiners (CB) are employed to connect and merge optical signals.

17.2.1 Optimization Techniques

One of the greatest problems shared by optical logic circuit is optical power depletion. For VG-based synthesis, each 2 × 1 splitter reduce the optical power by half. Due to the great number of

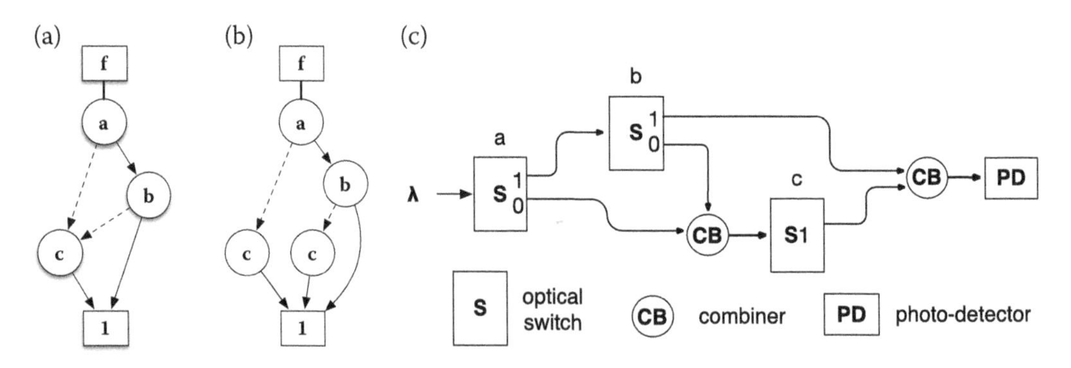

FIGURE 17.1 (a) Original 1-terminal BDD, (b) BDD after combiner elimination, and (c) optical implementation of (a) [9].

splitters, the loss cascaded inevitably leads to an extremely weak output signal indistinguishable from noises. For BDD-based synthesis, optical combiners bear the same problem. Due to BDD's single-path principle [10], only one input of the 2-input combiner has light. Mode mismatch causes half of the light escape from the waveguide to the free space. Consequently, there will still be a –3 dB (50%) power loss at the output [10]. To mitigate the issue, there are two techniques [10]: *combiner elimination* and *coupler assignment.*

Considering the example in Figure 17.1(a), we can see if we duplicate node c and obtain the equivalent BDD in Figure 17.1(b), the combiner at the original node c can be eliminated. Though at the terminal, the number of combiner inputs is increased, it is not cascaded. Furthermore, the number of optical switches also increases. This example shows the trade-off of using combiner elimination. As a general guidance, suppose the terminal combiner has *nTerm* inputs. Let γ_{org} be the worst-case efficiency factor from the top node to the node along a path before reaching the terminal combiner, and thus, the original efficiency becomes $\gamma_{org}/nTerm$. Given an *nIn*-input node on the path, with *nCornTerm* of the terminal combiner inputs belonging to its fan-out corn, we create *nCopy* copies of this node. The overhead of each copy is the number of the node in the fanout corn (including the corn root), denoted by *nCornSize*. The new efficiency factor becomes

$$\mathrm{nIn}/(\mathrm{nTerm} + \mathrm{nCopy}\cdot\mathrm{nCornTerm})\cdot\gamma_{org} \tag{17.1}$$

The ratio over the original efficiency is thus

$$r = \mathrm{nIn}/(1 + \mathrm{nCopy}\cdot\mathrm{nCornTerm}/\mathrm{nTerm}) \tag{17.2}$$

As can be calculated by the formula, the example using combiner elimination improves the worst-case path power efficiency from 1/4 to 1/3. The detailed flow is shown in Algorithm 17.1. Given a BDD, the expected minimum benefit ratio *Benefit* upon one copy, and the area overhead budget *Budget*, we have

The second technique is coupler assignment, which improves the power efficiency by redistributing the optical power directly using directional coupler. Figure 17.2 shows a plain combiner and directional coupler. The power efficiency for the former is always 1/2 for either of the inputs. For directional coupler, we can choose the coupling coefficient k (where $0 < k < 1$) so that the power efficiency for port a and b can be different: $\gamma_a = k$ and $\gamma_b = 1 - k$.

ALGORITHM 17.1 COMBINER ELIMINATION

1:*Overhead* $\leftarrow$ 0
 2:**for** each path p_i starting from the smallest path power efficiency **do**
 3:**for** each edge (u, v) on p_i bottom to top **do**
 4:Compute estimated benefit r and overhead *nCornSize*(v)
 5:**if***Overhead* + *nCornSize*(v) > *Budget***then**
 6:**continue**
 7:**else if**r > *Benefit***then**
 8:$v'$$\leftarrow$ Copy v and its fanout corn
 9:Reconnect u to v'
 10:*Overhead* $\leftarrow$ *Overhead* + *nCornSize*

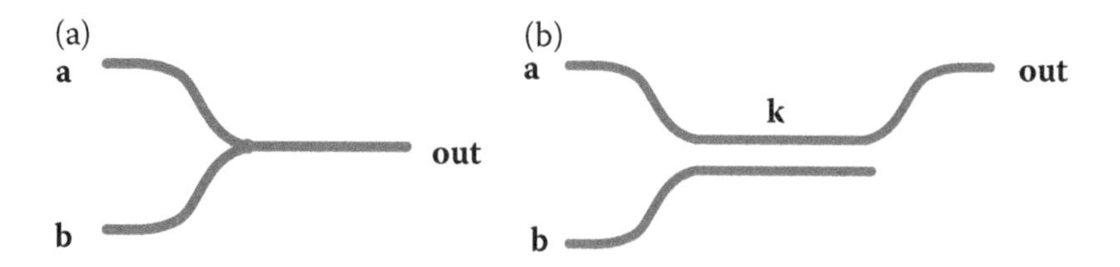

FIGURE 17.2 (a) Plain combiner and (b) directional coupler.

Basically, coupler assignment decides the efficiency factors to maximize the worst-case power efficiency. Note that the size of a typical coupler is almost two times the size of a typical microresonator-based switch. The trade-off of the flexibility of reassigning the efficiency factor and the induced area overhead is tackled in the synthesis flow.

For example, the implementation in Figure 17.1(c) simply uses plain combiners, which always half the incoming optical powers. However, if we leverage directional coupler, with a careful assignment of the coupling ratio, optical power can be better exploited and balanced among the all paths to improve the general, i.e., worse-case power efficiency of the paths. As an example, in the implementation in Figure 17.1(c), the worst-case ratio of the output optical power to the input power of all the three paths is 1/4 (path a → c → 1 and a → b → c → 1). If we replace the terminal combiner with a directional coupler with a coupling efficiency of 2/3 for port c and 1/3 for port b, then the worst-case power ratio can be calculated to be 1/3, with an improvement of 33%.

Coupler assignment is essentially a constrained polynomial programming problem, where the objective is to maximize the worst-case output optical power, the variables are the coupling ratio of each directional coupler, and the constraints are based on the law of conservation of energy of the couplers. The general coupler assignment problem can be formulated as a polynomial programming problem as follows

$$\begin{aligned} &\textbf{\textit{Maximize}}\ \gamma_{network} = min_i\ \{\gamma_i \cdot \Pi_{j:e_j \in :p_i} x_j\} \\ &\text{s. t. } \Sigma_{i \in In(n)} x_i = 1, \quad \forall\ n \\ &0 \le x_i \le 1, \forall\ i \end{aligned} \tag{17.3}$$

The objective is the efficiency factor of the whole PIC network, which is defined by the minimum of all the path efficiency factor. Each variable x_i represents an assignment of coupling efficiency for the edge e_j on path p_i. The path efficiency of p is calculated as the product of the coupling efficiency x_i's times γ, the efficiency factor related to the other parts, e.g., the switch efficiency. The first constraint represents the rule of power conservation; i.e., the sum of the efficiency factor of the input ports to a coupler is 1. The second constraint simply states the efficiency factor should be positive and smaller than 1. The max-min objective can be transformed by adding a dummy variable f and add the constraints that

$$f \le \gamma_i \cdot \prod x_j, \quad \forall\ i \tag{17.4}$$

Note that the polynomial programming problem can be solved by semidefinite programming (SDP) relaxation [20], but when the variable number is non-trivial, it usually takes long time to solve. A more scalable solution uses quadratically constrained programming (QCP), so that an efficient second-order cone programming solver can be adopted. In QCP, the number of variables in each constraint is limited to be less than or equal to 2, so that each constraint is quadratic. The QCP optimization is run iteratively until no further optimize can be achieved or the overhead limit is

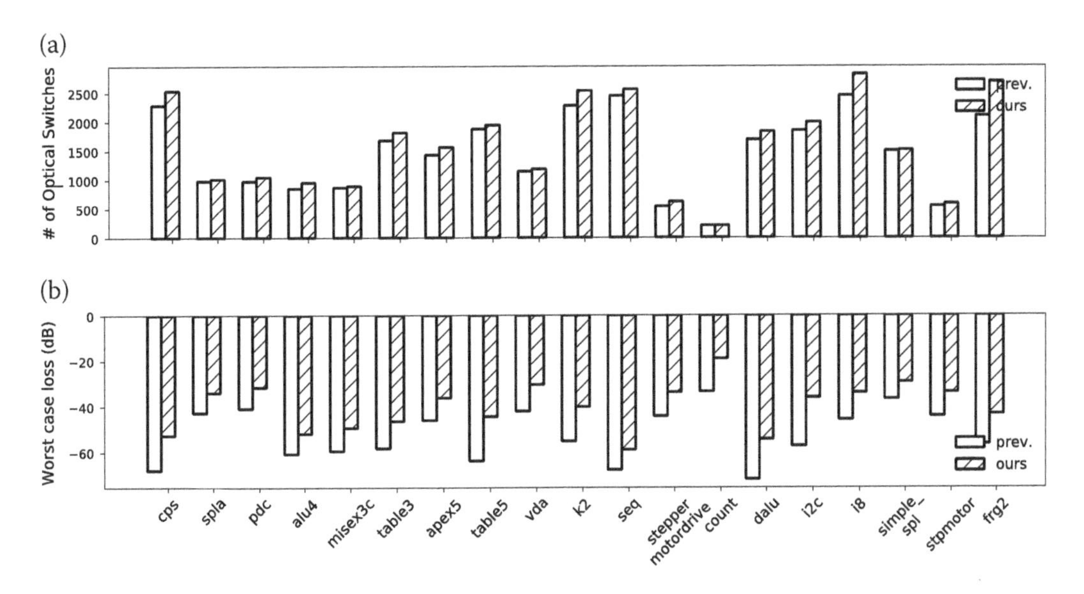

FIGURE 17.3 (a) Number of optical switches and (b) worst-case optical loss (dB) for different benchmark circuits.

reached. This also allows us to have a prior control of the overhead related to the couplers in each iteration, as the overhead is proportional to the number of variables in QCP.

17.2.2 Simulation Results

The techniques were implemented in C++ with CUDD package [21], and tested it on an 8-core 3.4 GHz Linux machine with a 32GB RAM. The experiments were conducted on Microelectronics Center of North Carolina (MCNC) [22] and International Workshop on Logic and Synthesis (IWLS) benchmarks [23]. The comparison reference (*prev.*) is the method without consideration of optical power loss [9]. The area measured by the number of optical switches are shown in Figure 17.3(a). The overhead percentage in terms of the increase of the number of optical switches (incl. the couplers) for each benchmark is limited to be less than 20%. The average is a small 7.63%. The network efficiency (in dB) is plotted in Figure 17.3(b). For all the benchmarks, the power efficiency has been improved by magnitudes. The average improvement ratio is 27.02X. The maximum CPU time is 14.5 s and the average is 1.88 s.

17.3 EXPLOITING WAVEGUIDE DIVISION MULTIPLEXING

The classical way of BDD-based synthesis has the limitation that multi-output functions are required to be separated to single-output function for implementation. Figure 17.4 shows a multi-primary output (PO) BDD containing two functions f_1 and f_2, where $f_1 = a'c + ab'c + ab$ and $f_2 = b'c + b$. In multi-PO BDD, common functional structures can be shared among different functions so that the number of BDD nodes is reduced. However, consider if we directly apply the synthesis method from Section 17.2 and let multiple light signals be sourced together into the circuit, it is impossible to differentiate the light signal received at the PD. Wavelength division multiplexing (WDM), at this point, naturally emerges as a promising solution. WDM is a technology that enables multiple optical signals to be transmitted independently and simultaneously in a single waveguide. Due to its advantages of high bandwidth, fast speed, low power and minimal crosstalk, WDM provides much higher bit rate for data transmission and is suitable for the high bandwidth requirement of current SoCs [24]. Using WDM, Figure 17.4(b) shows that if two

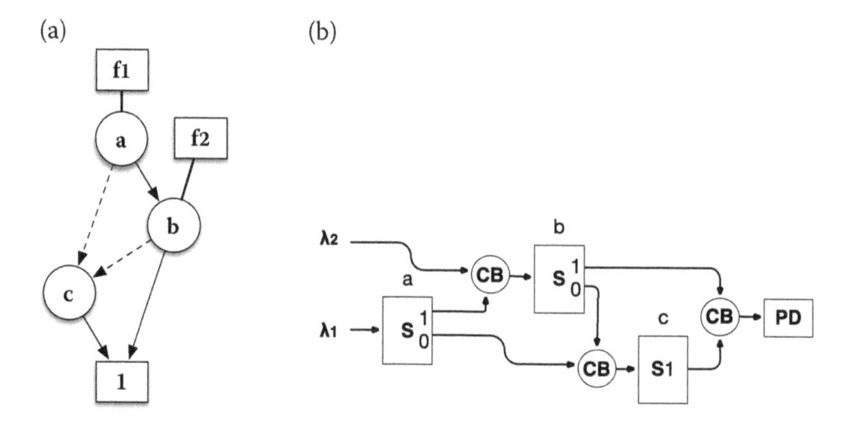

FIGURE 17.4 (a) Multi-function BDD and (b) the optical implementation.

wavelength λ_1 and λ_2 are sourced to represent the two PO functions, the structural sharing can be exploited. It can be calculated that by sharing instead of separation, two optical switches can be saved. The difference would be non-trivial for bigger logic functions. Generally, due to the limitation of integrated lasers on wavelength spacing, on-chip WDM has a capacity limit varying from 2 to 16 [2,25].

17.3.1 Proposed Synthesis Flow

The general flow of WDM-based synthesis methods is shown in Figure 17.5. Given an arbitrary multi-PO function and WDM capacity constraint, the BDD reordering engine is called to build a BDD with all the POs. The synthesis problem is first modeled to a hypergraph partitioning problem and solved to minimize the partitioning cost (*HyPart*). As the result of hypergraph partitioning contains infeasible partitions, a second resolving step is introduced (*ReFlow*). The resolving is achieved by modeling and solving a min-cost max-flow problem. The final result is hence guaranteed to be feasible. During the whole procedure, BDD reordering engine is called for multiple times for either the complete PO set or some PO subset in order to produce further improvement: at the beginning, at the end of HyPart and ReFlow.

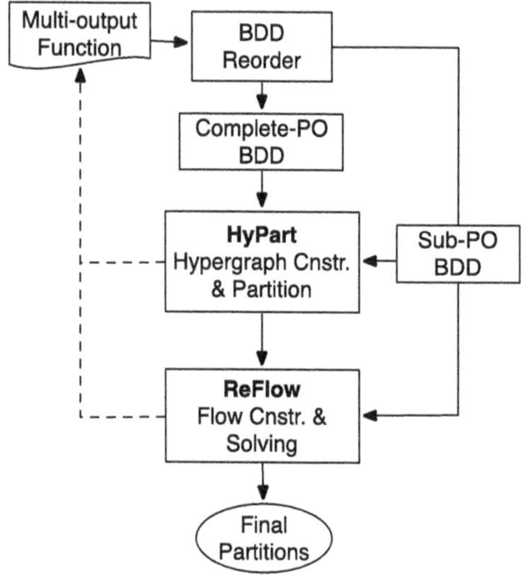

FIGURE 17.5 Proposed synthesis flow.

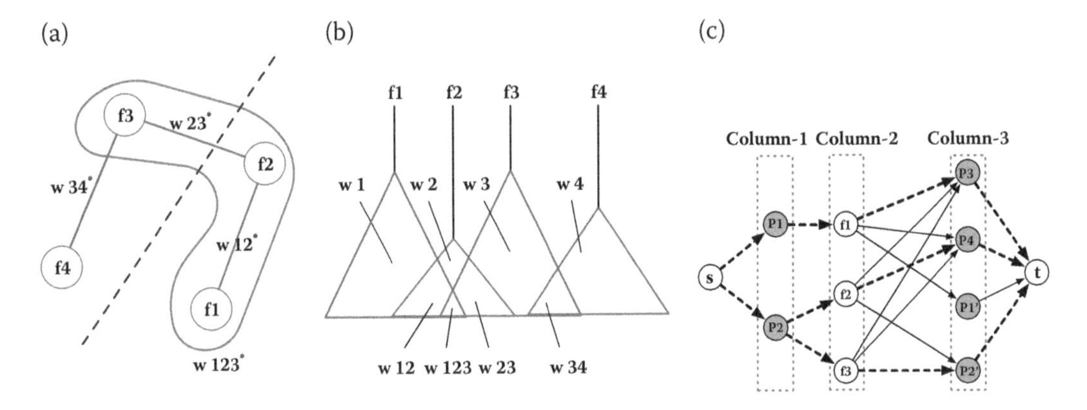

FIGURE 17.6 (a) Hypergraph and (b) corresponding BDD example. (c) ReFlow example.

Based on the equivalence of BDD and optical implementation, the BDD-based optical synthesis problem under WDM capacity constraint is presented in the following BDD-partitioning problem. We denote the WDM capacity by *Cap*.

Problem 17.1: *BDD partitioning problem (BPP). Given a BDD composed of multiple output functions and a constant Cap, BPP is to find a partition of the BDD that has the minimum number of BDD nodes. The number of functions of each partition is no greater than Cap.*

In the following, we approximate BPP by a hypergraph partitioning problem (HPP) [29]. A hypergraph is a generalization of a graph in which a hyperedge can join any number of vertices. Formally, a hypergraph G_h is a pair $G_h = (V_h, E_h)$ where V_h is a set of vertices, and E_h is a set of hyperedges. As an example, Figure 17.6(a) shows a hypergraph with four vertices labeled by f_1 to f_4, three 2-vertex hyperedges represented by solid lines: (f_1, f_2), (f_2, f_3), (f_3, f_4); and one 3-vertex hyperedge represented by a curved area: (f_1, f_2, f_3). Each hyperedge weight $w*$ is indexed by the indices of the nodes connected to the hyperedge. For example, the weights of the hyperedge (f_2, f_3) and the hyperedge (f_1, f_2, f_3) are w^*_{23} and w^*_{123} respectively. The HPP problem is defined as follows:

Problem 17.2: *Hypergraph partitioning problem (HPP). Given a hypergraph, the hypergraph partitioning problem is to assign the vertices into disjoint non-empty partitions so that the hyperedge cut weight is minimum.*

HPP is known to be NP-hard [26]. Given a multi-output BDD, we construct the hypergraph by Algorithm 17.2. For each PO function *f*, a hypergraph node is created and the fanout cone *Cone* in the BDD is collected (Lines 1–3). Each BDD node is then classified based on which PO cone it is contained. If BDD nodes are shared by different POs, they are merged to the same group *G* which is indexed by the corresponding output function set *s* (Lines 4–7). A hyperedge is created for each such group whose *s* has more than one output function. Let *k* be the actual number of partitions in a set *s*. The edge weight w_S is set to be the number of BDD nodes contained in the group ($|G_S|$) times a scaling factor $\alpha = k* - 1$, where $k*$: $=\lceil |S|/Cap \rceil$ as a straightforward approximation of *k*. If $k* = 1$, it is set to 1 to avoid zero weights (Lines 8–11).

Figures 17.6(b) and 17.6(a) show a 4-PO BDD and its corresponding hypergraph. There are four hypergraph nodes representing POs and four hyperedges, each corresponding to a shared region. The area of the regions in terms of the BDD nodes is denoted as w_{12}, w_{123}, w_{23}, and w_{34}, respectively. Each hyperedge cost is marked by a corresponding weight $w*$: $=w \cdot \alpha$, which is *w* weighed by the scaling factor α described above. In the hypergraph, the edge cost is the region area

ALGORITHM 17.2 HYPERGRAPH CONSTRUCTION FOR BPP

1:**for** each output function f_i of BDD **do**
  2:Create a hypergraph vertex marked by f_i
  3:$Cone_i$ ←{BDD nodes in its fanout cone}
  4:**for** each BDD node n_i**do**
    5:$s_{cur} \leftarrow \{f'_i s: \ n_i \in Cone_j\}, \ G_{s_{cur}} \leftarrow \{n_i\},$
    6:**if**s_{cur} is identical to an existing group G_s's index s**then**
      7:Merge $G_{s_{cur}}$ and G_s to $G_{s \leftarrow s_{cur} \cup s} \leftarrow G_s \cup G_{s_{cur}}$
    8:**for** each group G_s whose $|s_i| > 1$ **do**
      9:Create a hyperedge connecting all $f s \in s_i$, whose
      10:weight $w^*_{s_i}: =w_{s_i} \cdot \alpha_i$, where
      11:$w_{s_i}: =|G_{s_i}|, \ \alpha_i: =k^*_{s_i} - 1, \ k^*_{s_i}: =\lceil |s_i|/Cap \rceil$

weighted by a constant $\alpha_{s_i} = k^*_{s_i} - 1 = |S|/Cap$ can be viewed ii as an approximation of k_{s_i}: the actual number of partitions in set s_i. The fundamental idea of linking HPP and BPP is that, for HPP the lower cut cost meaning the fewer shared nodes are separated to different partitions, then for BPP, the less area overhead is introduced by a certain partition solution. Once the partitions are computed, BDD reordering engine is called again as a post-optimization.

HyPart does not always produce perfectly balanced partitions, in the sense that the numbers of elements in some partitions are smaller than the specified capacity of WDM and others are greater. The latter case is infeasible in the context of optical implementation. To resolve the problem, the following algorithm ReFlow is proposed to balance the element allocation in each partition. Basically, the above partition methodology has achieved a near-balanced solution but with a small number of capacity violations. The key idea of ReFlow is to transfer POs from the partitions over the WDM capacity to those under the capacity. Ultimately, we obtain a solution which satisfies the capacity constraints. ReFlow is achieved through a min-cost max-flow model, which demonstrated effective application in physical design problems [24]. Figure 17.6(c) is a pseudo-example illustrating how the network flow model. Suppose in the partition solution given by the previous step, P_1 and P_2 are the over-capacity partitions that require re-allocation and P_3 and P_4 are the under-capacity partitions that can accommodate extra elements. To construct the flow graph, we require the following flow nodes: (1) node P_1 and P_2, corresponding to over-capacity partitions (marked in Column-1); (2) node f_1 to f_3, corresponding to the POs contained in the over-capacity partitions (marked in Column-2); (3) node P_3, P_4, P'_1 and P'_2 corresponding to the target partitions (marked in Column-3); (4) node s and t representing the pseudo starting and terminating nodes. In Column-3, P_3 and P_4 are the under-capacity partitions and P'_1 and P'_2 mirror the original over-capacity partitions P_1 and P_2, which allows the possibility that nodes in P_1 and P_2 can still stay in the original partition as long as the flow is feasible, i.e., the capacity constraint is satisfied. As for the flow edges, one flow edge is created: (1) from an over-capacity P node to an f node, if $f \in P$ given by HyPart; (2) from each f node to each under-capacity P node; (3) from an f node to an mirroring P' node if $f \in P$ given by HyPart; (4) from s to each over-capacity P node; and from each under-capacity P node and mirroring P' node to t. Intuitively, if we observe a flow from f_i to partition P or P' in Column 3, it means f_i's eventual destination is the partition P or P'. For the example, the flow solution marked by the blue edges means $f_1 \in P_1$ is moved to P_3, $f_2 \in P_2$ is moved to P_4, and $f_3 \in P_2$ stays in P_2.

17.3.2 Simulation Results

The experimental setup follows Section 17.2. Figure 17.7 shows the first set of experiments, which mainly demonstrates the effect of the number of partitions on the number of BDD nodes. The x-axis is for the number of partitions, starting from *sep* meaning that each PO is separated to an individual partition, then the number of the partitions are set to 10, 5, 2, and ending with *all*, meaning all the POs are grouped in one partition, i.e., the complete-PO BDD. The y-axis is for the total number of BDD nodes. The solid curves represent the average of 10 random partition solutions. The dashed curves represent the results by using the flow in Figure 17.5. In general, the trend shows that for both methods, the number of BDD nodes would generally decrease with the number of partitions and the latter is more area-efficient. This is consistent to the idea that if the PO functions are implemented in one BDD instead of separated, it is more likely to enable the sharing of BDD nodes and sub-functions given a good BDD reordering engine. The second set of experiments compares the area measured by the number of optical switches (including WDM overhead) for separate method (*prev.*) [9] and our WDM-based flow [12]. The ratios of the switch number of our method compared to that of the previous method are 81.2%, 75.2%, and 72.0% on average for WDM capacity of 4, 8, and 16, respectively. Figure 17.8(b) depicts the average number of nodes given by the end of HyPart, re-synthesizing the HyPart (ReSyn), and ReFlow. It can be seen re-synthesizing can help to further reduce the number of BDD nodes. Further, the ReFlow resolves the infeasible partitions with a negligible overhead.

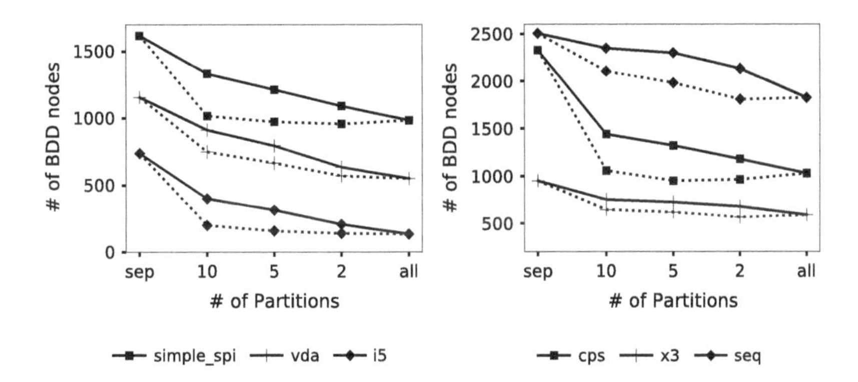

FIGURE 17.7 Relation of number of BDD nodes with number of partitions.

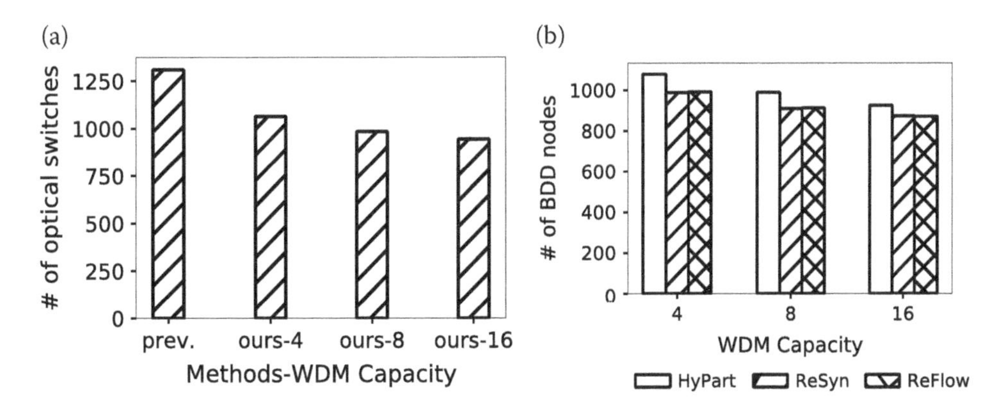

FIGURE 17.8 (a) Number of optical switches for different WDM capacities and (b) number of BDD nodes at different flow steps.

17.4 AREA-EFFICIENT OPTICAL NEURAL NETWORKS

17.4.1 Background of Optical Neural Networks

In this section, we embark on the optical analog computing paradigm, highlighting the optical neural network (ONN). We start with the introduction of the optical components and classic ONN architecture. Mach-Zehnder Interferometer (MZI) is the fundamental building block of the ONN. Figure 17.9(a) shows the schematic of a 2 × 2 MZI. The working principle of MZI is based on interference of light. Initially, two lights source from the inputs, entering the first coupling region and split into the two arms of the interferometer and re-combined. With the phase difference (φ) induced by the coupling region, MZI can provide constructive or destructive interference. Specifically, suppose the input modal amplitudes are y_1 and y_2; output modal amplitudes are y_1', y_2'; the transfer relation of a 2 × 2 MZI can be written as

$$\begin{pmatrix} y_1' \\ y_2' \end{pmatrix} = \begin{pmatrix} \cos\varphi & \sin\varphi \\ -\sin\varphi & \cos\varphi \end{pmatrix} \begin{pmatrix} y_1 \\ y_2 \end{pmatrix} \tag{17.5}$$

The phase difference φ is determined by the length of the coupling region or the temperature, so that the transfer behavior of MZI can be adjusted.

The classic ONN realizes the basic MLP neural network [17] as shown in Figure 17.10. In MLP, each hidden layer is a fully connected layer. The fully connected layer has two parts: the linear part which realizes a weight matrix W and the non-linear part which realizes an activation function. To implement a real weight matrix W for the linear part, it is first decomposed into three matrices using singular value decomposition $W = U\,\Sigma\,V^*$, where U and V^* are unitary matrices; Σ is a diagonal matrix whose diagonal values are non-negative real numbers called singular values; $(\cdot)^*$ denotes the conjugate transpose. A square matrix A is unitary if the product of it and its conjugate transpose is identity, i.e., $AA^* = I$. Each unitary matrix is implemented with a triangular planar array with 2 × 2 MZIs shown in Figure 17.9(b). As can be verified, the transfer matrix of a 2 × 2 MZI is a 2-dimensional unitary matrix $U(2)$. It is proved in [13,16] that, by joining multiple $U(2)$ blocks into the triangular array, we can realize any arbitrary unitary matrix. As for the real non-negative diagonal matrix Σ which simply performs a scaling operation, it can be implemented using optical attenuators or optical amplification materials. The area overhead of each attenuator is estimated to be the size of an MZI: indeed, if

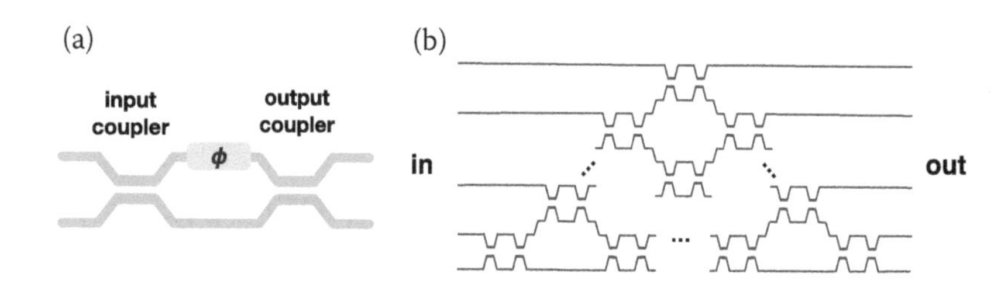

FIGURE 17.9 (a) MZI schematic and (b) MZI array for unitary implementation.

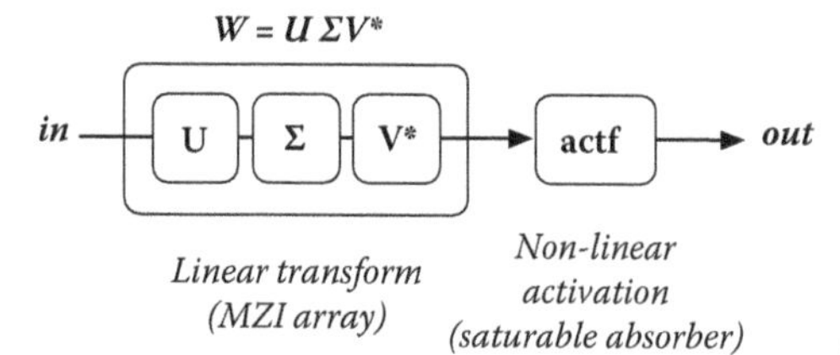

FIGURE 17.10 Basic ONN architecture layer.

the scaling factor is smaller than 1, we can readily use an MZI with one of its outputs to realize it. It should be noted that the nonlinear activation part is not easy to be fully integrated in optical domain. The authors of [17] offered the saturable absorber as a potential candidate to simulate the nonlinearity function in their experiment. The unitary matrices are the area-dominating blocks of the architecture.

17.4.2 Slimmed Optical Neural Network

To reduce the area, [18] proposes a slimmed ONN architecture using hardware-software co-design. The general architecture of a single fully connected layer is shown in Figure 17.11. The linear weight matrix (W) of each layer is implemented by three parts, a tree network (T), a unitary network (U) and a diagonal network (Σ). The transfer matrix of the linear part is then

$$W = TU\Sigma. \tag{17.6}$$

The tree network allows sparse connections of the inputs and outputs of the network. The next two blocks of the linear part, the unitary network and the diagonal network, share the same physical implementation as their counterparts of the classic neural network [17]. The difference from the basic architecture is, in the training, their parameters are directly encoded as training parameters and are decided by the training engine, so that the neural network accuracy is optimized. Since the area bottleneck comes from the unitary block, by skillfully eliminating one of the unitary matrices, we are able to achieve area reduction.

Tree Network Construction: A tree network example which contains three subtrees, is schemed in Figure 17.12. We can prove that, for each N-input *subtree* in the network, with arbitrary modal amplitude ratios satisfying the constraint

$$\alpha_1, \alpha_2, \ldots, \alpha_N, \quad \text{s. t.,} \sum_{i=1}^{N} \alpha_i^2 = 1, \; -1 \le \alpha_i \le 1, \; i = 1, \ldots, N \tag{17.7}$$

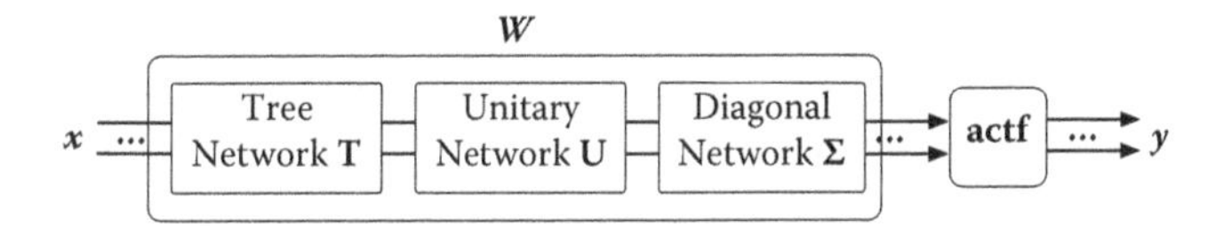

FIGURE 17.11 Proposed slimmed layer implementation.

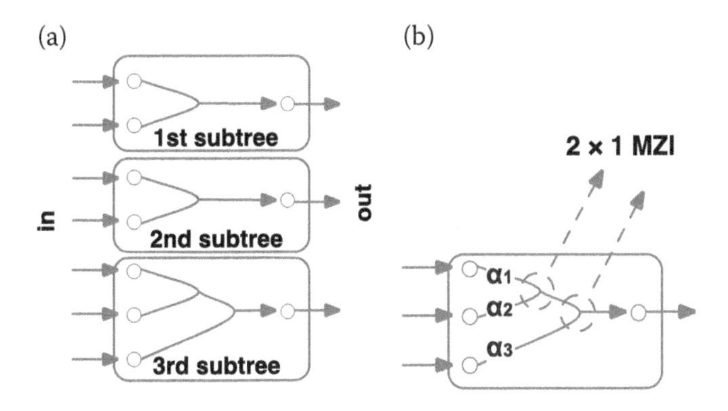

FIGURE 17.12 Tree network example.

we can implement it by cascading $(N - 1)$ (2×1) MZIs. Hence, the area of this construction is upper bounded by the input dimension. The best modal amplitude ratios α_is are determined by the training. The energy conservation is also encoded simultaneously.

Unitary and Diagonal Block Construction: Following the tree network, the second and the third blocks of the proposed architecture are the unitary block and the diagonal block. The optical implementations follow the same as the basic architecture. However, in order to save one unitary matrix, SVD is not reflected in our physical architecture. On that account, the software setup for the training is also changed. Basically, we will apply the gradient descent optimization engine to train all the parameters of the unitary block and the diagonal block as well. For the unitary block U, which is constrained by $UU^* = I$, we add the regularization term

$$reg = \|UU^* - I\|_F \tag{17.8}$$

to the training objective that originally embodies the accuracy objective, where $\|\cdot\|_F$ is the Frobenius norm. This regularization will impel the optimization engine to find a matrix that is close to a unitary matrix. As shown in the experiments, the trained result U_t can be sufficiently close to the real unitary. However, as only a true unitary can be implemented by the MZI array, we further find a closest unitary matrix by leveraging the software SVD-decomposition. Specifically, the trained matrix is decomposed as

$$U_t = PSQ^* \tag{17.9}$$

Since U_t is close to a unitary matrix, meaning that the column vectors being close to orthogonal, the decomposed singular value diagonal matrix S will be close to an identity matrix I. Therefore, U_t can be approximated by $U_a = PQ^*$. The matrices P and Q^* are true unitary and so is their product U_a. As for the effectiveness of this approximation, we have the following claim.

Claim 17.1: *To minimize the regularization term reg is equivalent to minimize the Frobenius norm of the difference* $\varepsilon = U_t - U_{aF}$.

Based on the true unitary matrix U_a, we parameterize the phases of the MZI array for constructing the unitary block. The last block before the activation function is a diagonal block. The dimension of the diagonal block is the same as the unitary block. The diagonal elements are also encoded in training to optimize the objective.

17.4.3 Simulation Results

We implemented the proposed architecture in Tensorflow and tested it on a machine with an Intel Core i9-7900X CPU and an NVIDIA TitanXp GPU. The experiments were conducted on the machine learning dataset MNIST [27]. We also implemented the neural network model of different sizes (labeled from N1 to N9). Figure 17.13 shows the comparison of the classic architecture and the slimmed architecture (using U_a), in terms of (a) testing accuracy and (b) area measured by MZI count. The greatest testing accuracy degradation is a negligible 0.0088; the average is 0.0058. The saving of the area varies from 15% to 38%. The average is 28.7%. The second set of experiments studies the robustness advantage of a slimmed architecture. Similar to other common neuromorphic computing hardwares, the ONN hardwares, especially the phases of the MZIs, are also exposed to the problem of noise, such as manufacturing imperfectness and temperature crosstalk [17]. As depicted in the box plots of Figure 17.14, for different random noise amplitude settings imposed upon the phases of MZI, the accuracy distribution of the slimmed architecture not only has higher average and geometric means but also a smaller variation range among all the samples.

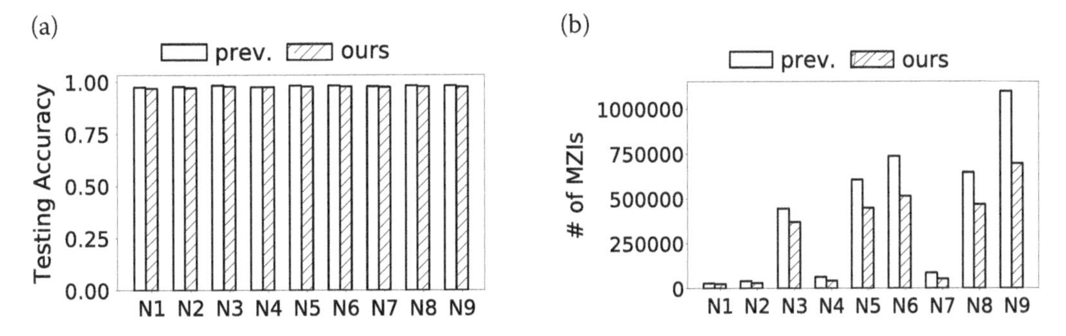

FIGURE 17.13 (a) Testing accuracy and (b) number of MZIs for different ONN configurations.

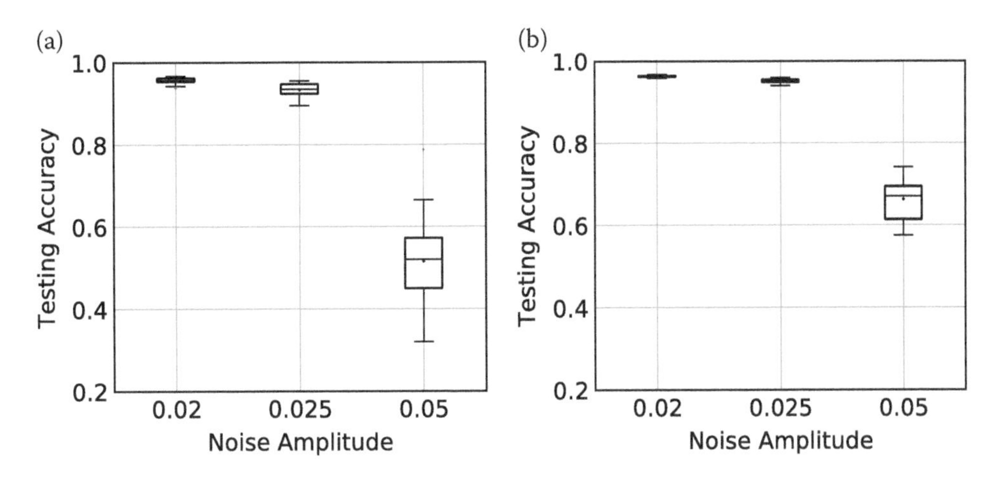

FIGURE 17.14 Noise robustness of (a) the previous architecture and (b) the proposed architecture.

17.5 CONCLUSION AND FUTURE DIRECTIONS

This chapter discusses two optical computing paradigms: digital computing and analog computing. Automated design methodologies are introduced to enhance different aspects including scalability and robustness. There are still many challenges and research opportunities for each of them.

Scalability: The optical power loss is a major obstacle to build complicated systems. As demonstrated in the previous sections, optical devices lack the capability of signal restoration and input-output isolation as CMOS counterparts, optical signal inevitably diminishes throughout the computation and may even become indistinguishable to environmental noise. Therefore, the scalability (as well robustness) of optical circuits is considerably limited by optical power loss. One practical direction is to introduce the idea of "buffer insertion", i.e., optical amplifier and OE/EO converter insertion into the synthesis flow.

Robustness: like many emerging technologies, computing optics may suffer from manufacturing defects and process/environmental variations. Robust and fault-tolerant designs that can endure these uncertainties are critical to bringing optical computing in practice. Importantly, as the integration advances, crosstalk noise may also become a critical problem of the signal integrity. The crosstalk noise has already been revealed in large-dimensional optical routers [28] and it would be worthwhile to revisit the solutions in the new context of computing.

REFERENCES

[1] D. A. Miller, "Attojoule optoelectronics for low-energy information processing and communications," Journal of Lightwave Technology, 35, no. 3, pp. 346–396, 2017.
[2] C. Sun, M. T. Wade, Y. Lee, J. S. Orcutt, L. Alloatti, M. S. Georgas, A. S. Waterman, J. M. Shainline, R. R. Avizienis, S. Lin, et al., "Single-chip microprocessor that communicates directly using light," Nature 528, no. 7583, pp. 534–538, 2015.
[3] Y. Tian, L. Zhang, J. Ding, and L. Yang, "Demonstration of electro-optic half-adder using silicon photonic integrated circuits," Optics Express22, no. 6, pp. 6958–6965, 2014.
[4] P. Zhou, L. Zhang, Y. Tian, and L. Yang, "10 GHz electro-optical OR/NOR directed logic device based on silicon micro-ring resonators," Optics Letters 39, no. 7, pp. 1937–1940, 2014.
[5] L. Yang, L. Zhang, C. Guo, and J. Ding, "XOR and XNOR operations at 12.5 Gb/s using cascaded carrier-depletion microring resonators," Optics Express 22, no. 3, pp. 2996–30122014.
[6] Y. Tian, L. Zhang, and L. Yang, "Electro-optic directed AND/NAND logic circuit based on two parallel microring resonators," Optics Express 20, no. 15, pp. 16794–16800, 2012.
[7] L. Zhang, J. Ding, Y. Tian, R. Ji, L. Yang, H. Chen, P. Zhou, Y. Lu, W. Zhu, and R. Min, "Electro-optic directed logic circuit based on microring resonators for XOR/XNOR operations," Optics Express 20, no. 11, pp. 11605–11614, 2012.
[8] C. Condrat, P. Kalla, and S. Blair, "Logic synthesis for integrated optics," in Proceedings of the 21st edition of the great lakes symposium on Great lakes symposium on VLSI, pp. 13–18. 2011.
[9] R. Wille, O. Keszocze, C. Hopfmuller, and R. Drechsler, "Reverse BDD-based synthesis for splitter-free optical circuits," In The 20th Asia and South Pacific Design Automation Conference, IEEE, pp. 172–177, 2015.
[10] Z. Zhao, Z. Wang, Z. Ying, S. Dhar, R. T. Chen, and D. Z. Pan, "Logic synthesis for energy-efficient photonic integrated circuits," in Proceedings of the 23rd Asia and South Pacific Design Automation Conference, IEEE, pp. 355–360, 2018.
[11] Z. Ying, Z. Wang, Z. Zhao, S. Dhar, D. Z. Pan, R. Soref, and R. T. Chen, "Silicon microdisk-based full adders for optical computing," Optics Letters 43, no. 5, pp. 983–986, 2018.
[12] Z. Zhao, D. Liu, Z. Ying, B. Xu, R. T. Chen, and D. Z. Pan, "Exploiting wavelength division multiplexing for optical logic synthesis," in Proceedings of the 2019 Design, Automation and Test in Europe Conference (DATE), IEEE, pp. 1567–1570, 2019.
[13] M. Reck, A. Zeilinger, H. J. Bernstein, and P. Bertani, "Experimental realization of any discrete unitary operator," Physical Review Letters 73, no. 1, pp. 58, 1994.
[14] D. A. Miller, "Perfect optics with imperfect components," Optica 2, no. 8, pp. 747–750, 2015.
[15] A. Ribeiro, A. Ruocco, L. Vanacker, and W. Bogaerts, "Demonstration of a 4 × 4-port universal linear circuit," Optica 3, no. 12, pp. 1348–1357, 2016.
[16] W. R. Clements, P. C. Humphreys, B. J. Metcalf, W. S. Kolthammer, and I. A. Walmsley, "Optimal design for universal multiport interferometers," Optica 3, no. 12, pp. 1460–1465, 2016.
[17] Y. Shen, N. C. Harris, S. Skirlo, M. Prabhu, T. Baehr-Jones, M. Hochberg, X. Sun, S. Zhao, H. Larochelle, D. Englund, et al., "Deep learning with coherent nanophotonic circuits," Nature Photonics 11, no. 7, pp. 441–446, 2017.
[18] Z. Zhao, D. Liu, M. Li, Z. Ying, L. Zhang, B. Xu, B. Yu, R. T. Chen, and D. Z. Pan, "Hardware-software co-design of slimmed optical neural networks," in Proceedings of the 24th Asia and South Pacific Design Automation Conference, ACM, pp. 705–710, 2019.
[19] R. E. Bryant, "Symbolic Boolean manipulation with ordered binary-decision diagrams," ACM Computing Surveys (CSUR) 24, no. 3, pp. 293–318, 1992.
[20] P. A. Parrilo and B. Sturmfels, "Minimizing polynomial functions," Algorithmic and quantitative real algebraic geometry, DIMACS Series in Discrete Mathematics and Theoretical Computer Science 60, pp. 83–99, 2003.
[21] F. Somenzi, "CUDD: Colorado University Decision Diagram package." http://vlsi.colorado.edu/fabio/CUDD/. Accessed: 2015-09-30.
[22] "MCNC benchmarks." https://s2.smu.edu/~manikas/Benchmarks/MCNC_Benchmark_Netlists.html. Accessed: 2015-09-30.
[23] "IWLS 2005 benchmarks." http://iwls.org/iwls2005/benchmarks.html. Accessed: 2015-09-30.
[24] D. Liu, Z. Zhao, Z. Wang, Z. Ying, R. T. Chen, and D. Z. Pan, "Operon: Optical-electrical power-efficient route synthesis for on-chip signals," in Proceedings of the 55th Annual Design Automation Conference, ACM, pp. 1–6, 2018.

[25] J. M. Kahn and K.-P. Ho, "Spectral efficiency limits and modulation/detection techniques for DWDM systems," IEEE Journal of Selected Topics in Quantum Electronics 10, no. 2, pp. 259–272, 2004.
[26] K. Andreev and H. Racke, "Balanced graph partitioning," Theory of Computing Systems 39, no. 6, pp. 929-939, 2006.
[27] Y. LeCun, "The MNIST database of handwritten digits." http://yann.lecun.com/exdb/mnist/, 1998. Accessed: 2016-09-30.
[28] Y. Xie, J. Xu, J. Zhang, Z. Wu, and G. Xia, "Crosstalk noise analysis and optimization in 5 × 5 hitless silicon-based optical router for optical networks-on-chip (ONoC)," Journal of Lightwave Technology 30, no. 1, pp. 198–203, 2012.

18 High-Performance Programmable MZI-Based Optical Processors

Farhad Shokraneh, Simon Geoffroy-Gagnon, and Odile Liboiron-Ladouceur

CONTENTS

18.1 INTRODUCTION AND MOTIVATION

Over the last few years, the considerable progress in silicon photonics technology has made it possible to fabricate multiport optical interferometric structures with small footprints to perform computations. The increa singly developed and standardized process design kit (PDK) libraries in silicon photonics [1,2], along with the large variety of photonic components have led to large-scale photonic integration capabilities, which fundamentally reduces the complexity in assembly, calibration, and configuration of such computational structures [3–6]. As a result, on-chip photonic systems are increasingly in demand for large-scale computational architectures. The interest in reconfigurable multiport linear optical interferometers is growing rapidly, due to the high speed and low power consumption. Thanks to the inherent parallelism in optics, these structures provide linear computational time complexity for matrix multiplications and thus can be exploited as computational accelerators. Interestingly, these structures can be theoretically analyzed and experimentally configured. The use of reconfigurable MZIs to construct various mesh topologies allows for implementation of unitary transformation matrices for different applications. In this regard, the computations are done by processing the field properties of the propagating light in the optical system. The unitary matrix linearly relates an input vector of optical signals to an output vector of coherent optical modes [3,4,7–13]. The phase shifters in such a mesh of Mach-Zehnder interferometers (MZIs) are employed for their simple experimental calibration and configuration, which

DOI: 10.1201/9780429292033-18

make the structure suitable candidate to serve as a reconfigurable linear optical processor. The programmed MZIs can redistribute the light to implement any arbitrary unitary transformation matrix [3,4,7–10]. The unitary section is of great importance as any linear matrix can be decomposed by SVD into two unitary matrices and one diagonal matrix. In other words, such an optical processor can be programmed to implement the linear transformation matrix of a given application by adjusting the phase shifters of the MZIs in the structure [7,10,14].

There have been several demonstrations of these multiport reconfigurable MZI-based structures performing matrix multiplications in computational systems using machine learning algorithms, e.g., ONNs [3,15,16]. In [9], a dual-drive tunable directional coupler was proposed to serve as a low-insertion-loss MZI in such reconfigurable optical processors. It was also demonstrated how to perform singular value decomposition (SVD) in optics to implement an arbitrary linear transformation matrix of a given application [17,18]. Although digital computers exploiting graphical processing units (GPUs) have been commonly used to efficiently perform computational tasks that use large amounts of matrix multiplications, such as NNs, optical processors have been recognized as an excellent alternative in terms of speed and energy efficiency. The implemented ONNs by $N \times N$ optical processors can perform multiply-accumulate (MAC) operations in matrix multiplications with a time complexity in the order of N, whereas digital platforms scale matrix multiplications by a computational time complexity of greater than N^2 [19,20]. Compared to state-of-the-art electronic-based counterparts with various platforms [21], photonic NNs are promising in MAC operations, exhibiting computational speed and efficiency above 10^8 MegaMAC (MMAC)/s/cm^2 and 10^4 GigaMAC (GMAC)/s/W, respectively. In electronic application-specific-integrated circuits (ASICs) [22], the power consumption can be reduced considerably by using a trained ONN, which requires a one-time passive configuration for implementing the weight matrix without additional power consumption [15]. In [3], an ONN is experimentally implemented for vowel recognition using a programmable mesh of MZIs that employs the Reck structure [7] to implement singular value decomposition (SVD) to perform fully optical matrix multiplications. To implement the 4×4 ONN, the silicon photonic chip was programmed to perform SVD in a two-layer ONN. The optical nonlinear activation function is simulated on a digital computer. There were 24 MZIs with 48 phase shifters programmed for experimental classification of the data samples to achieve a final classification accuracy of 76.7%. The implemented ONN is at least two orders of magnitude faster than state-of-the-art electronic counterparts. In terms of power efficiency, the optical processor performs at least five orders of magnitude better than conventional GPUs. According to [23], such structures provide 30 fJ/MAC energy efficiency and 0.56 TMACs/s/mm^2 computational density for matrix multiplications compared to that of modern GPUs [24] at 0.43 pJ/MAC and 580 GMACs/s/mm^2. Consequently, MZI-based optical processors are promising candidates for practical computational accelerators.

18.2 THEORY AND ANALYSIS

This section presents the theoretical and experimental analysis of multiport reconfigurable MZI-based optical processors. The fundamental theory of a 2×2 reconfigurable MZI is used to determine its transformation matrix which is a unitary matrix of degree two, i.e., U(2). The 2×2 reconfigurable MZI can be tuned with one thermo-optic phase shifter in one of its internal arms and another one at one of its output arms. The unitary transformation matrix of the reconfigurable MZI is determined by the product of the transformation matrices of its directional couplers and phase shifters. Similarly, it is demonstrated how the unitary transformation matrix of a reconfigurable multi-port optical processor can be obtained through the product of the unitary transformation matrices of its constituent MZIs. In the case of a 4×4 structure, it implements a unitary transformation matrix of degree four, i.e., U(4), which can be calculated by the successive multiplications of the MZIs unitary matrices. The programming process of the U(4) is carried out based on the decomposition of its unitary transformation matrix into that of its constituent MZIs. The decomposition process of a unitary transformation matrix provides the required phase shifts to implement it in optics. The experimental characterization results show that the MZIs in the device exhibit random phase offsets originating from fabrication process variations. Thus, all the MZIs in the device are characterized experimentally through a presented calibration scheme prior to programming it for a given application. Finally, as an example, an arbitrary transformation matrix is decomposed into the unitary transformation matrices of the MZIs in the 4 × 4 optical processor. It is explained how to determine the required phase shifts and their corresponding bias voltages to program the optical processor experimentally.

18.2.1 2 × 2 Reconfigurable MZI

A reconfigurable MZI-based linear optical processor is a compact and energy-efficient integrated structure that can perform fast and power efficient matrix multiplications. Figure 18.1 shows the schematic of a 2 × 2 reconfigurable MZI, which is the building block of the MZI-based optical processor. The 2 × 2 reconfigurable MZI consists of two 3-dB (50:50) couplers with one phase shifter (θ) on one of the internal arms of the MZI and an external phase shifter (ϕ) at one of the outputs of the MZI. The internal phase shifter controls the power at the MZI outputs. The external one determines the relative phase of the two MZI outputs. As a result, it can be reconfigured through adjusting its two phase shifters.

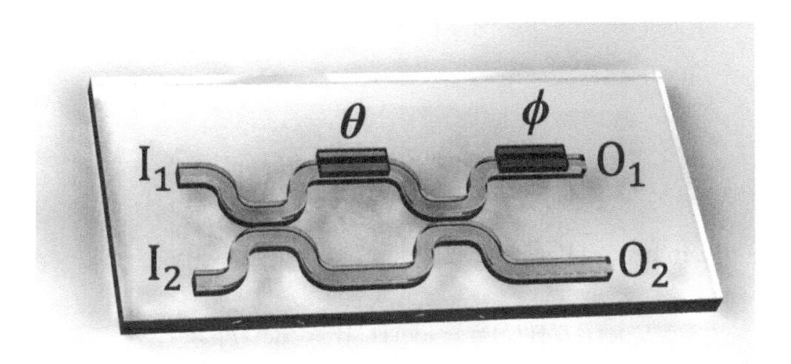

FIGURE 18.1 Schematic illustration of the 2 × 2 reconfigurable MZI. Adapted with permission from [32] The Optical Society.

The linear transformation matrix of a lossless reconfigurable MZI is determined by the product of the transformation matrices of its two 3-dB couplers with the field coupling coefficient of $\sqrt{\rho}$=0.5 and the phase shifters θ and ϕ as follows:

$$
\begin{aligned}
[D_{MZI}] = \begin{bmatrix} u_{11} & u_{12} \\ u_{21} & u_{22} \end{bmatrix} = \\
\begin{bmatrix} e^{j\phi} & 0 \\ 0 & 1 \end{bmatrix}\begin{bmatrix} \sqrt{\rho} & j\sqrt{1-\rho} \\ j\sqrt{1-\rho} & \sqrt{\rho} \end{bmatrix}\begin{bmatrix} e^{j\theta} & 0 \\ 0 & 1 \end{bmatrix}\begin{bmatrix} \sqrt{\rho} & j\sqrt{1-\rho} \\ j\sqrt{1-\rho} & \sqrt{\rho} \end{bmatrix} = \\
\frac{1}{2}\begin{bmatrix} e^{j\phi} & 0 \\ 0 & 1 \end{bmatrix}\begin{bmatrix} (e^{j\theta}-1) & j(1+e^{j\theta}) \\ j(1+e^{j\theta}) & (1-e^{j\theta}) \end{bmatrix} = \\
je^{j\left(\frac{\theta}{2}\right)}\begin{bmatrix} e^{j\phi} sin\left(\frac{\theta}{2}\right) & e^{j\phi} cos\left(\frac{\theta}{2}\right) \\ cos\left(\frac{\theta}{2}\right) & -sin\left(\frac{\theta}{2}\right) \end{bmatrix}.
\end{aligned}
\tag{18.1}
$$

$[D_{MZI}]$ is a unitary transformation matrix of degree two i.e., U(2), which relates the MZI input optical signals to the output ones as follows:

$$\begin{bmatrix} E_{O1} \\ E_{O2} \end{bmatrix} = [D_{MZI}] \begin{bmatrix} E_{I1} \\ E_{I2} \end{bmatrix}. \tag{18.2}$$

The unitary transformation matrix $[D_{MZI}]$ is a complex square matrix which can perform both amplitude transformations and phase rotations on the propagating fields to transform the input optical signals to output optical coherent modes. It should be noted that to implement an arbitrary unitary operation, an additional phase shifter is required on one of the input waveguides of the MZI. This additional phase shifter enables the unitary operation to combine the power on the two input waveguides to one output, for any specific relative phase of the two inputs. In other words, it allows for nulling out the power on the other output by adjusting the relative phase of the two inputs. In this chapter, the implemented unitary matrices are specifically generated for the presented application and thus the additional phase shifters at the inputs are not used.

An interesting property of a unitary transformation matrix $[U]$ is that its conjugate transpose $[U]^\dagger$ is equal to its inverse $[U]^{-1}$ [25] as given by

$$[U]^\dagger = [U]^{-1}. \tag{18.3}$$

As a result,

$$[U][U]^\dagger = [U][U]^{-1} = [I], \tag{18.4}$$

where $[I]$ is the identity matrix. Therefore,

$$\det([U][U]^\dagger) = |\det[U]|^2 = \det[I] = 1, \tag{18.5}$$

Consequently, a unitary matrix has an absolute value of determinant equal to one,

$$|\det[U]| = 1. \tag{18.6}$$

The reconfigurable MZI can be tuned to control the fields, or in other words the power levels, and the relative phase of the two output ports by adjusting its phase shifters (θ and ϕ). For a lossless MZI, if the input fields are given by $E_{I_1} = E_{in}\, e^{j0}$ and $E_{I_2} = 0$, using the transformation matrix of the MZI given in equation (18.1), the electric fields at the MZI outputs are given by

$$E_{O_1} = je^{j\left(\frac{\theta}{2}\right)} E_{in} sin\left(\frac{\theta}{2}\right) e^{j\phi}, \tag{18.7}$$

$$E_{O_2} = je^{j\left(\frac{\theta}{2}\right)} E_{in}\, cos\left(\frac{\theta}{2}\right) e^{j0}. \tag{18.8}$$

According to (18.7) and (18.8), the output field amplitudes are controlled by the phase shifter θ while the phase shifter ϕ determines the relative phase of the outputs. Figure 18.2 shows the electric field propagation after every constituent component of the reconfigurable MZI.

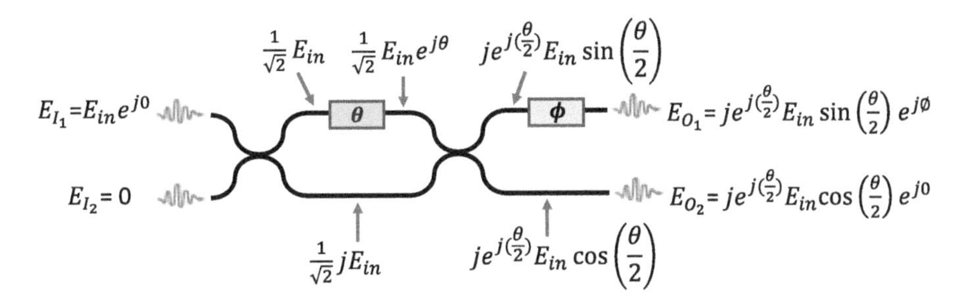

FIGURE 18.2 Field transmission of each constituent component of a 2 × 2 reconfigurable MZI.

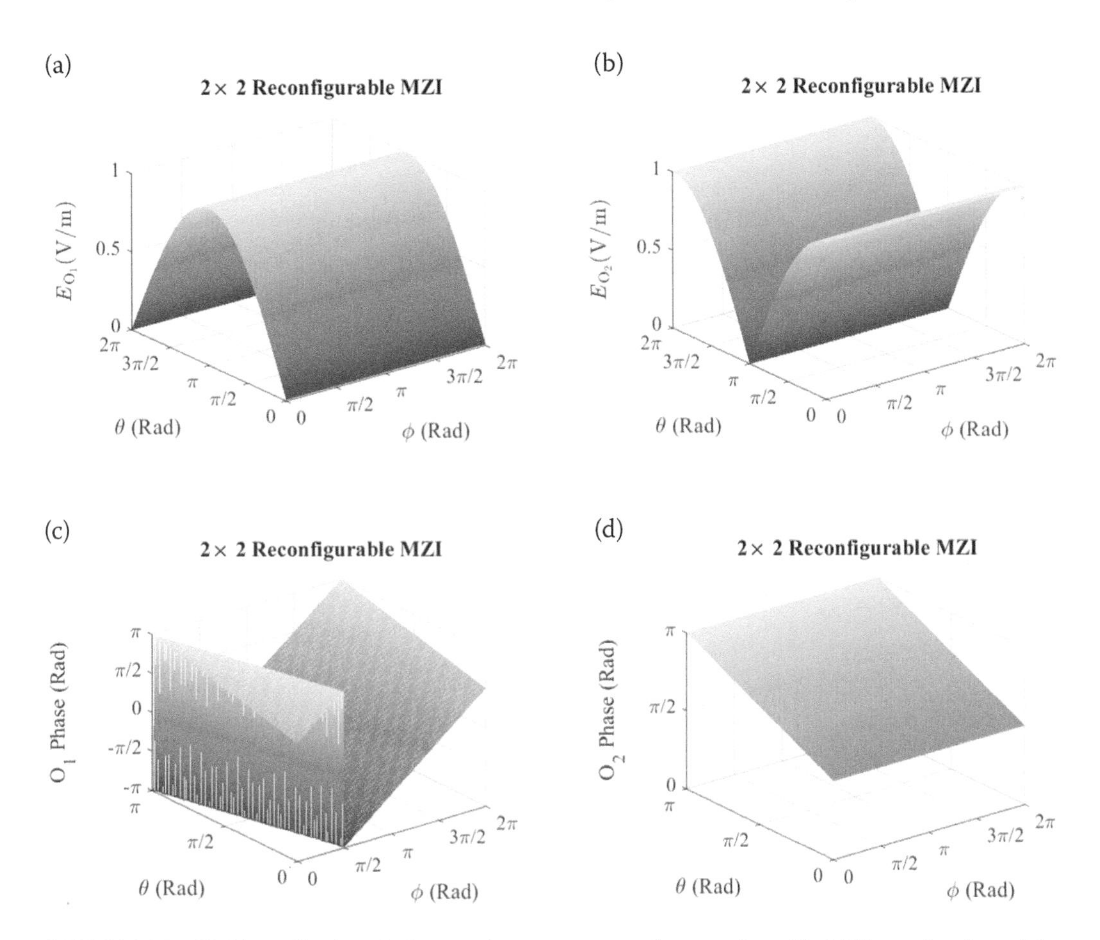

FIGURE 18.3 Field amplitude and phase at the output ports of a reconfigurable MZI as a function of the phase shifts θ and ϕ. (a) and (b) The field amplitude at the output ports O_1 and O_2, respectively, which is controlled by the phase shifter θ. (c) and (d) The phases at the output ports O_1 and O_2, respectively, which is adjusted by the phase shifter ϕ. The phase ranges were chosen for better clarity of the results.

Figures 18.3a and 18.3b demonstrate the transmitted electric field amplitudes, $|E_{O_1}|$ and $|E_{O_2}|$ as a function of θ and ϕ at the two outputs of the MZI for inputs $E_{I_1}=1\,e^{j0}$ and $E_{I_2}=0$. As can be seen in the figures and according to equations (18.7) and (18.8), the transmitted field amplitude is controlled by θ and the maximum field transmission at the output ports O_1 and O_2 corresponds to $\theta=\pi$ and $\theta=0$, respectively. According to equations (18.7) and (18.8), and as can be inferred from Figures 18.3c and 18.3c,

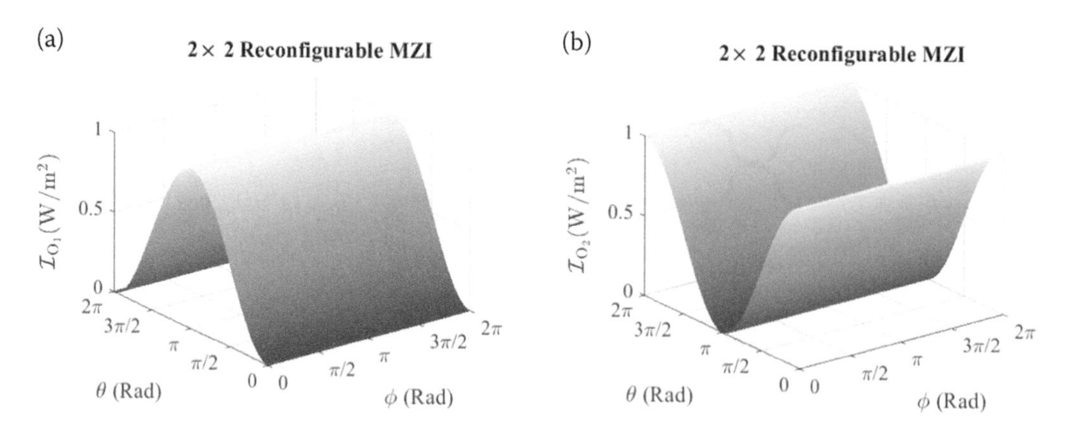

FIGURE 18.4 Optical field intensity at the output ports, (a) O_1 and (b) O_2, of the reconfigurable MZI as a function of the phase shifts θ and ϕ.

both output fields have a phase shift $je^{\left(\frac{\theta}{2}\right)}$ in common and the phase difference at the output ports is determined by the phase shift ϕ. For instance, in the case of $\theta = 0$ and $\phi = \left(\frac{\pi}{2}\right)$ the field phases at the outputs O_1 and O_2 are π and $\frac{\pi}{2}$, respectively.

In this context, the field intensity $\mathcal{I}$ is expressed as

$$\mathcal{I} \propto (E \cdot \overline{E}) = |E|^2, \tag{18.9}$$

where $\overline{E}$ denotes the complex conjugate of electric field E. Consequently, the field intensities at the two outputs are obtained by

$$\mathcal{I}_{O_1} = |E_{in}^2|\, sin^2\left(\frac{\theta}{2}\right) = \mathcal{I}_{in}\, sin^2\left(\frac{\theta}{2}\right), \tag{18.10}$$

$$\mathcal{I}_{O_2} = |E_{in}^2|\, cos^2\left(\frac{\theta}{2}\right) = \mathcal{I}_{in}\, cos^2\left(\frac{\theta}{2}\right), \tag{18.11}$$

where $\mathcal{T}_{O_1} = sin^2\left(\frac{\theta}{2}\right)$ and $\mathcal{T}_{O_2} = cos^2\left(\frac{\theta}{2}\right)$ represent the power transmission coefficients of the MZI outputs O_1 and O_2, respectively, such that $\mathcal{T}_{O_1} + \mathcal{T}_{O_2}=1$. $\theta=0$ leads to the total transmission at O_2, whereas $\theta = \pi$ results in maximum transmission at O_1. Figure 18.4 shows the field intensities at the two outputs of the reconfigurable MZI, $\mathcal{I}_{O_1}$ and $\mathcal{I}_{O_2}$, for the example $E_{I_1}=1\, e^{j0}$ and E_{I_2}.

As can be inferred from Figure 18.4 and equations (18.10) and (18.11), the internal phase shifter can be used to control the power levels at the outputs of the MZI. It is also expected that the input light intensity is equal to the total light intensity at the two outputs of an ideal lossless MZI, which can be concluded from equations (18.10) and (18.11) as follows:

$$\mathcal{I}_{in} = \mathcal{I}_{O_1} + \mathcal{I}_{O_2}. \tag{18.12}$$

It is essential to note that although the phase shifter ϕ of a single MZI does not control the light intensities at its two outputs, it affects the optical power at the outputs of the subsequent MZIs in a mesh of MZIs.

18.3 4 × 4 RECK MZI-BASED OPTICAL PROCESSOR

An $N \times N$ Reck structure shown in Figure 18.5 consists of $n = N(N-1)/2$ MZIs, i.e., $N^2 - N$ phase shifters [7,12]. To represent a universal structure that can implement an arbitrary unitary matrix using the triangular mesh, N additional phase shifters should be added at its inputs. The additional phase shifters complement the required N^2 degrees of freedom to construct the N-dimensional arbitrary unitary matrix with N^2 matrix elements. Adding N additional phase shifters at the inputs enables the Reck mesh to combine the power on its input waveguides to one output port, for any specific relative phase of its inputs. Therefore, it can perform any rotation in a unitary group of degree N, U(N). In this chapter, the unitary matrices implemented by the optical processors are generated for the presented specific application, i.e., ONNs. As a result, the N additional phase shifters at the inputs of the structure were not used.

In this respect, N denotes the number of main channels from N input ports to N output ports, i.e., $I_i - O_i$ for $i = 1, 2, 3, \ldots, N$. Each MZI can be reconfigured individually by controlling its two phase shifters. To construct the transformation matrix of an U(N), the unitary matrix of each MZI is diagonally presented on a two-dimensional subspace within an N-dimensional Hilbert space ($H_{N\times N}$) leaving ($N-2$)-dimensional subspace unchanged [7]. In this regard, the diagonal elements of $[D_{MZI}]$ are in line with the 1 elements in the Hilbert vector space. Therefore, the unitary transformation matrix of the n MZIs within U(N) can be expressed as

$$[D^{(n)}_{MZI\,(s,t)}]_{H_{N\times N}} = \begin{bmatrix} 1 & 0 & & \cdots & \cdots & & \cdots & 0 \\ 0 & 1 & \ddots & \cdots & \cdots & & \cdots & 0 \\ \vdots & \ddots & \ddots & & & & \vdots & \vdots \\ 0 & \cdots & & u_{11} & u_{12} & & \cdots & 0 \\ 0 & \cdots & & u_{21} & u_{22} & & \cdots & 0 \\ \vdots & \vdots & & & & \ddots & \ddots & \vdots \\ 0 & \cdots & & \cdots & \cdots & \ddots & 1 & 0 \\ 0 & 0 & & \cdots & \cdots & & 0 & 1 \end{bmatrix}_{N\times N}. \quad (18.13)$$

In other words, $[D_{MZI}]$ is defined in an N-dimensional Hilbert space using a two-dimensional subspace of it, which is along with the diagonal elements of 1s while the unchanged off-diagonal elements are 0s, i.e., similar to an identity matrix of degree N ($[I]_{(N)}$). The transformation matrix of an U(N) with N main channels can be determined by the multiplication of the transformation matrices of the MZIs each of which is connected to two adjacent channels, s and t, i.e., $t = 2, 3, \ldots, N$ and $s = t - 1$.

As can be seen in Figure 18.5, the 4 × 4 unitary structure has four main channels, i.e., $I_1 - O_1$, $I_2 - O_2$, $I_3 - O_3$ and $I_4 - O_4$, representing channels one to four, respectively. For instance, MZI (3) is connected to channels one and two ($s = 1$ and $t = 2$). MZIs (1) to (6) construct a unitary transformation matrix denoted by $[T_{U(4)}]$. Each MZI in the structure can be configured by controlling the power and the relative phase of its output ports through tuning its internal and external

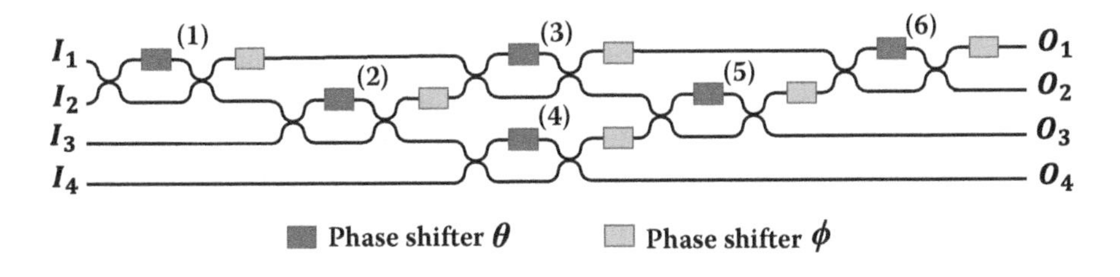

FIGURE 18.5 Schematic of the 4 × 4 Reck MZI-based reconfigurable optical processor consisting of MZIs (1) to (6) to implement the unitary section U(4) [7,12].

phase shifters, respectively. It is essential to note that, although, the external phase shifter ϕ of each MZI does not affect the optical power at its output ports, it changes the optical power splitting ratio of the subsequent MZIs in the optical processor.

To construct $[T_{U(4)}]$, the unitary matrices of MZIs (1) to (6) are defined on a two-dimensional subspace within a four-dimensional Hilbert space based upon their location in the structure being connected to channels s and t such that $t = 2, 3, 4$ and $s = t - 1$. For instance, MZI (1) is connected to channels one and two, i.e., $s = 1$, and $t = 2$. In this respect, the unitary transformation matrices of MZIs (1) to (6), $[D_{MZI}^{(n)}]_{H_{4\times4}}$, in the the 4×4 optical processor are given by

$$[D_{MZI}^{(1)}]_{H_{4\times4}} = \begin{bmatrix} u_{11}^{(1)} & u_{12}^{(1)} & 0 & 0 \\ u_{21}^{(1)} & u_{22}^{(1)} & 0 & 0 \\ 0 & 0 & 1 & 0 \\ 0 & 0 & 0 & 1 \end{bmatrix}, \quad [D_{MZI}^{(2)}]_{H_{4\times4}} = \begin{bmatrix} 1 & 0 & 0 & 0 \\ 0 & u_{11}^{(2)} & u_{12}^{(2)} & 0 \\ 0 & u_{21}^{(2)} & u_{22}^{(2)} & 0 \\ 0 & 0 & 0 & 1 \end{bmatrix}, 0.3cm$$

$$[D_{MZI}^{(3)}]_{H_{4\times4}} = \begin{bmatrix} u_{11}^{(3)} & u_{12}^{(3)} & 0 & 0 \\ u_{21}^{(3)} & u_{22}^{(3)} & 0 & 0 \\ 0 & 0 & 1 & 0 \\ 0 & 0 & 0 & 1 \end{bmatrix}, \quad [D_{MZI}^{(4)}]_{H_{4\times4}} = \begin{bmatrix} 1 & 0 & 0 & 0 \\ 0 & 1 & 0 & 0 \\ 0 & 0 & u_{11}^{(4)} & u_{12}^{(4)} \\ 0 & 0 & u_{21}^{(4)} & u_{22}^{(4)} \end{bmatrix}, 0.3cm \tag{18.14}$$

$$[D_{MZI}^{(5)}]_{H_{4\times4}} = \begin{bmatrix} 1 & 0 & 0 & 0 \\ 0 & u_{11}^{(5)} & u_{12}^{(5)} & 0 \\ 0 & u_{21}^{(5)} & u_{22}^{(5)} & 0 \\ 0 & 0 & 0 & 1 \end{bmatrix}, \quad [D_{MZI}^{(6)}]_{H_{4\times4}} = \begin{bmatrix} u_{11}^{(6)} & u_{12}^{(6)} & 0 & 0 \\ u_{21}^{(16)} & u_{22}^{(6)} & 0 & 0 \\ 0 & 0 & 1 & 0 \\ 0 & 0 & 0 & 1 \end{bmatrix}.$$

Consequently, the unitary transformation matrix of the U(4) is defined as follows:

$$[T_{U(4)}] = [D_{MZI}^{(6)}]_{H_{4\times4}} \cdot [D_{MZI}^{(5)}]_{H_{4\times4}} \cdot [D_{MZI}^{(4)}]_{H_{4\times4}} \cdot [D_{MZI}^{(3)}]_{H_{4\times4}} \cdot [D_{MZI}^{(2)}]_{H_{4\times4}} \cdot [D_{MZI}^{(1)}]_{H_{4\times4}}. \tag{18.15}$$

Each MZI contributes to the linear optical wave interactions in the physical device represented by $[T_{U(4)}]$ [17] given by

$$[T_{U(4)}] = \begin{bmatrix} U_{11} & U_{12} & U_{13} & U_{14} \\ U_{21} & U_{22} & U_{23} & U_{24} \\ U_{31} & U_{32} & U_{33} & U_{34} \\ U_{41} & U_{42} & U_{43} & U_{44} \end{bmatrix}, \tag{18.16}$$

where U_{kl} (k and $l \in \{1, 2, 3, 4\}$) are elements in $[T_{U(4)}]$, which can be determined by the product of matrices $[D_{MZI}^{(n)}]$ with $u_{pq}^{(n)}$, for p, $q \in \{1,2\}$ and $n \in \{1, 2, ..., 6\}$. The unitary transformation matrix is the core of Reck-mesh-based optical processors. It has been demonstrated that it is possible to implement efficient ONNs using the unitary section of the MZI-based optical meshes. The use of unitary section implemented by the Reck mesh allows for efficient computational tasks and better training in such ONNs [26[30]]. Additionally, the unitary section is of great importance in general ONNs since a general weight matrix, which is linear as opposed to simply unitary, can be decomposed through SVD into two unitary matrices and one diagonal matrix.

18.3.1 Programming the 4 × 4 Reck-Based Optical Processor

The programming process of an $N \times N$ Reck-topology-based reconfigurable optical processor is carried out based on the decomposition of the linear transformation matrix of $[T_{U(N)}]$ into the unitary matrices of the corresponding MZIs. In this regard, the unitary transformation matrix $[T_{U(N)}]$ given by an application, such as the weight matrix in ONNs, is successively multiplied by the inverse unitary transformation matrices of the MZIs defined on the corresponding Hilbert space, $[D_{MZI}^{(n)}]_{H_{N\times N}}^{-1}$. Due to the unitary property of the transformation matrix of each reconfigurable MZI, its inverse $[D_{MZI}]^{-1}$ is equivalent to its conjugate transpose, which can be given by

$$[D_{MZI}]^{-1} = -je^{-j\left(\frac{\theta}{2}\right)} \begin{bmatrix} e^{-j\phi} sin\left(\frac{\theta}{2}\right) & cos\left(\frac{\theta}{2}\right) \\ e^{-j\phi} cos\left(\frac{\theta}{2}\right) & -sin\left(\frac{\theta}{2}\right) \end{bmatrix}. \tag{18.17}$$

The decomposition process of $[T_{U(N)}]$ is equivalent to a reverse propagation, i.e., the light is coupled to the device from the right side of the device, where the output ports are used as inputs to couple the light and the input ports are used as outputs. As a result, the successive products should be done in the corresponding order starting from the left first layer of MZIs on the left side of the optical processor shown in Figure 18.5. It is essential to note that in each step of the successive multiplications of $[T_{U(N)}]$ by $[D_{MZI}^{(n)}]_{H_{4\times4}}^{-1}$, an off-diagonal element in the lower triangle of the resultant matrix becomes zero, a method similar to Gaussian elimination. Due to the unitary property of the resultant matrices in every step, once an off-diagonal element becomes zero, it will not be changed by the subsequent transformations. Moreover, in each row, when all off-diagonal elements become zero, the off-diagonal elements in the corresponding column also become zero in accordance with the unitary property. Thus, after the successive multiplications in a row, the effective dimension of the resultant matrix is

$$[T_{U(N)}] \cdot [D_{MZI(N,1)}]_{H_{N\times N}}^{-1} \cdot [D_{MZI(N,2)}]_{H_{N\times N}}^{-1} \cdots [D_{MZI(N,N-1)}]_{H_{N\times N}}^{-1} = \begin{bmatrix} [T_{U(N-1)}] & 0 \\ 0 & 1 \end{bmatrix}. \tag{18.18}$$

Eventually, this successive product results in the identity matrix $I_{(N)}$. Consequently, the required phase shifts for programming the device can be determined with a limited number of multiplications in the decomposition process.

To decompose $[T_{U(4)}]$ for programming the optical processor, it is successively multiplied by $[D_{MZI}^{(n)}]_{H_{4\times4}}^{-1}$ for n=1, 2, 3, …, 6, from the right. These successive products are carried out in a specific order based on the equivalent reverse propagation. As can be seen in Figure 18.6, the

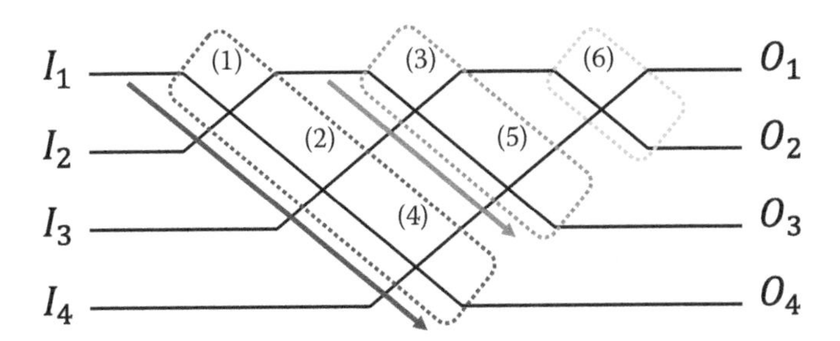

FIGURE 18.6 Schematic illustration of the MZIs order in the U(4) section of the 4 × 4 Reck-based unitary optical processor for decomposing $[T_{U(4)}]$ to experimentally program the device [12].

decomposition process starts with MZIs (1), (2), and (4) (blue box), followed by MZIs (3) and (5) (red box) and MZI (6) (green box), respectively. As a result of the successive multiplications of the matrices in this given order, the final resulting matrix is an identity matrix of degree four, $[I]_{(4)}$.

$$[T_{U(4)}]\cdot[D_{MZI}^{(1)}]_{H_{4\times4}}^{-1}\cdot[D_{MZI}^{(2)}]_{H_{4\times4}}^{-1}\cdot[D_{MZI}^{(4)}]_{H_{4\times4}}^{-1}\cdot[D_{MZI}^{(3)}]_{H_{4\times4}}^{-1}\cdot[D_{MZI}^{(5)}]_{H_{4\times4}}^{-1}\cdot[D_{MZI}^{(1)}]_{H_{4\times4}}^{-1} = [I]_{(4)}. \quad (18.19)$$

By setting the off-diagonal elements to zero, step by step, the required phase shifts in the inverse transformation matrix of the corresponding MZI ($[D_{MZI}^{(n)}]_{H_{4\times4}}^{-1}$) can be calculated. These steps for decomposing the linear transformation matrix of a given application can be expressed as the followings:

Step 1: Considering the left first layer of MZIs (blue box) in Figure 18.6, the product of $[T_{U(4)}]$ from the application with $[D_{MZI}^{(1)}]_{H_{4\times4}}^{-1}$ nulls the first element in the fourth row of the resultant matrix,

$$[T_{U(4)}]\cdot[D_{MZI}^{(1)}]_{H_{4\times4}}^{-1}=$$

$$\begin{bmatrix} U_{11} & U_{12} & U_{13} & U_{14} \\ U_{21} & U_{22} & U_{23} & U_{24} \\ U_{31} & U_{32} & U_{33} & U_{34} \\ U_{41} & U_{42} & U_{43} & U_{44} \end{bmatrix} \cdot -je^{-j\left(\frac{\theta_1}{2}\right)} \begin{bmatrix} e^{-j\phi_1} sin\left(\frac{\theta_1}{2}\right) & cos\left(\frac{\theta_1}{2}\right) & 0 & 0 \\ e^{-j\phi_1} cos\left(\frac{\theta_1}{2}\right) & -sin\left(\frac{\theta_1}{2}\right) & 0 & 0 \\ 0 & 0 & 1 & 0 \\ 0 & 0 & 0 & 1 \end{bmatrix} = \underbrace{\begin{bmatrix} * & * & * & * \\ * & * & * & * \\ * & * & * & * \\ 0 & * & * & * \end{bmatrix}}_{[A]}, \quad (18.20)$$

where (*) denotes a matrix element, which is determined in a later step. Therefore, A_{41}, which is a function of phase θ_1 is set to zero, resulting in

$$U_{41}\, sin\left(\frac{\theta_1}{2}\right) + U_{42}\, cos\left(\frac{\theta_1}{2}\right)=0. \quad (18.21)$$

Thus, θ_1 in MZI (1) can be calculated by

$$\theta_1=2\, tan^{-1}\left(\frac{-U_{42}}{U_{41}}\right). \quad (18.22)$$

The experimental realization of this fact is equivalent to applying the required phase shift to the phase shifter of MZI (1) to implement its inverse transformation matrix in the reverse propagation. Thus, the effect of MZI (1) is eliminated from the constructed $[T_U(4)]$ in the forward propagation. In this sense, the multiplication of the resultant matrix [A] by the input vector $[O_1, O_2, O_3, O_4]$ must lead to an output vector $[I_1, I_2, I_3, I_4]$ where $I_4 = 0$. Thus,

$$\underbrace{\begin{bmatrix} * & * & * & * \\ * & * & * & * \\ * & * & * & * \\ 0 & * & * & * \end{bmatrix}}_{[A]} \cdot \begin{bmatrix} O_1 \\ O_2 \\ O_3 \\ O_4 \end{bmatrix} = \begin{bmatrix} I_1 \\ I_2 \\ I_3 \\ I_4 \end{bmatrix}. \quad (18.23)$$

To this end, light is only injected from O_1 and not from O_2, O_3 and O_4, i.e., setting $O_2 = O_3 = O_4 = 0$. In this sense,

$$I_4 = (0 \cdot O_1) + (* \cdot O_2) + (* \cdot O_3) + (* \cdot O_4) = 0, \quad (18.24)$$

can be achieved if and only if the phase shifter of MZI (1) is properly adjusted to implement its inverse transformation matrix such that its contribution from $[T_U(4)]$ is eliminated in reverse propagation.

Step 2: The product of [A] with $[D_{MZI}^{(2)}]_{H_{4\times4}}^{-1}$ nulls the second element on the fourth row of the resultant matrix [B] as follows:

$$[A]\cdot[D_{MZI}^{(2)}]_{H_{4\times4}}^{-1} = \underbrace{\begin{bmatrix} * & * & * & * \\ * & * & * & * \\ * & * & * & * \\ 0 & 0 & * & * \end{bmatrix}}_{[B]}. \quad (18.25)$$

Similarly, $B_{42}(\theta_2) = 0$ leads to the determination of the phase shift of θ_2 in MZI (2).

Step 3: Likewise, the product of [A] with $[D_{MZI}^{(4)}]_{H_{4\times4}}^{-1}$ sets the last off-diagonal elements in the last row of the resultant matrix [C] to zero, given by

$$[B]\cdot[D_{MZI}^{(4)}]_{H_{4\times4}}^{-1} = \underbrace{\begin{bmatrix} * & * & * & 0 \\ * & * & * & 0 \\ * & * & * & 0 \\ 0 & 0 & 0 & 1 \end{bmatrix}}_{[C]}. \quad (18.26)$$

As a result, $C_{43}(\theta_4) = 0$ allows for obtaining the phase shift of θ_4 in MZI (4). It is essential to note that due to the unitary property of the resultant matrices, once all off-diagonal elements in a row become zero, the diagonal element is set to one and the off-diagonal elements in the corresponding column also become zero. Consequently, by determining θ_4 in MZI (4) in this step, $C_{44} = 1$ and $C_{14} = C_{24} = C_{34} = 0$, which means the effective dimension of the resultant matrix in this step is reduced by one as shown in (18.18).

Step 4: The next step is to multiply [C] by $[D_{MZI}^{(3)}]_{H_{4\times4}}^{-1}$ corresponding to the red box in Figure 18.6, which nulls the first element in the third row of the resultant matrix [D] given by

$$[C]\cdot[D_{MZI}^{(3)}]_{H_{4\times4}}^{-1} = \underbrace{\begin{bmatrix} * & * & * & 0 \\ * & * & * & 0 \\ 0 & * & * & 0 \\ 0 & 0 & 0 & 1 \end{bmatrix}}_{[D]}, \quad (18.27)$$

$$D_{31}(\theta_3, \phi_1, \phi_2) = 0. \quad (18.28)$$

Step 5: The product of [D] with $[D_{MZI}^{(5)}]_{H_{4\times4}}^{-1}$ sets the second element in the third row of the resultant matrix [E] expressed as

$$[D]\cdot[D_{MZI}^{(5)}]_{H_{4\times4}}^{-1} = \underbrace{\begin{bmatrix} * & * & 0 & 0 \\ * & * & 0 & 0 \\ 0 & 0 & 1 & 0 \\ 0 & 0 & 0 & 1 \end{bmatrix}}_{[E]}, \tag{18.29}$$

$$E_{32}(\theta_3, \theta_5, \phi_1, \phi_2, \phi_4) = 0. \tag{18.30}$$

Since all the off-diagonal elements in the third row of the resultant matrix [E] are set to zero, it can be concluded that

$$E_{33}(\theta_3, \theta_5, \phi_1, \phi_2, \phi_4) = 1, \tag{18.31}$$

$$E_{23}(\theta_3, \theta_5, \phi_1, \phi_2, \phi_4) = 0, \tag{18.32}$$

$$E_{13}(\theta_3, \theta_5, \phi_1, \phi_2, \phi_4) = 0. \tag{18.33}$$

Using the five equations (18.28), (18.30), (18.31), (18.32), and (18.33) the phases shifts θ_3, θ_5, ϕ_1, ϕ_2 and ϕ_4 of the corresponding MZIs in the device can be determined.

Step 6: The multiplication of [E] by $[D_{MZI}^{(6)}]_{H_{4\times4}}^{-1}$ nulls the first element of the second row in [F], expressed by

$$[E]\cdot[D_{MZI}^{(6)}]_{H_{4\times4}}^{-1} = \underbrace{\begin{bmatrix} 1 & 0 & 0 & 0 \\ 0 & 1 & 0 & 0 \\ 0 & 0 & 1 & 0 \\ 0 & 0 & 0 & 1 \end{bmatrix}}_{[F]}, \tag{18.34}$$

$$F_{21}(\theta_6, \phi_3, \phi_5) = 0. \tag{18.35}$$

In this step, all the off-diagonal elements in the second row of [F] are set to zero, thus, similar to previous rows,

$$F_{22}(\theta_6, \phi_3, \phi_5) = 1, \tag{18.36}$$

$$F_{12}(\theta_6, \phi_3, \phi_5) = 0, \tag{18.37}$$

$$F_{11}(\theta_6, \phi_3, \phi_5, \phi_6) = 1. \tag{18.38}$$

In other words, [F] is the identity matrix, and using the four equations (18.35), (18.36), (18.37), and (18.38) the phases shifts θ_6, ϕ_3, ϕ_5 and ϕ_6 can be obtained. Therefore, all the required phase shifts to experimentally implement the linear transformation matrix by the 4×4 optical processor are determined. The next section gives an example for determining the required phase shifts for experimentally implementing an arbitrary unitary matrix of U(4).

18.3.2 An Example for the Decomposition of an Arbitrary Unitary Transformation Matrix U(4)

In this section, an example is given for the experimental construction of an arbitrary unitary transformation matrix $[T_{U(4)}]$ generated by (18.15) using arbitrary phases for the phase shifters of the MZIs in the 4 × 4 optical processor. The MZIs (1) to (6) shown in Figure 18.5 should be tuned using the corresponding phase shifts θ_n and ϕ_n $(n = 1, 2, \ldots 6)$ to implement the unitary transformation $[T_{U(4)}]$ given by

$$[T_{U(4)}] = \begin{bmatrix} U_{11} & U_{12} & U_{13} & U_{14} \\ U_{21} & U_{22} & U_{23} & U_{24} \\ U_{31} & U_{32} & U_{33} & U_{34} \\ U_{41} & U_{42} & U_{43} & U_{44} \end{bmatrix} =$$

$$\begin{bmatrix} -0.0575 + 0.0037i & 0.0217 + 0.0487i & 0.0388 - 0.0260i & 0.9904 - 0.1041i \\ 0.0642 + 0.0446i & 0.0293 - 0.0652i & 0.5842 - 0.8035i & -0.0352 + 0.0256i \\ 0.0179 - 0.0873i & 0.2908 - 0.9487i & -0.0253 + 0.0644i & 0.0467 + 0.0233i \\ 0.2230 - 0.9659i & -0.0195 + 0.0845i & 0.0776 + 0.0109i & 0.0037 - 0.0609i \end{bmatrix}.$$

To implement this unitary transformation matrix $[T_{U(4)}]$ using the optical structure, the required phase shifts are applied to the phase shifters of the MZIs (1) to (6), respectively. For the given example, the phase shifts can be determined through the decomposition protocol given in Section 18.3. The successive multiplication of $[T_{U(4)}]$ from the right side by the corresponding inverse unitary matrices of the six MZIs, i.e., $[D_{MZI}^{(n)}]_{H_{4\times4}}^{-1}$, allows for nullification of all off-diagonal elements in the resultant matrices step by step. Eventually, these consecutive multiplications result in an identity matrix of degree four, $I_{(4)}$. In each step, once an element becomes zero, they will not be affected by the transformation of the next steps. For the given unitary matrix these steps are expressed by

Step 1:

$$[T_{U(4)}]\cdot[D_{MZI}^{(1)}]^{-1} =$$

$$\begin{bmatrix} 0.2665 + 0.0454i & -0.0298 + 0.1490i & 0.5216 + 0.4673i & 0.1664 + 0.6210i \\ -0.2413 - 0.6546i & 0.3999 + 0.3555i & 0.1030 + 0.3555i & -0.2120 - 0.2120i \\ -0.2269 - 0.6234i & -0.4778 - 0.3561i & -0.1671 - 0.0954i & 0.0357 + 0.4080i \\ \mathbf{0.0000 - 0.0000i} & -0.1006 - 0.5704i & 0.1006 + 0.5704i & 0.3290 - 0.4698i \end{bmatrix},$$

from which θ_1 can be determined as follows:

$$\theta_1 = 1.75\ \text{Rad}.$$

Step 2:

$$[T_{U(4)}]\cdot[D_{MZI}^{(1)}]^{-1}\cdot[D_{MZI}^{(2)}]^{-1} =$$

$$\begin{bmatrix} 0.2665 + 0.0454i & -0.0349 - 0.5565i & 0.1165 + 0.4349i & 0.1664 + 0.6210i \\ -0.2413 - 0.6546i & -0.0029 - 0.6158i & -0.1484 - 0.1484i & -0.2120 - 0.2120i \\ -0.2269 - 0.6234i & 0.1904 + 0.5231i & 0.0250 + 0.2857i & 0.0357 + 0.4080i \\ \mathbf{0.0000 - 0.0000i} & \mathbf{0.0000 - 0.0000i} & -0.4698 + 0.6710i & 0.3290 - 0.4698i \end{bmatrix},$$

which results in

$$\theta_2 = 1.57\,\text{Rad}.$$

Step 3:

$$[T_{U(4)}]\cdot[D_{MZI}^{(1)}]^{-1}\cdot[D_{MZI}^{(2)}]^{-1}\cdot[D_{MZI}^{(4)}]^{-1} = \begin{bmatrix} 0.2665+0.0454i & -0.0349-0.5565i & 0.2684-0.7376i & 0.0000+0.0000i \\ -0.2413-0.6546i & -0.0029-0.6158i & 0.0636+0.3604i & 0.0000+0.0000i \\ -0.2269-0.6234i & 0.1904+0.5231i & 0.2500-0.4330i & \mathbf{0.0000-0.0000i} \\ \mathbf{0.0000-0.0000i} & \mathbf{0.0000-0.0000i} & \mathbf{0.0000-0.0000i} & \mathbf{1.0000-0.0000i} \end{bmatrix},$$

Step 3 results in the determination of θ_4 given by

$$\theta_4 = 1.22\ \ \text{Rad}.$$

Step 4:

$$[T_{U(4)}]\cdot[D_{MZI}^{(1)}]^{-1}\cdot[D_{MZI}^{(2)}]^{-1}\cdot[D_{MZI}^{(4)}]^{-1}\cdot[D_{MZI}^{(3)}]^{-1} = \begin{bmatrix} -0.3462+0.2424i & 0.1550-0.4258i & 0.2684-0.7376i & \mathbf{0.0000-0.0000i} \\ -0.3830+0.8214i & 0.0367+0.2081i & 0.0636+0.3604i & \mathbf{0.0000-0.0000i} \\ \mathbf{0.0000-0.0000i} & -0.4330+0.7500i & 0.2500-0.4330i & \mathbf{0.0000-0.0000i} \\ \mathbf{0.0000-0.0000i} & \mathbf{0.0000-0.0000i} & \mathbf{0.0000-0.0000i} & \mathbf{1.0000-0.0000i} \end{bmatrix},$$

Using the equations obtained through Steps 4, and 5, the phase shifts θ_3, θ_5, ϕ_1, ϕ_2, and ϕ_4 can be obtained as follows:

$$\begin{aligned} \theta_3 &= 1.40\,\text{Rad}, \\ \theta_5 &= 1.05\,\text{Rad}, \\ \phi_1 &= 0.09\,\text{Rad}, \\ \phi_2 &= 0.17\,\text{Rad}, \\ \phi_4 &= 0.35\,\text{Rad}. \end{aligned}$$

Step 5:

$$[T_{U(4)}]\cdot[D_{MZI}^{(1)}]^{-1}\cdot[D_{MZI}^{(2)}]^{-1}\cdot[D_{MZI}^{(4)}]^{-1}\cdot[D_{MZI}^{(3)}]^{-1}\cdot[D_{MZI}^{(5)}]^{-1} = \begin{bmatrix} -0.3462+0.2424i & -0.7424+0.5198i & \mathbf{0.0000-0.0000i} & \mathbf{0.0000-0.0000i} \\ -0.3830+0.8214i & 0.1786-0.3830i & \mathbf{0.0000-0.0000i} & \mathbf{0.0000-0.0000i} \\ \mathbf{0.0000-0.0000i} & \mathbf{0.0000-0.0000i} & \mathbf{1.0000-0.0000i} & \mathbf{0.0000-0.0000i} \\ \mathbf{0.0000-0.0000i} & \mathbf{0.0000-0.0000i} & \mathbf{0.0000-0.0000i} & \mathbf{1.0000-0.0000i} \end{bmatrix},$$

Step 6:

$$[T_{U(4)}]\cdot[D_{MZI}^{(1)}]^{-1}\cdot[D_{MZI}^{(2)}]^{-1}\cdot[D_{MZI}^{(4)}]^{-1}\cdot[D_{MZI}^{(3)}]^{-1}\cdot[D_{MZI}^{(5)}]^{-1}\cdot[D_{MZI}^{(6)}]^{-1} = \begin{bmatrix} \mathbf{1.0000+0.0000i} & \mathbf{0.0000-0.0000i} & \mathbf{0.0000+0.0000i} & \mathbf{0.0000+0.0000i} \\ \mathbf{0.0000-0.0000i} & \mathbf{1.0000-0.0000i} & \mathbf{0.0000-0.0000i} & \mathbf{0.0000+0.0000i} \\ \mathbf{0.0000+0.0000i} & \mathbf{0.0000+0.0000i} & \mathbf{1.0000-0.0000i} & \mathbf{0.0000-0.0000i} \\ \mathbf{0.0000-0.0000i} & \mathbf{0.0000+0.0000i} & \mathbf{0.0000+0.0000i} & \mathbf{1.0000+0.0000i} \end{bmatrix}.$$

Eventually, the transformation in Step 6 allows for calculating the phase shifts θ_6, ϕ_3, ϕ_5, and ϕ_6 as follows:

$$\theta_6 = 0.87\,\text{Rad},$$
$$\phi_3 = 0.26\,\text{Rad},$$
$$\phi_5 = 0.44\,\text{Rad},$$
$$\phi_6 = 0.52\,\text{Rad}.$$

Consequently, all the required phase shifts for experimental implementation of $[T_{U(4)}]$ are determined. According to (18.15), these phase shifts can be applied to the MZIs in the optical processor to experimentally construct the unitary transformation matrix of the given application. The next section is allocated to the performance analyses of a single layer ONN implemented by different sizes of the Reck-topology-based and the Diamond-topology-based optical processors.

18.4 THE DIAMOND MESH, A PHASE-ERROR- AND LOSS-TOLERANT MZI-BASED OPTICAL PROCESSOR FOR ONNS

This section presents how the topology of the MZI-based optical processor can effect the performance of the system in a given application. The optical components imperfections play a key role in the performance of the optical devices, which necessitate a more robust architecture against these performance degrading factors. In this respect, a diamond topology of field-programmable MZI-based optical processors is introduced. The presented performance analysis confirms that the single layer ONNs implemented by the Diamond mesh exhibit more robustness to fabrication faults and experimental imperfections compared to that of the Reck mesh studied in the previous section and reported in [12,15]. Both structures can be experimentally calibrated and programmed for a given application. Compared to the Reck structure, the proposed Diamond mesh employs $(N - 1)(N - 2)/2$ additional reconfigurable MZIs for a more symmetric topology with more loss-balanced optical paths. Furthermore, the additional MZIs in the Diamond mesh allows the structure to be programmed in a way that optimal light intensities are directed towards the output ports for a better classification performance of the implemented ONNs. This is achieved by excluding the light intensity that causes incorrect classifications through the tapered-out waveguides of these MZIs.

The analytical results presented in this section show the effects of experimental uncertainties on the classification accuracy of the ONNs implemented by different sizes of the two topologies. The obtained results on the classification performance and scalability of the two structures show that the additional MZIs in the proposed Diamond mesh provides extra degrees of freedom in optimizing the weight matrix of the implemented optical NNs (ONNs). Consequently, the Diamond mesh is more robust to fabrication process variations and experimental imperfections, i.e., insertion loss (IL) of the MZIs and phase errors.

18.4.1 Diamond Topology and Its Programming Process

As explained in Section 18.3, a multiport field-programmable MZI-based optical processor is a mesh of 2 × 2 reconfigurable MZIs. As shown in Figure 18.1, a 2 × 2 reconfigurable MZI is composed of two 3-dB couplers with one phase shifter (θ) on the top arm in between, and another one (ϕ) on the top output of the MZI.

Figure 18.7 demonstrates a 4 × 4, i.e., $N = 4$, Diamond MZI-based optical processor which is composed of $n = (N - 1)^2 = 9$ reconfigurable MZIs.

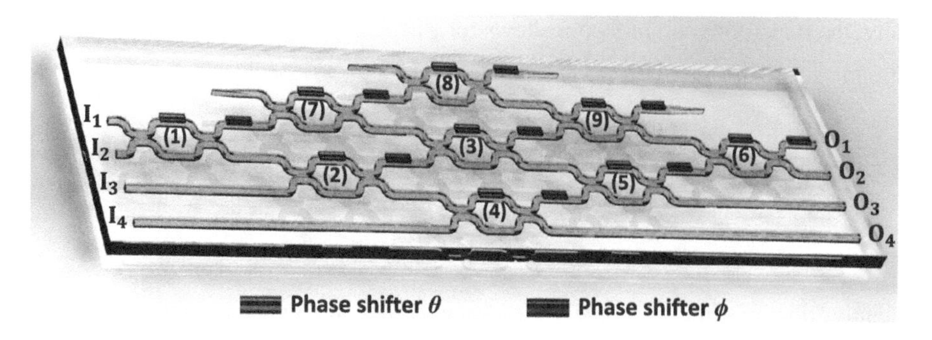

FIGURE 18.7 Schematic layout of the 4 × 4 Diamond reconfigurable optical processor. Adapted with permission from [32] The Optical Society.

As can be seen, it employs $(N-1)(N-2)/2 = 3$ additional MZIs, i.e., MZIs (7), (8), and (9), as compared to the Reck structure studied in Section 18.3. Both structures can be experimentally calibrated and programmed for a given application. The optical processor linearly relates the four inputs to the four outputs through the field interactions between its nine MZIs. To construct the unitary transformation matrix of the 4 × 4 Diamond structure shown in Figure 18.7, the 2 × 2 unitary transformation matrix of each MZI is defined on a two-dimensional subspace within a $(2N-2=6)$-dimensional Hilbert space ($H_{6\times6}$) based on its location in the mesh. The matrices of the MZIs are denoted by $[D^{(n)}]_{H_{6\times6}}$, where $n = 1, 2, 3, \ldots, 9$ represents the labels of the MZIs in the structure. As a result, the contribution of each MZI is reflected into the unitary transformation matrix of the whole structure with respect to its connections to the optical paths within the mesh. In this regard, the unitary transformation matrix of the Diamond mesh can be calculated by

$$\begin{aligned}[T_{U(6)}] = {} & [D^{(6)}]_{H_{6\times6}} \cdot [D^{(5)}]_{H_{6\times6}} \cdot [D^{(4)}]_{H_{6\times6}} \cdot [D^{(9)}]_{H_{6\times6}} \cdot [D^{(3)}]_{H_{6\times6}} \cdot \\ & [D^{(2)}]_{H_{6\times6}} \cdot [D^{(8)}]_{H_{6\times6}} \cdot [D^{(7)}]_{H_{6\times6}} \cdot [D^{(1)}]_{H_{6\times6}} .\end{aligned} \tag{18.39}$$

To determine the required phases for implementing the unitary transformation matrix $[T_{U(6)}]$ given by an application, it is essentially decomposed based on the protocol given in [12]. Figure 18.8

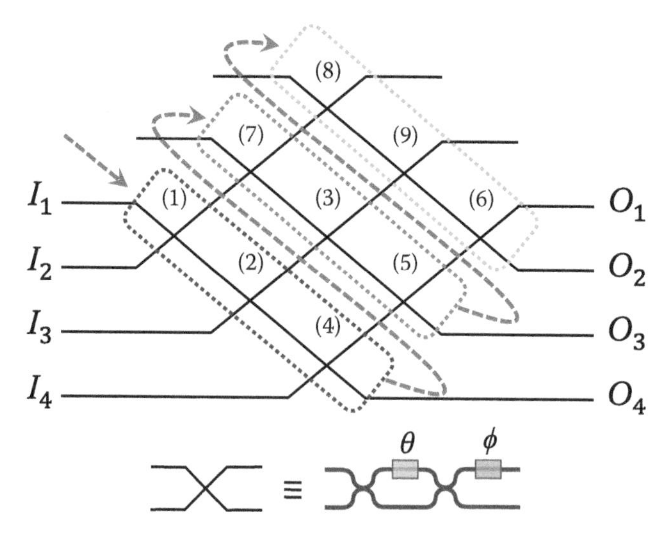

FIGURE 18.8 Schematic of the decomposition order of the MZIs in the 4 × 4 Diamond reconfigurable MZI-based optical processor. Adapted with permission from [32] The Optical Society.

depicts the decomposition process order of $[T_{U(6)}]$ which is carried out through its successive product with the inverse transformation matrices of the MZIs defined in a six-dimensional Hilbert space.

Due to the unitary property of U(2) matrix given by equation (18.1), the inverse transformation matrix of each reconfigurable MZI in the structure is equal to its conjugate transpose. The decomposition process is carried out based on the fact that the structure implementing $[T_{U(6)}]$ in the forward propagation, is setup to eliminate the effects of the MZIs, one by one, in the reverse direction, i.e., when the light propagates within the structure from the right to the left. As shown in Figure 18.8, the decomposition of $[T_{U(6)}]$ starts from the layer of MZIs on the far left part of the structure indicated by the blue box (MZIs (1), (2), and (4), respectively) followed by the second layer of MZIs, i.e., the red box (MZIs (7), (3), and (5), in the named order) and finally, the green box (MZIs (8), (9), and (6), respectively). Therefore, the decomposition processes can be carried out based on

$$\begin{aligned}&[T_{U(6)}]\cdot[D^{(1)}]^{-1}_{H_{6\times6}}\cdot[D^{(2)}]^{-1}_{H_{6\times6}}\cdot[D^{(4)}]^{-1}_{H_{6\times6}}\cdot[D^{(7)}]^{-1}_{H_{6\times6}}\cdot[D^{(3)}]^{-1}_{H_{6\times6}}\\&\cdot[D^{(5)}]^{-1}_{H_{6\times6}}\cdot[D^{(8)}]^{-1}_{H_{6\times6}}\cdot[D^{(9)}]^{-1}_{H_{6\times6}}\cdot[D^{(6)}]^{-1}_{H_{6\times6}}=[I]_{(6)}.\end{aligned}\tag{18.40}$$

In each step of the decomposition process, an off-diagonal matrix element in the resultant matrix becomes zero, and thanks to the unitary property, it will not be changed by the transformations in the following steps. Furthermore, once all off-diagonal elements in each row become zero, the corresponding diagonal elements is set to one and all off-diagonal elements in the corresponding column become zero. Eventually, this process results in an identity matrix of order six ($[I]_{(6)}$). Thus, the required phases to implement $[T_{U(6)}]$ can be calculated [12]. The resultant matrices in different steps of the decomposition process for $[T_{U(6)}]$ is expressed by $[T^{(n)}]_{H_{6\times6}}$, which is multiplied by the inverse transformation matrix of the nth MZI, $[D^{(n)}]^{-1}_{H_{6\times6}}$, to null the off-diagonal matrix element in the corresponding step, given by

$$[T^{(n)}]_{6\times6}=\begin{bmatrix} * & * & * & * & 0 & 0\\ * & * & * & * & * & 0\\ * & * & * & * & * & *\\ *^{(8)} & *^{(9)} & *^{(6)} & * & * & *\\ 0 & *^{(7)} & *^{(3)} & *^{(5)} & * & *\\ 0 & 0 & *^{(1)} & *^{(2)} & *^{(4)} & * \end{bmatrix},\tag{18.41}$$

where $*$ in different colors denotes unknown matrix elements in the resultant matrices during the decomposition process. For comparison between the two topologies, the Reck mesh matrix elements within $[T_{U(6)}]$ of the Diamond mesh are shown by $*^{(n)}$ and $*$. Furthermore, the matrix elements $*^{(n)}$ and $*$ in the resultant matrix $[T^{(n)}]_{6\times6}$ of the Diamond mesh are determined by the contributions of MZIs (7), (8) and (9). These matrix elements are set to zero or one during the decomposition process. As explained in Section 18.3, during this process, the resultant matrix in each step is multiplied by the inverse transformation matrix of the related MZI, i.e., $[D^{(n)}]^{-1}_{H_{6\times6}}$ where $n = 1, 2, 3, \ldots, 9$ for the Diamond mesh. As a result, one can determine the required phases for implementing $[T_{U(6)}]$ using the Diamond optical processor chip.

Based on equation (18.41) and the related information given above, the Diamond mesh shown in Figure 18.7 can be viewed as a 6 × 6 Reck mesh from which six MZIs are excluded from the

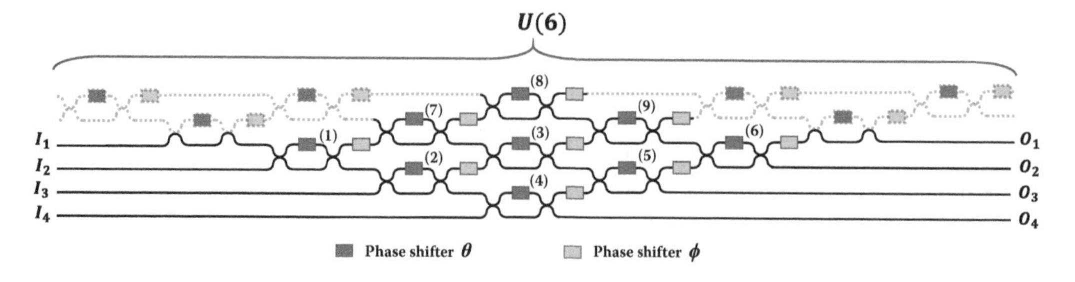

FIGURE 18.9 Schematic of the 4 × 4 Diamond reconfigurable MZI-based optical processor embedded in a 6 × 6 Reck mesh.

structure. The Diamond mesh is represented by a unitary transformation matrix of degree six from which a 4 × 6 subspace is used for its performance analysis. Figure 18.9 demonstrates the 4 × 4 Diamond MZI-based optical processor embedded in a 6 × 6 Reck mesh.

The excluded six MZIs could have affected the matrix elements $T_{2,1}$, $T_{3,1}$, $T_{3,2}$, $T_{5,1}$, $T_{6,1}$, and $T_{6,2}$ in the lower triangle of $[T^{(n)}]_{6\times6}$ given by equation (18.41). In other words, to implement the transformation matrix of the Diamond mesh, the effects of these six MZIs are eliminated from the unitary transformation matrix of the corresponding Reck mesh. Thus, the programming protocol for the Reck mesh presented in Section 18.3 can be similarly used to program the Diamond mesh, as shown in Figure 18.8. This allows to compare the two topologies within the same programming framework. According to Figure 18.8, and equations (18.40) and (18.41), the phase extraction equations for programming the diamond mesh are given as follows:

Step 1:

$$[T_{U(6)}]\cdot[D^{(1)}]^{-1}_{H_{6\times6}} = [T^{(1)}]_{6\times6}, \tag{18.42}$$

where $T^{(1)}_{6,3} = 0$, and thus, the phase shift θ_1 can be determined.

Step 2:

$$[T^{(1)}]_{6\times6}\cdot[D^{(2)}]^{-1}_{H_{6\times6}} = [T^{(2)}]_{6\times6}, \tag{18.43}$$

where $T^{(2)}_{6,4} = 0$, which allows for obtaining the phase shift θ_2.

Step 3:

$$[T^{(2)}]_{6\times6}\cdot[D^{(4)}]^{-1}_{H_{6\times6}} = [T^{(4)}]_{6\times6}, \tag{18.44}$$

where $T^{(4)}_{6,5} = 0$, from which the phase shift θ_4 can be calculated. In this step, due to the unitary property, $T^{(4)}_{6,6} = 1$, $T^{(4)}_{5,6} = 0$, $T^{(4)}_{4,6} = 0$, and $T^{(4)}_{3,6} = 0$.

In the second row of multiplications:

Step 4:

$$[T^{(4)}]_{6\times6}\cdot[D^{(7)}]^{-1}_{H_{6\times6}} = [T^{(7)}]_{6\times6}, \tag{18.45}$$

which results in $T^{(7)}_{5,2} = 0$.

Step 5:

$$[T^{(7)}]_{6\times6}\cdot[D^{(3)}]^{-1}_{H_{6\times6}} = [T^{(3)}]_{6\times6}, \tag{18.46}$$

which leads to $T^{(3)}_{5,3} = 0$.

Step 6:

$$[T^{(3)}]_{6\times6}\cdot[D^{(5)}]^{-1}_{H_{6\times6}} = [T^{(5)}]_{6\times6}, \tag{18.47}$$

which yields $T^{(5)}_{5,4} = 0$. The unitary property of the matrix leads to $T^{(5)}_{5,5} = 1$, $T^{(5)}_{4,5} = 0$, $T^{(5)}_{3,5} = 0$, and $T^{(5)}_{2,5} = 0$. Using the equations obtained in the second row of transformations, i.e., Steps 4, 5, and 6, the phase shifts θ_7, θ_3, θ_5, ϕ_1, ϕ_2, and ϕ_4 can be determined.

In the last row of transformations:

Step 7:

$$[T^{(5)}]_{6\times6}\cdot[D^{(8)}]^{-1}_{H_{6\times6}} = [T^{(8)}]_{6\times6}, \tag{18.48}$$

which leads to $T^{(8)}_{4,1} = 0$.

Step 8:

$$[T^{(8)}]_{6\times6}\cdot[D^{(9)}]^{-1}_{H_{6\times6}} = [T^{(9)}]_{6\times6}, \tag{18.49}$$

which results in $T^{(9)}_{4,2} = 0$.

Step 9:

$$[T^{(9)}]_{6\times6}\cdot[D^{(6)}]^{-1}_{H_{6\times6}} = [T^{(6)}]_{6\times6}, \tag{18.50}$$

where $T^{(6)}_{4,3} = 0$. The unitary property of the matrix leads to $T^{(6)}_{4,4} = 1$, $T^{(6)}_{3,4} = 0$, $T^{(6)}_{2,4} = 0$, and $T^{(6)}_{1,4} = 0$. Similarly, $[T^{(6)}]_{6\times6}$ results in $T^{(6)}_{3,1} = 0$, $T^{(6)}_{3,2} = 0$, $T^{(6)}_{3,3} = 1$, $T^{(6)}_{2,3} = 0$, $T^{(6)}_{1,3} = 0$, $T^{(6)}_{2,1} = 0$, $T^{(6)}_{2,2} = 1$, $T^{(6)}_{1,2} = 0$ and $T^{(6)}_{1,1} = 1$. Eventually, the obtained equations in the last layer of transformations, i.e., Steps 7, 8, and 9, allow for calculating the phase shifts θ_8, θ_9, θ_6, ϕ_7, ϕ_3, and ϕ_5.

As can be inferred from the decomposition process, the matrix elements in $[T_{U(6)}]$ reflect the contribution of several MZIs. By eliminating the effect of each MZI step by step, all the phases of the MZIs in the Diamond mesh can be determined, which can be used for programming the optical processor in a similar fashion as the algorithm presented in [12]. As a result, the performance comparison between the two practical topologies can be made within the same programming framework. The Diamond mesh constructs a unitary transformation matrix of degree six from which a 4×6 subspace is used for its performance analysis. The next section, presents the performance analyses of the Diamond mesh in ONNs and the results are compared with that of the 4×4 Reck mesh studied in Section 18.3 and reported in [12,15]. The next section shows how the proposed Diamond structure is employed to implement single layer ONNs, and the obtained classification accuracies are compared to that of the Reck mesh studied in [15].

18.4.2 Optical Neural Networks for Classification

NNs are machine learning models that can be used for performing various computational tasks. Figure 18.10 illustrates a single layer of a NN where a series of matrix multiplications denoted by $\mathbf{W}$ are interleaved with nonlinear activation functions $H(\cdot)$ to perform the required machine learning tasks, such as classification mechanisms in voice and image recognition [33,34], and autonomous driving control systems [35]. Such a classification process involves taking multi-dimensional inputs, $\mathbf{I}^0$, and sorting them into their respective classes, c. Each input $\mathbf{I}^0 \in \mathbb{R}^f$ is a multi-dimensional sample that represents multiple features, where f is the number of features. The output vector $\hat{\mathbf{O}}$ in the NN is classified based on the index of the element with the maximum value.

The size of the weight matrix $\mathbf{W}^k$ in a NN represented by m^k can vary from one layer to another, where m is the size of the matrix in the kth layer, $k \in \{1, ..., K\}$. Therefore, the output of layer k is expressed as $\mathbf{O}^k = H(\mathbf{W}^k \cdot O^{k-1})$. Consequently, the NN with an input $\mathbf{I}^0 \in \mathbb{R}^f$ produces a final output $\hat{\mathbf{O}} \in \mathbb{R}^c$, where c is the number of possible classes. In other words, the output of layer 1 represented by $\mathbf{O}^1 = H(\mathbf{W}^1 \cdot \mathbf{I}^0)$ passes through the rest of the layers, being transformed by $\mathbf{W}^k$ and $H^k(\cdot)$ to generate the final output $\hat{\mathbf{O}} = H(\mathbf{W}^K \cdot \mathbf{O}^{K-1})$, where the maximum argument in the final output of the NN $\hat{\mathbf{O}}$ designates the class of the input sample.

The optimal weight matrix in ONNs is obtained by performing backpropagation all the way down to the phases in the MZIs phase shifters using interferometric measurements to obtain the loss gradient using the Neuroptica Python package [36]. This permits the backpropagation to the phases, as opposed to simply to the NN matrix weights. The algorithm in Neuroptica uses the adjoint electric field method to implement the *in-situ* backpropagation routine, where the loss gradients are obtained through interferometric measurements to perform the gradient descent algorithm [31,37]. It should be noted that while the results are similar to the standard backpropagation algorithm described in this work, the phases themselves are modified rather than the matrix weights [36]. During the training process of the NN, the weight matrix elements are optimized through backpropagation, where the output vector is compared to the true class of the sample, called the ground truth $\mathbf{O}$ by using a loss function $\mathcal{L}$ [38]. The loss function is a distance metric such as Mean Squared Error (MSE) between the obtained values $\hat{\mathbf{O}}$ and $\mathbf{O}$, given by

$$\mathcal{L}_{(\text{MSE})} = \frac{1}{S} \sum_{p=1}^{S} (\hat{\mathbf{O}}_p - \mathbf{O}_p)^2, \tag{18.51}$$

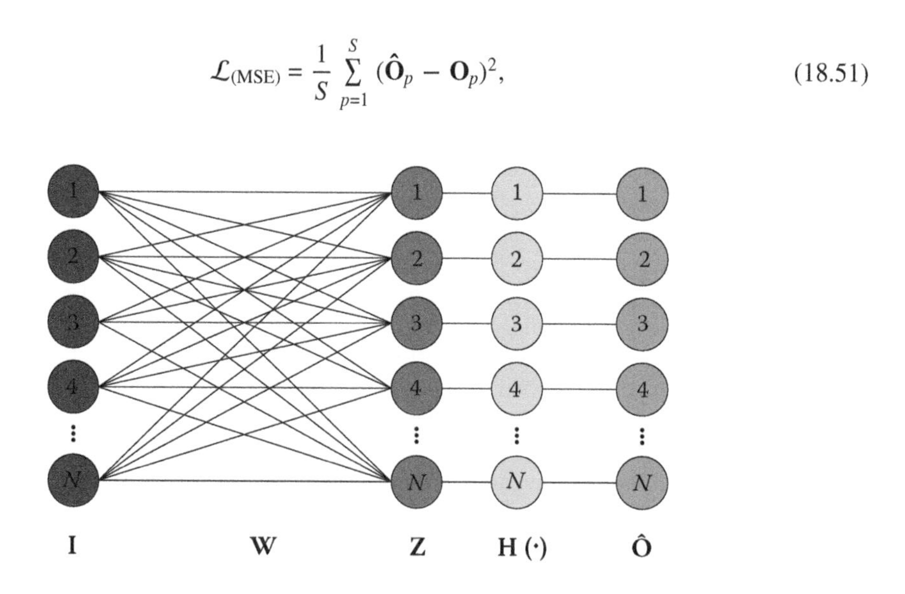

FIGURE 18.10 Schematic of a single-layer NN.

where S is the total number of samples in the training data set. The ground truth $\mathbf{O}$ is a one-hot-encoded vector with one element being set to 1 depending on the class of the sample, while every other element being set to 0 [39]. The use of MSE for classification in this work is only to train a linear ONN, and MSE was used for its simplicity. In a single layer NN, which is the case of this research work, there is no difference between using categorical cross-entropy or MSE as the loss function, since they both achieve a final validation classification accuracy of 100%. An optimal $\mathbf{W}$ results in a lower loss function value, which will increase the training classification accuracy [40,41]. The backpropagation algorithm modifies the weight matrix elements in the NN based on the product of a learning rate α with the negative gradient of the loss function $\mathcal{L}$ with respect to the weights $\mathbf{W}$ in the NN. If this process is repeated enough times, the loss function decreases and as a result, the classification accuracy will increase [38,42].

$$\mathbf{W}^{t+1} = \mathbf{W}^{t} - \alpha \cdot \nabla_{\mathbf{W}^{t}} \mathcal{L}, \tag{18.52}$$

where ∇ denotes the gradient operator and t is the current epoch, i.e., a single training cycle of the entire training process. It should be noted that in a single layer NN, a linearly separable data set can be classified perfectly with no need for a nonlinear activation function at the output [15,43]. The reason lies in the fact that most nonlinear activation functions are monotonic and do not affect the classification of a sample in a single layer NN. The optimal weight matrix in ONNs is obtained by performing backpropagation to find optimal phases in the MZIs phase shifters. According to equations (18.1) and (18.39), the phase shifters θ and ϕ of the MZIs determine the weight matrix elements of the structure. The use of linearly separable data set allows for evaluating the proposed Diamond structure itself and investigate the device performance improvement compared to the Reck structure, rather than enhancing the NN algorithm. Additionally, the data sets were separated in an 80:20 ratio between the training set and validation set. The validation set is used to test the ONN on data that was never seen during training. The data set is 100% classifiable by a digital NN such that the ONNs implemented by the two structures can be assessed in the presence of experimental imperfections, i.e., phase errors and IL of each MZI.

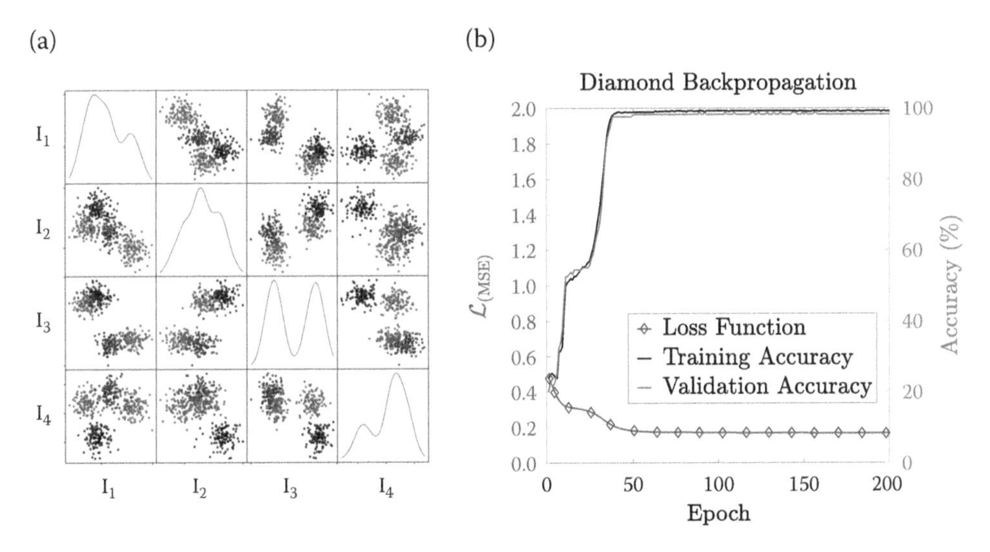

FIGURE 18.11 Dataset and training algorithm of the single layer ONN implemented by the 4 × 4 Diamond mesh with 0 dB loss per MZI; (a) multivariate Gaussian data set, (b) backpropagation (training) process of the single layer ONN for weight matrix optimization through minimizing the loss function value. Adapted with permission from [32] The Optical Society.

18.4.3 4 × 4 Diamond-Topology- and Reck-Topology-Based Single Layer ONNs

Figure 18.11a shows the data set used to characterize a 4 × 4 single layer ONN. The data set consists of four multivariate Gaussian distributions represented by four different colors, each of which composed of a set of four-dimensional points, i.e., $\mathbf{I} \in \mathbb{R}^4$. The four different Gaussian distributions are classified using a one-hot encoding scheme such that the correct output vector for a single sample denoted by $\mathbf{O} \in \mathbb{R}^4$ has one element set to 1, depending on the class of the sample, and the rest set to zero [39]. Figure 18.11b shows the backpropagation results for the Diamond mesh with 0 dB loss per MZI used in the single layer ONN. As can be seen, the loss function reaches a minimum value of approximately 0.18, while the ONN achieves a final validation accuracy of 98.75%. The backpropagation process optimizes the weight matrix to minimize the loss function [38].

According to Figure 18.11b, the training and validation classification accuracies are well matched, implying that the ONN is still generalizing well after 200 epochs. If the training data set classification accuracy remains high while the validation data set classification accuracy dropped, the ONN would be over-fitting and therefore not generalizing well. The validation data set is composed of data-points not seen during training, allowing for an unbiased classification metric. Using the validation data set, the validation accuracy is determined and compared with the training accuracy of the ONN. It should be noted that after around epoch 50, while the accuracy does not change, the difference between the output values of the ONN still increase. This means that the network becomes more resilient to the experimental uncertainties, up to a certain point.

It should be noted that in these analyses, the training process of the ONN is carried out for ideal scenarios with perfect MZIs, i.e., no loss nor phase uncertainties. The standard deviation values of the phases θ and ϕ and the loss per MZI are then added to perform the analysis for the given mesh of MZIs.

Figure 18.12 shows the classification accuracy of the single layer ONN implemented by the 4 × 4 structure size of the Reck mesh and the Diamond mesh as a function of the phase error

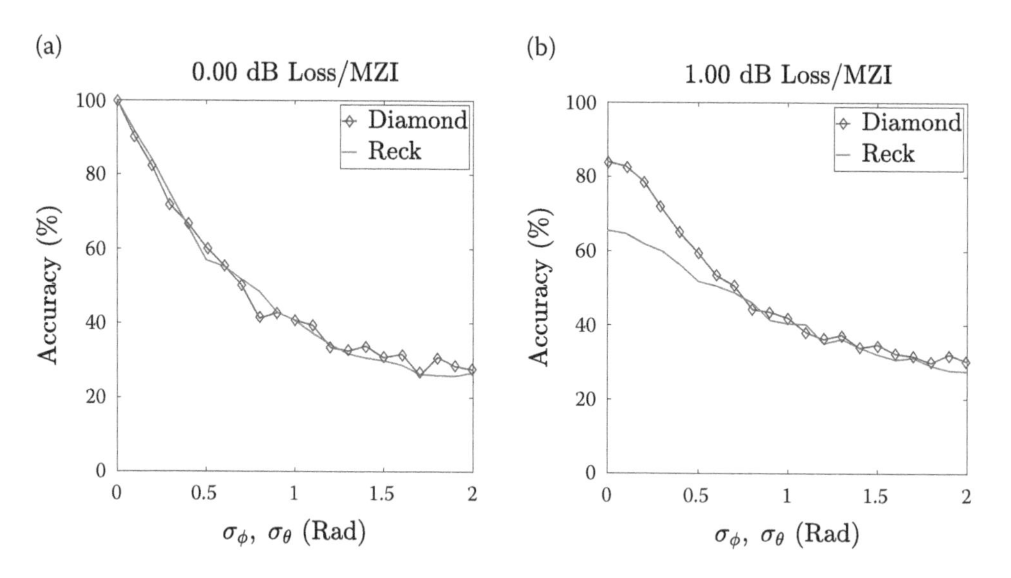

FIGURE 18.12 Classification performance of the single layer ONNs implemented by the 4 × 4 structure size of the Reck mesh and the proposed Diamond mesh. The results are obtained as a function of the phase error standard deviation, σ_θ and σ_ϕ, for 0 dB and 1 dB loss per MZI with 0.5 dB standard deviation. Adapted with permission from [32] The Optical Society.

standard deviation, σ_θ and σ_ϕ, for 0 dB and 1 dB loss per MZI with a standard deviation of 0.5 dB. The results in this work are obtained based on adding different normally distributed error terms to the phases and insertion loss per MZI with a specific standard distribution for every sample.

In this regard, the phases and the insertion loss of the MZIs were defined by

$$\theta = \theta_{true} + \mathcal{N}(0, \sigma_\theta^2), \tag{18.53}$$

$$\phi = \phi_{true} + \mathcal{N}\left(0, \sigma_\phi^2\right), \tag{18.54}$$

$$IL = IL_{mean} + \mathcal{N}(0, \sigma_{IL}^2), \tag{18.55}$$

where $\mathcal{N}$ is the added Gaussian noise with zero mean and a standard deviation of σ for the different parameters. This entire process is, repeated multiple times and the mean classification accuracy is taken as the final classification accuracy [36]. The simulated standard deviations of the phase errors and the loss per MZI are implemented only in the testing phase of the ONN. This is done by adding a different normally distributed error term to the phases and insertion losses per MZI with a specific standard distribution for every sample, repeating the entire testing phase multiple times and taking the mean of the classification accuracies as the final classification accuracy [36].

According to Figure 18.12a, as expected in perfect condition, i.e., 0 dB loss per MZI and zero Rad phase uncertainty, both the Reck mesh and the Diamond mesh provide 100% classification accuracy. However, the trend is different in practice since the phase error and the IL of the MZIs are almost inevitable. As shown in Figure 18.12b, in the case of 1 dB loss per MZI with a standard deviation of 0.5 dB and zero Rad phase error, the ONN implemented by a 4 × 4 diamond mesh classifies the data samples with 85% accuracy, where as the classification accuracy obtained by that of the Reck structure is approximately 75%. As can be seen in this figure, the diamond mesh tends to outperform the Reck mesh when the phase uncertainty is less than 0.5 Rad, which confirms its better candidacy for practical scenarios. It should be noted that for large phase uncertainties, the classification accuracy of both the Reck-mesh-based and the Diamond-mesh-based ONNs decay to that of a random classifier. The better performance of the Diamond mesh can be associated with its possibility to pass out the excess light intensity which can affect its classification performance destructively, through the tapered-out waveguides of its three additional MZIs. Moreover, the symmetric topology of the Diamond mesh allows for more phase error robustness and better loss balanced optical paths from its inputs to outputs.

Figure 18.13 demonstrates how the classification performance of the 4 × 4 Reck-topology-based and Diamond-topology-based single layer ONNs are degraded by phase errors σ_θ and σ_ϕ in the constituent MZIs. As can be seen in the case of 4 × 4 mesh size of the two structures, the classification accuracy is more sensitive to σ_θ than σ_ϕ. This can be associated with the fact that in a single MZI, only the phase shifter θ determines the power splitting ratio at the outputs. However, in a mesh of MZIs, ϕ phase shifters also affect the power splitting ratios of the subsequent MZIs.

In a small size of meshes, such as the 4 × 4 Reck mesh or Diamond Mesh with only three layers of MZI, the impact of ϕ phase shifters on the power at the outputs is less than that of θ phase shifters, being more similar to that of a single MZI. This is mainly because, as the light propagates though the waveguides of the structure from the input ports towards the outputs,

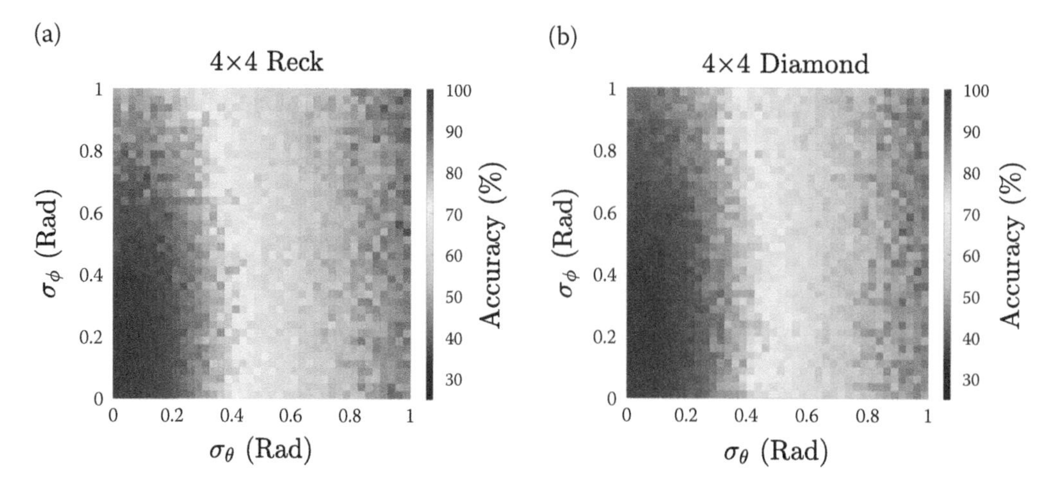

FIGURE 18.13 Classification accuracy of the single layer ONNs implemented by the 4 × 4 (a) Reck mesh, and (b) Diamond mesh with respect to phase errors σ_θ and σ_ϕ. Adapted with permission from [32] The Optical Society.

the ϕ phase shifters of the MZIs in the first layer, i.e., MZIs (1), (2), and (4), do not play a role in the adjustment of the power levels at the outputs of the corresponding MZIs. However, they affect the power splitting ratio of the MZIs in the next two layers. Additionally, the ϕ phase shifters in the last layer of MZIs, i.e., MZI (6) in the case of the Reck mesh and MZIs (9), (8), and (6) in the Diamond mesh, do not have an impact on the power splitting ratio of the corresponding MZIs. Consequently, σ_ϕ is less significant than that of the σ_θ in the classification accuracy of the 4 × 4 ONNs. As the mesh size increases, the number of MZI layers in the structure becomes larger, and thus, the impact of σ_ϕ on the classification accuracy of the structures becomes on par with that of σ_θ. This fact will be investigated in the next section of the chapter.

Figure 18.14 shows the classification accuracy of the ONN exploiting the Reck and the Diamond topologies as a function of phase uncertainty, σ_θ and σ_ϕ, and loss per MZI. In this

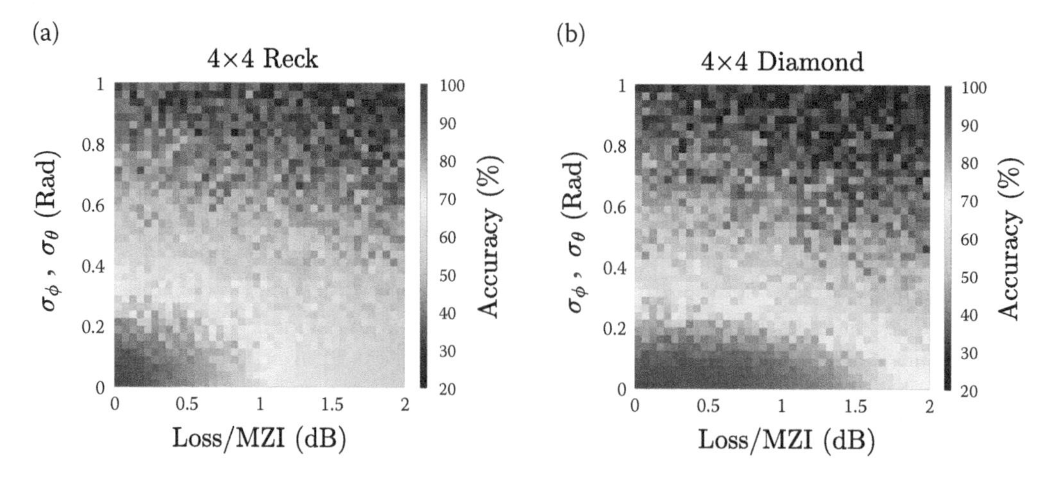

FIGURE 18.14 Classification accuracy of the single layer ONNs implemented by the 4 × 4 (a) Reck mesh, and (b) Diamond mesh with respect to phase error, σ_θ and σ_ϕ, and loss per MZI. The standard deviation of the loss per MZI is set to 0.5 dB. Adapted with permission from [32] The Optical Society.

analysis, the standard deviation of the loss per MZI is set to 0.5 dB. It can be seen that compared to the Reck mesh, the proposed Diamond topology is less sensitive to the IL of the MZIs for the data sample classification.

According to Figures 18.12, 18.13, and 18.14, the 4 × 4 Diamond-topology-based single layer ONN outperforms that of the Reck topology when classifying the data samples in the presence of inevitable fabrication and experimental imperfections. Te more symmetric topology of the Diamond mesh makes its optical paths more loss-balanced compared to the Reck mesh. The additional MZIs in the Diamond structure also provide extra degrees of freedom in the ONN weight matrix optimization during the backpropagation process. Additionally, the extra set of waveguides tapered-out from the additional MZIs in the Diamond mesh allows to exclude the excess light intensity which would have otherwise degraded the ONN classification performance. The triangular mesh, however, maintains the entire optical power propagating through its waveguides.

18.4.4 Scalability Investigation of the Reck and Diamond Topologies in Single Layer ONNs

This section is allocated to the performance analysis of a single layer ONN with appropriate data sets implemented by different sizes of the Reck and the proposed Diamond structures. It is investigated that as the size of the ONN increases, the classification accuracy becomes more sensitive to phase errors and IL of the constituent MZIs in the optical device due to a larger number of reconfigurable MZIs. The results shown in this section confirm that the $N \times N$ Diamond mesh, with $(N-1)(N-2)/2$ additional MZIs, outperforms the $N \times N$ Reck structure. The better scalability of the Diamond topology compared to the Reck mesh is attributed to the more symmetry present in the Diamond topology, resulting in more loss-balanced optical paths. Furthermore, the additional MZIs in the Diamond structure provides more degrees of freedom allowing the ONN to provide better differentiation between the correct class and the other classes. Figure 18.15compares the backpropagation process of a 32 × 32 ONN implemented by the Reck and the Diamond meshes with 0 dB loss per MZI.

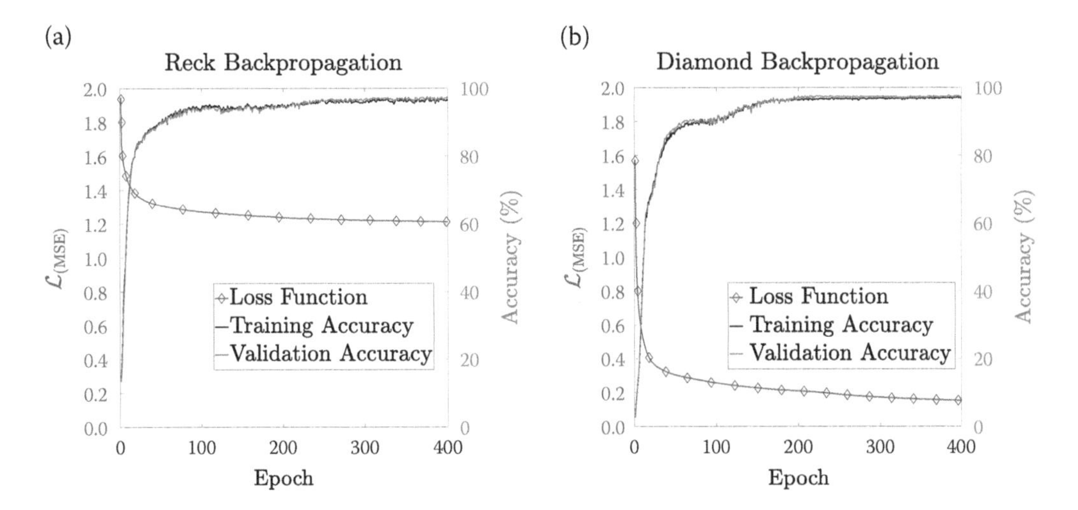

FIGURE 18.15 Back-propagation of the single layer ONNs implemented by the Reck and the proposed Diamond meshes of 32 × 32 structure size with 0 dB loss per MZI. The unitary transformation (weight) matrices in both scenarios are optimized through minimizing the $L_{(MSE)}$ value. The Diamond mesh achieves a loss function value as low as 0.153 compared to that of the Reck scenario at 1.21. Adapted with permission from [32] The Optical Society.

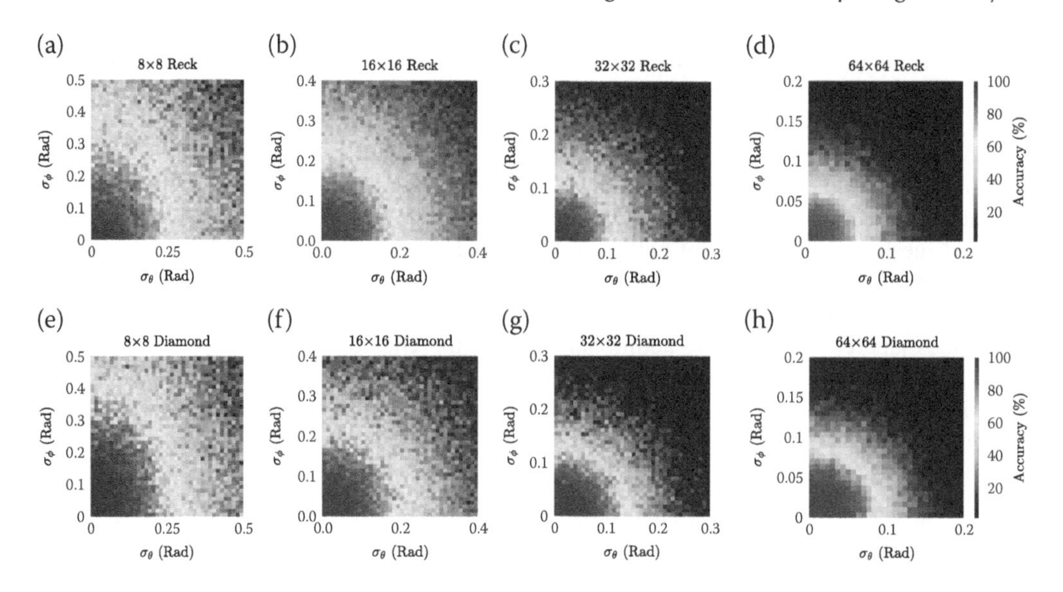

FIGURE 18.16 Classification accuracy of the single layer ONNs implemented by different sizes of the Reck structure and the proposed Diamond mesh. The results are obtained as a function of phase uncertainties σ_θ and σ_ϕ. Adapted with permission from [32] The Optical Society.

As can be seen in Figures 18.15a and 18.15b, the weight matrix is optimized through minimizing the loss function, where the Diamond mesh achieves a validation classification accuracy of 97.71% with a final $L_{(\mathrm{MSE})}$ value as low as 0.15 compared to that of the Reck mesh achieving 97.40% with a final $L_{(\mathrm{MSE})}$ value of 1.21. The lower loss function value in the Diamond scenario is translated to a smaller difference between the output vector and the ground truth vector for each sample during classification. In fact, the Diamond mesh has the possibility to optimally direct the proper amount of light intensity to the outputs for correct classification by excluding the degrading portion of the light intensity from the classification process of the ONN through the tapered-out waveguides of its additional MZIs.

Figure 18.16 demonstrates the classification accuracy of the single layer ONNs implemented by various sizes of the two structures as a function of phase errors σ_θ and σ_ϕ. As shown in the figure, the area of high classification accuracy in both cases decreases with the size of the structures i.e., the size of the ONN.

As can be inferred from these figures, the Diamond topology exhibits more robustness in terms of possible experimental phase uncertainties which makes it more suited for large scale ONNs. The reason lies in the fact that compared to the Reck mesh, the Diamond topology has the possibility to optimally direct the required light intensity towards its outputs for more correct classification. Furthermore, as shown in Figure 18.15, the additional MZIs in the Diamond mesh yields a set of extra degrees of freedom during the training process of the ONNs, allowing for a better minimization of the loss function value. It should be noted that in this analysis, every mesh size has its own data set, where the mesh size is equal to the dimensionality of the data set, i.e., the $N = 64$ meshes have a 64 dimensional multivariate Gaussian data set, while the $N = 32$ meshes have a 32 dimensional one.

Figure 18.17 depicts the performance analysis of different sizes of the single layer ONNs constructed by the Reck and the Diamond meshes in regards to the phase errors, σ_θ and σ_ϕ, and the IL of each MZI in the structures. According to Figure 18.17, the Figure of Merit (FoM) in Rad·dB,

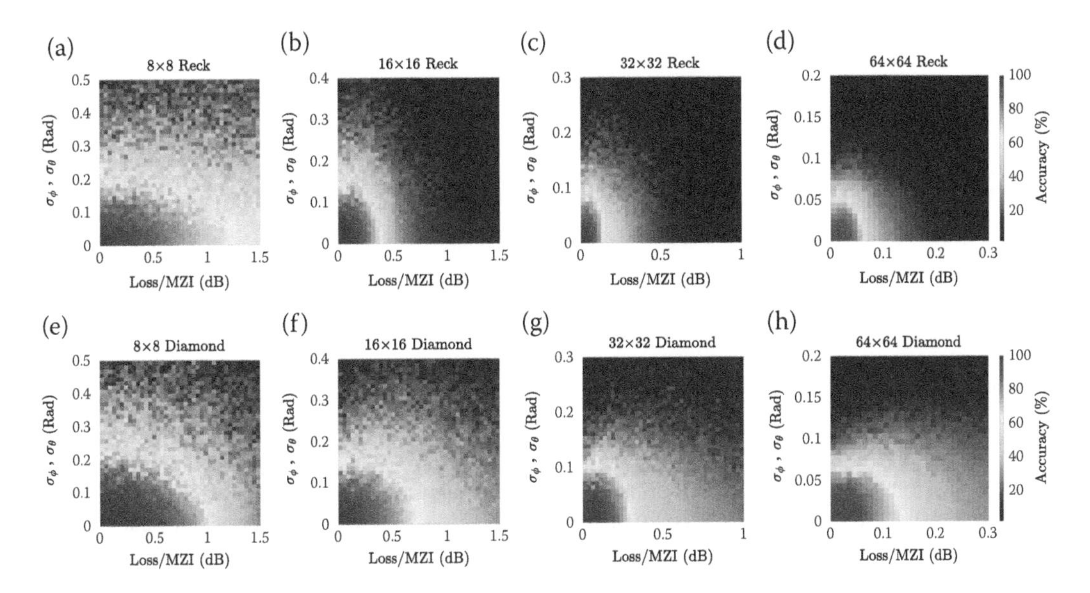

FIGURE 18.17 Classification accuracy of the single layer ONNs implemented by different sizes of the Reck mesh and the proposed Diamond mesh. The results are obtained as a function of the IL (loss) per MZI and phase uncertainty, σ_θ and σ_ϕ. Adapted with permission from [32] The Optical Society.

representing the classification accuracy area above 75%, is reduced in larger structures implying a higher sensitivity to the possible experimental phase errors and IL of the MZIs. According to Figure 18.17, as the structure size of the Reck mesh and the Diamond mesh increases, the effect of phase error on the classification performance of the single layer ONNs is more significant, particularly, in the Reck-topology-based scenarios. For instance, according to [15], a voltage noise standard deviation of 8.67 mV corresponding to σ_θ and σ_ϕ of 0.013 Rad, which results in a classification accuracy degradation of 3% in the 64 × 64 Reck-topology-based ONN compared to 0.56% in the corresponding Diamond-topology-based one.

Figure 18.18 summarizes the performance analysis of the corresponding ONNs implemented by the Reck and the proposed Diamond meshes with different structure sizes. Figure 18.18a compares the related FoM values of the classification accuracy area above 75% and Figure 18.18b shows the final loss function values (MSE) obtained through the corresponding backpropagation processes. It can be seen that the diamond mesh is expected to experimentally outperform the Reck mesh as it exhibits higher FoM and lower final loss function values for various sizes of ONNs. The higher value of FoM in any given size of the Diamond structure can be translated to higher classification accuracies under experimental conditions, where the phase errors and the IL of the MZIs are inevitable. Additionally, the lower final loss function value (MSE) in the Diamond mesh is due to the extra degrees of freedom provided by its $(N-1)(N-2)/2$ additional MZIs, leading to a more optimal weight matrix in the ONN. Contrary to the Reck mesh, the Diamond topology has the possibility of eliminating the excess light intensity that degrades its classification accuracy through the tapered-out waveguides of its additional MZIs.

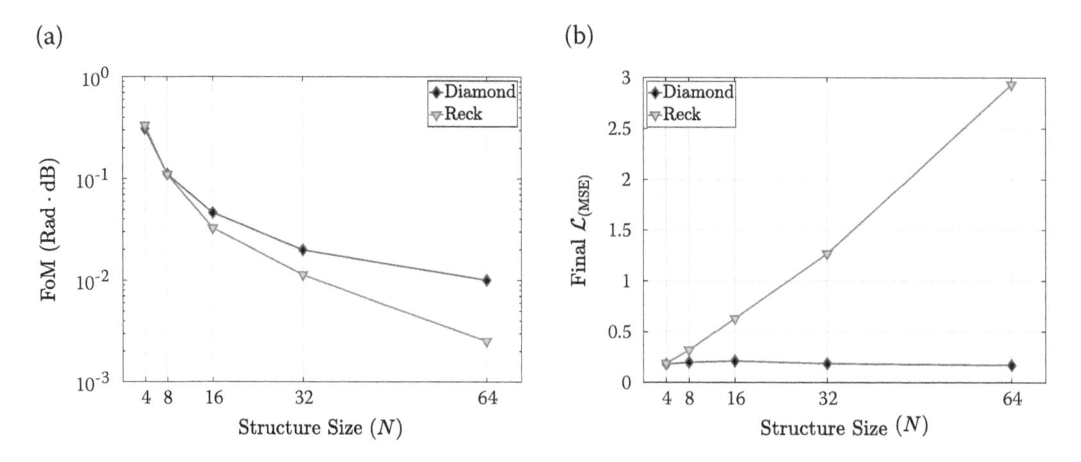

FIGURE 18.18 Performance comparison between the Reck and the proposed Diamond structures in different sizes (N) of the single layer ONNs; (a) figure of merit (FoM), and (b) final loss function values (MSE). Adapted with permission from [32] The Optical Society.

18.5 CONCLUSION

The chapter presented the theoretical principles of multiport reconfigurable MZI-based optical processors. It was shown how to determine the unitary transformation matrix of the triangular (Reck) mesh of MZIs, and how to decompose it to extract the required phase shifts to program the optical device for a given application. The presented analyses in this chapter confirm that the Reck topology is sensitive to the insertion loss of the MZIs and the phase errors due to its loss-unbalanced optical paths. The investigation in this respect has led to the introduction of a phase-error- and loss-tolerant field-programmable MZI-based optical processor with Diamond mesh topology. The Diamond mesh was used to implement various sizes of single layer ONNs and its classification performance was compared to that of the Reck mesh. The obtained results confirm that the Diamond structure is a good practical candidate for ONNs. This is associated with the $(N-1)(N-2)/2$ additional MZIs in the Diamond mesh, which makes its topology more symmetric with more loss-balanced optical paths compared to the Reck topology. Additionally, the extra set of MZIs in the Diamond structure provides additional degrees of freedom in the ONN weight matrix optimization for optimal classification accuracy. As a result, it yield a higher robustness to phase error and the IL of the MZIs.

REFERENCES

[1] A. E. Lim, J. Song, Q. Fang, C. Li, X. Tu, N. Duan, K. K. Chen, R. P. Tern, and T. Liow, "Review of Silicon Photonics Foundry Efforts," *IEEE Journal of Selected Topics in Quantum Electronics*, vol. 20, no. 4, pp. 405–416, 2014.

[2] A. Rahim, T. Spuesens, R. Baets, and W. Bogaerts, "Open-Access Silicon Photonics: Current Status and Emerging Initiatives," *Proceedings of the IEEE*, vol. 106, no. 12, pp. 2313–2330, 2018.

[3] Y. Shen, N. C. Harris, S. Skirlo, M. Prabhu, T. Baehr-Jones, M. Hochberg, X. Sun, S. Zhao, H. Larochelle, D. Englund *et al.*, "Deep Learning with Coherent Nanophotonic Circuits," *Nature Photonics*, vol. 11, no. 7, p. 441, 2017.

[4] A. Ribeiro, A. Ruocco, L. Vanacker, and W. Bogaerts, "Demonstration of a 4×4-Port Self-

Configuring Universal Linear Optical Component," in *2016 Progress in Electromagnetic Research Symposium (PIERS)*, 2016, pp. 3372–3375.

[5] X. Qiang, X. Zhou, J. Wang, C. M. Wilkes, T. Loke, S. O'Gara, L. Kling, G. D. Marshall, R. Santagati, T. C. Ralph *et al.*, "Large-Scale Silicon Quantum Photonics Implementing Arbitrary
Two-Qubit Processing," *Nature Photonics*, vol. 12, no. 9, pp. 534–539, 2018.

[6] A. N. Tait, T. F. De Lima, E. Zhou, A. X. Wu, M. A. Nahmias, B. J. Shastri, and P. R. Prucnal, "Neuromorphic Photonic Networks Using Silicon Photonic Weight Banks," *Scientific Reports*, vol. 7, no. 1, pp. 1–10, 2017.

[7] M. Reck, A. Zeilinger, H. J. Bernstein, and P. Bertani, "Experimental Realization of Any Discrete Unitary Operator," *Physical Review Letters*, vol. 73, pp. 58–61, Jul 1994. [Online]. Available: https://link.aps.org/doi/10.1103/PhysRevLett.73.58.=0pt.

[8] W. R. Clements, P. C. Humphreys, B. J. Metcalf, W. S. Kolthammer, and I. A. Walmsley, "Optimal Design for Universal Multiport Interferometers," *Optica*, vol. 3, no. 12, pp. 1460–1465, Dec 2016. [Online]. Available: http://www.osapublishing.org/optica/abstract.cfm.=0pt.

[9] D. Perez, E. S. Gomariz, and J. Capmany, "Programmable True-Time Delay Lines Using Integrated Waveguide Meshes," *Journal of Lightwave Technology*, vol. 36, no. 19, pp. 4591–4601, 2018.

[10] D. A. Miller, "Self-Aligning Universal Beam Coupler," *Optics Express*, vol. 21, no. 5, pp. 6360–6370, 2013.

[11] N. C. Harris, G. R. Steinbrecher, M. Prabhu, Y. Lahini, J. Mower, D. Bunandar, C. Chen, F. N. Wong, T. Baehr-Jones, M. Hochberg *et al.*, "Quantum Transport Simulations in a Programmable Nanophotonic Processor," *Nature Photonics*, vol. 11, no. 7, p. 447, 2017.

[12] F. Shokraneh, M. S. Nezami, and O. Liboiron-Ladouceur, "Theoretical and Experimental Analysis of a 4×4 Reconfigurable MZI-Based Linear Optical Processor," *Journal of Lightwave Technology*, vol. 38, no. 6, pp. 1258–1267, Mar 2020.

[13] J. Carolan, C. Harrold, C. Sparrow, E. Martn-López, N. J. Russell, J. W. Silverstone, P. J. Shadbolt, N. Matsuda, M. Oguma, M. Itoh *et al.*, "Universal Linear Optics," *Science*, vol. 349, no. 6249, pp. 711–716, Aug 2015.

[14] R. Burgwal, W. R. Clements, D. H. Smith, J. C. Gates, W. S. Kolthammer, J. J. Renema, and I. A. Walmsley, "Using an Imperfect Photonic Network to Implement Random Unitaries," *Optics Express*, vol. 25, no. 23, pp. 28 236–28 245, Nov 2017. [Online]. Available: http://www.opticsexpress.org/abstract.cfm?URI=oe-25-23-28236=0pt.

[15] F. Shokraneh, S. Geoffroy-Gagnon, M. S. Nezami, and O. Liboiron-Ladouceur, "A Single Layer Neural Network Implemented by a 4×4 MZI-Based Optical Processor," *IEEE Photonics Journal*, vol. 11, no. 6, pp. 1–12, Dec 2019.

[16] Q. Cheng, J. Kwon, M. Glick, M. Bahadori, L. P. Carloni, and K. Bergman, "Silicon Photonics Codesign for Deep Learning," *Proceedings of the IEEE*, pp. 1–22, 2020.

[17] D. A. Miller, "Self-Configuring Universal Linear Optical Component," *Photonics Research*, vol. 1, no. 1, pp. 1–15, 2013.

[18] D. A. Miller, "Establishing Optimal Wave Communication Channels Automatically," *J. Lightwave Technol.*, vol. 31, no. 24, pp. 3987–3994, 2013.

[19] R. A. Athale and W. C. Collins, "Optical Matrix–Matrix Multiplier Based on Outer Product Decomposition," *Applied Optics*, vol. 21, no. 12, pp. 2089–2090, 1982.

[20] N. H. Farhat, D. Psaltis, A. Prata, and E. Paek, "Optical Implementation of the Hopfield Model," *Applied Optics*, vol. 24, no. 10, pp. 1469–1475, 1985.

[21] M. A. Nahmias, B. J. Shastri, A. N. Tait, T. F. de Lima, and P. R. Prucnal, "Neuromorphic Photonics." OSA Optical Society, Jan 2018.

[22] V. Sze, Y.-H. Chen, T.-J. Yang, and J. S. Emer, "Efficient Processing of Deep Neural Networks: A Tutorial and Survey," *Proceedings of the IEEE*, vol. 105, no. 12, pp. 2295–2329, 2017.

[23] T. F. de Lima, A. N. Tait, A. Mehrabian, M. A. Nahmias, C. Huang, H.-T. Peng, B. A. Marquez, M. Miscuglio, T. El-Ghazawi, V. J. Sorger *et al.*, "Primer on Silicon Neuromorphic Photonic Processors: Architecture and Compiler," *Nanophotonics*, vol. 1, no. ahead-of-print, 2020.

[24] N. P. Jouppi, C. Young, N. Patil, D. Patterson, G. Agrawal, R. Bajwa, S. Bates, S. Bhatia, N. Boden, A. Borchers *et al.*, "In-datacenter Performance Analysis of a Tensor Processing Unit," in *Proceedings of the 44th Annual International Symposium on Computer Architecture*, 2017, pp. 1–12.
[25] B. Hall, *Lie Groups, Lie Algebras, and Representations: An Elementary Introduction*. Springer, 2015, vol. 222.
[26] L. Jing, Y. Shen, T. Dubcek, J. Peurifoy, S. Skirlo, Y. LeCun, M. Tegmark, and M. Soljačić, "Tunable Efficient Unitary Neural Networks (EUNN) and Their Application to RNNs," in *International Conference on Machine Learning*, 2017, pp. 1733–1741.
[27] R. Hamerly, A. Sludds, L. Bernstein, M. Prabhu, C. Roques-Carmes, J. Carolan, Y. Yamamoto, M. Soljacic, and D. Englund, "Towards Large-Scale Photonic Neural-Network Accelerators," in *2019 IEEE International Electron Devices Meeting (IEDM)*, 2019, pp. 22.8.1–22.8.4.
[28] I. A. D. Williamson, T. W. Hughes, M. Minkov, B. Bartlett, S. Pai, and S. Fan, "Reprogrammable Electro-Optic Nonlinear Activation Functions for Optical Neural Networks," *IEEE Journal of Selected Topics in Quantum Electronics*, vol. 26, no. 1, pp. 1–12, Jan 2020.
[29] L. Chrostowski, H. Shoman, M. Hammood, H. Yun, J. Jhoja, E. Luan, S. Lin, A. Mistry, D. Witt, N. A. Jaeger *et al.*, "Silicon Photonic Circuit Design Using Rapid Prototyping Foundry Process Design Kits," *IEEE Journal of Selected Topics in Quantum Electronics*, 2019.
[30] J. M. Shainline, S. M. Buckley, R. P. Mirin, and S. W. Nam, "Superconducting Optoelectronic Circuits for Neuromorphic Computing," *Physical Review Applied*, vol. 7, no. 3, p. 034013, Mar 2017.
[31] T. W. Hughes, M. Minkov, Y. Shi, and S. Fan, "Training of Photonic Neural Networks Through In Situ Backpropagation and Gradient Measurement," *Optica*, vol. 5, no. 7, pp. 864–871, Jul 2018. [Online]. Available: http://www.osapublishing.org/optica/abstract.cfm?URI=optica-5-7-864=0pt.
[32] F. Shokraneh, S. Geoffroy-Gagnon, and O. Liboiron-Ladouceur, "The Diamond Mesh, a Phase-Error- and Loss-Tolerant Field-Programmable MZI Based Optical Processor for Optical Neural Networks," *Optics Express*, vol. 28, no. 16, pp. 23 495–23 508, Aug 2020. [Online]. Available: http://www.opticsexpress.org/abstract.cfm?URI=oe-28-16-23495.
[33] J. Fu, H. Zheng, and T. Mei, "Look Closer to See Better: Recurrent Attention Convolutional Neural Network for Fine-Grained Image Recognition," in *2017 IEEE Conference on Computer Vision and Pattern Recognition (CVPR)*, July 2017, pp. 4476–4484.
[34] G. K. Venayagamoorthy, V. Moonasar, and K. Sandrasegaran, "Voice Recognition Using Neural Networks," in *Proceedings of the 1998 South African Symposium on Communications and Signal Processing-COMSIG '98 (Cat. No. 98EX214)*, Sep 1998, pp. 29–32.
[35] B. Wu, A. Wan, F. Iandola, P. H. Jin, and K. Keutzer, "SqueezeDet: Unified, Small, Low Power Fully Convolutional Neural Networks for Real-Time Object Detection for Autonomous Driving," in *2017 IEEE Conference on Computer Vision and Pattern Recognition Workshops (CVPRW)*, July 2017, pp. 446–454.
[36] S. Geoffroy-Gagnon, "Neuroptica: Towards a Practical Implementation of Photonic Neural Networks," GitLab Repository, 2020. [Online]. Available: https://gitlab.com/simongg/neuroptica.
[37] G. Veronis, R. W. Dutton, and S. Fan, "Method for Sensitivity Analysis of Photonic Crystal Devices," *Optics Letters*, vol. 29, no. 19, pp. 2288–2290, Oct 2004. [Online]. Available: http://ol.osa.org/abstract.cfm?URI=ol-29-19-2288=0pt.
[38] M. A. Nielsen, *Neural Networks and Deep Learning*. Determination Press, 2018. [Online]. Available: http://neuralnetworksanddeeplearning.com/=0pt.
[39] F. Pedregosa, G. Varoquaux, A. Gramfort, V. Michel, B. Thirion, O. Grisel, M. Blondel, P. Prettenhofer, R. Weiss, V. Dubourg *et al.*, "Scikit-learn: Machine Learning in Python," *Journal of Machine Learning Research*, vol. 12, no. Oct, pp. 2825–2830, 2011.
[40] I. Goodfellow, Y. Bengio, and A. Courville, *Deep Learning*. MIT Press, 2016. [Online]. Available: http://www.deeplearningbook.org=0pt
[41] Y. LeCun, L. Bottou, G. Orr, and K. Muller, "Efficient BackProp," in *Neural Networks: Tricks of the Trade*, G. Orr and K. Muller, Eds. Springer, 1998.

[42] Y. LeCun, "A Theoretical Framework for Back-Propagation," in *Artificial Neural Networks: Concepts and Theory*, P. Mehra and B. Wah, Eds. Los Alamitos, CA: IEEE Computer Society Press, 1992.

[43] H. Wu, "Stability Analysis for Periodic Solution of Neural Networks with Discontinuous Neuron Activations," *Nonlinear Analysis: Real World Applications*, vol. 10, no. 3, pp. 1717–1729, 2009. [Online]. Available: http://www.sciencedirect.com/science/article/pii/S1468121808000540=0pt.

19 High-Performance Deep Learning Acceleration with Silicon Photonics

Febin P. Sunny, Asif Mirza, Mahdi Nikdast, and Sudeep Pasricha

CONTENTS

19.1 INTRODUCTION

Many emerging applications such as self-driving cars, autonomous robotics, fake news detection, pandemic growth and trend prediction, and real-time language translation are increasingly being powered by sophisticated machine learning models. With researchers creating deeper and more complex deep neural network (DNN) architectures, including multi-layer perceptron (MLP) and convolution neural network (CNN) architectures, the underlying hardware platform must consistently deliver better performance while satisfying strict power dissipation limits. Such an endeavor to achieve higher performance-per-watt has driven hardware architects to design custom accelerators for deep learning, e.g., Google's TPU [1] and Intel's Movidius [2], with much higher performance-per-watt than conventional CPUs and GPUs.

Electronic accelerators architectures, unfortunately face fundamental limits in the post Moore's law era where processing capabilities are no longer improving as they did over the past several decades [3]. In particular, moving data electronically on metallic wires in these accelerators creates a major bandwidth and energy bottleneck [4]. Silicon photonics is a promising technology to enable energy-efficient, ultra-high bandwidth, and low-latency communication solutions [5]. CMOS-compatible photonic interconnects have already replaced metallic ones for light-speed

DOI: 10.1201/9780429292033-19

data transmission at almost every level of computing, and are now actively being considered for chip-scale integration [6].

Remarkably, it is also possible to use optical components to perform computation, e.g., matrix-vector multiplication [7]. By employing on-chip waveguides, electro-optic modulators, photodetectors, and lasers to build photonic interconnects and photonic integrated circuits, it is now possible to conceive a new class of DNN accelerators which are effective for low-latency and energy-efficient optical domain data transport and communication. Not only can such photonics-based accelerators address the fan-in and fan-out problems with linear algebra processors, but their operational bandwidth can approach the photodetection rate (typically in the hundreds of GHz), which is orders of magnitude higher than electronic systems today that operate at a clock rate of a few GHz [8].

Despite the above benefits, a number of obstacles must be overcome before viable photonic DNN accelerators can be realized. Fabrication process and thermal variations can adversely impact the robustness of photonic accelerator designs by introducing undesirable crosstalk noise, tuning overheads, resonance drifts, optical phase shifts, and photo-detection current mismatches. For example, experimental studies have shown that micro-ring resonator (MR) devices used in chip-scale photonic interconnects can experience significant resonant drifts (e.g., ~9 nm reported in [9]) within a wafer due to process variations. This matters because even a 0.25 nm drift can cause the bit-error-rate (BER) of photonic data traversal to degrade from 10-12 to 10-6. Moreover, thermal crosstalk in silicon photonic devices such as MRs can significantly reduce DNN model accuracy by limiting the achievable precision (i.e., resolution) of weight and bias parameters to a few bits. Common tuning circuits that rely on thermo-optic phase-change effects to control photonic devices, e.g., when imprinting activations or weights on optical signals, also place a limit on the achievable throughput and parallelism in photonic accelerators. Lastly, at the architecture level, there is a need for a scalable, adaptive, and low-cost computation and communication fabric that can handle the demands of diverse MLP and CNN models.

In this chapter, we introduce CrossLight, novel silicon photonic neural network accelerator that addresses the challenges highlighted above through a cross-layer design approach. By cross-layer, we refer to the design paradigm that involves considering multiple layers in the hardware-software design stack together, for a more holistic optimization of the photonic accelerator. CrossLight involves device-level engineering for resilience to fabrication-process variations and thermal crosstalk, circuit-level tuning enhancements for inference latency reduction, and an optimized architecture-level design that also integrates the device- and circuit-level improvements to enable higher resolution, better energy-efficiency, and improved throughput compared to prior efforts on photonic accelerator design. Our novel contributions in this chapter include:

- Improved silicon photonic device designs that we fabricated to make our architecture more resilient to fabrication-process variations;
- An enhanced tuning circuit to simultaneously support large thermal-induced resonance shifts and high-speed, low-loss device tuning;
- Consideration of thermal crosstalk mitigation methods to improve the weight resolution achievable by CrossLight architecture;
- Increased throughput and energy-efficiency by improving wavelength reuse and further use of matrix decomposition at the architecture-level;
- A comprehensive comparison with state-of-the-art accelerators that shows the efficacy of our cross-layer optimized solution.

19.2 RELATED WORK

Silicon-photonics based DNN accelerator architectures represent an emerging paradigm that can immensely benefit the landscape of deep learning hardware design [10–14]. A photonic neuron in these architectures consists of three components: a weighting, a summing, and a nonlinear unit which is analogous to an artificial neuron. Noncoherent photonic accelerators, such as [11–13], typically employ the Broadcast and Weight (B&W) protocol [10] to manipulate optical signal power for setting and updating weights and activations. The B&W protocol is an analog networking protocol that uses wavelength-division multiplexing (WDM), photonic multiplexors, and photodetectors to combine outputs from photonic neurons in a layer. Coherent photonic accelerators, such as [8], [14], typically use only a single wavelength to manipulate the electrical field amplitude rather than signal power. Weighting occurs with electrical field amplitude attenuation proportional to the weight value, and phase modulation that is proportional to the sign of the weight. The weighted signals are then coherently accumulated with cascaded Y-junction combiners. For both types of accelerators, non-linearity can be implemented with devices such as electro-absorption modulators [8].

Due to the scalability, phase encoding noise, and phase error accumulation limitations of coherent accelerators [15], there is growing interest in designing efficient noncoherent photonic accelerators. In particular, the authors of DEAP-CNN [11] have described a noncoherent neural network accelerator that implements the entirety of the CNN layers using connected convolution units. The tuned MRs in these units assume the kernel values by using phase tuning to manipulate the energy in their resonant wavelengths. Holylight [12] is another noncoherent architecture that uses microdisks (instead of MRs) for its lower area and power consumption. It utilizes a "whispering gallery mode" resonance for microdisk operation, which unfortunately is inherently lossy due to a phenomenon called tunneling ray attenuation [16]. More generally, these noncoherent architectures suffer from susceptibility to process variations and thermal crosstalk, which are not addressed in these architectures. Microsecond-granularity thermo-optic tuning latencies further reduce the speed and efficiency of optical computing [17]. We address these shortcomings as part of our proposed cross-layer optimized noncoherent photonic accelerator architecture in this chapter.

19.3 NONCOHERENT PHOTONIC COMPUTATION OVERVIEW

As mentioned earlier, noncoherent photonic accelerators typically utilize the Broadcast and Weight (B&W) photonic neuron configuration with multiple wavelengths. Figure 19.1 shows an example of this B&W configuration with n neurons in a layer where the colored-dotted box represents a single neuron. Each input to a neuron is imprinted onto a unique wavelength (λ_i) emitted by a laser diode (LD) using a Mach–Zehnder modulator (MZM). The wavelengths are multiplexed (MUXed) into a single waveguide using arrayed waveguide grating (AWG), and split into n branches that are each weighted with a micro-ring resonator (MR) bank that alters optical signal power proportional to weight values. Summation across positive and negative weight arms at each branch is performed

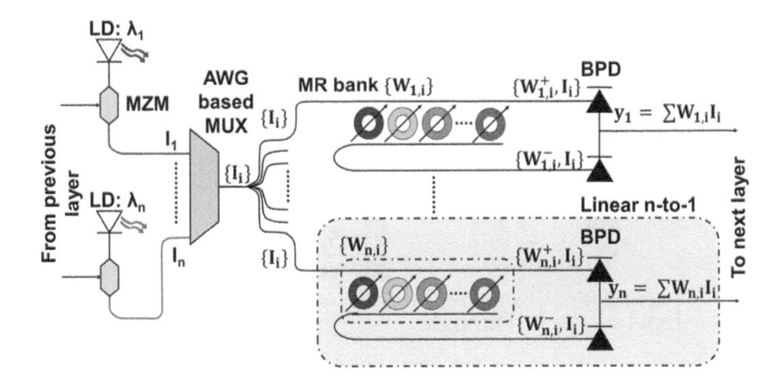

FIGURE 19.1 Noncoherent Broadcast-and-Weight (B&W) based photonic neuron [38].

using a balanced photodetector (BPD). Optoelectronic devices such as electro-absorption modulators (not shown for brevity) introduce non-linearity after the multiplication and summation operations.

MRs are the fundamental components that impact the efficiency of this configuration. Weights (and biases) are altered by tuning MRs so that the losses experienced by wavelengths—on which activations have been imprinted—can be modified to realize matrix-vector multiplication. MR-weight banks have groups of these tunable MRs, each of which can be tuned to drain energy from a specific resonant wavelength so that the intensity of the wavelength reflects a specific value (after it has passed near the MR). As an example of performing computation in the optical domain, consider the case where an activation value of 0.8 must be weighted by a value of 0.5 as part of a matrix-vector multiplication in a DNN model inference phase. Let us assume that the red wavelength (λ_1) is imprinted with the activation value of 0.8 by using the MZM in Figure 19.1 (alternatively, MRs can be used for the same goal, where an MR will be tuned in such a way that 20% of the input optical signal intensity is dropped as the wave traverses the MR). When λ_1 passes through an MR bank, e.g., the one in the dotted-blue box in Figure 19.1, the MR in resonance with λ_1 can be tuned to drop 50% of the input signal intensity. Thus, as λ_1 passes this MR, we will obtain 50% of the input intensity at the through port, which is 0.4 (= 0.8 × 0.5). The BPD shown in Figure 19.1 then converts the optical signal intensity from that wavelength (and other wavelengths) into an electrical signal that represents an accumulated single value.

An MR is essentially an on-chip resonator which is said to be in resonance when an optical wavelength on the input port matches with the resonant wavelength of the MR, generating a Lorentzian-shaped signal at the through port. An all-pass MR and its output optical spectrum is shown in Figure 19.2. The free-spectral range (FSR) and extinction ratio (ER) are two primary characteristics of an MR. These depend on several physical properties in the MR, including its width, thickness, radius, and the gap between the input and ring waveguide [18]. Changing any of these properties changes the effective index (n_{eff}) of the MR, which in turn causes a change in the output optical spectrum. It is crucial to maintain the central wavelength at the output optical spectrum for reliable operation of MRs. However, MRs are sensitive to fabrication-process variations (FPVs) and variations in surrounding temperature. These cause the central wavelength of the MR to deviate from its original position, causing a drift in the MR resonant wavelength ($\Delta\lambda_{MR}$) [19]. Such a drift (due to FPV or thermal variations) can be compensated using electro-optic (EO) or thermo-optic (TO) tuning mechanisms. Both of these have their own advantages and disadvantages. EO tuning is faster (~ns range) and consumes lower power (~4 μW/nm) but with a smaller tuning range [20]. In contrast, TO tuning has a larger tunability range, but consumes higher power (~27 mW/FSR) and has higher (~μs range) latency [17].

A large number of MRs must be used at the architecture-level to support complex MLP and CNN model executions. As the number of MRs increase, so does the length of the waveguide which hosts the banks. Unfortunately, this leads to an increase in the total optical signal propagation, modulation, and through losses experienced, which in turn increases the laser power

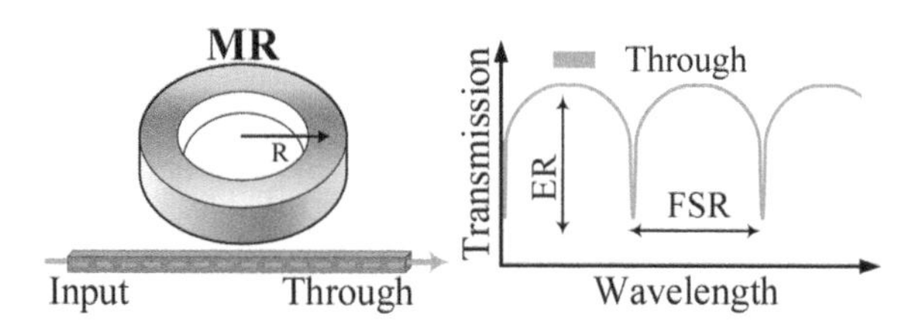

FIGURE 19.2 An all-pass MR with output spectral characteristics at the through port with extinction ratio (ER) and free spectral range (FSR) specified in the figure [38].

required to drive the optical signals through the weight banks, so that they can be detected error-free at the photodetector. An excessive number of parallel arms with MR weight banks (the dotted box in Figure 19.1 represents one arm working in parallel with other arms) also increases optical splitter losses. Moreover, without considering crosstalk mitigation strategies (as is the case with previously proposed photonic accelerators), the weight resolution of the architecture goes down with increased crosstalk noise in optical signals.

In summary, to design efficient photonic accelerators, there is a need for (1) improved MR device design to better tolerate variations and crosstalk; (2) efficient MR tuning circuits to quickly and reliably imprint activation and parameter values; and (3) a scalable architecture design that minimizes optical signal losses. Our novel CrossLight photonic accelerator design addresses all of these concerns and is discussed next.

19.4 CROSSLIGHT ARCHITECTURE

Figure 19.3 shows a high-level overview of our CrossLight noncoherent silicon photonic neural network accelerator. The photonic substrate performs vector dot product (VDP) operations using silicon photonic MR devices, and summation using optoelectronic photodetector (PD) devices over multiple wavelengths. An electronic control unit is required for the control of photonic devices, and for communication with a global memory to obtain the parameter values, partial sum buffering, and for mapping of the vectors. We use digital to analog converter (DAC) arrays to convert buffered signals into analog tuning signals for MRs. Analog to digital converter (ADC) arrays are used to map the output analog signals generated by PDs to digital values that are sent back for post-processing and buffering. We break down the discussion of this accelerator into three parts (Sections 19.4.1–19.4.3), corresponding to the contributions at the device, tuning circuit, and architecture levels, as discussed next.

19.4.1 MR Device Engineering and Fabrication

Process variations are inevitable in CMOS-compatible silicon photonic fabrications, causing undesirable changes in resonant wavelength of MR devices ($\Delta\lambda_{MR}$). A 1.5 × 0.6 mm^2 chip was fabricated using high-resolution Electron Beam (EBeam) lithography and we performed a comprehensive design-space exploration of MRs to compensate for FPVs while improving MR device insertion loss and Q-factor. In this exploration, we varied the input and ring waveguide widths to find an MR device design that was tolerant to FPVs. We found that in an MR design of any radii and gap, when the input waveguide is 400 nm wide and the ring waveguide is 800 nm wide at room temperature (300 K), the undesired $\Delta\lambda_{MR}$ due to FPVs can be reduced from 7.1 to 2.1 nm (70% reduction). This is a significant result, as these engineered MRs require less compensation for FPV-induced resonant wavelength shifts, which can reduce the power consumption of architectures using such MRs.

Unfortunately, the impact of FPVs is not completely eliminated, even with such optimized MR designs, and there is still a need to compensate for FPVs. Thermal variations are another major factor to cause changes in MR n_{eff} which also leads to undesirable $\Delta\lambda_{MR}$. Thermo-optic (TO) tuners are used to compensate for such deviations in $\Delta\lambda_{MR}$. These TO tuners use microheaters to change the temperature in the proximity of an MR device, which then alters the n_{eff} of the MR, changing the device resonant wavelength, and correcting the $\Delta\lambda_{MR}$. High temperatures from such heaters can unfortunately cause thermal energy dissipation, creating thermal crosstalk across MR devices placed close to each other. One can avoid such thermal crosstalk by placing devices at an appropriate distance from each other, typically 120 μm to 200 μm (depending on the number of MR devices in proximity within an MR bank). But such a large spacing hurts area efficiency and also increases waveguide length, which increases propagation losses and its associated laser power overhead. We propose to address this challenge at the circuit level, as discussed in the next section.

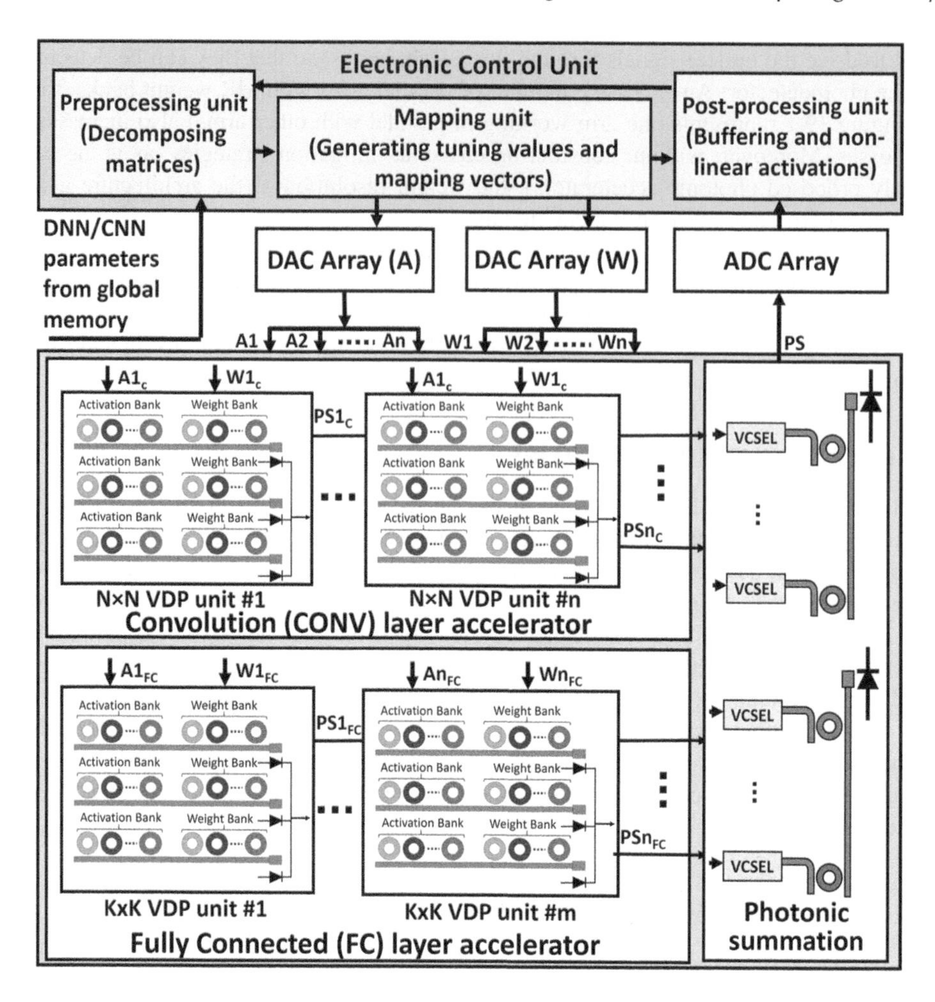

FIGURE 19.3 An overview of CrossLight, showing dedicated vector dot product (VDP) units for CONV and FC layer acceleration, and the internal architecture [38].

19.4.2 Tuning Circuit Design

To reduce thermal crosstalk, we must reduce the reliance on TO tuning, an approach that is used in all prior photonic neural network accelerators, but one that entails high overheads. We propose to use a hybrid tuning circuit where both thermo-optic (TO) and electro-optic (EO) tuning are used to compensate for $\Delta\lambda_{MR}$. Such a tuning approach has previously been proposed in [21] for silicon photonic Mach–Zehnder Interferometers with low insertion loss. Such an approach can be easily transferred to an optimized MR for hybrid tuning in our architecture. The hybrid tuning approach supports faster operation of MRs with fast EO tuning to compensate for small $\Delta\lambda_{MR}$ shifts and, using TO tuning when necessary to compensate for large $\Delta\lambda_{MR}$ shifts. To further reduce the power overhead of TO tuning in this hybrid approach, we adapt a method called Thermal Eigen Decomposition (TED), which was first proposed in [22]. Using TED, we can collectively tune all the MRs in an MR bank to compensate for large $\Delta\lambda_{MR}$ shifts. By doing so, we can cancel the effect of thermal crosstalk (i.e., an undesired phase change) in MRs with much lower power consumption. The TO tuning power can be calculated by the amount of phase shift necessary to apply to the MRs in order for them to be at their desired resonant wavelength. The extent of phase crosstalk ratio (due to thermal crosstalk) as a function of the distance between an MR pair is shown in Figure 19.4,

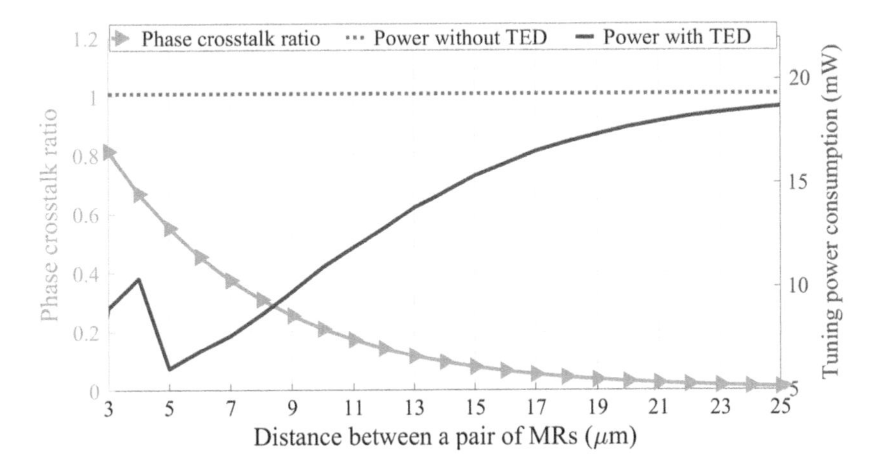

FIGURE 19.4 Phase crosstalk ratio and tuning power consumption in a block of 10 fabricated MRs with variable distance between adjacent pair of MRs [38].

for our fabricated MR devices. The results are based on detailed analysis with a commercial 3D heat transport simulation EDA tool for silicon photonic devices (Lumerical HEAT [23]). It can be seen from the orange line that the amount of phase crosstalk reduces exponentially as the distance between an MR pair increases. Such a trend has also been observed in [24]. To find a balance between tuning power savings while having reduced crosstalk, we perform a sensitivity analysis based on the distance between two adjacent MRs in our architecture. We placed the optimized MRs (described in the previous section) in such a manner that maximum tuning power is saved when they are close to each other while compensating for thermal crosstalk. Results from our analysis (the solid-blue line in Figure 19.4) indicate that placing each MR pair at a distance of 5 μm is optimal, as decreasing or increasing such a distance causes an increase in power consumption of individual TO heaters in the MRs. Figure 19.4 also shows the tuning power required without using the TED approach (blue dotted line), which can be seen to be notably higher.

The workflow of our circuit-level hybrid tuning approach can be summarized as follows. When the accelerator is first booted at runtime, a one-time compensation for design-time FPVs is applied using TO tuning. The extent of compensation for crosstalk is calculated offline during the test phase, where the required phase shift in each of the MRs is calculated, and once the system is online, the respective phase shift values are applied to cancel the impact of thermal crosstalk. Subsequently, we apply EO tuning due to its extremely low latency to represent vector elements in each vector operation with MRs (discussed in more detail in the next section). If large shifts in temperature are observed at runtime, we can perform a one-time calibration with TO tuning to compensate for it. In our analysis, runtime TO tuning would be required rarely beyond its first use after the initial bootup of the photonic accelerator platform.

19.4.3 Architecture Design

The optimized MR devices, layouts, and tuning circuits are utilized within optical vector dot product (VDP) units, which are shown in Figure 19.3. We use banks (groups) of MRs to imprint both activations and weights onto the optical signal. At the architecture level, we compose multiples of VDP units into two architectural sub-components: one to support convolution (CONV) layer acceleration and the other to support fully connected (FC) layer acceleration. We focus on these two types of layers as they are the most widely used and consume the most significant amount of latency and power in computational platforms that execute DNNs. In contrast, other layer types (e.g., pooling, batch normalization) can be implemented very efficiently in the electronic domain. Note also that we focus on inference acceleration, as done in all photonic DNN accelerators, and almost all electronic DNN accelerators.

19.4.3.1 Decomposition Vector in CONV/FC Layers

To map CONV and FC layers from DNN models to our accelerator, we first need to decompose large vector sizes into smaller ones. In CONV layers, a filter performs convolution on a patch (e.g., 2 × 2 elements) of the activation matrix in a channel to generate an element of the output matrix. The operation can be represented as follows:

$$K \otimes A = Y \tag{19.1}$$

For a 2 × 2 filter kernel and weight matrices, (19.1) can be expressed as:

$$\begin{bmatrix} k_1 & k_2 \\ k_3 & k_4 \end{bmatrix} \otimes \begin{bmatrix} a_1 & a_2 \\ a_3 & a_4 \end{bmatrix} = k_1a_1 + k_2a_2 + k_3a_3 + k_4a_4 \tag{19.2}$$

Rewriting (19.2) as a vector dot product, we have:

$$[k_1 \quad k_2 \quad k_3 \quad k_4]. \begin{bmatrix} a_1 \\ a_2 \\ a_3 \\ a_4 \end{bmatrix} = k_1a_1 + k_2a_2 + k_3a_3 + k_4a_4 \tag{19.3}$$

Once we represent the operation as a vector dot product, it is easy to see how it can be decomposed into partial sums. For example:

$$\begin{aligned} &[k_1 \quad k_2\ ]. \begin{bmatrix} a_1 \\ a_2 \end{bmatrix} = k_1a_1 + k_2a_2 = PS_1 \\ &[k_3 \quad k_4\ ]. \begin{bmatrix} a_3 \\ a_4 \end{bmatrix} = k_3a_3 + k_4a_4 = PS_2 \\ &PS_1 + PS_2 = Y \end{aligned} \tag{19.4}$$

In FC layers, typically much larger dimension vector multiplication operations are performed between input activations and weight matrices:

$$AW = \begin{bmatrix} a_1 \\ a_2 \\ \vdots \\ a_n \end{bmatrix} [w_1 \quad w_2 \ldots w_n] \tag{19.5}$$

$$AW = \begin{bmatrix} a_1{\cdot}w_1 & +a_1{\cdot}w_2 & +\cdots & a_1{\cdot}w_n \\ a_2{\cdot}w_1 & +a_2{\cdot}w_2 & +\cdots & a_2{\cdot}w_n \\ & \vdots & & \\ a_n{\cdot}w_1 & +a_n{\cdot}w_2 & +\cdots & a_n{\cdot}w_n \end{bmatrix} \tag{19.6}$$

In (19.5), a_1 to a_n represent a column vector of activations (A) and w_1 to w_n represent a row vector of weights (W). The resulting vector is a summation of dot products of vector elements (19.6). Much like with CONV layers, these can be decomposed into lower dimensional dot products.

19.4.3.2 Vector Dot Product (VDP) Unit Design

We separated the implementation of CONV and FC layers in CrossLight due to different orders of vector dot product computations required to implement each layer. For instance, typical CONV layer kernel sizes vary from 2×2 to 5×5, whereas in FC layers it is not uncommon to have 100 or more neurons (requiring 100×100 or higher order multiplication). State-of-the-art photonic DNN accelerators, e.g., [11], only consider the scales involved at the CONV layer, and either only support CONV layer acceleration in the optical domain, or use the same CONV layer implementation to accelerate FC layers. This leads to increased latencies and reduced throughput as the larger vectors involved with FC layer calculation must be divided up into much smaller chunks, in the order of the filter kernel size of the CONV layer.

For improved efficiency, we separately support the unique scale and requirements of vector dot products involved in CONV and FC layers. For CONV layer acceleration, we consider n VDP units, with each unit supporting an $N \times N$ dot product. For FC layer acceleration, we consider m units, with each unit supporting a $K \times K$ dot product. Here $n > m$ and $K > N$, as per the requirements of each of the distinct layers. In each of the VDP units, the original vector dimensions are decomposed into N or K dimensional vectors, as discussed above. We performed an exploration to determine the optimal values for N, K, n, and m. The results of this exploration study are presented in Section 19.5.

19.4.3.3 Optical Wavelength Reuse in VDP Units

Prior work on photonic DNN accelerator design typically considers a separate wavelength to represent each individual element of a vector. This approach leads to an increase in the total number of lasers needed in the laser bank (as the size of the vectors increases) which in turn increases power consumption. Beyond employing the decomposition approach discussed above, we also consider wavelength reuse per VDP unit to minimize laser power. In this approach, within VDP units, the N or K dimensional vectors are further decomposed into smaller sized vectors for which dot products can be performed using MRs in parallel, in each arm of the VDP unit. The same wavelengths can then be reused across arms within a VDP to reduce the number of unique wavelengths required from the laser. PDs perform summation of the element-wise products to generate partial sums from decomposed vector dot products. The partial sums from the decomposed operations are then converted back to the photonic domain by VCSELs (bottom right of Figure 19.3), multiplexed into a single waveguide, and accumulated using another PD, before being sent for buffering. Thus, our approach leads to an increase in the number of PDs compared to other accelerators but significantly reduces both the number of MRs per waveguide and the overall laser power consumption.

In each arm within a VDP unit, we used a maximum of 15 MRs per bank for a total of 30 MRs per arm, to support up to a 15×15 vector dot product. The choice of MRs per arm considers not only the thermal crosstalk, layout spacing issues (discussed earlier), and the benefits of wavelength reuse (discussed in previous para), but also non-negligible optical splitter losses as the number of MRs per arm increases, which in turn increases laser power requirements. Thus, the selection of MRs per arm within a VDP unit was carefully adjusted to balance parallelism within/across arms, and laser power overheads.

19.5 EVALUATION AND SIMULATION SETUP

19.5.1 Simulation Setup

To evaluate the effectiveness of our CrossLight accelerator, we conducted several simulation studies. These studies were complemented by our MR-device fabrication and optimization efforts

TABLE 19.1
Models and datasets considered for evaluation

Model No.	CONV Layers	FC Layers	Parameters	Datasets
1	2	2	60,074	Sign MNIST
2	4	2	890,410	CIFAR10
3	7	2	3,204,080	STL10
4	8	4	38,951,745	Omniglot

TABLE 19.2
Parameters considered for analyses of photonic accelerators

Devices	Latency	Power
EO tuning [20]	20 ns	4 μW/nm
TO tuning [17]	4 μs	27.5 mW/FSR
VCSEL [33]	10 ns	0.66 mW
TIA [34]	0.15 ns	7.2 mW
Photodetector [35]	5.8 ps	2.8 mW

on real chips, as discussed in Section 19.4. We considered the four DNN models shown in Table 19.1 for execution on the accelerator. Model 1 is Lenet5 [25] and models 2 and 3 are custom CNNs with both FC and CONV layers. Model 4 is a Siamese CNN utilizing one-shot learning. The datasets used to train these models are also listed in the table. We designed a custom CrossLight accelerator simulator in Python to estimate its performance and power/energy. We used Tensorflow 2.3 along with Qkeras [26], for analyzing DNN model accuracy across different parameter resolutions.

We compared CrossLight with the DEAP-CNN [11] and Holylight [12] photonic DNN accelerators from prior work. Table 19.2 shows the optoelectronic parameters considered for this simulation-based analysis. We considered photonic signal losses due to various factors: signal propagation (1 dB/cm [6]), splitter loss (0.13 dB [27]), combiner loss (0.9 dB [28]), MR through loss (0.02 dB [29]), MR modulation loss (0.72 dB [30]), microdisk loss (1.22 dB [31]), EO tuning loss (6 dB/cm [20]), and TO tuning loss (1 dB/cm [17]). We also considered the 1-to-56-Gb/s ADC/DAC-based transceivers from recent work [32]. To calculate laser power consumption, we use the following laser power model:

$$P_{laser} - S_{detector} \geq P_{photo\_loss} + 10 \times \log_{10} N_{\lambda} \tag{19.7}$$

where P_{laser} is laser power in dBm, $S_{detector}$ is the PD sensitivity in dBm, and P_{photo_loss} is the total photonic loss encountered by the optical signal, due to all of the factors discussed above.

19.5.2 Results: CrossLight Resolution Analysis

We first present an analysis of the resolution that can be achieved with CrossLight. We consider how the optical signals from MRs impact each other due to their spectral proximity, also known as inter-channel crosstalk. For this, we use the equations from [36]:

$$\varphi(i, j) = \frac{\delta^2}{(\lambda_i - \lambda_j)^2 + \delta^2} \tag{19.8}$$

In (19.8), $\varphi(i, j)$ describes the noise content from the *j*th MR present in the signal from the *i*th MR. As the noise content increases, the resolution achievable with CrossLight will decrease. Also, $(\lambda_i - \lambda_j)$ is the difference between the resonant wavelengths of *i*th MR and *j*th MR, while δ ($= \lambda_i/2Q$) denotes the 3 dB bandwidth of the MRs, with Q being the quality factor (Q-factor) of the MR being considered. The noise power component can thus be calculated as:

$$P_{noise} = \sum_i^{n-1} \varphi(i, j) P_{in}[i] \tag{19.9}$$

For unit input power intensity, resolution can then be computed as:

$$Resolution = \frac{1}{max|P_{noise}|} \tag{19.10}$$

From this analysis, we found that with the FSR value of 18 nm, the Q-factor value of ~8000 in our optimized MR designs, and the wavelength reuse strategy in CrossLight, which allows us to have large $(\lambda_i - \lambda_j)$ values (>1 nm), our MR banks will be able to achieve a resolution of 16 bits for up to 15 MRs per bank (Section 19.4.3.2). This is much higher than the resolution achievable by many photonic accelerators. For instance, DEAP-CNN can only achieve a resolution of 4 bits, whereas Holylight can only achieve a 2-bit resolution per microdisk (they however combine 8 microdisks to achieve an overall 16-bit resolution). Higher resolution ensures better accuracy in inference, which can be critical in some applications. Figure 19.5 shows the impact of varying the resolution across the weights and activations from 1 bit to 16 bits (we used quantization-aware training to maximize accuracy), for the four DNN models considered (Table 19.1). A crucial observation is that model inference accuracy is sensitive to the resolution of weight and activation parameters. Models such as the one for STL10 are particularly sensitive to the resolution. Thus, the high resolution afforded by CrossLight can allow achieving higher accuracies than other photonic DNN accelerators, such as DEAP-CNN.

19.5.3 Results: CrossLight Sensitivity Analysis

We performed a sensitivity analysis by varying the number of VDP units in the CONV layer accelerator (*n*) and FC layer accelerator (*m*), along with the complexity of the VDP units (*N* and *K*, respectively).

Figure 19.6 shows the frames per second (FPS; a measure of inference performance) vs. energy per bit (EPB) vs. area of various configurations of CrossLight. We selected the best configuration as the one which had the highest value of FPS/EPB. In terms of (*N, K, n, m*), the values of the four parameters for this configuration are (20, 150, 100, 60). This configuration also ended up being the one with the highest FPS value, but had a higher area overhead than other configurations. Nonetheless, this area is comparable to that of other photonic accelerators. This configuration was used for comparisons with prior work, as discussed next.

19.5.4 Results: Comparison with State-of-the-Art Accelerators

We compared our CrossLight accelerator against two well-known photonic accelerators: DEAP-CNN and Holylight, within a reasonable area constraint for all accelerators (~16–25 mm^2). We present results for four variants of the CrossLight architecture: (1) *Cross_base* utilizes

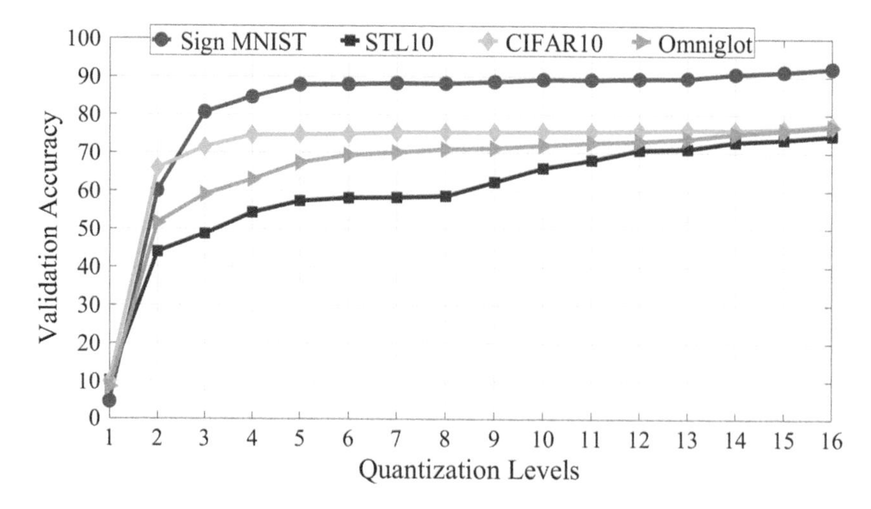

FIGURE 19.5 Inference accuracy of the four DNN models considered, across quantization (resolution) range from 1 bit to 16 bits (for both weights and activations) [38].

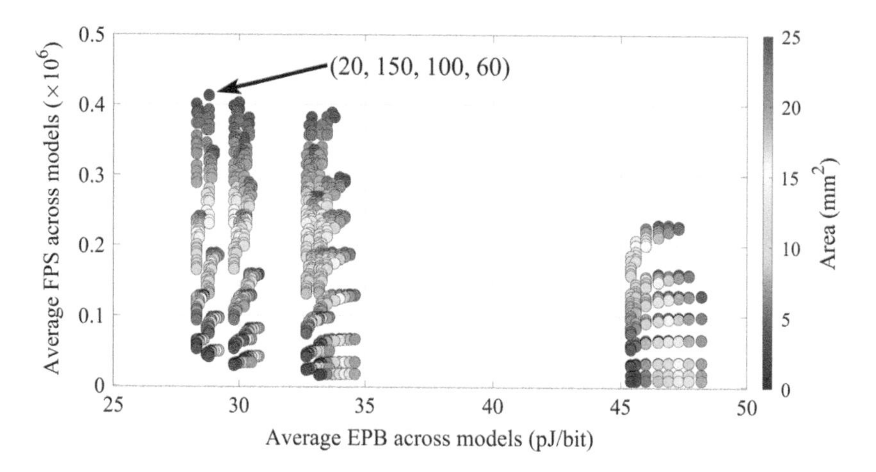

FIGURE 19.6 Scatterplot of average FPS vs. average EPB vs. area of various CrossLight configurations. The configuration with highest FPS/EPB (and FPS) is highlighted [38].

conventional MR designs (without FPV resilience) and traditional TO tuning; (2) *Cross_opt* utilizes the optimized MR designs from Section 19.4.1, and traditional TO tuning; (3) *Cross_base_TED* utilizes the conventional MR designs with the hybrid TED-based tuning approach from Section 19.4.2; and (4) *Cross_opt_TED* utilizes the optimized MR designs and the hybrid TED-based tuning approach.

Figure 19.7 shows the power consumption comparison across the four CrossLight variants and the two photonic accelerators from prior work. We also included numbers for comparing electronic platforms: three deep learning accelerators (DaDianNao, Null Hop, and EdgeTPU), a GPU (Nvidia Tesla P100), and CPUs (Intel Xeon Platinum 9282 denoted as IXP9282, and AMD Threadripper 3970 × denoted as AMD-TR) [37]. The difference in power values between the CrossLight variants arises due to the optimization approaches adopted in each of the variant. The variants which considered conventional MR design instead of the optimized designs have larger power consumption for compensating for FPV. This value becomes non-trivial as the number of MRs increase, and thus having reduced tuning power requirement per MR (in *Cross_opt* and *Cross_opt_TED*) becomes a significant advantage. Using the TED based hybrid tuning approach

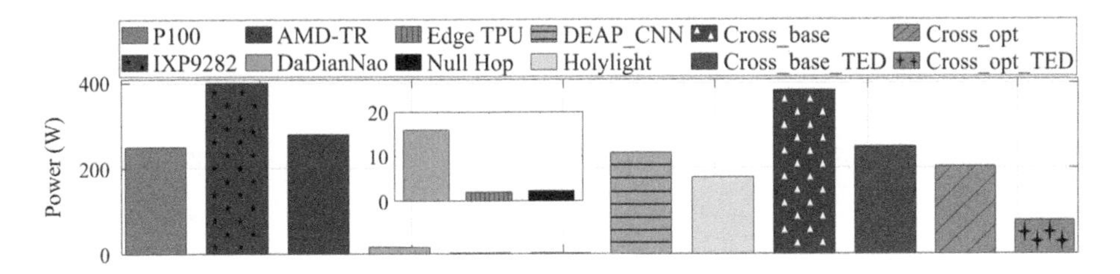

FIGURE 19.7 Power consumption comparison among variants of CrossLight vs. photonic accelerators (DEAP-CNN, Holylight), and electronic accelerator platforms (P100, Xeon Platinum 9282, Threadripper 3970x, DaDianNao, EdgeTPU, Null Hop) [38].

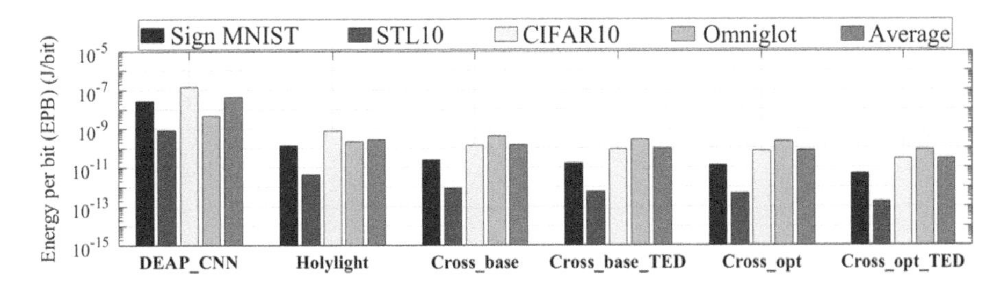

FIGURE 19.8 Comparison of EPB values of the photonic DNN accelerators [38].

provides further significant power benefits for *Cross_opt_TED* over *Cross_opt*, which uses conventional TO tuning. *Cross_opt_TED* can be seen to have lower power consumption than both photonic accelerators, as well as the CPU and GPU platforms, although this power is higher than that of the edge/mobile electronic accelerators (Figure 19.7).

Figure 19.8 shows a comparison of energy-per-bit (EPB) across all of the photonic accelerators, for the four DNN models. On average, our best CrossLight configuration (Cross_opt_TED) has 1544× and 9.5× lowerEPB compared to DEAP-CNN and Holylight, respectively. CrossLight has lower EPB is because we comprehensively took into consideration various losses and crosstalk that a photonic DNN accelerator would experience, and put in place novel approaches at the device, circuit, and architecture layers to counteract their impact in CrossLight. The utilization of TED-based thermal crosstalk management allows us to have MRs placed much closer together, which in turn reduces propagation losses. Additionally, CrossLight considers a combination of TO and EO tuning which enables the reduction of power and EPB as well. The use of EO tuning in our hybrid tuning approach also provides the advantage of lower latencies, which is apparent in the EPB values.

Table 19.3 summarizes the average values of EPB (in pJ/bit) and performance-per-watt (in kiloFPS/watt) of the photonic accelerators as well as the electronic accelerators considered in this chapter. It can be observed that the best *CrossLight* configuration (*Cross_opt_TED*) achieves significantly lower EPB and higher performance-per-watt values than all of the accelerators considered. Specifically, against Holylight, which is the best out of the two photonic DNN accelerators considered, CrossLight achieves 9.5× lower energy-per-bit and 15.9× higher performance-per-watt. Our chapter demonstrates the effectiveness of cross-layer design of deep learning accelerators with the emerging silicon photonics technology. With the growing maturity of silicon photonic device fabrication in CMOS-compatible processes, it is expected that the energy costs of device tuning, losses, and laser power overheads will go further down, making an even stronger case for considering optical-domain accelerators for deep learning inference.

TABLE 19.3
Average EPB and kiloFPS/watt values across accelerators

Accelerator	Avg. EPB (pJ/bit)	Avg. KiloFPS/Watt
P100	971.31	24.9
IXP 9282	5099.68	2.39
AMD-TR	5831.18	2.09
DaDianNao	58.33	0.65
Edge TPU	697.37	17.53
Null Hop	2727.43	4.48
DEAP_CNN	44453.88	0.07
Holylight	274.13	3.3
Cross_base	142.35	10.78
Cross_base_TED	92.64	16.54
Cross_opt	75.58	20.25
Cross_opt_TED	28.78	52.59

19.6 CONCLUSION

In this chapter, we presented a novel cross-layer optimized photonic neural network accelerator called CrossLight. Utilizing silicon photonic device-level fabrication-driven optimizations along with circuit-level and architecture-level optimizations, we demonstrated 9.5× lower energy-per-bit and 15.9× higher performance-per-watt compared to state-of-the-art photonic DNN accelerators. CrossLight also shows improvements in these metrics over several CPU, GPU, and custom electronic accelerator platforms considered in our analysis. CrossLight shows the promise of cross-layer optimization strategies in countering various challenges such as crosstalk, fabrication-process variations, high laser power, and excessive tuning power. The results presented in this chapter demonstrates the promise of photonic DNN accelerators in addressing the need for energy-efficient and high performance-per-watt DNN acceleration.

REFERENCES

[1] N. P. Jouppi, C. Young, N. Patil, D. Patterson, G. Agrawal, R. Bajwa, S. Bates, S. Bhatia, N. Boden, A. Borchers, and R. Boyle, "In-datacenter performance analysis of a tensor processing unit," in Proceedings of the 44th International Symposium on Computer Architecture, pages 1–12, 2017.
[2] Intel Movidius VPU, 2020. [Online]. https://www.intel.com/content/www/us/en/products/processors/movidius-vpu/movidius-myriad-x.html.
[3] Waldrop and M. Mitchell, "The chips are down for Moore's law," in *Nature News*, vol. 530, no. 7589, 2016.
[4] S. Pasricha and N. Dutt, "On-Chip Communication Architectures," xxx Morgan Kauffman, IS BN 978-0-12-373892-9, Apr2008.
[5] A. K. Ziabari, J. L. Abellán, R. Ubal, C. Chen, A. Joshi, and D. Kaeli, "Leveraging silicon-photonic NoC for designing scalable GPUs," in *ACM International Conference on Supercomputing*, 2015.
[6] S. Pasricha and M. Nikdast, "A survey of silicon photonics for energy efficient manycore computing," in *IEEE Design and Test*, vol. 37, no. 4, pages 60–81, 2020.
[7] D. A. Miller, "Silicon photonics: Meshing optics with applications," in *Nature Photonics*, vol. 11, no. 7, pages 403–404, 2017.
[8] Y. Shen, N. C. Harris, S. Skirlo, M. Prabhu, T. Baehr-Jones, M. Hochberg, X. Sun, S. Zhao, H. Larochelle, D. Englund, and M. Soljačić, "Deep learning with coherent nanophotonic circuits," in *Nature Photonics*, vol. 11, no. 7, pages 441–446, 2017.

[9] W. A. Zortman, D. C. Trotter, and M. R. Watts, "Silicon photonics manufacturing," in *Optics Express*, vol. 18, no. 23, pages 23598–23607, 2010.

[10] A. N. Tait, T. F. De Lima, E. Zhou, A. X. Wu, M. A. Nahmias, B. J. Shastri, and P. R. Prucnal, "Neuromorphic photonic networks using silicon photonic weight banks," in *Scientific Reports*, vol. 7, no. 7430, 2017.

[11] V. Bangari, B. A. Marquez, H. Miller, A. N. Tait, M. A. Nahmias, T. F. De Lima, H. T. Peng, P. R. Prucnal, and B. J. Shastri, "Digital electronics and analog photonics for convolutional neural networks (DEAP-CNNs)," in *IEEE Journal of Selected Topics in Quantum Electronics*, vol. 26, no. 1, 2020.

[12] W. Liu, W. Liu, Y. Ye, Q. Lou, Y. Xie, and L. Jiang, "HolyLight: A nanophotonic accelerator for deep learning in data centers," in *IEEE/ACM Design, Automation & Test in Europe Conference & Exhibition*, 2019.

[13] K. Shiflett, D. Wright, A. Karanth, and A. Louri, "PIXEL: Photonic neural network accelerator," in *IEEE International Symposium on High Performance Computer Architecture (HPCA)*, 2020.

[14] Z. Zhao, D. Liu, M. Li, Z. Ying, L. Zhang, B. Xu, B. Yu, R. T. Chen, and D. Z. Pan, "Hardware-software co-design of slimmed optical neural networks," in *IEEE/ACM Asia and South Pacific Design Automation Conference*, 2019.

[15] G. Mourgias-Alexandris, A. Totović, A. Tsakyridis, N. Passalis, K. Vyrsokinos, A. Tefas, and N. Pleros, "Neuromorphic photonics with coherent linear neurons using dual-IQ modulation cells," in *Journal of Lightwave Technology*, vol. 38, no. 4, pages 811–819, 2020.

[16] C. Pask, "Generalized parameters for tunneling ray attenuation in optical fibers," in *Journal of the Optical Society of America*, vol. 68, no. 1, pages 110–116, 1978.

[17] P. Pintus, M. Hofbauer, C. L. Manganelli, M. Fournier, S. Gundavarapu, O. Lemonnier, F. Gambini, L. Adelmini, C. Meinhart, C. Kopp, and F. Testa, "PWM-driven thermally tunable silicon microring resonators: Design, fabrication, and characterization," in *L&P Reviews*, vol. 13, no. 9, 2019.

[18] W. Bogaerts, P. De Heyn, T. Van Vaerenbergh, K. De Vos, S. Kumar Selvaraja, T. Claes, P. Dumon, P. Bienstman, D. Van Thourhout, and R. Baets, "Silicon microring resonators," in *L&P Reviews*, vol. 6, no. 1, pages 47–73, 2012.

[19] M. Nikdast, G. Nicolescu, J. Trajkovic, and O. Liboiron-Ladouceur, "Chip-scale silicon photonic interconnects: A formal study on fabrication non-uniformity," in *Journal of Lightwave Technology*, vol. 34, no. 16, pages 3682–3695, 2016.

[20] S. Abel, T. Stöferle, C. Marchiori, D. Caimi, L. Czornomaz, M. Stuckelberger, M. Sousa, B. J. Offrein, and J. Fompeyrine, "A hybrid barium titanate–silicon photonics platform for ultraefficient electro-optic tuning," *Journal of Lightwave Technology*, vol. 34, no. 8, pages 1688–1693, 2016.

[21] L. Lu, X. Li, W. Gao, X. Li, L. Zhou, and J. Chen, "Silicon non-blocking 4 × 4 optical switch chip integrated with both thermal and electro-optic tuners," in *IEEE Photonics*, vol. 11, no. 6, pages 1–9, 2019.

[22] M. Milanizadeh, D. Aguiar, A. Melloni, and F. Morichetti, "Canceling thermal cross-talk effects in photonic integrated circuits," in *Journal of Lightwave Technology*, vol. 37, no. 4, pages 1325–1332, 2019.

[23] Lumerical Solutions Inc. Lumerical HEAT. [Online]. http://www.lumerical.com/tcad-products/heat/.

[24] S. De, R. Das, R. K. Varshney, and T. Schneider, "Design and simulation of thermo-optic phase shifters with low thermal crosstalk for dense photonic integration," in *IEEE Access*, vol. 8, pages 141632–141640, 2020.

[25] Y. LeCun, L. Bottou, Y. Bengio, and P. Haffner, "Gradient-based learning applied to document recognition," in *Proceedings of the IEEE*, 1998.

[26] QKeras. https://github.com/google/qkeras.

[27] L. H. Frandsen, P. I. Borel, Y. X. Zhuang, A. Harpøth, M. Thorhauge, M. Kristensen, W. Bogaerts, P. Dumon, R. Baets, V. Wiaux, and J. Wouters, "Ultralow-loss 3-dB photonic crystal waveguide splitter," in *Optics letters*, vol. 29, no. 14, pages 1623–1625, 2004.

[28] Y. C. Tu, P. H. Fu, and D. W. Huang, "High-efficiency ultra-broadband multi-tip edge couplers for integration of distributed feedback laser with silicon-on-insulator waveguide," in *IEEE Photonic Journal*, vol. 11, no. 4, pages 1–13, 2019.

[29] S. Bahirat and S. Pasricha, "OPAL: A multi-layer hybrid photonic NoC for 3D ICs," in *IEEE/ACM Asia and South Pacific Design Automation Conference*, 2011.

[30] H. Jayatilleka, M. Caverley, N. A. Jaeger, S. Shekhar, and L. Chrostowski, "Crosstalk limitations of microring-resonator based WDM demultiplexers on SOI," in *IEEE Optical Interconnects Conference*, 2015.

[31] E. Timurdogan, C. M. Sorace-Agaskar, E. S. Hosseini, G. Leake, D. D. Coolbaugh, and M. R. Watts, "Vertical junction silicon microdisk modulator with integrated thermal tuner," in *CLEO: Science and Innovations, OSA*, 2013, paper CTu2F.2.

[32] M. Pisati, F. De Bernardinis, P. Pascale, C. Nan, M. Sosio, E. Pozzati, N. Ghittori, F. Magni, M. Garampazzi, G. Bollati, and A. Milani, "A sub-250 mW 1-to-56Gb/s continuous-range PAM-4 42.5 dB IL ADC/DAC-based transceiver in 7 nm FinFET," in *IEEE International Solid-State Circuits Conference*, 2019.

[33] Z Ruan, Y. Zhu, P. Chen, Y. Shi, S. He, X. Cai, and L. Liu, "Efficient hybrid integration of long-wavelength VCSELs on silicon photonic circuits," *Journal of Lightwave Technology*, vol. 38, no. 18, pages 5100–5106, 2020.

[34] A. D. Güngördü, G. Dündar, and M. B. Yelten, "A high performance TIA design in 40 nm CMOS," in *IEEE International Symposium on Circuits and Systems*, 2020.

[35] B. Wang, Z. Huang, W. V. Sorin, X. Zeng, D. Liang, M. Fiorentino, and R. G. Beausoleil, "A low-voltage Si-Ge avalanche photodiode for high-speed and energy efficient silicon photonic links," in *Journal of Lightwave Technology*, vol. 38, no. 12, pages 3156–3163, 2020.

[36] L. H. Duong, M. Nikdast, S. Le Beux, J. Xu, X. Wu, Z. Wang, and P. Yang, "A case study of signal-to-noise ratio in ring based optical networks-on-chip," *IEEE Design & Test*, vol. 31, no. 5, pages 55–65, 2014.

[37] M. Capra, B. Bussolino, A. Marchisio, M. Shafique, G. Masera, and M. Martina, "An updated survey of efficient hardware architectures for accelerating deep convolutional neural networks," in *Future Internet*, vol. 12, no.7, page 113, 2020.

[38] F. Sunny, A. Mirza, M. Nikdast and S. Pasricha, "CrossLight: A cross-layer optimized silicon photonic neural network accelerator," *IEEE/ACM Design Automation Conference (DAC)*, San Francisco, CA, December 2021.

Index

N

O

P

Q

R

S

T